Green Building: An Engineering Approach to Sustainable Construction

Green Building: An Engineering Approach to Sustainable Construction

Christian M. Carrico
Associate Professor in Civil and Environmental Engineering, New Mexico Institute of Mining and Technology, Socorro, NM, United States

Butterworth-Heinemann is an imprint of Elsevier
125 London Wall, London EC2Y 5AS, United Kingdom
50 Hampshire Street, 5th Floor, Cambridge, MA 02139, United States

Copyright © 2025 Elsevier Inc. All rights are reserved, including those for text and data mining, AI training, and similar technologies.

For accessibility purposes, images in electronic versions of this book are accompanied by alt text descriptions provided by Elsevier. For more information, see https://www.elsevier.com/about/accessibility.

Publisher's note: Elsevier takes a neutral position with respect to territorial disputes or jurisdictional claims in its published content, including in maps and institutional affiliations.

No part of this publication may be reproduced or transmitted in any form or by any means, electronic or mechanical, including photocopying, recording, or any information storage and retrieval system, without permission in writing from the publisher. Details on how to seek permission, further information about the Publisher's permissions policies and our arrangements with organizations such as the Copyright Clearance Center and the Copyright Licensing Agency, can be found at our website: www.elsevier.com/permissions.

This book and the individual contributions contained in it are protected under copyright by the Publisher (other than as may be noted herein).

Notices

Knowledge and best practice in this field are constantly changing. As new research and experience broaden our understanding, changes in research methods, professional practices, or medical treatment may become necessary.

Practitioners and researchers must always rely on their own experience and knowledge in evaluating and using any information, methods, compounds, or experiments described herein. In using such information or methods they should be mindful of their own safety and the safety of others, including parties for whom they have a professional responsibility.

To the fullest extent of the law, neither the Publisher nor the authors, contributors, or editors, assume any liability for any injury and/or damage to persons or property as a matter of products liability, negligence or otherwise, or from any use or operation of any methods, products, instructions, or ideas contained in the material herein.

ISBN: 978-0-12-824365-7

For information on all Butterworth-Heinemann publications visit our website at https://www.elsevier.com/books-and-journals

Publisher: Peter Linsley
Acquisitions Editor: Stephen Merken
Editorial Project Manager: Helena Beauchamp
Production Project Manager: Prasanna Kalyanaraman
Cover Designer: Mark Rogers

Typeset by STRAIVE, India

Contents

Preface xiii
List of acronyms and symbols xvii

CHAPTER 1 Introduction to green building 1
    1.1 Green and sustainable building definitions 1
    1.2 Evolution of the field 1
    1.3 Intended audience: Engineering students and beyond 6
    1.4 Approach and differentiators of this textbook 6
    1.5 Trajectory of this textbook 8
    1.6 Engineering problem-solving approach and dimensional homogeneity 8
    1.7 Graphical presentation of data 10
    1.8 Accreditation programs: ABET: Gold standard for engineering programs 10
    1.9 Summary and additional resources 11

CHAPTER 2 Energy science: Key underlying physics 13
    2.1 Energy as a fundamental physical quantity 13
    2.2 Forms of energy 14
    2.3 Power (W) and power density or flux (W/m^2) 15
    2.4 Essential electricity relationships 17
    2.5 Thermodynamic properties, systems, and processes 18
    2.6 Heat and work 19
    2.7 Heat transfer: Conduction, convection, and radiation 22
    2.8 Heat capacity of materials, latent and sensible heat, stored chemical energy 24
    2.9 Mass & energy balance 26
    2.10 The laws of thermodynamics 27
    2.11 Heat engines, combustion, and heat pumps 27
    2.12 Combustion and reacting systems 30
    2.13 Ideal gas law and other essential fluids relationships 30
    2.14 Water vapor and relative humidity 31
    2.15 Chapter summary and conclusions 33
    2.16 End of chapter exercises 34

CHAPTER 3 Energy and building history: How have we arrived here? 39
    3.1 Human population and exponential growth 39
    3.2 Global population and environmental impacts 42
    3.3 Historical United States and global energy use 44

v

3.4	Sustainable global energy use	50
3.5	Anatomy of centralized electrical generating units	51
3.6	Historical and current materials use	53
3.7	Energy, materials, and sustainability	56
3.8	Chapter summary and conclusions	57
3.9	End of chapter exercises	58

CHAPTER 4 Environmental problems from the local to global scale: Motivation for why we need to build better ... 61

4.1	Infrastructure and environment	61
4.2	Hydrological resources on planet earth	62
4.3	Land use, soils, and materials extraction	65
4.4	Materials extraction and processing	67
4.5	Solid, hazardous, and construction and demolition waste	67
4.6	Microplastics and marine debris	69
4.7	Biodiversity, habitat loss, and species extinction	71
4.8	Ambient and indoor air pollution	72
4.9	Anthropogenic climate change and the climate crisis	81
	4.9.1 Overview of fundamental climate system physics	81
	4.9.2 Radiative forcing of climate	85
	4.9.3 The carbon cycle and a human fingerprint	88
	4.9.4 Overwhelming evidence for anthropogenic climate change	92
	4.9.5 Modeling current and future climate disruption and relevance to green building	96
	4.9.6 Case study: Wildland fire and the built environment	98
4.10	Environmental policy and environmental justice	102
4.11	Chapter summary and conclusions	103
4.12	End of chapter exercises	104

CHAPTER 5 Global solutions to the climate crisis: Our friends with cobenefits ... 109

5.1	Climate change mitigation approaches: Enhancing sustainability	109
5.2	Decarbonization of our energy infrastructure	112
5.3	Energy conservation, efficiency, and curtailment: A building-centric decarbonization	115
5.4	Additional built environment solutions to the climate crisis	117
5.5	Agriculture, forestry, and land use changes: Reinventing our food system	117
5.6	Carbon capture, use, and sequestration	118
5.7	Development of a hydrogen economy	123

5.8	Adaptation to an altered climate	124
5.9	Geoengineering approaches: Purposely altering the radiative balance	125
5.10	Chapter summary and conclusions	126
5.11	End of chapter exercises	127

CHAPTER 6 Green design, delivery, commissioning, and auditing: Starting the project right 131

6.1	Site sustainable development	131
6.2	Global scale concerns to integrated design process	134
6.3	Building renovation projects	138
6.4	Case study: Ghirardelli Square, San Francisco, California	139
6.5	Building sustainable design: Appropriate sizing	140
6.6	Building as a system and net zero	142
6.7	Building construction and manufacturing	143
6.8	Design for sustainable transportation options	144
6.9	Case study: Fuel & Iron redevelopment in Pueblo, Colorado	146
6.10	Building orientation and passive solar considerations	147
6.11	Building codes, commissioning, and auditing of buildings	149
6.12	Design for future adaptability	149
6.13	Chapter summary and conclusions	150
6.14	End of chapter exercises	151

CHAPTER 7 Materials use in buildings: Choosing the best options 153

7.1	Overview of materials	153
7.2	Materials attributes and sustainability	154
7.3	Embodied energy and embodied carbon	155
7.4	Solid waste, construction waste, hazardous waste and material reuse	158
7.5	Concrete and cement: Scale, composition, and impacts	163
7.6	Steel and other metals	169
7.7	Structural wood products: Mass timber and cross laminated timber (CLT)	170
7.8	Bamboo, hemp straw bale, and other structural plant fibers	172
7.9	Advanced manufacturing and construction methods	174
7.10	Tall buildings and green building	176
7.11	Case study: Green renovated tall building Empire State Building in New York	179
7.12	Case study: Newly constructed green tall tower in Austin, TX	181
7.13	Chapter summary and conclusions	183
7.14	End of chapter exercises	184

CHAPTER 8 Building thermal envelope: Constructing a tight barrier between the indoors and outdoors 185

- 8.1 Defining the thermal envelope of a building 185
- 8.2 Heat and mass transfer in buildings 186
- 8.3 Insulating materials and R-value in series 188
- 8.4 Calculating parallel pathway R-values 192
- 8.5 Estimating heat loss through the thermal envelope 193
- 8.6 New innovations in insulating materials 195
- 8.7 Windows and doors: Minimizing heat transfer while allowing access, light, ventilation, visibility, and views 196
- 8.8 Roofs and attics including "cool roof" technologies 199
- 8.9 Basements and crawlspaces 200
- 8.10 Thermal mass and thermal bridging 201
- 8.11 Other perils, problems, and possibilities in insulating the building envelope 204
- 8.12 Tools for testing the building envelope and performance metrics 206
- 8.13 Chapter summary and conclusions 208
- 8.14 End of chapter exercises 208

CHAPTER 9 Electro-mechanical systems and appliances in buildings: Evaluating alternatives 215

- 9.1 Overview of energy systems in buildings 215
- 9.2 Climate control systems for conditioned spaces 216
    - 9.2.1 Advent of refrigerated air heat pumps 218
    - 9.2.2 Refrigerants, stratospheric ozone depletion, and global warming potential 219
    - 9.2.3 HVAC system sizing and ducting 221
    - 9.2.4 High-efficiency combustion systems 221
    - 9.2.5 Air source heat pumps and mini-split systems 223
    - 9.2.6 Evaporative coolers 225
    - 9.2.7 Ventilation systems for space cooling and heating 226
    - 9.2.8 Heat and energy recovery ventilators (HRVs and ERVs) 230
    - 9.2.9 Combined systems and cogeneration 230
- 9.3 Case Study: Laboratory building district chiller systems control 232
- 9.4 Water heating systems-storage tank and tankless on-demand systems 233
- 9.5 Lighting 235
- 9.6 Electrical appliances and phantom loads 239
- 9.7 Demand management and load shifting 240
- 9.8 Chapter summary and conclusions 241
- 9.9 End of chapter exercises 242

CHAPTER 10 Water, infrastructure, and buildings 249
- 10.1 Water quality parameters and pollution 249
- 10.2 Potable water treatment 252
- 10.3 Wastewater treatment 256
- 10.4 Stormwater management 258
- 10.5 Systems promoting more sustainable water use 260
- 10.6 Onsite black and graywater systems 262
- 10.7 Water management on the building scale 263
- 10.8 Other water problems in buildings 265
- 10.9 Water-energy nexus: Water treatment requires energy and vice versa 265
- 10.10 Chapter summary and conclusions 267
- 10.11 End of chapter exercises 268

CHAPTER 11 Green outdoor spaces: Harmonizing the built and natural environments 273
- 11.1 History and importance of landscape design 273
- 11.2 Demographic trends and green sites 275
- 11.3 Site planning, engineering, and minimizing construction impacts 276
- 11.4 Community, commercial, and other shared scale green spaces 278
- 11.5 Case study High Line rail line repurposed into a linear park in New York city 279
- 11.6 Appropriate plants and their benefits to site and soil health 281
- 11.7 Xeriscape: Low-water-use landscaping 283
- 11.8 Case study: Landscaping of the "Hellstrip" of an arid region 286
- 11.9 Outdoor rooms: A low energy extension of the indoors 287
- 11.10 Permaculture, regenerative landscaping, and producing food onsite 288
- 11.11 Site water management systems: Rainwater, graywater, constructed wetlands 289
- 11.12 Case study xeriscaped residential landscapes 291
- 11.13 Landscape maintenance and equipment use 292
- 11.14 Green roofs 292
- 11.15 Chapter summary and conclusions 293
- 11.16 End of chapter exercises 294

CHAPTER 12 Healthy indoor spaces: Maintaining indoor environmental quality 297
- 12.1 Human comfort and hygro-thermal properties 297
- 12.2 Indoor environmental quality (IEQ) 300
- 12.3 Lighting characteristics 300

12.4	Indoor sound characteristics	300
12.5	Indoor air quality	301
12.6	Mitigation approaches for indoor air pollution	304
12.7	Radon and its mitigation	306
12.8	Biohazard concerns: Ventilation, air cleaning and disease transmission	306
12.9	Interior furnishings	310
12.10	Chapter summary and conclusions	311
12.11	End of chapter exercises	312

CHAPTER 13 Renewable energy systems: Advancing distributed energy production ... 313

13.1	Overview of renewable energy systems	313
13.2	Spectrum of solar energy and electromagnetic energy	317
13.3	Behind the meter renewable systems	320
13.4	Solar photovoltaic cells and PV systems	320
	13.4.1 Solar cell physics and efficiency	320
	13.4.2 Electricity generation with solar PV panels	323
	13.4.3 Solar PV costs	324
13.5	Solar daylighting and passive solar space heating	328
13.6	Case study: CSU powerhouse research facility	331
13.7	Active solar thermal systems	331
13.8	Geothermal energy systems	333
13.9	Wind power systems	335
13.10	Biomass energy	338
13.11	Micro hydro applications	340
13.12	Energy storage systems and transmission	340
13.13	Constraints and opportunities in a renewable energy transition	343
13.14	Case study part I—Mid-century schoolhouse/church repurposed into a home	345
13.15	Chapter summary and conclusions	348
13.16	End of chapter exercises	349

CHAPTER 14 Socioeconomic context and equity in green building: Policy, tools, codes, and certifications ... 355

14.1	Economic factors, financing, and payback times	355
14.2	Federal policy and financial incentives	357
14.3	Zoning and building codes	358
14.4	Utilities and rate structures	359
14.5	Environmental justice issues and the building industry	360
14.6	Incorporating social, environmental, and economic costs over the life cycle	363

14.7	LEED: Leadership in energy and environmental design	364
14.8	ISI & ASCE ENVISION sustainability program	366
14.9	Indoor environmental quality certification programs	367
14.10	Wider World of green certification programs	367
14.11	Career directions in green building: Energy systems engineer	370
14.12	Chapter summary and conclusions	372
14.13	End of chapter exercises	373

CHAPTER 15 Climate adaptation, resiliency, and the built sector: Designing for future disruptions 375

15.1	Overview of global change effects on the built environment and adaptation	375
15.2	Extreme weather events & the built environment	376
15.3	Adaptation and resiliency: Planning for current and anticipated changes	381
15.4	Resilience organizations and tools	385
15.5	Coastal construction and sea level rise	387
15.6	Inland flooding and mitigation of its effects	389
15.7	Case study: Babcock Ranch Community in Florida	390
15.8	Drought risks	392
15.9	Wildland fire and the built environment	392
15.10	Microgrids, smart grid, and grid resiliency	395
15.11	Case study: Oregon state treasury building	398
15.12	Managed retreat: The ultimate adaptation	399
15.13	Case study part II—Resilience in mid-century schoolhouse/church repurposed into a home	400
15.14	Chapter summary and conclusions	401
15.15	End of chapter and end of book summation exercises	403

References 421
Index 439

Preface

Societal environmental challenges

I like to think our current eco-status lies somewhere in the middle of an accelerating crescendo of environmental tragedies and a resounding list of environmental problems tackled, as examples of both exist. The challenge of the climate crisis is among the most important humans have faced, and our buildings are a key contributor in both operations and construction. The estimated contributions to total greenhouse gas emissions from the built environment approach half of our greenhouse gas emissions when the embodied carbon and associated emissions are included. The focus of this book is a hopeful look at solving our problems. My area of specialization, air quality, offers models for fixing our problems (improvements in urban air quality, acid rain, and stratospheric ozone depletion, among others).

The engineering field has a growing recognition that it will have an important role to play in efforts to improve the sustainability of our infrastructure. This is increasingly emphasized with the prominence of the Leadership in Energy and Environmental Design (LEED) program, the sustainability focus in the American Society of Civil Engineers (ASCE) through its Envision program, and more emphasis on the Accreditation Board for Engineering and Technology (ABET) criteria. Though LEED, Envision, and other programs are highlighted, this is not a test prep book; rather, I intended it as a book to help engineering students start problem-solving in this area.

The need is great for engineers trained in green building and broader sustainability issues. Likewise, engineering students need a book focused on the sustainability of the built sector. Civil and environmental engineers of course play a leading role in construction, roads, buildings, and environmental infrastructure. However, all engineers have to consider the sustainability aspects of their projects. At the same time, engineering students often do not realize the career opportunities within green building.

Evolution of this textbook

As many textbooks begin, this book grew from not finding the right match for teaching a green building engineering class at the New Mexico Institute of Mining and Technology (NMT). It started as a pandemic project and has been slowly progressing for a few years. It likely started earlier, and perhaps it was destined that I would arrive at this calling. As a kid, I marveled at Chicago brick that had been reused in our house and marble slabs purportedly reused from a bathroom renovation in the Old State Capitol building in Springfield, Illinois. The (unconfirmed) stories about Abraham Lincoln possibly traversing (or worse) our marble flooring continue to this day. Such efforts were as much about limited resources, thriftiness, and using what is available vs environmental concerns or material's reuse.

Why not an energy book? Energy use features prominently, but it is necessary but not sufficient. Materials, water, and embodied carbon are all important parameters in their own right. There exist many well-regarded books on energy and sustainability. One of the early motivators is a book by the late David J. Rose of MIT, Learning about Energy, that I used circa 1990. More recently, Dunlap, Hinrichs and Kleinbach, McKay, Jelley and Andrews, and Jacobson all come to mind as sustainable energy-related textbooks that I have used, but there are many more, some of which are referenced.

I wrote this for use in my class on green building and hope you find it useful. I found that more civil engineers than environmental take the class at NMT. They are not required to take thermodynamics or heat transfer, and hence I give a crash course on energy mechanics in one chapter. This and other early chapters could be skipped in cases where students have had this background. We will either solve our planetary energy imbalance or, in time, an increasingly perturbed climate system will solve its human problem; I prefer the former. My hope is it will play a minuscule role in helping solve our climate crisis and generally improve our built and natural environments. At least it will help train the generation that is inheriting the problem from the current and previous generations.

Focus and unique features

The book is an introductory and broad overview of green building and the broader sustainability space. It targets engineering students who are uninitiated to the field, and thus it is also an introductory textbook for engineers who may not have taken any other sustainability coursework. The focus here is developing the students' understanding, problem-solving abilities, and how to quantify aspects of green building rather than attempting to keep up with the dizzying progression of new materials and techniques. A healthy dose of energy calculations, the underpinnings of understanding energy use in buildings.

You will notice I tend to the "back of the envelope" scale of the problem-type exercises. Understanding the scale of the problem and the scalability is vital to effective solutions. This is not a book of textbook problems but rather seeks to develop problem-solving skills and engineering intuition for real-world solutions. In many cases, though, it comes down to fundamental quantities like joules and kilograms. The students sometimes blow a fuse on these problems as they expect all the information to be provided, and sometimes it requires appropriate engineering approximations or for them to find info to solve these problems (and problems do not always have one exact solution of course depending on assumptions). These problems usually start as questions or new stories that make me go "hmmm," and so I try calculating these to satisfy my curiosity.

What differentiates this book from what perhaps an AI-generated textbook might look like? I would say mostly the problems, examples, figures/photos, and case studies, mainly from my own experiences. Many examples from my quest to be more energy and water-efficient, with plenty of mistakes made along the way! Combination of key concepts and on-the-ground practical applications from the living laboratory of building renovations (sometimes to the chagrin of the other occupants). I do not think the chatbots have mastered building a low-carbon stone wall (yet). I also tried to make the book visually informative and engaging with 200 or so figures, likely related to my own visual learning. I encourage instructors and other readers to freely use, modify, and adapt the problems and the images (which are mostly mine) to their classes.

Teaching ancillaries for this title, including solutions and an image bank, are available to instructors by visiting https://educate.elsevier.com/book/details/9780128243657 and following instructions to log in or register.

Acknowledgments

There were extremely helpful reviewers, mostly anonymous, of both the proposal and individual chapters of the book. I appreciate their feedback and incorporated much of that to the extent my brain's bandwidth could handle it.

I am indebted to my colleagues at NMT and beyond who gave feedback and encouragement. In particular, NMT department colleagues provided encouragement and feedback, including Profs. Richardson, Huang, Harb, Cook, Morris, and Wilson. I will also thank NMT professional staff, especially Bedelia Apodaca, the HR office, Sponsored Projects office, and Academic Affairs office, for support during the sabbatical year. Jason Hebert and Alex Garcia in facilities at NMT often gave enlightening tours of construction projects happening around campus. Students who have taken or TA'd the green building class and/or reviewed drafts, notably Ryan Himes, Mercy Ajigah, and Samantha Gomez, helped lend a student perspective and feedback. Longer-term colleagues from Illinois including Drs. Mark Rood and Mark Cal are thanked for their influence.

Publicly available data was extremely helpful including Our World in Data for amazing data products which are freely available and they are providing a service to the world. USEIA, EPA, NREL, DOE, and other federal sources were also used heavily as well for images and data. Shutterstock images were used in a few cases where I lacked images. Grammarly and MS Word spelling and grammar checkers were both used for spelling and grammar checks (apologies for any remaining such errors).

I will also offer an up-front apology to international readers as this book uses both US customary and SI units. I also, in moving many things around, have neglected to fully spell out acronyms at first use, so I provide a table of acronyms/definitions below. I attempt to provide examples throughout the United States and beyond, but my experience is mostly in the southwestern United States, and these biases show through.

The Elsevier editorial team, most notably Helena Beauchamp and Steve Merken, who stuck with me and the project over its ~4 years, should be commended. Doug Swartz and John Phelan, who were energy engineers at Fort Collins Utilities, helped to reignite my long-term interest in buildings and efficiency. My friends, family, and my spouse Liz, whose experience in book publishing was extremely helpful, as was the fact that we were both tackling book projects at the same time. Please feel free to contact me with feedback or suggestions for the next iteration (email to: kip.carrico@nmt.edu).

Christian M. 'Kip' Carrico

List of acronyms and symbols

Acronym or symbol	Parameter or term	Notes
A	Amp	Unit of electrical current
A	Area	
ABET	Accreditation Board for Engineering and Technology	
AC	Alternating current	
ACH50	Air changes per hour	Often measured at a pressure difference of 50 Pa
AEE	Association of energy engineers	
AF	Air to fuel ratio	
A-h	Amp-hour	A unit of battery energy storage
AI	Artificial intelligence	
AQI	Air quality index	Composite air quality index reported to the public to quantify air pollution hazard
ASCE	American Society of Civil Engineers	
ASHRAE	American Society for Heating, Refrigerating and Air-Conditioning Engineers	
BECCS	Bioenergy with carbon capture and sequestration	
BIFMA	Business and Institutional Furniture Manufacturers' Association	
BOD	Biochemical oxygen demand	
BPI	Building Performance Institute	
BREEAM	Building Research Establishment Environmental Assessment Methodology	
BTU	British Thermal Unit	
c	Heat capacity	Sensible heat
C	Runoff coefficient	
C	Concentration	
°C	Degree Celsius	Unit of temperature
CDD	Cooling degree days	
$CaCO_3$	Calcium carbonate	Limestone
CADR	Clean air delivery rate	
cal	calorie	The energy needed to raise the temperature of 1 g of water by 1°C
Cal	Calorie	Food Calorie equivalent to 1 kcal
CaO	Calcium oxide	Lime
CCUS	Carbon capture, utilization, and storage	Also known as CCS

Continued

List of acronyms and symbols

Acronym or symbol	Parameter or term	Notes
CEA	Certified Energy Auditor	
CEER	Combined energy efficiency ratio	
CEM	Certified Energy Manager	
C&D	Construction and Demolition	
CFC	Chlorofluorocarbons	Ozone-depleting compounds that are also potent greenhouse gases
CFM	Cubic feet per minute	
CH_4	Methane	A greenhouse gas and the main component of natural gas
CHP	Combined heat and power	
CMU	Concrete masonry unit	
CO	Carbon monoxide	Incomplete combustion product and EPA criteria air pollutant
CO_2	Carbon dioxide	One of the products of combustion and a key atmospheric greenhouse gas
CO_2,e	Carbon dioxide equivalent	Greenhouse gas of interest multiplied by its GWP
COP	Coefficient of performance	Quantifies effectiveness of a refrigeration system
CPI	Consumer price index	
CRI	Color rendering index	
CSU	Colorado State University	
D	Diameter	
Δ	Delta	Change in a given parameter
dB	Decibel	
DC	Direct current	
dW/dt or P	Rate of work	In power units
ϵ	Emissivity	
EER	Energy efficiency ratio	
EIFS	Exterior insulation and finish systems	
EOR	Enhanced oil recovery	
η (eta)	Efficiency	Unitless
Embodied carbon	Embodied carbon	The emissions associated from extraction to delivery for materials
Embodied energy	Embodied energy	The energy use associated from extraction to delivery for materials
ERV	Energy recovery ventilator	
ESB	Empire state building	
ESG	Environmental, social, and governance	
ΔF	Radiative forcing	Perturbation of the atmospheric radiative balance
°F	Degree Fahrenheit	Unit of temperature

Continued

List of acronyms and symbols

Acronym or symbol	Parameter or term	Notes
FEMA	Federal Emergency Management Agency	
FSC	Forestry Stewardship Council	
FSEC	Florida Solar Energy Center	
g	Gravitational constant	$9.8\,\text{ms}^{-2}$ at earth's surface
gal	Gallon	
GBA	Green Building Alliance	
GBCI	Green Business Certification, Inc.	
GCM	General circulation model	
GDP	Gross domestic product	
GHG	Greenhouse gas	
GPF	Gallons per flush	
GSA	Government Services Administration	
GT	Gigatonne	
GWP	Global warming potential	The effectiveness at serving as a greenhouse gas scaled to CO_2, which has GWP = 1
h	Convective heat transfer coefficient	
h	Enthalpy	
HAB	Harmful algae bloom	
H_2	Hydrogen gas	An energy carrier
H_2S	Hydrogen sulfide	
H_2SO_4	Sulfuric acid	
HDD	Heating degree days	
HDI	Human development index	
HERS	Home index rating system	
HFC	Hydrofluorocarbon	An interim replacement compound for CFCs
HOA	Homeowners' Association	
HRV	Heat recovery ventilator	
HRT	Hydraulic residence time	
HSPF	Heat seasonal performance factor	
HVAC		
I	Current of electricity	
i	Discount rate	
IAQ	Indoor air quality	
IBC	International building code	
ICF	Insulating concrete form	
ICLUS	Integrated climate and land-use scenarios	
IDP	Integrated design process	
IPCC	Intergovernmental panel on climate change	
IR	Infrared	Radiation with wavelengths larger than visible
ISI	Institute for sustainable infrastructure	
IWBI	International well building institute	

Continued

Acronym or symbol	Parameter or term	Notes
J, kJ, MJ	Joule, kilojoules, megajoules	Units of energy
k	Reaction rate constant	
K	Kelvin	Unit of temperature
K	Hydraulic conductivity	
κ	Conductive heat transfer coefficient	
KE	Kinetic energy	Energy of motion
KM	Kilometer	
kWh	Kilowatt hour	Unit of energy consumption
L	Latent heat	
L	Remaining biochemical oxygen demand (BOD)	
L_o	Ultimate BOD	
λ	Wavelength	
LASSO	Locating and selecting scenarios online	
LCA	Life cycle analysis	
LCOE	Levelized cost of energy	
LEED	Leadership in energy and environmental design	Administers by USGBC
LED	Light emitting diode	
LID	Low impact development	
M	Mass	
Manual J	Calculation method for building heat loss	
MCL	Maximum contaminant level	
MEP	Mechanical, electrical, and plumbing	
MERV	Minimum efficiency reporting value	Rating for the effectiveness at an air filter for removing particulate matter
MGD	Million gallons per day	
Mol	Moles	
MSW	Municipal solid waste	
μ (Mu)	Dynamic viscosity	
µm (Mu)	Micrometer	1E-6 m
N	Newton	Unit of force ($1\,kgms^{-2}$)
NAAQS	National Ambient Air Quality Standards	
NASEM	National Academies of Sciences, Engineering, and Medicine	
NCEP	National Center for Environmental Prediction	
NEPA	National Environmental Policy Act	
NMT	New Mexico Institute of Mining and Technology	
NOAA	National Oceanic and Atmospheric Administration	

Continued

Acronym or symbol	Parameter or term	Notes
NO, NO_2, NO_x	Nitric oxide, nitrogen dioxide ($NO_x = NO + NO_2$)	NO_2 is an EPA criteria air pollutant
NPV	Net present value	
NREL	National Renewable Energy Lab	
NRI	National Risk Index	
ω	Rotational velocity	Unit of rotation speed (radians/s)
Ω	Ohms	Unit of electrical resistance
O&M	Operations and maintenance	
ODP	Ozone depleting potential	
OPC	Ordinary Portland Cement	
O_3	Ozone	EPA criteria air pollutant
P	Pressure	
PCi/L	PicoCurie per liter	Unit of radiation
PFAS	Per- and polyfluoroalkyl substances	
pH	pH scale	Quantifies the acidity of a system
P_{sat}	Saturation vapor pressure	
P_v	Vapor pressure	
P	Power	Rate of work or energy use
PE	Potential energy	Stored gravitational, chemical, or other energy
$PM_{2.5}$, PM_{10}	Particulate matter with diameter less than 2.5 and 10 micrometers, respectively	EPA criteria air pollutants
Pt	Penetration through a device	
PV	Photovoltaic	
Q	Heat transfer quantity	
Quad	Quadrillion BTU	Unit of energy equal to 1E15 BTU
R	Radius	
R_t	Cash flow	
R&D	Research and development	
RESNET	Residential Energy Service Network	
R-value	Resistance to heat transfer parameter	
RH	Relative humidity	
ρ	Density	
RCRA	Resource Conservation and Recovery Act	
RMI	Rocky Mountain Institute	
σ (sigma)	Stefan-Boltzmann constant	5.67E-8 $Wm^{-2}K^{-4}$
SCM	Supplementary cementitious material	
SDG	Sustainable development goal	
SEER	Seasonal energy efficiency ratio	
SF_6	Sulfur hexafluoride	Greenhouse gas
SHGC	Solar heat gain coefficient	
SO_2	Sulfur dioxide	EPA criteria air pollutant
SPL	Sound pressure level	

Continued

Acronym or symbol	Parameter or term	Notes
SRCC	Solar Rating and Certification Corporation	
STEM	Science, technology, engineering and mathematics	
STP	Standard temperature and pressure	Typically 1 atm and 298 K
T	Temperature	
TOD	Time of day	
U	Internal energy	
U	Thermal transmittance	
UEF	Uniform energy factor	
UHI	Urban heat island	
UNEP	United Nations Environment Programme	
US DOE	United States Department of Energy	
US EIA	United States Energy Information Agency	
US EPA	United States Environmental Protection Agency	
USGBC	US Green Building Council	Administers the LEED program
UV	Ultraviolet	
ν	Specific volume	Reciprocal of density
V	Volts	Unit of electrical voltage
v	Velocity	
V	Volume	
V_o	Overflow velocity	
V_s	Settling velocity	
VOC	Volatile organic compound	
VT	Visible transmittance	
W, kW, MW	Watt, kilowatt, megawatt	Units of power (J/s)
WUI	Wildland-Urban Interface	
WWTP	Wastewater treatment plant	
Xeriscape	Xeriscape	Using native and adapted plant species and landscaping to minimize water use

CHAPTER 1

Introduction to green building

Learning objectives

(1) Integrate a big-picture view of the green building space and where it fits in the greater world of sustainability.
(2) Examine the trajectory of this textbook, its audience, and its applicability.
(3) Set the stage with key fundamentals, concepts, and terms of relevance.

1.1 Green and sustainable building definitions

Green building is an emerging interdisciplinary field at the intersection of environmental, civil, and construction engineering as well as architecture. A first order of business is the distinction between sustainable and green, often used interchangeably. Green of course considers the impacts on the environment taken into consideration and hopefully minimized. Green building can be reduced to the concept of minimizing the impacts of buildings on the planet while maintaining the health of the occupants inside of them. According to the EPA webpage on the topic "Green engineering is the design, commercialization, and use of processes and products that minimize pollution, promote sustainability, and protect human health without sacrificing economic viability and efficiency (https://www.epa.gov/green-engineering/about-green-engineering)." Applying this to the built sector, green building seeks to design and build both buildings and other constructed infrastructure that minimizes environmental impacts. Moreover, human health, economics, viability, and other aspects combine with the environmental protection aspect to make a more holistic, sustainable approach. Though the focus is environment here both in construction and operations of the built environment, the need to consider broader sustainability aspects is recognized. Green building applies this to the building industry from planning and procurement to ultimate demolition. The co-benefits of green building are important as well. A more sustainable building often results in more satisfied and productive occupants, lower utility bills, and more durable and esthetic indoor environments.

1.2 Evolution of the field

The building part needs clarification as well. The built environment is distinguished from the natural environment and thus includes buildings—commercial, industrial, and residential—as well as the interconnecting, supporting infrastructure such as roads and bridges. The focus is firstly on the buildings and their sites, with some connections made to the larger infrastructure as well. The focus extends beyond the construction process to the operations and the entire life cycle.

Why do we need to build better? According to the US Energy Information Association (EIA), buildings account for roughly 37% in 2023 of direct energy use through the residential and commercial sectors, including upstream electrical system losses (Frequently Asked Questions (FAQs)—US Energy Information Administration (EIA)). The United Nations Environment Program estimated approximately 38% of both global energy use as well as CO_2 emissions in 2015 were attributed to the built sector (UNEP, 2021). Energy use and emissions dropped modestly in 2020 due to the pandemic but have resumed increasing as the world emerged from the pandemic. Some energy economists suggest we are at or near our peak CO_2 emissions, which would be good news, albeit several decades after when the science was sufficiently clear to act.

The embodied energy and embodied carbon of a material represent the energy and emissions, respectively, associated with the production and delivery of materials. This includes everything from extraction to disposal of the materials. For materials, upfront emissions from extraction to production dominated because of their immediate impacts and our short timeframe to decarbonize. They are typically measured in MJ/kg for the embodied energy and kg of CO_2 equivalents (CO_2,e) per kg of material for embodied emissions.

The overall contributions of the built sector to climate-altering greenhouse gas emissions are estimated as high as 48% when all embodied emissions, leakage, and all contributions from the residential, commercial, and industrial buildings are included (Mazria, 2003; Fig. 1.1). In cities, it is an even higher

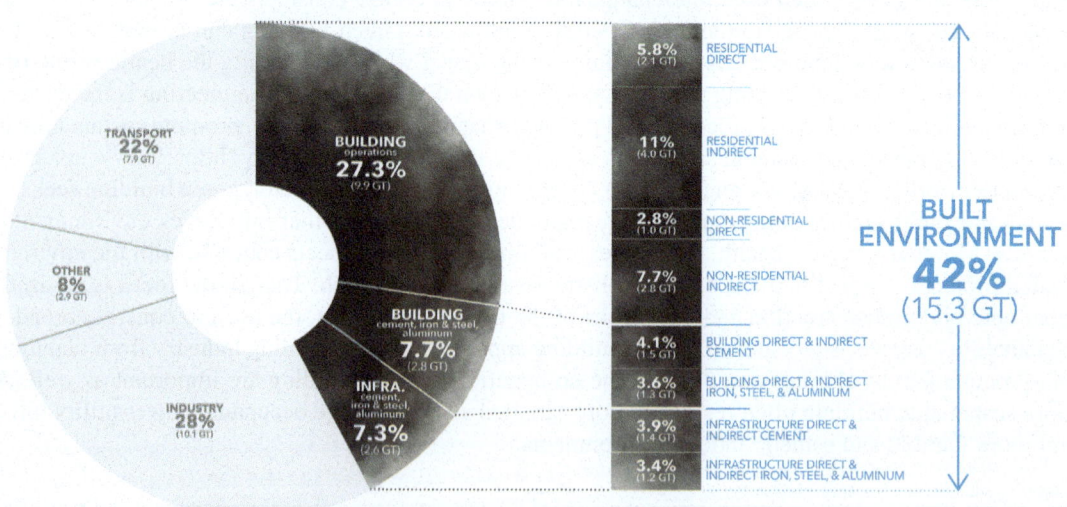

FIG. 1.1

Breakdown of total global greenhouse emissions associated with building-related emissions.

Architecture 2030, used with permission.

fraction. When accounting for the energy use in the manufacturing of building materials and that associated with the transportation of materials and workers to building locations, it approaches half of global energy use by some estimates. Clearly, from any perspective, buildings are central to addressing the climate crisis, as will be discussed in more detail. Building energy and materials use offer one of the monumental opportunities for civilization to reduce greenhouse gas emissions and help solve the climate crisis, all while improving the performance of the built sector.

The history of green building could occupy a book on its own so only a few highlights can be offered. In terms of integrating the built environment with the natural world, the concept of modern green building can be traced back a century or more. It could be described in part as a reaction to the utilitarian and brutalist buildings of the Industrial Era. Ancient civilizations such as the Anasazi, ancient Greece, Persia, and others used principles of building design orientation, passive solar, wind towers, evaporative cooling, thermal mass, and daylighting in their buildings (Fig. 1.2). Although many efforts that could be characterized as "green" came before the 20th century, the whimsical architecture of Antoni Gaudí was noted for the reuse of waste materials (Fig. 1.3). The buildings of architect Frank Lloyd Wright are the modern epitome of what was then termed organic architecture, although not without flaws. Fallingwater may be the most obvious example of the integration of constructed buildings with their natural surroundings (Fig. 1.4).

An overview figure showing numerous aspects of a green building and site is given in Fig. 1.5. The key concepts include thoughtful use of low-impact materials, minimizing building energy use and/or

FIG. 1.2

Cliff dwellings of Mesa Verde in southwestern Colorado from ∼ a millennium ago. The buildings were oriented to collect the winter sun and remain shaded in the summer sun.

FIG. 1.3
Parc Guell in Barcelona, Spain, was designed by Antoni Gaudí ~1900 using organic shapes and integrated waste materials into the construction.

Image used with permission Jenifoto/Shutterstock.

FIG. 1.4
Frank Lloyd Wright's Fallingwater home was built in ~1937 for a private client in Pennsylvania and integrated it with the surrounding environment.

Image used with permission Will Ashley/Shutterstock.

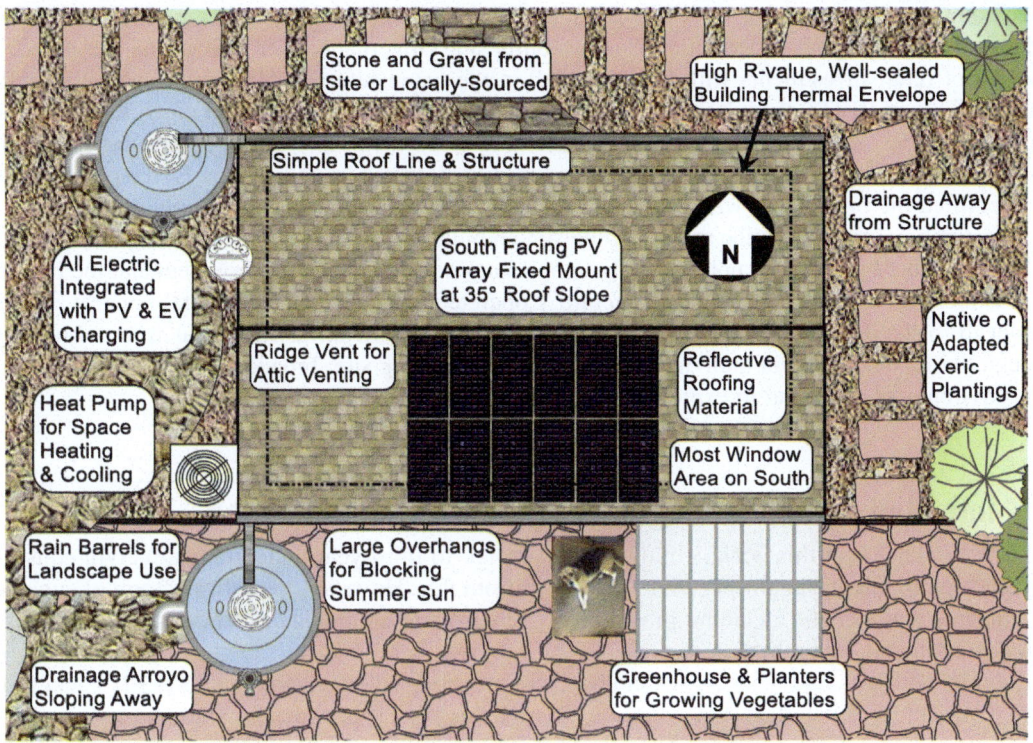

FIG. 1.5

Key features of a green-built residential structure intended for an arid, sunny US location.

producing energy on-site, and a building that is well insulated from atmospheric extremes though is well integrated with the site otherwise.

According to the Brundtland Commission, which first defined sustainability, sustainability is meeting the needs of the present generation without compromising those of future generations. This is an elegant statement of the notion of incorporating minimizing resource use, maximizing efficiency in materials and energy resources, and fostering the use of materials and energy that can be rapidly renewed. Sustainability is defined by the American Society of Civil Engineers (ASCE) by the following: "a set of economic, environmental, and social conditions (aka the 'Triple Bottom Line') in which all of society has the capacity and opportunity to maintain and improve its quality of life indefinitely without degrading the quantity, quality, or the availability of economic, environmental, and social resources. Sustainable development is the application of these resources to enhance the safety, welfare, and quality of life for all of society (ASCE)."

Sustainable is green but is larger, as it also incorporates these broader economic aspects and social aspects. A building could be green but not cost-effective, reliant on child labor, and thus not truly sustainable. The triple bottom line in a sustainable project considers the economics, social aspects, and environmental impacts of a project.

The economics of green building is also transforming. Capital and financing markets for both construction and more widely have responded in kind, and there are now numerous options, including green bonds, stocks, exchange-traded funds, and other vehicles for investing in green building efforts and even more having a broader Equity, Sustainability, and Governance (ESG) focus. As of April 2022, a search of retail investments with "sustainable" in the name returned 500 options, mostly mutual funds, ETFs, and REITs, though the focus on environmental sustainability is not clear. Investment research firm MSCI established a global green building stock index that tracks companies that derive 50% or more of their revenues from products and services in green building. Clearly, the private sector has identified the problem as well as the emerging players working in the green building space.

A last important clarification regarding the "green building" term, which will mostly be used here. It should be noted that it does not just apply to buildings but to the built environment and the building process more broadly, e.g., green infrastructure. Concrete with lower environmental impacts is as important to roads, dams, water treatment facilities, and other built infrastructure as it is to buildings. Though the focus here is the individual building at its building site, broader applicability to a wider built environment is natural to explore.

1.3 Intended audience: Engineering students and beyond

The primary target for this book is mid-stream (second- to third-year undergraduate) civil, construction, and environmental engineers who will benefit from this. However, the full spectrum of engineering students may benefit including most notably: energy, environmental, architectural, mechanical, material, chemical, and others.

Having said that, the science, technology, engineering, and mathematics (STEM) background needed to tackle this material is not extensive. Though this book is intended for engineers, it is certainly accessible and potentially useful to physical scientists, architects, sustainability practitioners, constructors, construction managers, builders, green building and sustainability specialists, and contractors. The sciences of most relevance include architecture, building science, physics, and chemistry, and students in these disciplines may also find the book useful. Calculus is used but is done so sparingly and is not mandatory to understand the book. Algebra, geometry, and arithmetic are reasonably sufficient. A well-versed, numerically oriented nonscientist could successfully take this on.

Problem-solving is key to science and engineering. The book has ample examples and exercises for students invoking both calculations (only some calculus-based) as well as conceptual problems. Both are important; science is about diagnosing the problems we face, and engineering is about designing the solutions to these (using scientific principles). Discussing one without the other is a less fulfilling approach.

1.4 Approach and differentiators of this textbook

Why another textbook on green building? There has been a profusion of books on green building, and there is now even an entry in the series "Green Building and Remodeling for Dummies." A broader field, sustainable engineering is an emerging field, and several new textbooks and other learning aids have been released in recent years (Reddy et al., 2019). The key differentiators here are (a) a wide introductory focus on the green-built environment from construction to operations and (b) an engineering and science approach.

There are great textbook options for architects, construction practitioners, and sustainability experts. The Sustainable Construction book by Kibert is an excellent and comprehensive introduction to the field aimed at architecture students and a broader audience. Likewise, the book by Kruger and Seville is also an excellent and well-diagrammed book aimed at green building practitioners, residential builders, and sustainability experts. What I found lacking was an introductory book targeting engineers and scientists, e.g., a quantitative and problem-solving approach that resonates with engineering students (albeit with considerable qualitative parts as well).

This book takes an engineering problem-solving approach to evaluate green building techniques and materials, their environmental impacts, and consider the economics. We examine the built environment from inside to outside in looking at minimizing environmental impacts. Energy and water use and efficiency are key concerns. Although technologies are described and referenced, the focus is a more fundamental understanding and analysis approach that enhances the students' underlying curiosity and problem-solving skills. This can then be applied to many technologies and problems. Gaining a physical (and at times chemical) understanding of building science is important as it affects energy use, water use, materials, and other aspects of relevance. The approach is as much about energy and materials' literacy and an intuitive feel for their relevance and use.

The proliferation of materials, components, and systems over the last several decades is a boon to green building but bewildering to neophytes. It has made the critical jobs of the architect, engineer, and most importantly, constructors and contractors, more complex. This book does not seek to enumerate or describe all materials, systems, specifications, or techniques on the forefront. For example, the book reviews the accreditation and certification programs available to students, including the Leadership in Energy and Environmental Design (LEED) program and ASCE's Envision program. However, coverage and details of any particular technology or certification program such as LEED are thus not extensive. Many other resources that can be kept current are better suited for keeping pace with the latest and greatest. Many are referenced where applicable.

This book pursues an overarching framework that will help the engineer or scientist make these evaluations. The approach hereto builds a technical intuition for scale and the ever-important mass and energy balance concepts that can be applied in analyzing any technology or system. Practical, everyday descriptions, examples, and analogies are used often because this is how I understand it. The scale of the problems we face—and the challenge of the scale-up of solutions—are emphasized. What does 40 gigatons/year of CO_2 mean? It is hard to picture this mass, particularly when it is gas-phase. Equating with other large objects sometimes helps, e.g., "How many Empire State Buildings does this represent?" can make it more palpable. A few "capstone" student group projects are suggested in the exercises at the end of the chapter that could serve or be modified for semester course projects.

An engineer's job combines words and numbers, and both are vital to the profession, and engineers need to be literate and numerate in their work. Throughout there are writing-oriented exercises, as engineers will almost invariably be involved in writing reports, summaries, white papers, journal articles, reviews, and the list goes on. We also borrow from the architectural approach to present a dozen or so case studies spanning a range of building and project types. These help to bring some life and real-world feeling to the material. The use of photos, diagrams, and illustrations is important here as well. The majority of these are my own but I have resorted to public domain and images licensed from Shutterstock in cases where I lack images.

As a forewarning and apology, my experience and biases give the book a strong tilt toward the United States and more specifically the southwestern region. This is not to diminish the efforts throughout the rest of the United States and the world. It is my experience, and much of it is on the street knowledge from trying to implement these, or the "school of hard knocks" rather than the ivory tower. Perhaps a second edition can broaden my perspective with the help of an international co-author.

1.5 Trajectory of this textbook

The book is intended as an introduction for engineering students to environmental threats, and in particular climate crisis, its impacts, and the contributions and role of the built environment in responding to this threat. This could serve as a stand-alone book for those engineers who take a single course focused on environment and sustainability. The primary focus is on applications to buildings, a secondary focus on the site, and a tertiary focus on the broader built infrastructure. The site and broader infrastructure are important contexts for the building design itself. At times, particularly in the first few chapters, there are examples and problems beyond buildings, applicable to understanding the climate emergency we face. Students must have a fundamental understanding of the problems that need to be addressed.

The book takes the reader through big-picture topics that are key to understanding the field first and the fundamental concepts of physical and chemical principles that can be applied. Exploration of green building requires an understanding of some key physical relationships, including the concepts of energy balance and mass balance. A subset of the physical laws familiar to scientists and engineers—ideal gas law, key fluid dynamics relationships, heat transfer mechanisms—also need review as they apply to green buildings. A historical look at both energy and materials use follows along with a chapter motivating why we care about building greener as a solution to environmental problems, most notably climate change. This approach is to highlight a selection of products, but it is not at all comprehensive. The focus here is problem-solving and thinking analytically—the whole house, systems engineering approach.

Green building is framed as one of many pieces of the puzzle to solving climate change and a myriad of environmental problems plus big picture climate change solutions, both of which are given as an overview for the uninitiated (and which can be skipped for well-versed students). Following, the green planning and design approach is discussed, as many of the most impactful decisions are made before any project is shovel-ready. Next, the material choices are discussed, and then their application to the thermal envelope of the building. Some of the key engineering systems in a building such as HVAC systems are then covered, as this is integral to the building energy use and engineering design. Water use in buildings is next discussed and connected to the outdoor environment of the building. Maintaining healthy indoor environments and the use of on-site renewable energy sources are the next topics. The various certification programs and the socioeconomic context, including cost and equity issues, are next discussed. Finally, in the concluding chapter, the concept of resilience engineering is introduced, underscoring how we must future-proof our infrastructure.

1.6 Engineering problem-solving approach and dimensional homogeneity

The comprehension of the key elements of sustainability in the built environment can be quantified in joules (of energy use, of embodied energy), kilograms (of material use), and liters (of water use). The problem-solving process relies on units' management and assuring the units of the solution make sense (a power rating should be in kilowatts and not in kilograms). To compound the problem, in the United States, we often deal with both SI and customary US units primarily derived from the imperial system, making unit management even more vital. Take, for instance, pressure that has a myriad of units including Pa, mm Hg, in. liquid H_2O, mBar, atmospheres, psi, torr, and the list goes on.

An engineering approach to problem-solving requires some methodical forethought and time invested. The solution should have the following (and I require students to use these steps):

1. Summarize the problem to make sure it is grasped.
2. State the given parameters in the problem.

1.6 Engineering problem-solving approach and dimensional homogeneity

3. State what is sought or the solution you are pursuing.
4. State assumptions that are required to get to a solution.
5. Diagram the process as appropriate.
6. Provide legible and logical solutions with appropriate detail.
7. Give the final solution in a box and with proper units.
8. Comment in a sentence or two on the meaning and reasonableness of the solution.
9. Make sure all terms in an additive equation have the same units—they must!
10. Make sure they all have the same system (SI) and order of magnitude (kW vs W vs MW)!
11. Units management—work through the units in a problem.
12. Check your final answer—are the units logical?
13. Is the magnitude and sign of the final answer logical?
14. Significant figures: usually two to three digits past the decimal point is plenty precise for the final solution!
15. The use of proper units for a solution is key to engineering. My advice to students is that if you get an answer that seems of the right magnitude and is the expected unit, there is at least a 50% probability it was correctly solved. Nonsensical units or orders of magnitude guarantee it was not.

A final exercise scientists and engineers will frequently encounter is the graphical display of information. This is often the most effective and efficient means to convey a concept and show evidence for an assertion. An example is given in Fig. 1.6 from the water quality world, namely displaying the biochemical oxygen demand (BOD_r) of wastewater remaining over time.

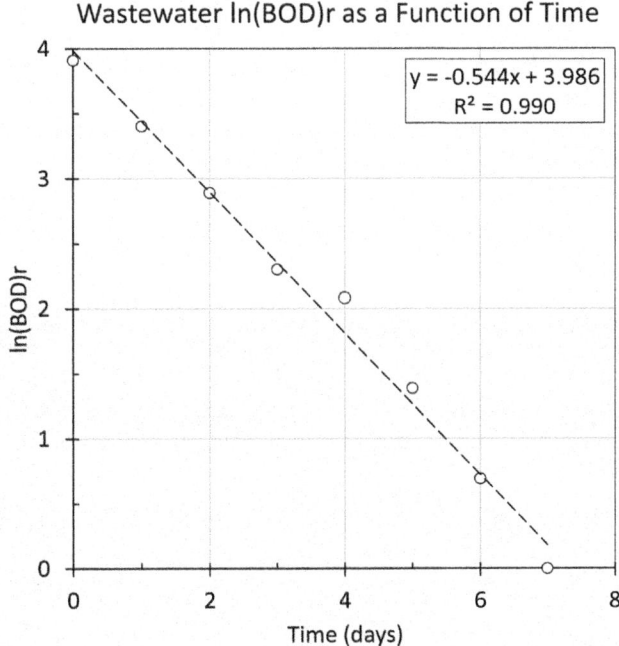

FIG. 1.6

Effective graphical presentation of data is important to engineering. The plot is the natural logarithm of remaining biochemical oxygen demand (BOD_r) versus time and shows that the BOD_r decays exponentially and thus is a first-order removal process.

1.7 Graphical presentation of data

A challenge for developing engineers and scientists to effectively communicate technical information. Thus, graphing data is a focus here as well. An engineer needs to be able to graphically convey information succinctly. This is a key form of technical communication scientists and engineers need to be able to accomplish effectively. The series of books by Edward Tufte on graphical data presentation gives a deeper dive into effectively displaying data. An effective graph should have (Fig. 1.6):

(1) Labeled axes with parameters and units.
(2) Axes with a logical number and size of increments.
(3) A reasonable number of data series on a plot and differentiated appropriately with colors, symbols, and lines.
(4) A legend or differentiation of datasets displayed.
(5) A readable font and size.
(6) Lack of visual clutter.

Numerous end-of-chapter exercises involve the graphing of data.

1.8 Accreditation programs: ABET: Gold standard for engineering programs

The Accreditation Board for Engineering and Technology, Inc. (ABET) accreditation program is essential to science and engineering programs and provides the norm for maintaining a viable engineering program in the US and beyond. ABET is a nonprofit nongovernmental organization (NGO) that has served as a third-party reviewer to assess and accredit higher education programs in the sciences and engineering since 1932. Throughout the book, problems related to ABET criteria are given. The most essential part is encapsulated by the following Criterion 3 for Student Outcomes (www.ABET.org):

> "The program must have documented student outcomes that support the program's educational objectives. Attainment of these outcomes prepares graduates to enter the professional practice of engineering. ABET specified Student Outcomes (1) through (7) below as the goals for graduates of an accredited engineering program, plus any additional outcomes that may be articulated by the program.
>
> 1. An ability to identify, formulate, and solve complex engineering problems by applying principles of engineering, science, and mathematics.
> 2. An ability to apply engineering design to produce solutions that meet specified needs with consideration of public health, safety, and welfare, as well as global, cultural, social, environmental, and economic factors.
> 3. An ability to communicate effectively with a range of audiences.

> **4.** An ability to recognize ethical and professional responsibilities in engineering situations and make informed judgments, which must consider the impact of engineering solutions in global, economic, environmental, and societal contexts.
> **5.** An ability to function effectively on a team whose members together provide leadership, create a collaborative and inclusive environment, establish goals, plan tasks, and meet objectives.
> **6.** An ability to develop and conduct appropriate experimentation, analyze, and interpret data, and use engineering judgment to draw conclusions.
> **7.** An ability to acquire and apply new knowledge as needed, using appropriate learning strategies."

The student outcome ABET criteria may be summarized as (1) problem-solving, (2) design and optimization, (3) communications, (4) ethics and environment, (5) teaming and leadership, (6) experimentation, and (7) constant learning. These seven outcomes are the focus of the approach and problem-solving nature of this book.

Related to climate change, the engineering profession is putting much greater emphasis on ethics, sustainability, and the broader context of the engineering profession. Civil and environmental engineers are duty-bound to consider the sustainability of the projects they undertake (e.g., the Professional Engineers Exam, the ASCE code of ethics, and other credentials mandate this more commonly). This shows up both as the focus of the ASCE Envision program, which will be detailed in a later chapter, as well as ABET Student Outcome 2 being directly related.

1.9 Summary and additional resources

Green building is quickly transforming from a niche sector of the construction market to the dominant approach. At some point, it will just become "building" shortly without the need to specify green or sustainable (a positive development). The world of green and sustainable construction is changing quickly, particularly concerning the technologies and products available in the marketplace. I would be remiss not to point the reader to a fraction of the other outstanding resources available. Many of the books are referenced throughout, as are some of the websites. It is impossible to keep up-to-date completely with these changes in a textbook format. Here we provide some resources for the aspiring student to consult about the latest and greatest developments in this arena.

The "Pretty Good House" is a green building and renovation movement that focuses on achievable results rather than ultra-high-performance temples to efficiency that usually are high-end homes that are only affordable by a very small subpopulation (Kolbert et al., 2022). Rather than perfect solutions, it focuses on achievable, affordable buildings that are healthy for the occupants and environmentally minimized. Likewise, the book "A House Needs to Breathe, or Does It" by Allison Bailes is a contemporary look at scientific knowledge applied to green building (Bailes, 2022).

The highly recommended website of Green Building Advisor (https://www.greenbuildingadvisor.com/) is best in class for in-depth online discussions of green building science, technology, and real-world experience. Likewise, the Energy Vanguard website moderated by Allison Bailes is an indispensable website for building efficiency discussion from a physicist. The reader will find extensive discussion ranging from building physics to on-the-ground experience with techniques and technologies from a passionate group of thinkers and doers. The Building Green website is another excellent resource to consult for more up-to-date information on innovations, technologies, and methods than a textbook like this one. Environmental Building News is another key resource that has shown longevity.

The Rocky Mountain Institute (RMI) is a nonprofit started by Amory and Hunter Lovins in the early 1980s and focused on sustainable energy generation and use (www.rmi.org). RMI established a highly efficient laboratory in the Rocky Mountains of Colorado that could be heated passively from solar input due to its high efficiency.

Architects have many venues, including the Architecture 2030 organization, which was a pioneer in quantifying and working to minimize the contributions of the built sector to climate change (https://architecture2030.org/). Building Science Corporation is another excellent resource for technical information related to building performance. They maintain a large library of white papers on a specific topic related to building performance.

The challenge is great. We have become wealthier, but this has had ramifications for the sustainable development of our resources. With the climate crisis, we have a limited timeframe to figure out how to sustainably rebuild our energy infrastructure to be more resilient and provide sufficient energy access, affordably and equitably, and with limited environmental liabilities (NASEM, 2021). It has been said that there is no silver bullet to solving our energy and environmental problems. Improving the built environment certainly represents a significant, though tarnished, piece of the silver buckshot that is available and necessary.

CHAPTER 2

Energy science: Key underlying physics

Learning objectives

(1) Grasp the fundamental aspects of energy science, including thermodynamics, heat transfer, and fluid mechanics, as it applies to the built environment.
(2) Apply the concepts of energy and mass balance in understanding the flows of energy and materials associated with a building.
(3) Differentiate energy, power, and efficiency as they apply to building systems, including heat engines and heat pumps.

Applying fundamental thermodynamics and heat transfer principles is central to evaluating the viability of green building materials and systems. This chapter is a broad (and admittedly cursory and incomplete) overview of the key scientific concepts that the engineering or science student learns in their physics, chemistry, and mathematics background. Important concepts and terms that will be reviewed include energy, work, power, the laws of thermodynamics, methods of heat transfer, enthalpy, internal energy, entropy, basics of fluid flow, latent and sensible heat, ideal gas law, and the heat capacity of materials.

The intent here is an overview (or preferably a review from completed semester-long courses on thermodynamics and heat transfer), and it is highly recommended that the students have prior exposure. The focus is on the key relationships, and their derivation is beyond the scope of this book.

It is difficult to distill the key concepts of thermodynamics and heat transfer into a single chapter as they are normally spread over two or more semester courses. The uninitiated may find it difficult to digest this all in reading one chapter; nonetheless, the reader is encouraged to bear with the process in this chapter even in the face of the most equation-dense of the chapters. Similarly, students who are well familiar with these concepts could easily spend minimal time reviewing this or skip this chapter entirely.

2.1 Energy as a fundamental physical quantity

Though it is not the sole metric relevant to green building, energy is one of the most universal parameters that relates to how a building impacts the environment. It is of relevance to climate change, building heat loss and gain, systems energy consumption, and materials embodied energy, among other important implications. An energy imbalance leading to the increase in global mean temperature is the central driver of climate change as well.

Energy is typically measured in joules (force of 1 N applied over 1 m) in the SI system or British thermal units (BTU) in the British system (energy required to raise 1 lb of liquid water by 1°F at standard temperature and pressure, STP). As will be seen, the joules associated with building energy use, embodied energy of materials, costs of producing treated water, the climate system, and heat loss in a building are all important facets of green building. A summary of important energy units is given in Table 2.1.

Table 2.1 Key energy units.

Unit	Definition and description	Notes
Joule (J)	A force of 1 Newton applied over 1 m	There are 4.18 J/cal, and thus the heat capacity of liquid water is approximately 4.18 J/(g K)
Calorie (cal)	The thermal energy required to raise the temperature of 1 g of water by 1°C	1 food Calorie is a kilocalorie (kcal) or 1000 cal
British thermal unit (BTU)	1 BTU is the amount of energy needed to raise the temperature of 1 lb of water by 1°F and is the force of 778 lb applied over 1 ft	A BTU is approximately the energy released from striking one kitchen match and is equal to 1055 J
Electron-volt (eV)	The energy change is accompanied by the acceleration of an electron when subject to a voltage difference of 1 V	Of use for particle physics where 1 eV is equal to 1.602E−19 J of energy

2.2 Forms of energy

Energy is an abstract parameter, and quotes from Dr. Richard Feynman underscore that it is hard to define. Even the physics definition of the "capacity to do work" is somewhat mysterious and vague. It does not help matters either that the myriads of energy units (joule, BTU, calorie, erg, electron-volt), both SI and imperial, that energy is measured. In a particular disservice to clarity, we invented the kilowatt-hour (kWh) energy unit that your electric bill uses (energy used with running 1 kW of electrical load for 1 h).

The reader is encouraged to conceptualize energy as to what form it takes—chemical, gravitational, kinetic, electromagnetic, nuclear, light, and sound. Finally, no energy is created nor destroyed, only transformed from one of these types to another. This forms the basis of the First Law of Thermodynamics. A summary of major forms of energy is given in Table 2.2.

Table 2.2 Forms of energy and examples.

Energy form	This form of energy is associated with	Example
Kinetic-translational	A mass in bulk translational movement	A car traveling along the highway
Kinetic-rotational	A mass in rotation	The car wheel rotating on its axle
Gravitational potential	A mass located in a gravitational field	The car at the top of a hill versus the bottom of the hill
Chemical potential	The energy associated with the chemical bonds in a substance	The gasoline containing H—C and C—C bonds versus the product oxidized form containing C—O and H—O bonds
Thermal	Molecular motion of an object or substance	The waste heat from the vehicle exhausted from its radiator
Electromagnetic	Transmission of electromagnetic energy such as light	The electromagnetic radiation transmitted from the sun to Earth growing the flora, fauna, and microbes forming the crude oil
Nuclear	The strong force that holds nuclei together	The nuclear bonds in the hydrogen and helium isotopes in the great fusion reactor in the sky, the sun
Internal	The molecular kinetic and potential energy	Combustion of methane releases internal energy that is stored chemical potential energy and transfers this to the kinetic internal energy of the combustion products

2.3 Power (W) and power density or flux (W/m²)

It is vital to understand the difference between energy and power, which are convoluted, even at times by engineers and scientists. Although used interchangeably in conversation, power is the rate of energy use. Power is typically in units of joules/second, which is a Watt, or in BTU/h in the imperial system. Using a fluid flow analogy, power is the flow rate of the river while energy is the accumulated water in the lake (Fig. 2.1). In another analogy, the odometer is to the speedometer as the energy used is to the power. It does not help that we use kWh as a billable energy unit (energy consumption related to the use of 1 kW of power such as a typical hair dryer for 1 h). Thus, the energy consumed in a process can be found from the integral over time of power (P) (Eq. 2.1).

$$E = \int P dt \qquad (2.1)$$

This of course becomes $E = P \times t$ if power is constant, or conversely, $P = E/t$.

The scales of our power consumption are of relevance to have an "energy intuition" and context. The listing in Table 2.3 gives power consumption from a wristwatch (microwatts) to the sun's power output (approaching the Exawatt scale) and spans ~24 orders of magnitude. Relevant electrical devices that consume power in each factor of 1000 are given (Table 2.3). The following example gives the reader a scale for human total power consumption from all sectors (Example 2.1).

Occasionally, "power density" (P/A) or power flux is a useful concept metric. The power scale of a facility as compared to its land use footprint. The latter can become a fuzzy concept in terms of what

16 Chapter 2 Energy science: Key underlying physics

FIG. 2.1

The analogy is power as the flow rate of the river and energy as the water accumulates in the lake.

Table 2.3 Scales of power consumption in Watts and corresponding devices.			
Number	Power of 10	Prefix	Device of this approximate power scale (Watts)
Millionth	$1E-6 = 1 \times 10^{-6}$	Micro-	Quartz wristwatch
Thousandth	$1E-3 = 1 \times 10^{-3}$	Milli-	Laser in CD or DVD player
One	$1E0 = 1 \times 10^{0}$	–	Nightlight (LED)
Thousand	$1E3 = 1 \times 10^{3}$	Kilo-	Residential photovoltaic systems
Million	$1E6 = 1 \times 10^{6}$	Mega-	Large wind turbine
Billion	$1E9 = 1 \times 10^{9}$	Giga-	Large coal or nuclear power station
Trillion	$1E12 = 1 \times 10^{12}$	Tera-	Electric power consumption in the United States
Quadrillion	$1E15 = 1 \times 10^{15}$	Peta-	The most powerful laser power output
Quintillion	$1E18 = 1 \times 10^{18}$	Exa-	$0.174 EW =$ Solar power received by Earth

> **EXAMPLE 2.1 Energy and power**
>
> **Problem:** The world currently uses ~524 Quads of energy for all uses in a year. Convert this to a steady-state Terawatt equivalent. What mass flow rate of coal is needed if entirely provided by 25 MJ/kg coal and typical power station efficiency?
> **Given:** 524 Quads/y worldwide energy use.
> **Find:** Steady-state (SS) power equivalent.
> **Assume:** Constant use. 33% efficient power plant.
>
> **Solution:**
>
> $$\text{Power} = \frac{\text{Energy}}{\text{Time}}$$
>
> $P = 524\text{E}15\,\text{BTU/year} * (1055\,\text{J/BTU}) * \text{year}/(24*3600*365)\,\text{s}.$
> $P = 1.75\text{E}13\,\text{J/s}$ where 1 TW is 1E12 W.
> $P = 17.5\,\text{TW}$ equivalent constant power usage globally.
> $dM/dt = (1.75\text{E}13\,\text{J/s} * 1/0.33) * \text{kg}/25\text{E}6\,\text{J} * 1\,\text{tonne}/1000\,\text{kg} = 2100\,\text{tonne/s}.$
>
> If all were provided from coal, it would take ~21 train cars/second or a 1-mile trainload every 5 s! Our global rate of energy "use" (more accurately energy conversion as we are merely converting it into a more useful form) is staggering.

upstream facilities to include in the footprint (e.g., mining for fuels or construction materials). P/A is also important for benchmarking power use in a structure, in which case the steady-state power usage is divided by the floor area of the building. It also becomes important for sizing and land use for energy facilities, particularly solar and wind applications that are distributed energy resources.

2.4 Essential electricity relationships

The key electrical relationships of relevance are expressed in Ohm's law and Joule's law as given below. Ohm's law states that the voltage drop across a device is equal to the current multiplied by the resistance, while power consumption is equal to the voltage times current (Eqs. 2.2 and 2.3).

$$V = IR \tag{2.2}$$

$$P = VI = I^2 R \tag{2.3}$$

One other aspect is the storage of electrical energy. Battery storage is quantified by the Amp-hours, kilowatt-hours, or joules of storage capacity. An example of the equivalencies is given (Example 2.2).

> **EXAMPLE 2.2 Battery storage**
>
> **Problem:** Find the equivalency in MJ and Amp-hours for a 12 VDC battery storage system that stores 100 kWh.
> **Given:** 12 VDC, 100 kWh.
> **Find:** Equivalent MJ and Amp-hours.
>
> **Solution:**
>
> 100 kWh is the approximate storage currently in the longest-range electric vehicles.
> Energy = 100 kWh = 100 kJ/s (3600 s) = 360,000 kJ = 360 MJ.
> At 12 VDC, 1 Ah is equivalent to:
> Electrical energy = $VI * t$ = 12 V (1 Ah) (3600 s/h) = 43,200 J = 0.0432 MJ.
> Thus, 360 MJ is equivalent to 8333 Ah stored in a 12 V battery.

2.5 Thermodynamic properties, systems, and processes

The proper treatment of thermodynamics requires a minimum of a semester-long course in science or engineering and the introductory physics and chemistry classes that precede it. Here, the key concepts related to buildings and energy use are discussed in brief.

The universe can be separated into a user-defined system and the surroundings, which are separated by a control surface (Fig. 2.2). A system may be open (allowing mass flow between the surroundings and itself) or closed. The closed system prohibits the exchange of mass with the surroundings, whereas the isolated system prohibits the exchange of mass or energy with the surroundings. An open system (such as a pump) is best modeled with a control volume approach where the mass flows can be clearly defined (e.g., pump inlet and outlet) across the control surface of the defined control volume (Kroos and Potter, 2014; Turns and Pauley, 2020).

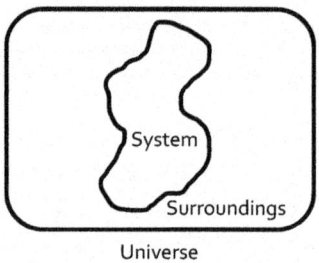

FIG. 2.2

In thermodynamics, the universe is divided by a control surface into the system of interest and its surroundings.

A system will occupy a singular, specified thermodynamic state at any point in time, which is defined by two state variables. Temperature is a fundamental thermodynamic state variable and is something for which we have an intuition of hot and cold. It represents the average kinetic energy of the molecules of a substance where the higher the molecular kinetic energy, the higher the temperature (usually in Kelvins which is an absolute scale starting at zero K or in °C which equals Kelvins-273.15). Molecular kinetic energy includes translational, rotational, and vibrational forms. Pressure is another key state variable that is defined as the force per unit area normal to a control surface due to the system.

Internal energy (u) is a thermodynamic state variable and is a function of temperature, itself a measure of the energy associated with the molecular scale kinetic energy and the potential energy associated with the bonding and molecular structure (Turns and Pauley, 2020; Kroos and Potter, 2014). The National Institute of Standards and Technology (NIST) steam tables and other related compilations provide the values for internal energy and other thermodynamic state variables for different substances as a function of T and P.

Enthalpy (h) of an ideal gas takes the (microscopic) internal energy of the system and adds the macroscopic energy associated with pressure and volume for the gas (Kroos and Potter, 2014; Tosun, 2015). For a liquid or solid, these two parameters are equivalent but differ for ideal gases due to the P-v characteristics of the gaseous system.

For these purposes, entropy is the state variable that quantifies the molecular scale randomness of a system. Generally, the gas phase > liquid phase > solid phase in terms of entropy.

A system goes through a process when it changes from one state to another as defined by the state variables discussed earlier. A few of the key concepts and processes that are covered in the field of thermodynamics are given in Table 2.4 (Example 2.3).

Table 2.4 Key thermodynamic terms and processes.

Process	Description	Example(s)
Open system	A system that can exchange mass with its surroundings	A pump, turbine, heat exchanger, or other device
Closed system	A system that does not exchange mass with its surroundings	A sealed piston-cylinder arrangement
Adiabatic	A process where the system changing does not exchange heat with the surroundings	A perfectly insulated system. Often power turbines in a power plant are considered adiabatic as the heat loss is negligible
Isothermal	The temperature of a system remains constant during a process	An indoor room temperature controlled with a thermostat (approximately)
Isobaric	A constant pressure process	A pressure cooker where the relief valve has opened; a frictionless piston-cylinder with a constant piston weight
Isentropic	A constant entropy process	Expansion through a throttle device is often modeled as isentropic
Isochoric	A constant volume process (also called isometric)	A rigid container that does not expand or shrink leading to no boundary work
Reversible process	A cyclic process that can be run in the reverse direction without any additional energy input	Carnot cycle
Phase change process	A change in phase from solid, liquid, or gas to another, where latent heat is absorbed or released depending on the direction of the process	Vaporization of steam in a power plant boiler

EXAMPLE 2.3 Open vs closed systems

Concepts: You have the following devices: Steam Turbine, Planet Earth, Heat Exchanger, Weather Balloon, Arctic Ice mass, and Automobile Piston. Determine if they are better represented by a control volume approach or a closed system.

Solution: Devices: Steam Turbine (control volume), Planet Earth (closed system essentially), Heat Exchanger (control volume), Weather Balloon (closed system until it ruptures), Arctic Ice mass (control volume), Automobile Piston (depends on time frame-closed system during the brief power stroke, control volume during exhaust stroke).

2.6 Heat and work

Two more tangible forms of energy transfer are denoted by the concepts of heat and work. Both are forms of energy as it is transferred from a system to the surroundings or vice versa. They are not properties of a system or thermodynamic state variables; they represent a boundary transfer process.

Heat is the flow of thermal energy from one body to another due to a temperature difference and occurs via conduction, convection, or radiative transfer. Heat can be the intended end-use (e.g., a gas furnace heating an interior space) or an intermediate form of energy converted into something more useful (e.g., a thermal power plant converting thermal energy into electrical energy). A table describing some of the units of heat and work is equivalent to the units for energy given previously (Table 2.1).

Energy systems involve transformations from heat to work (or vice versa for the heat pump). The transformations are sometimes many. The example of the energy transformations involved in the human body as well as transportation options illustrates this (Example 2.4).

EXAMPLE 2.4 Stored chemical energy in food

Problem: You are getting juiced up to pull an all-nighter to study for your exam (please do not do this). You go to the qwik-e-mart to pick up a 128-fluid-ounce mega-ultra super gulp of sugary soft drink. In doing so, you walk 2 km each way. Do you work off the Calories ingested?

(a) Estimate how much energy you ingest via drinking the sugary soft drink versus how much you burn in walking 4 km (expressed in joules, BTU, and calories, and don't overlook the difference in chemist's calories vs food Calories).

(b) Instead, you hop in the Hummer (assume 6 mpg) and drive to pick up the beverage. How much chemical energy did this burn in joules, BTUs, and calories for the roundtrip?

(c) Instead, you take your Airbus A380 which consumes 13.78 kg/km of fuel with a fuel energy content of 43.3 MJ/kg. Neglect takeoff and landing losses. How much chemical energy did this burn in joules, BTUs, and calories for the roundtrip?

(d) Make a bar chart comparison of the energy scales in SI units and comment.

(e) Show a box diagram of the transformations of energy from one form to another stepping back to the source of energy for the gasoline to the end product of the Hummer. There are at least 5 transformations.

Given: Soft drink, 4 km via walk, Hummer, A380.
Find: Energy scales of each, are the calories input from the drink consumed?
Assume: 4 km/h burns 236 Cal; Hummer 6 mpg; A380 13.8 kg/km jet fuel @ 43 MJ/kg.

Solution:

Sugary soft drink in:
14.5 kcal/oz × 128 oz × 1000 cal/kcal = 1,856,000 cal.
1.856 Mcal × 4.18 J/cal = 7.8 MJ.
7.8E6 J × BTU/1055 J = 7339 BTU.

Walking energy consumed:
236 Cal burned for 4 km (1 h) of walking at 4 km/h = 236,000 cal.
236,000 cal × 4.18 J/cal = 9.86E5 J = 0.99 MJ.
86E5 J × BTU/1055 J = 935 BTU.

Hummer gasoline energy consumed:
2.5 miles/6 mpg = 0.42 gal × 124,000 BTU/gallon = 51,700 BTU.
51,700 BTU × 1055 J/BTU = 5.45E7 J = 54.5 MJ.
5.45E7 J × cal/4.18 J = 13.0E6 cal.

A380 energy consumed:
13.78 kg/km * 4 km * 43.3 MJ/kg = 2387 MJ.
2387 MJ × cal/4.18 J = 5.71E8 cal = 571 Mcal.
2387 MJ × BTU/1055 J = 2.26 MBTU.

You do not work off the calories ingested. The range of energy scales is vast, from 1 MJ expended from walking to over 1000 MJ for the aircraft. Recommend: Walk, have water or coffee if necessary but get your sleep!! A comparison of energy scales and transformations is given in Fig. 2.3.

Continued

2.6 Heat and work

EXAMPLE 2.4 Stored chemical energy in food—cont'd

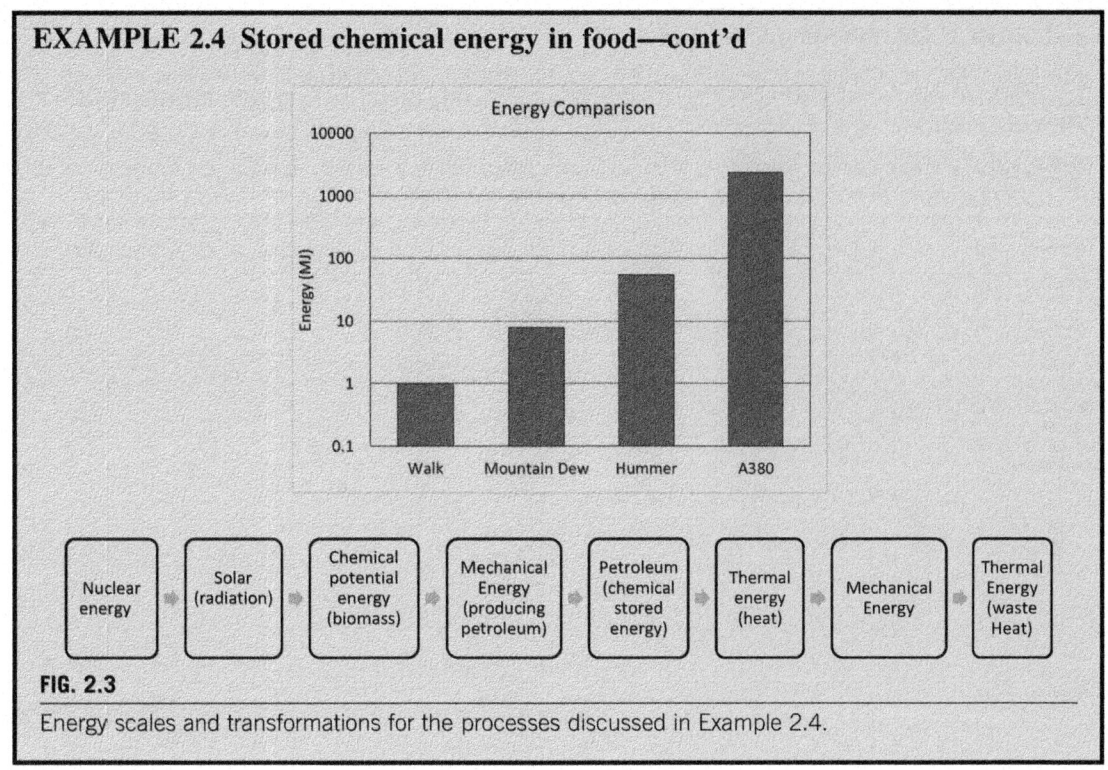

FIG. 2.3

Energy scales and transformations for the processes discussed in Example 2.4.

Work results from a force applied over a distance and can thus be thought of as macroscopic kinetic energy. Displacement work and work associated with gravity's resistance are given in differential form in Eqs. (2.4), (2.5), respectively.

$$dW = F \cdot dx = mA \cdot dx \tag{2.4}$$

$$\Delta PE = mg \cdot \Delta z \tag{2.5}$$

$$\text{Translational } KE = \frac{1}{2}mv^2 \tag{2.6}$$

Rotational energy is given by one-half the rotational velocity ω (radians/s) multiplied by the square of the moment of inertia I (which for a simple rotation mass on a connecting rod is mr^2).

$$v = r\omega$$

$$\text{Rotational } KE = \frac{1}{2}I\omega^2 \tag{2.7}$$

A useful diagram for an ideal gas is a *P-v* diagram plotting the pressure (*P*) and specific volume (*v*) relationship for a process or cycle. The area under the curve is the work done based on the relationship given for work in Eq. (2.8) (Example 2.5).

$$W = \int P \, dv \tag{2.8}$$

> **EXAMPLE 2.5 P-v diagrams and work**
>
> **Problem:** A heat engine using an ideal gas is working cyclically (Fig. 2.4). You are trying to maximize work out and minimize work into the device. When you are (a) going through a process from point A to point B, and when you are (b) returning from B to A, what are the preferred pathways to maximize cycle efficiency?
>
> **Solution:** The area under the *P-v* diagram should be maximized. Thus, pathway 5 (maximizing positive work or work out of the system) should be followed from A to B and pathway 1 from B to A (minimizing negative work or work into the system). This cycle encompasses the largest area internal to the cycle and thus the largest network out of the cycle from A to B and back again.
>
>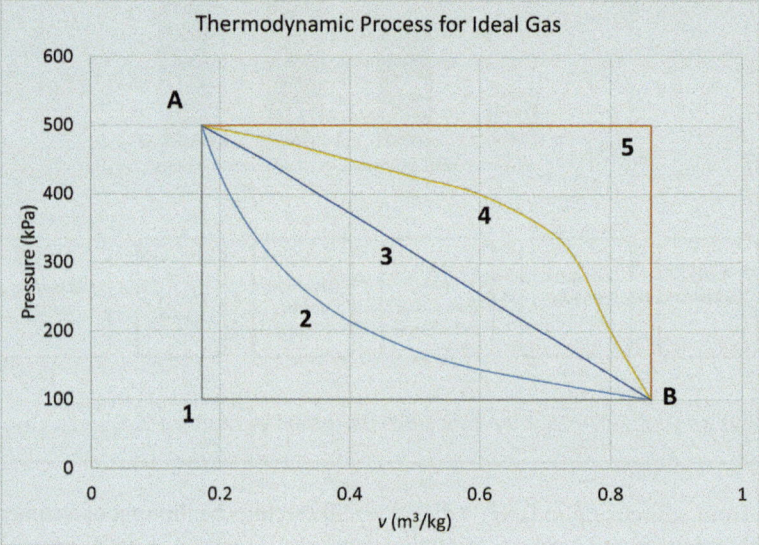
>
> **FIG. 2.4**
> Processes from point A to B are plotted on a *P-v* diagram as discussed in Example 2.5.

2.7 Heat transfer: Conduction, convection, and radiation

Heat transfer mechanisms are critical to the energy use of a structure, particularly related to the space heating and cooling of a building for indoor comfort. Heat is transferred between objects at differing temperatures, and the net heat flow is from the warmer to the cooler body. The three means of heat transfer are conduction, convection, and radiation, and all are dependent on the temperature difference ΔT between the objects.

Conduction is the direct transfer of heat through a heat-conducting body (e.g., the hot pot on the stove conducting heat through the handle to one's hand). Convection is the transfer of heat via a moving fluid (e.g., air or water typically). The transfer of heat via the water vapor motion above the pot is an example. The similar conduction and convection equations for the rate of heat transfer are given in

2.7 Heat transfer: Conduction, convection, and radiation

Eqs. (2.9), (2.10), respectively, where both processes are characterized by a constant (k or h) and are driven by a ΔT and an area of interface (A).

$$\dot{Q}_{cond} = \frac{\kappa A}{L}(T_2 - T_1) = UA(T_2 - T_1) \tag{2.9}$$

$$\dot{Q}_{conv} = hA(T_2 - T_1) \tag{2.10}$$

The key difference between the two equations is that the conductivity coefficient (κ) and conductor length (L) over which heat transfer occurs come into play for conduction, while the convection coefficient (h) characterizes convection. Of note, building components such as windows are often rated with a conductivity (U) that considers the conductivity κ and fixed length of the conduction path L such that $U = \kappa/L$. Also of note, the R-value is the reciprocal of U and is often specified for insulation (e.g., a 4″ thick fiberglass batt insulation has an R-value of approximately 12).

The third heat transfer mechanism is via radiation. A body, dependent on its temperature, emits radiation at a spectrum of energies (e.g., the flames below the boiling pot transferring energy to the surroundings) (Fig. 2.5). This mechanism requires no matter for it to occur (e.g., the sun's radiation intercepted by the Earth). This is the most temperature-sensitive heat transfer pathway (T^4 dependence) (Eq. 2.11).

$$\dot{Q}_{rad} = \sigma A \varepsilon (T_2^4 - T_1^4) \tag{2.11}$$

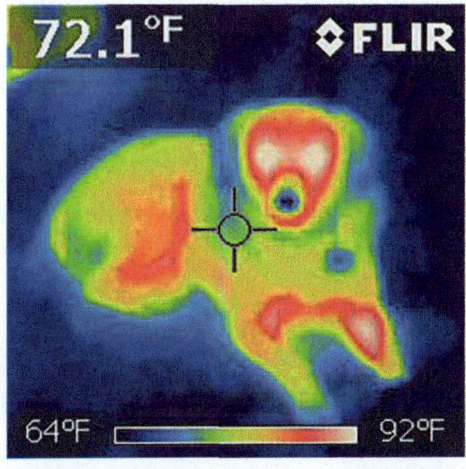

FIG. 2.5

Thermal image (infrared wavelength) of a healthy domestic canine. The effective body temperature as compared to the surroundings determines the infrared radiation emission magnitude and spectrum. The dog's extremities such as tail, ears, and nose are closer to the temperature of the surrounding room. The nose is particularly cool due to evaporative cooling. The less insulated areas near the core are warmer.

The constants in the equation are the Stefan-Boltzmann constant, which is $\sigma = 5.67E-8\,W/(m^2\,K^4)$, and the emissivity ε that takes on a value from 0 to 1 depending on how closely the body approximates a true blackbody. Lacking any other information, one can presume an emissivity approaching 1, a perfect blackbody (Example 2.6).

> **EXAMPLE 2.6 Radiative transfer**
>
> **Problem:** A typical person has a body surface temperature of 30°C. How much heat does he lose to the surroundings at 20°C? Assume the person has a surface area of 2 m² and is a perfect blackbody radiator as are the surroundings ($\varepsilon = 1$).
> **Given:** Human $2\,m^2$, $\varepsilon = 1$, $T = 303\,K$ and $T_{room} = 293\,K$.
> **Find:** dQ/dt via radiation.
> **Assume:** Steady state (SS).
>
> **Solution:**
>
> $$\dot{Q} = \sigma \times A \times \varepsilon (T_2^4 - T_1^4)$$
>
> $$\dot{Q} = 5.67 \times 10^{-8}\,Wm^{-2}K^{-4} \times 2\,m^2 \times 1(303^4 - 293^4)K^4$$
>
> $$\dot{Q} = 120\,W$$
>
> Each human radiates close to the input power of a 100 W lightbulb, plus or minus. This might be a small extra cooling load in a residence but is certainly worth considering in an arena with 25,000 people.

2.8 Heat capacity of materials, latent and sensible heat, stored chemical energy

Materials, by way of their molecular properties, including density, can store heat as internal energy. The simplest relationship, when the heat capacity may be considered constant, relates the quantity of heat transferred (Q) to the mass (m), the heat capacity of the material (c), and the temperature change (ΔT) (Eq. 2.12).

$$Q = mc\Delta T \tag{2.12}$$

For applications near room temperature, the heat capacities of liquid water and air are approximately 4184 and 1004 J/(kg K), respectively (slightly dependent on pressure and temperature conditions) (Table 2.5). The much larger heat capacity of water and its thermal inertia have numerous implications

Table 2.5 Heat capacities of common building materials.

Material	Heat capacity (J/g K)	Material density (kg/m³)	Heat capacity per unit volume (MJ/m³ K)
Water	4.18	1000	4.18
Basalt	0.84	3011	2.529
Limestone	0.84	2611	2.193
Granite	0.79	2691	2.125
Concrete	0.88	2371	2.086
Brick	0.84	2301	2.018
Gypsum	1.09	1602	1.746
Sand (dry)	0.835	1602	1.337
Soil	0.80	1522	1.217
Wood	0.42	550	0.231
Air	1.0035	1.204	0.0012

Data from Build Green Canada, http://www.buildgreen.ca/2008/09/an-explanation-of-thermal-mass/.

2.8 Heat capacity of materials, latent and sensible heat, stored chemical energy

from climate to thermal mass in green buildings. The climate system, due to the large heat reservoir represented by the ocean and its thermal mass, takes a long time to come to equilibrium given perturbation. For green buildings, thermal storage is much more viable with water than air, particularly when considering the much higher density as well. Water has the highest heat capacity of common construction materials per both unit mass and volume (Table 2.5).

Biological systems also use stored chemical energy in their tissues, most notably the hydrocarbon compounds dominant in fat tissue. A rough equivalence is that 0.454 kg (1 lb) of fat tissue stores approximately 4050 Calories or 17 million joules. The Calorie content of food is also a measure of this biological stored energy, where 1 nutritional Calorie (with a capital C) is 1000 cal or 1 kcal (small c). Recall the calorie is the amount of heat to raise the temperature of 1 g of water by 1°C, which is equivalent to 4.186 joules of energy. The sugary soft drink that listed 200 Calories on the label thus is 200,000 cal of stored energy.

An equally important consideration is the heat exchange associated with the phase change of materials. The latent heat of a substance involves the heat absorbed or released during a change of phase state. For example, it takes energy added to liquid water to vaporize (and this energy is released if it recondenses from water vapor to liquid water). The same applies to the solid-to-liquid transition. Upon a freeze risk, citrus growers often spray liquid water on the trees to (a) release heat upon its freezing and (b) provide a layer of insulating ice to protect it from colder air temperatures (Fig. 2.6). The equation of note is given in Eq. (2.13), where the latent heat of the material is given as L.

FIG. 2.6

During a hard freeze, spraying liquid water on fruit trees can help protect the fruit by releasing the latent heat of fusion upon freezing and creating an insulating ice layer from the yet colder air. The author attempted this with a blooming apricot tree with mixed results (the fragile blossoms did not hold up to the ice in all cases).

$$Q = mL \tag{2.13}$$

Of note, the temperature remains constant during an ideal phase change and thus does not appear in Eq. (2.13) (e.g., liquid water at 1 atm boils at a constant ~100°C). The heat of vaporization for water (2230 kJ/kg) is much larger than the heat of fusion (334 kJ/kg).

2.9 Mass & energy balance

Energy and mass are conserved quantities and thus always amenable to the conservation of energy and mass calculations, also known as energy or mass balances. As a result, both quantities are neither created nor destroyed but only changed in form (kinetic energy converted to potential energy). With energy, there is one exception where the conversion of mass to energy during nuclear reactions according to Einstein's well-known relationship. This results in the mass before a reaction being slightly larger than the mass after a nuclear reaction and in doing so releases thermal energy proportional to the mass loss and the speed of light squared ($E = mc^2$). This typically does not come into play for building systems, and a straight mass balance approach is appropriate. A mass balance equation based on the general continuity equation is given in Eq. (2.14) (Example 2.7).

> **EXAMPLE 2.7 Writing a mass balance on Antarctic ice mass**
>
> **Problem:** Write a mass balance equation incorporating the key terms for gain and loss of ice mass from the Antarctic ice sheet shown in Fig. 2.7.
>
>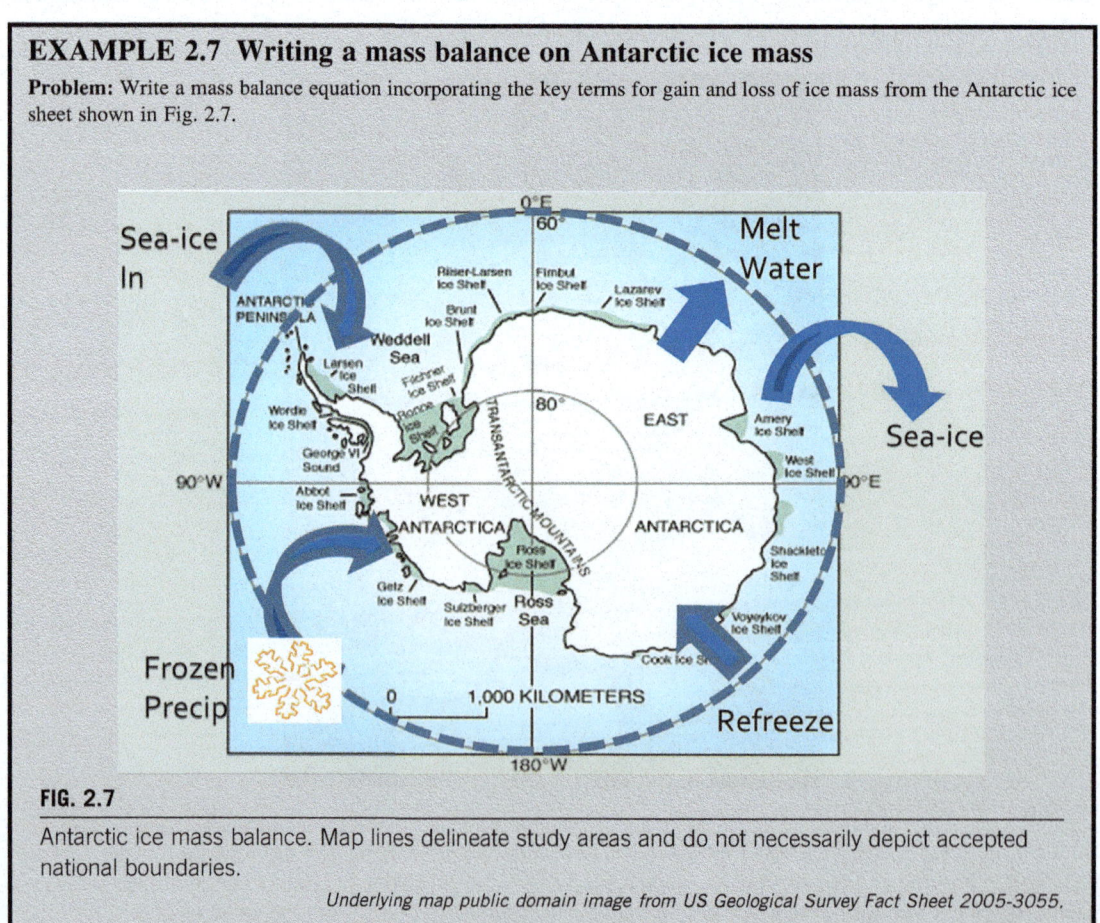
>
> **FIG. 2.7**
>
> Antarctic ice mass balance. Map lines delineate study areas and do not necessarily depict accepted national boundaries.
>
> *Underlying map public domain image from US Geological Survey Fact Sheet 2005-3055.*

Continued

> **EXAMPLE 2.7 Writing a mass balance on Antarctic ice mass—cont'd**
> **Solution:**
> Other small terms may contribute (sublimation). For simplicity, the rate of change of ice mass in the Arctic can be expressed as the amount of frozen precipitation (snow, ice, freezing rain) plus the difference between icebergs floating in vs those floating out plus the internal generation from phase change, which is the difference in water freezing to ice minus the thaw rate of the Arctic. The rate of change of ice mass (though varying with season) has been increasingly negative globally, especially in the Arctic.
>
> $$\frac{dM}{dt} = \dot{M}_{in} - \dot{M}_{out} + \dot{M}_{gen} = \dot{M}_{frozenprecip} + \left(\dot{M}_{icebergs,in} - \dot{M}_{icebergs,out}\right) + \left(\dot{M}_{freeze} - \dot{M}_{thaw}\right).$$

$$\text{Accumulation} = \text{Quantity In} - \text{Quantity Out} + \text{Quantity Generated Internally}$$
$$\Delta M = M_{in} - M_{out} + M_{gen}$$
$$\frac{dM}{dt} = \dot{M}_{in} - \dot{M}_{out} + \dot{M}_{gen} \tag{2.14}$$

2.10 The laws of thermodynamics

The First Law of Thermodynamics is that energy is neither created nor destroyed, i.e., it is conserved. The energy balance of a closed system (i.e., no mass flows in or out) may be expressed as the change of the system's total energy state E. E is the sum of its internal energy (U), its kinetic energy (KE), and its potential energy (PE). This is equal to the heat added to the system minus the work out of the system. The closed system form of the First Law is shown in Eq. (2.15), where ΔE is the change in the energy state of the system, which is the difference between heat (Q) into the system minus work (W) out of the system.

$$\Delta E = \Delta(U + KE + PE) = Q_{net,in} - W_{net,out} \tag{2.15}$$

For the open system where there are flows into and out of the control volume, the First Law may be expressed as in Eq. (2.16) where the inlet and outlet are balanced with the mass flow rate, $\dot{m}$.

$$\dot{Q}_{net,in} - \dot{W}_{net,out} = \dot{m}\Delta(h + ke + pe) \tag{2.16}$$

The above First Law equations can be expressed as quantities (J) or rates (W) if used consistently. The small "ke" and "pe" terms designate the specific kinetic and potential energy terms without the mass multiplier since it has been moved outside of the parentheses.

2.11 Heat engines, combustion, and heat pumps

Heat engines (Fig. 2.8) are innumerable throughout our built environment and everyday life. These are devices that harvest useful work (W_{out}) from the transfer of heat from a hot reservoir (Q_{hot}, e.g., the combustion temperature in an engine) and exhaust waste heat to a cold reservoir (Q_{cold}, often the

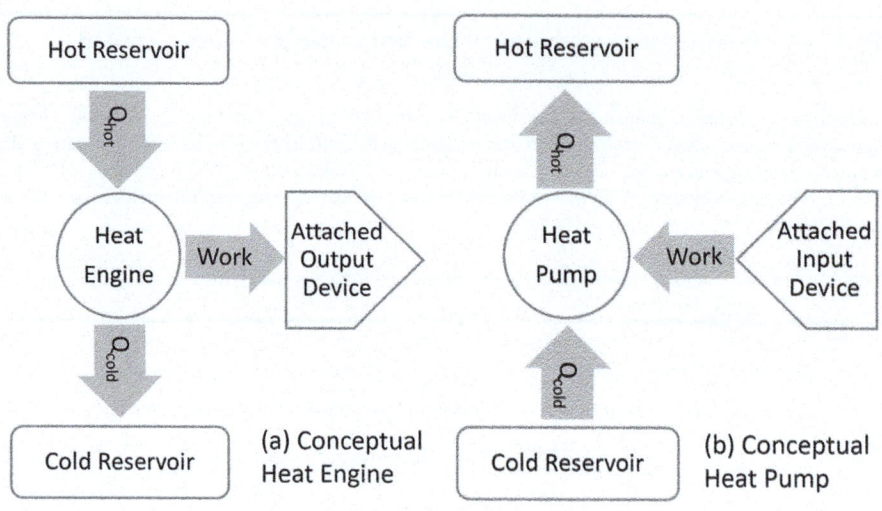

FIG. 2.8

Diagram of (A) a heat engine and (B) a heat pump. Both operate across a ΔT between a hot reservoir (often at the engine temperature) and a cold reservoir (often at atmospheric temperature). A heat pump simply reverses the cycle directions and thus has reversed energy flows compared with the heat engine.

atmosphere). The First Law is applicable, and thus the energy balance of the heat engine running at a steady state is given in Eq. (2.17).

$$Q_{hot} = Q_{cold} + W_{out} \qquad (2.17)$$

The above equations can be expressed as quantities (J) or rates (W) if used consistently.

End-use energy efficiency, as it applies to energy conservation efforts, is intrinsically related to the engineering concept of efficiency and will be discussed later in the book as one of the solutions to our climate crisis most important to the built environment. The thermal efficiency of a given heat engine is defined in words as what you get out of the process, e.g., useful work output, divided by what you put into the process, often thermal energy. It can be expressed fractionally (0 to 1) or as a percentage (0%–100%). When drawing a control surface around a given system and properly accounting for the energy flows into and out of it, the efficiency of a process cannot exceed 100%. It is typically far less than that (though it cannot be negative).

$$\eta_{heat\,=\,engine} = \frac{\text{Useful work output}}{\text{Required heat input}} = \frac{W_{out}}{Q_{hot}} = \frac{Q_{hot} - Q_{cold}}{Q_{hot}} \qquad (2.18)$$

$$\text{For Processes in Series}: \eta_{overall} = \prod \eta_i \qquad (2.19)$$

A detailed discussion of the Second Law of Thermodynamics—which has several statements—is beyond the scope here, though a few consequences are important. Heat cannot be completely converted to work in a cyclical heat engine (Kelvin-Planck Statement). Each heat engine requires a heat sink to dump waste heat. Think of a vehicle and its radiator or a power plant and its cooling towers. Though we seek to minimize waste heat, it is unavoidable. All work is eventually converted to dispersed, low-grade heat. Another consequence of the Second Law is that net heat transfer occurs spontaneously only from the hotter body to the cooler body (Clausius statement). The coffee cup sitting open in the room does not get hotter; it cools till it attains equilibrium with its surroundings.

For purposes here, entropy (S), another thermodynamic state variable, is a measure of the internal disorder in a system and is measured in units of J/K or J/(kg K) for specific entropy (s). The Second Law of Thermodynamics says that the entropy of the universe cannot decrease as we proceed along in time (entropy statement). Any entropy decrease in a given system must be accompanied by an equal or greater entropy increase in the surroundings. Phase state is an indicator of increasing entropy where the transitions from solid to liquid to gas to plasma all involve increases in entropy as each is a successively more disordered state.

Related to the Second Law, the upper limit on the efficiency of a heat engine is given by the idealized Carnot cycle composed of two adiabatic processes and two isothermal heat transfers. Carnot showed that the ratio of heat transfers is equivalent to the ratio of temperatures of the reservoirs (Eq. 2.20).

$$\eta_{Carnot} = 1 - \frac{Q_{cold}}{Q_{hot}} = 1 - \frac{T_{cold}}{T_{hot}} = \frac{T_{hot} - T_{cold}}{T_{hot}} \quad (2.20)$$

Heat pumps are the analogue of the heat engine and reverse the cycle of the heat engine. They are used in refrigeration-based systems, including heat pumps for space heating, kitchen refrigerators, and refrigerated air-based air conditioning systems, both central and window units. These systems require work input from a compressor to cause the flow of heat from the cold reservoir to the heat pump and a larger flow of heat to the hot reservoir. The associated energy balance equation and the coefficient of performance (COP) are given in Eqs. (2.21), (2.22), respectively.

$$Q_{hot} = Q_{cold} + W_{in} \quad (2.21)$$

$$COP_{heat\ pump} = \frac{\text{Heat output}}{\text{Required work input}} = \frac{Q_{hot}}{W_{in}} = \frac{Q_{hot}}{Q_{hot} - Q_{cold}} \quad (2.22)$$

A heat pump used as a chilled air system for space cooling is the reverse cycle from that for space heating, and the COP for the air conditioner is given in Eq. (2.23).

$$COP_{air\ conditioner} = \frac{\text{Heat extracted}}{\text{Required work input}} = \frac{Q_{cold}}{W_{in}} = \frac{Q_{hot}}{Q_{hot} - Q_{cold}} \quad (2.23)$$

An important variety of heat pumps is a refrigeration system. It removes heat from a colder place and dumps it into a warmer place. This applies to both chilled air systems (air conditioners) and refrigeration systems for chilling food and beverages. According to the Intergovernmental Panel on Climate Change (IPCC), approximately 1.5 billion refrigerators are in operation in the world, roughly one for every four people on the planet and increasing. The desired effect of the refrigeration system is removing heat from the chilled area, in other words maximizing Q_{cold} (Eq. 2.24) (Example 2.8).

$$COP_{refrigerator} = \frac{\text{Heat extracted}}{\text{Required work input}} = \frac{Q_{cold}}{W_{in}} = \frac{Q_{cold}}{Q_{hot} - Q_{cold}} \quad (2.24)$$

Our energy-climate problems can be viewed with these laws. The entire energy conversion system on the planet seeks an equilibrium where radiation energy in (solar) is balanced by radiation energy out (terrestrial), so the Earth achieves an equilibrium temperature. Extracting solids, liquids, and gases from the Earth and combusting those hydrocarbons creates products in a gas state that are more entropic.

> **EXAMPLE 2.8 Finding the Carnot COP of a refrigerator**
> **Problem:** The Carnot COP of a heat pump for space heating is $1/(1-T_{cold}/T_{hot})$. The desired output is different for a refrigerator, which means the COP expression is different. Find an expression for the Carnot COP of a refrigerator in terms of T_{cold} and T_{hot}. Of note, Carnot found that in the ideal cycle $Q_{hot}/Q_{cold} = T_{hot}/T_{cold}$.
> **Given:** Refrigerator.
> **Find:** Carnot COP for refrigerator.
> **Assume:** Steady state (SS).
>
> **Solution:**
>
> $COP_{ref} = Q_{cold}/W$.
> Energy must balance so $Q_{cold} + W_{in} = Q_{hot}$ or rearranging $W_{in} = Q_{hot} - Q_{cold}$.
> $COP_{ref} = Q_{cold}/(Q_{hot} - Q_{cold}) = 1/(Q_{hot}/Q_{cold} - 1)$.
> Using the Carnot equivalence between heat transfer ratio and absolute temperature ratio, the ideal Carnot refrigerator would have a Carnot $COP_{ref} = 1/(T_{hot}/T_{cold} - 1)$.
> Of note, the heat pump does not violate the Laws of Thermodynamics as the energy flows into and out of the heat pump still balance and the heat transfer to the hotter body is not spontaneous.

2.12 Combustion and reacting systems

The discussion above has been applied to a nonreacting system. Heat engines often rely upon the combustion of a fuel (gasoline in an internal combustion engine or natural gas in a furnace). Generalized combustion in air reaction is given in Eq. (2.25) for a generic hydrocarbon. For this equation, b is $x + y/4$.

$$C_xH_y + bO_2 + 3.76(b)N_2 \rightarrow xCO_2 + \left(\frac{y}{2}\right)H_2O + 3.76(b)N_2 \qquad (2.25)$$

N_2 in the above does not participate in the reaction but comes along for the ride with the air introduced to the combustion process and must be accounted for in the mass balance aspect. The important mass-based air-to-fuel ratio is given (Eq. 2.26) by the mole ratio multiplied by the molecular weight ratio of air:fuel.

$$AF = \frac{28.97 \,(\text{mol Air})}{(MWC_xH_y)(\text{mol HC})} \qquad (2.26)$$

2.13 Ideal gas law and other essential fluids relationships

Fluid flow is key to many systems in green building such as HVAC systems covered later in the electromechanical systems chapter. An extensive treatment of fluid mechanics requires a dedicated course. Here we will focus on a few key relationships relevant to building operations.

Water and air are the two substances of most relevance to green building. An approximate rule of thumb: water is 1000 times as dense as air. Liquid water is approximately 1 kg/L, and air is roughly 1.2 g/L at sea level. The heat capacity of water at 4186 J/(kg K) is ~4 times as large as air.

For these purposes, air is always approximated as an ideal gas invoking the ideal gas law (Seinfeld and Pandis, 2016). This states that the product of absolute gas pressure and gas volume (V) is equal to the moles of gas (n) multiplied by the absolute temperature (T) and universal gas constant (R which is 0.08206 L atm/(mol K)) (Eq. 2.27).

$$PV = nRT \qquad (2.27)$$

Recall that n is the moles of a given gas, and the molecules may be obtained by multiplying by Avogadro's Number, $6.022(10)^{23}$.

As a compressible fluid, the conversion of volumes (V) or volumetric flow rates (Q) from one temperature and pressure state to a second state is given in Eq. (2.28), where T and P must be in absolute scales. Conservation of mass of course still applies, but air volumes or volumetric flow rates can be expanded or compressed depending on the temperature and pressure conditions.

$$\dot{V}_2 = \dot{V}_1 \times \left(\frac{P_1}{P_2}\right)\left(\frac{T_2}{T_1}\right) \tag{2.28}$$

The mass flow rate of a fluid, gas, or liquid is found from its volumetric flow rate and fluid density (Eq. 2.29). The same relationship can be used to find the mass flow rate of a constituent in a flow by replacing the density with the mass concentration (e.g., µg/m³) of the constituent.

$$\dot{M} = \rho \, \dot{V} \tag{2.29}$$

For purposes here, the flow of a fluid through a pipe may be approximated as "plug flow" (picture a small plug of fluid the shape of the pipe flowing at a uniform velocity) (Fig. 2.9). The typical cross-sectional areas of a cylindrical tube (πR^2) or a square tube (D^2) apply, although any profile is possible. The flow rate of fluid through a pipe is calculated based on the mean fluid velocity (v), pipe cross-sectional area (A), and fluid density (ρ).

$$\dot{M} = \rho A v \tag{2.30}$$

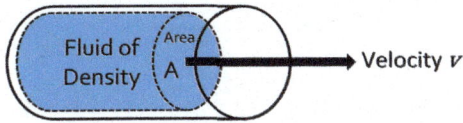

FIG. 2.9

The fluid flow through a pipe can often be represented as "plug flow" where a fluid element with cross-sectional area A flows at a velocity v through a pipe.

The Reynold Number (Re) characterizes the flow regime of the fluid sample passing through a pipe or tube of diameter D having fluid density, viscosity, and velocity of ρ, μ, and v (Eq. 2.31). Depending on the geometry of the flow system, Re indicates laminar (smooth) vs turbulent flow. For cylindrical pipe flow, the laminar region is for $Re < 2300$ and the turbulent region begins at about $Re = 3000$ with a transition regime in between.

$$Re = \frac{\rho D v}{\mu} \tag{2.31}$$

Although less energy-intensive than heating or cooling air, pumping a fluid requires power input. This power requirement is proportional to its volumetric flow rate, the pressure drop that must be overcome, and inversely proportional to the pump's efficiency (Eq. 2.32).

$$\text{Power} = \frac{\dot{V} \Delta P}{\eta} \tag{2.32}$$

2.14 Water vapor and relative humidity

Water vapor is a vital parameter for many environmental processes as well as for building physics. It plays a fundamental role in human comfort as well as building processes related to moisture and energy transport. The vaporization and condensation of water vapor in the atmosphere are central to the energetics of the atmosphere and thus the climate system via the latent heat effects as well. Relative humidity (*RH*) is a key parameter related to the drying of building materials and potential rot issues. A humid climate versus a dry climate manifests quite different effective indoor climate control options. Relative humidity also plays a key role in atmospheric processes, including the aerosol impacts on haze as well as climate change via water uptake by hydrophilic species in aerosol particles (Carrico et al., 2003).

The relative humidity in the air compares the actual partial pressure of water vapor (P_v) in the air to the saturation water vapor pressure (P_{sat}), which is a function of the dry bulb temperature (Eq. 2.33).

$$RH = \frac{P_v}{P_{sat}} \tag{2.33}$$

The Antoine equation gives the relationship between ambient dry bulb temperature and saturation vapor pressure for water vapor (Eq. 2.34). The equation gives the saturation water vapor pressure (P_v, bar) for pure water as a function of dry bulb temperature (K) and constants for water A_i (6.20963), B_i (2354.731), and C_i (7.559). The relationship between saturation water vapor pressure and the dry bulb temperature is shown in Fig. 2.10. The Clausius-Clapeyron relationship gives another approximate relationship between the two parameters (Eq. 2.35) where P_v is found in hPa and T is input in degrees C (Eq. 2.35) (Example 2.9).

$$\log P_v = A_i - \frac{B_i}{C_i + T} \tag{2.34}$$

$$P_v \cong 6.1094 \exp\left(\frac{17.625T}{T + 243.04}\right) \text{ where } T \text{ is in } °C \tag{2.35}$$

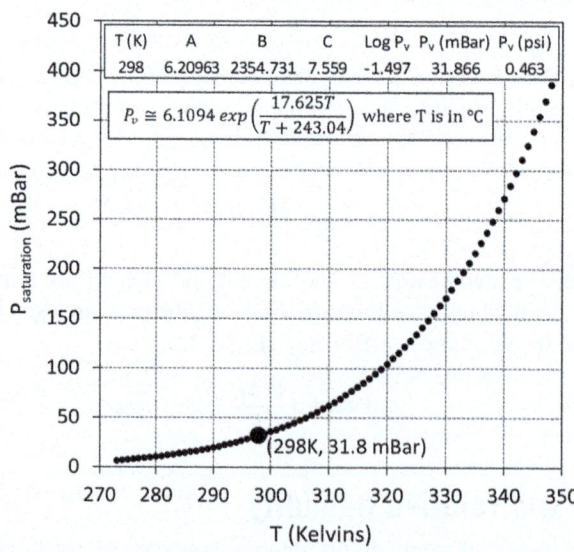

FIG. 2.10

Antoine Equation relationship between dry bulb temperature and saturation water vapor pressure for pure water.

EXAMPLE 2.9 Antoine relationship

Problem: Use the Antoine relationship. How much additional water is stored in the atmosphere if RH has remained constant vs the dry bulb temperature increase of 1°C? Compare this to Earth's total freshwater resources. Assume a preindustrial global average temperature of 13.7°C.

Given: Antoine relationship.
Find: ΔP of water vapor with $\Delta T = 1\,K$.
Assume: $T_{start} = 13.7°C$, steady state (SS) at start and finish.

Solution:

$$\log P_{vi} = A_i - \frac{B_i}{C_i + T} = 6.20963 - \frac{2354.731}{7.559 + 286.85} = -1.7885$$

$$P_{vi} = 0.01627\,\text{Bar}$$

$$\log P_{vi} = A_i - \frac{B_i}{C_i + T} = 6.20963 - \frac{2354.731}{7.559 + 287.85} = -1.7615$$

$$P_{vi} = 0.01732\,\text{Bar}$$

Thus the increase in water vapor pressure is 1.05 mBar or ~7%

This increase at the surface is equal to the increased water vapor mass

$$\text{Force} = P \times SA_{earth} = mg$$

$$m = \frac{P \times SA_{earth}}{g} = \frac{0.00105\,\text{Bar}\left(\frac{101325\,\text{Pa}}{\text{Bar}}\right) \times 4\pi(6.378 \times 10^6 m)^2}{9.8 \frac{m}{s^2}} = 5.55 \times 10^{15}\,\text{kg} = 5.55 \times 10^6\,\text{km}^3$$

This quantity of water is far more than the total freshwater in lakes and rivers. The atmosphere is a growing reservoir of water as the global average temperature increases as its capacity to hold water increases, and surface evaporation is enhanced at higher temperatures.

2.15 Chapter summary and conclusions

Energy and power (rate of energy use) are key concepts in the physics of buildings. Balances of mass and energy are fundamental concepts of engineering and applicable to building systems. Working knowledge of energy mechanics, thermodynamics, and heat transfer is necessary to understand energy and water flows in a building. Thermodynamics explores the conversion of heat into useful work. Thermodynamic state properties describe the state of a system at a given time. They include temperature, pressure, internal energy, enthalpy, and entropy. Knowing two independent state parameters specifies the state of the system.

When a constant heat capacity can be assumed, the familiar $MC\Delta T$ equation can be used to calculate the temperature change of a system due to heat gain or loss. This is called sensible heating, while the additional heat needed for a phase change (e.g., boiling), which happens at a constant temperature, is latent heat. Latent heat is important to atmospheric thermodynamics and building cooling loads.

The two boundary processes that allow energy to be transferred across a system boundary are heat transfer and work (a force applied over a distance for mechanical work). The output of useful work is the goal of a heat engine (e.g., vehicle engine), while the movement of heat is the goal of a heat pump (e.g., air conditioner). No system is 100% efficient, and the Carnot efficiency and coefficient of performance describe the ideal operations of a heat engine and pump, respectively.

Heat is transferred due to a temperature difference and occurs via conduction, via convective heat transfer from the movement of a fluid due to temperature and thus density difference, and via electromagnetic radiation. The latter is a function of T^4.

Entropy is the concept of disorder at the molecular level, and the Second Law of Thermodynamics states that this continuously increases across the universe. The consequences include that all machines will have systematic losses and an efficiency less than 100%. Other relationships including the Ideal Gas Law and Ohm's Law are important concepts for analyzing buildings and their energy use.

2.16 End of chapter exercises

(1) **Concepts:** On a hot day, the liquid on one's forehead and condensation on one's cold beverage have opposite effects. Explain.

(2) **Concepts:** In Table 2.3, the power consumption of the world's largest laser exceeds that of the United States. How does this not bring down the grid or at least dim the lights a bit when it is operated?

(3) **Concepts:** Design and illustrate an ideal passive to-go cup to keep your hot beverage hot. Discuss 4 features. Think about heat transfer mechanisms. How would you change it to keep your cold beverage cold?

(4) **Concepts:** Specify the best answer for the predominant means of heat transfer in each of these situations:
   (a) Heat transferred from a heated air dryer to one's hands.
   (b) Heat transferred from a warm front moving into New Mexico from the south.
   (c) Standing in front of a fireplace with a roaring fire behind a glass door.
   (d) Burning one's hand on a hot surface.
   (e) A cat warming itself in a sunbeam.

(5) **Problem:** Draw the conceptual heat engine diagram showing the flow directions of heat and work as well as the hot and cold temperature reservoirs. Express the work output and efficiency in terms of the hot and cold heat flows, Q_h and Q_c.

(6) **Problem:** Convert the global average energy use of ~500 Quads to (a) an equivalent ExaJoules (10^{18}) per year, (b) an equivalent TWh per year, and (c) a steady-state horsepower.

(7) **Problem:** The first steam engine worked between the upper and lower temperatures of 10°C and 100°C. What is the maximum efficiency possible?

(8) **Problem:** Estimate the mass of the atmosphere. Assume surface $P = 1$ atm (Earth's radius is 6370 km).

(9) **Problem:** You are using a heated 1-m diameter mass of iron at 500°C to provide warmth in a room. At the outset, what is the rate of heat transfer via radiation in a room temperature environment? Assume a perfect blackbody.

(10) **Problem:** A heat engine is generating useful work at a rate of 50 kW. The waste heat loss rate to the environment in the heat engine is 30 kW. What is the thermal efficiency of this engine? What is the hp. rating of the engine? What is the rate of gasoline consumption (120,000 BTU/gal) in gal/h?

2.16 End of chapter exercises

(11) **Problem:** Your power plant has an output of 2000 MW. Its flame temperature is 1227°C and it is exhausting to ambient conditions at 20°C.
   (a) Find Carnot efficiency.
   (b) For the idealized Carnot cycle heat engine, what is the minimum rate of input heat (Q_{hot})?
   (c) What is the waste heat (Q_{cold}) being dumped into the surroundings?
   (d) Draw a diagram showing the energy flows in and out and that they balance.

(12) **Problem:** Power consumed by an engine as it ramps up is given by $P(t) = 100\,kW + 10\,kW/min * t$. Find the energy consumed in kWh in 1 h.

(13) **Problem:** You successfully lift a 25 kg stone to the top of the Empire State Building (1250 ft to the top floor). How much work did you do on the stone assuming constant g? If you burned the calories you ingested from drinking a cola while accomplishing this (cola has 182 food Calories), what is the thermodynamic efficiency of the process (your body!)?

(14) **Problem:** An inventor has a new power plant design that is claimed to have a 2000 MW output with a feed rate of 100 kg/s of fuel with an energy content of 25 MJ/kg. The plant is operating at hot and cold reservoir temperatures of 1000°C and 20°C. Is this possible?

(15) **Problem:** A human can be approximated as a puny little machine outputting $\sim$1/10 hp. steady state. (a) If the person is maintaining a constant temperature, what is the rate of heat transfer for this system? Think about the heat coming from the stored fat tissues into the human, machine which then produces some useful work. (b) Assume that the human's useful work output can be converted into electricity with no loss and at a value of $0.12/kWh. Calculate the value of the work output over a year and lifetime if the same output can be maintained around the clock.

(16) **Problem:** Your bicycle tire gauge reads the following: 33.5 psi. You are at sea level ambient pressure. Find the absolute pressure in kPa. Estimate the mass of air in the tire using reasonable assumptions for a typical bike tire.

(17) **Problem:** A 32-fluid-ounce jug of maple syrup is very energy-dense. It contains 16 servings with 210 food Calories per 2-oz serving.
   (a) How many BTUs of energy does the jug contain?
   (b) How many degrees C would that raise the temperature of 50 kg of water (assume no heat loss and a reasonable heat capacity)?
   (c) Alternately, how high with it lift that 50 kg person in Earth's gravitational field (again assume no losses)?

(18) **Problem:** Calculate the power contained in Niagara Falls at a volumetric flow rate of 84,760 cubic feet per second 167-ft drop. What is the maximum power that could be generated from this resource? How many typical large-scale power plants does this equate to?

(19) **Problem:** An LNG tanker or a coal train can be thought of as stored chemical energy. An LNG tanker contains 130,000 m^3 of compressed liquefied natural gas at -163°C and 22 kPa (assume it is pure liquid methane with a density of 0.425 g/cm^3 and 55 MJ/kg). LNG tanker length is 1000 ft, and it travels at 10 m/s. A coal train (traveling at 25 m/s) has 120 cars, each 40 ft long, each containing 90 short tons of coal. We'll assume 25 MJ/kg for coal energy content.
   (a) Find the energy content of each in MJ.
   (b) What power flows do each represent?

(20) **Problem:** A weight management expert claims you will lose an extra $\sim$6 lb a year by ingesting an extra 2 L of chilled water (50°F) every day. Confirm or refute the claim. Ignore effects on

metabolism and other complicating factors, just a straight energy basis. Think about the sensible heating. A rough guideline is 3500 Calories of energy content per pound of fat tissue.

(21) Problem: 1 g of methane is burned, and the heat is used to raise the temperature of 1 kg of water. If the initial temperature of the water is 25°C, what is the final temperature?

(22) Problem: An internal combustion engine in a car burns ethanol at a rate of 42 kg/h and produces 74.6 kW of power delivered to the car. What percentage of the chemical energy of ethanol is being converted to work? Assume complete combustion of ethanol.

(23) Problem: We are going to do some super-hot yoga and have pressurized the classroom to 1.5 atm and raised the temperature to 120°F. (a) Calculate the mass of air in this classroom (assume 10 m × 15 m × 15 m). (b) How many molecules of air does this represent? (c) How many molecules of carbon dioxide are in this room assuming the current global average ambient concentration?

(24) Problem: At high speeds such as 75 mph on the highway, air drag force is a major component of the power consumption of a vehicle. The drag force on a vehicle is approximated as $0.5 * C_d * A * \rho * V^2$. What is the rate of work in overcoming the aerodynamic drag for a 2013 Toyota Prius vs a 2003 Hummer H2? Compared to their highway fuel consumption, does the ratio at least make sense with their MPG ratings? Calculate the fuel input power rates in kW based on fuel consumption. You will have to do some research.

(25) Problem: The Four Corners power plant was operated by Arizona Public Service and was located in New Mexico. It was one of the largest power plants in the world with a generating capacity of 2040 MWe. Assuming the coal it burns is 75% carbon with a heating value of 13,000 BTU/lb and has a sulfur content of 0.8%. It has a thermal efficiency of 38%, and it runs annually at 80% capacity (20% downtime for maintenance). Find the mass generation of CO_2 and SO_2 in metric tonnes/year. Assume complete combustion and no emission controls. Compare each to the Empire State Building, which has a mass of 331,000 tonnes.

(26) Problem: A Eurobus A380F, the world's largest passenger aircraft (fully loaded mass = 592,000 kg; length = 72.73 m), is traveling at full speed of 634 mph (1020 km/h).
(a) What is the kinetic energy of this aircraft?
(b) Think of it as passing a given point in a time increment. What power flow does this represent?

(27) Problem: How much would a 1 L cold drink that is at a temperature of 3°C drop the temperature of a person on a hot day vs how much would sweating this 1 L drop T. Assume a 70 kg person with a body temperature of 37°C. Assume no other gains or losses of heat.

(28) Problem: How much energy (in MJ) does it take to lift a person and their luggage (250 lb total weight) to a height of 30,000 ft and accelerate them from zero to a speed of 500 mph? Consider both the person's kinetic energy and potential energy but neglect friction, wind resistance, and inefficiencies. How much jet fuel does this consume?

(29) Problem: A heat engine is operating at T_{hot} = 932°F. It's exhausting in an environment at freezing temperature, 32°F. What is the maximum Carnot efficiency of this heat engine?

(30) Problem: You travel from an altitude of 4600 to 7400 ft for a hike. To lift your vehicle (3100 lb) and 2 humans and 2 dogs (350 lb), how much gasoline would you burn? Assume 25% engine efficiency. You accelerate from rest to 65 mph for the kinetic energy contribution. Neglect frictional and aerodynamic losses which would increase this consumption.

(31) Problem: You get ready to stick some delicious queso in your face-o. You have 1.5 L of grade-A prime cheese product (sg = 1.08) with a heat capacity of 3 kJ/kg/K. You are using a microwave oven with electrical power consumption of 1100 W. It has the typical efficiency of a microwave oven of 64%. You start from a room temperature of 20°C and heat it to 150 F, just below the melting point. How long should you microwave it, ignoring any other losses?

(32) **Problem:** A 25 lb beagle jumps to a height of 2 ft once every 3 s. The beagle continues this for 5 min. If the beagle is 25% efficient at converting kibble calories into motion, how much kibble will it need to fuel this? Kibbles contain 50 food Calories in a 30 g piece.

(33) **Problem:** Potential energy
  (a) How much energy does it take to lift a person (i.e., how much potential energy would the person gain) going from the base to the top of Long's Peak in Colorado? Assume starting $h1 = 7000$ ft, ending $h2 = 14,000$ ft, weight = 175 lbs. Assume 100% efficiency.
  (b) How many food calories is this (remember a food calorie = 1 kcal)?
  (c) How much gasoline would need to be burned to generate this amount of thermal energy, again assuming 100% efficiency?

CHAPTER 3

Energy and building history: How have we arrived here?

Learning objectives

(1) Identify the use of energy and materials as key inputs to our civilization.
(2) Recognize the contribution of population, its recent growth, as well as population density as amplifiers of environmental degradation.
(3) Understand the history of energy and materials used in civilization and as related to buildings.

It is useful to look at the big picture of energy, materials, and water use to put in context the building-related consumption. Buildings account for roughly 40% of global energy end use and ~30% of CO_2 emissions (Omer, 2008; Yang et al., 2014). Buildings notably also consume about 40% of materials and 14% of water use, according to the US Green Building Council (USGBC), as discussed below.

3.1 Human population and exponential growth

Humans have more or less come to dominate the planet for better and for worse. The scale of our population, activities, and the associated impacts has become staggeringly clear, particularly in the last century. Essentially, the larger the human population and size of the economy, the larger the consumption of materials, water, and energy on the global scale. Recent efforts in several advanced economies have been somewhat successful at dissociating the growth of the economy with ever-increasing materials and energy use.

At a sufficiently long-time scale, the human population curve looks very much like a J-shaped exponential growth curve. Around 1800, the human population moved from numbering in the hundreds of millions to billions. Most of the additions to the global population have occurred in the last several centuries, coincident with our industrialization and the agricultural revolution (Fig. 3.1). The advent of modern medicine and the lengthening of human lifetimes—both good things—have both undoubtedly contributed to population issues as well.

Population, its growth, and its density have all contributed in many ways to numerous environmental problems. In a closed petri dish with finite resources, a bacterial colony, no matter how clever, will eventually decline if it continues growing exponentially. It is hard to argue a planet with a population of 0.8 billion would not face far simpler ecological problems than one with 8 billion (and the likely several billion more humans by mid-century) (Fig. 3.2).

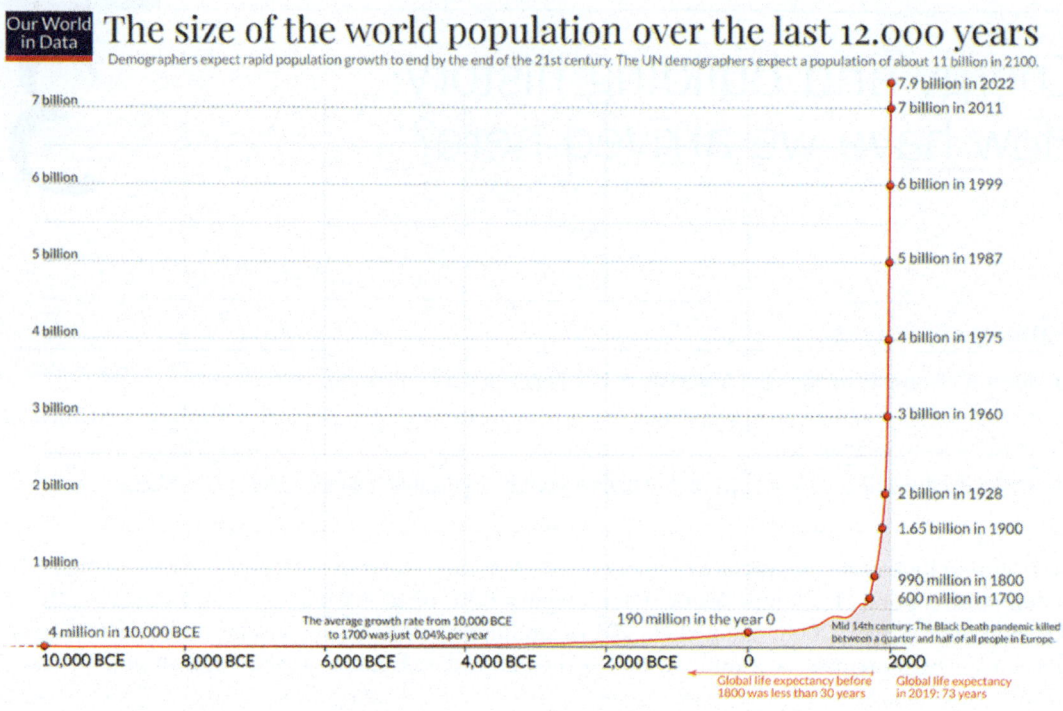

FIG. 3.1

Global human population.

Creative Commons image by Max Roser, Our World in Data.

The global human population continues to increase (for real-time updates: https://www.worldometers.info/world-population/). However, the rate of increase is declining and is now ~1% annually after exceeding 2% annually in the mid-1960s. This declining growth rate leads demographers to project that we will reach a global "steady state" population of ~12 ± 2 billion around the turn of the next century (Fig. 3.2).

There will likely be socioeconomic problems with stabilizing the global population, such as is seen in "aging" countries such as Japan. Here the increasing elderly population is being supported by a shrinking working population. However, it is not a sustainable justification to keep growing our population to support the growth of the past century. Moreso, it calls for a prudent, rational approach to managing our global population versus risking a major population decline due to resource constraints, climate change, or other problems brought about by rapid growth.

It is also noteworthy that much of the current and projected population growth is in developing and newly industrializing countries. The prospect of rapidly growing economies, energy, and material use in these nations is a huge challenge clearly linked to the building sector.

The global economy has also grown exponentially with a larger growth rate than population, averaging ~3% annually over the last 50 years. We are getting more numerous and wealthier on average as measured by the gross domestic product (GDP), which puts a monetary value on all goods and

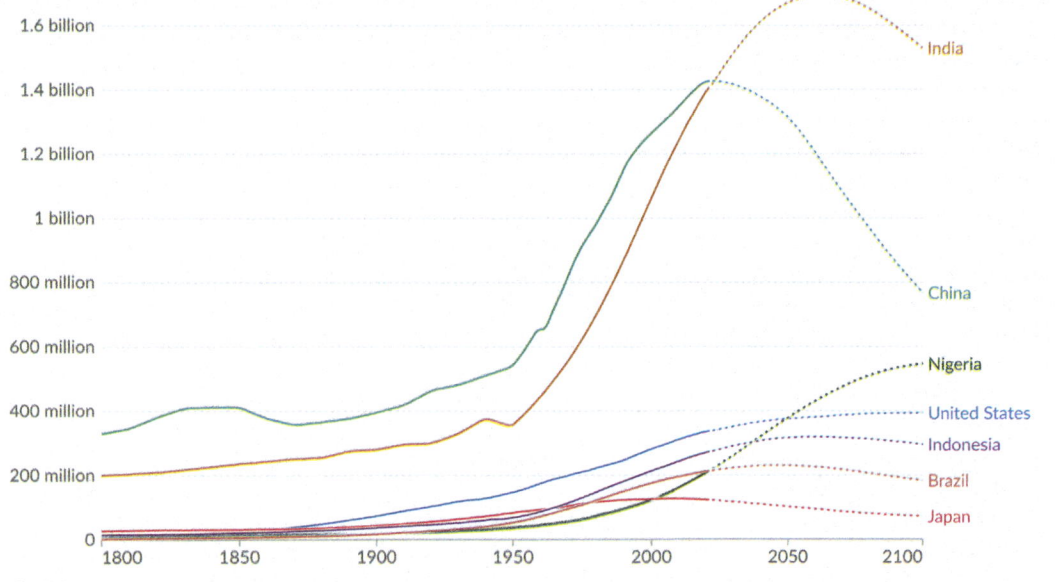

FIG. 3.2

Global regional human population as a function of year (projected to 2100) as well as geographical region.

Creative Commons image from OurWorldInData.org using United Nations Population Division data. Data from HYDE, 2017. Utrecht University/PBL Netherlands Environmental Assessment Agency – History Database of the Global Environment (HYDE v 3.2, 2017); Klein Goldewijk, C.G.M., Beusen, A., Doelman, J., Stehfest, E., 2017. Anthropogenic land use estimates for the Holocene – HYDE 3.2, Earth Syst. Sci. Data 9, 927–953; United Nations, Department of Economic and Social Affairs, Population Division, 2022. World Population Prospects 2022, Online Edition; Gapminder – Systema Globalis, 2022. Retrieved from https://github.com/open-numbers/ddf–gapminder–systema_globalis.

services produced. Year over year, GDP has been much more variable, though, than population and at times venturing negative (e.g., a negative 3% growth rate in 2020 with the global pandemic). There have also been large disparities in terms of the populations who have benefited and those who have been left out of this economic expansion. A key question is: Can economic growth be decoupled from ever-increasing environmental impacts and resource use on the global scale?

Our economic and population growth has been coupled with the increasing urbanization of our population, now over half of the global population. The urbanization trend has led to the development of the urban ecology field, which tries to maintain connections to nature with increasingly urbanized landscapes (Halliday, 2008). Population density in cities is increasing, and thus the incidence of increasing numbers and heights of tall buildings. This has led to many more "tall" buildings, supertall buildings exceeding 1000 ft, and now even mega-tall buildings of 2000 ft or higher. The mile-high building envisioned by Frank Lloyd Wright in the mid-20th century is over 50% a reality with the Burj Khalifa building in Dubai at 830 m (2722 ft).

It should be emphasized that at some point the gains in quality of life with increasing prosperity become little to gain with only increasing consumption. The Human Development Index (HDI) combines social indicators (e.g., education) with health indicators (e.g., lifespan) with economic metrics (e.g., GDP) to measure the quality of life of a population. A comparison of the HDI vs the strictly economic indicator of GDP/capita shows an expected strong correlation as countries begin development. However, this curve flattens out as electricity consumption exceeds 5000 kWh/year or GDP exceeds $30,000/capita with little additional gains in quality of life as measured by the HDI with increasing economic activity (Fig. 3.3).

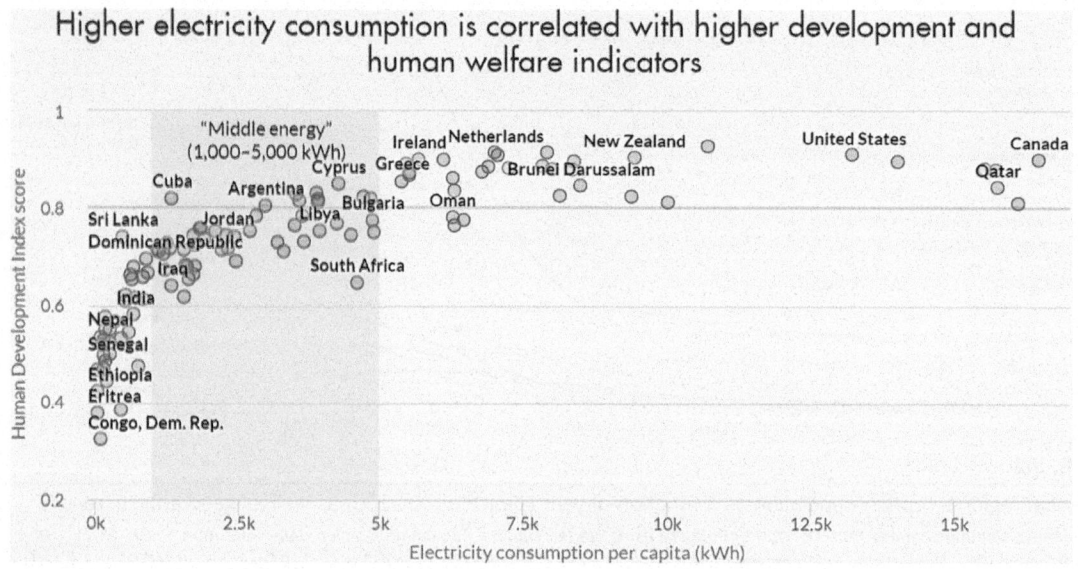

FIG. 3.3

Relationship between the Human Development Index (combining social, health, and economic well-being) and the electric use per capita (kWh) for the world's nations.

>Creative Commons image from Our World in Data (CGD, 2016. More than a lightbulb: five recommendations to make modern energy access meaningful for people and prosperity. A Report of the Energy Access Targets Working Group Center for Global Development Washington, DC, USA 35.).

3.2 Global population and environmental impacts

Even when minimized, by nature humans will have impacts from seeking food, clothing, and shelter, much less the biological processes of eating, drinking, breathing, and waste production. Several studies have shown that the most consequential action a person can take concerning the environment is limiting the number of new mouths to feed (Wynes and Nicholas, 2017), particularly in high-resource-use countries. Though there are nations that are facing a dearth of youth to support a large and aging population (e.g., Japan), supporting past population growth seems like a flimsy justification for promoting unabated continued human expansion.

An economist categorizes the environmental effects of our activities as an "externality," a negative effect or cost of an economic activity that is borne by those not directly benefiting from the activity (e.g., the general public). One entity's benefits (e.g., industrial output) result in costs (degraded environmental quality) borne by the public at large. This "tragedy of the commons" involves the degradation of a shared resource, e.g., the atmosphere or the hydrosphere. It results from the multiple activities of many individuals who do not recognize the small incremental contribution to that degradation and derive more utility from the activity than this social cost. Anyone who has shared a kitchen at a workplace has experienced the degradation associated with the use and abuse of such shared resources!

As a big-picture model, the classic "*IPAT*" equation (Eq. 3.1) captures the large-scale contributions to environmental impacts (I) as a product of population (P), an affluence factor (A), and a technology factor (T). This can be applied to varying scales or problems from the local to the global scale, as shown in the next example for greenhouse gas emissions (Eq. 3.1). The Kaya Identity takes this one step further by breaking down the technology factor into two independent terms: the energy use per unit of economic output (E, MJ/$GDP) and the pollutant intensity (C, kg/MJ) for the energy mix in use (Kaya and Yokobori, 1998) (Eq. 3.2).

$$I = PAT \tag{3.1}$$

$$I = PACE \tag{3.2}$$

The P in the *IPAT* equation implies no other alternative than making sure the population stabilizes sooner rather than later. The population factor could conceivably be the global population, though it often makes sense to break this down into country-specific populations due to the variability in the affluence and technology factors.

The level of development and thus the material and energy footprints of our lives are represented by an affluence factor (A). This factor is larger for wealthier nations or populations. However, it also should be noted that a level of affluence is often associated with an expanded ability to address environmental problems, at some level a "luxury" to those who are barely subsisting.

The importance of all four factors in the Kaya Identity (and many others folded into this simple model) cannot be overstated. The magnitude and nature of insults we have committed against nature have increased with both population, level of affluence, energy consumption, and the technology choices we make. The following example looks at the IPAT and Kaya factors for global CO_2 emissions (Example 3.1).

EXAMPLE 3.1 Global greenhouse gas emissions

Problem: Specify the IPAT and Kaya Identity contributors related to global greenhouse gas emissions.
  Given: IPAT, Kaya.
  Find: global CO_2 emission factors.
  Assume: globally homogeneous.
  Solution:
  The GHG impact (I) in mass units results from the annual quantity of emissions in CO_2 equivalents ($CO_{2,e}$), which is in turn linked to its climate forcing. The affluence factor (A) can be taken as the GDP/capita, while the technology factor is the carbon intensity of the economy under consideration, i.e., the mass of $CO_{2,e}$ emitted per unit of economic output GDP.
  (Global CO_2 emissions in 2020 is 36 GT CO_2 = 36E12 kg CO_2 = 36E15 g CO_2)

Continued

> **EXAMPLE 3.1 Global greenhouse gas emissions—cont'd**
>
> Global primary energy consumption in 2020 is 580E6 TJ = 580E12 MJ.
> Global CO_2 energy intensity is approximately 62 g CO_2/MJ.
> The size of the global economy in 2020 is \$84.7E12/year.
> Energy intensity of economy is (580E12 MJ/year)/(\$84.7E12/year) = 6.85 MJ/\$GDP.
> CO_2 intensity of economy is (36 GT CO_2/year)/(\$84.7E12/year) = 425 g CO_2/\$GDP.
> IPAT: $I_{GHG\ Emissions} = PAT =$ Population × GDP/Capita × CO_2/GDP.
> GHG Emissions = 8E9 person × \$11,350/person/year × 0.425 kg CO_2/\$GDP = 38.6 GT CO_2/year.
> $I_{GHG\ Emissions} = PACE =$ 8E9 person × \$11,350/person/year × 0.062 kg CO_2/MJ × 6.85 MJ/\$GDP = 38.5 GT CO_2/year.
> These are slightly different than the current 36GT CO_2/year (covariances between factors likely affect this).

Generally, the developed world has emerged from the mid-20th century era of polluted rivers catching fire and air pollution episodes like the London smog episodes, where hundreds to thousands of excess deaths occurred acutely. The developing world still grapples with some of these acute issues, and the issues more globally have become chronic environmental issues with larger time and spatial scales. We all desire the improvements in lifestyle that come with prosperity and the mitigation of pain, difficulty, and shorter lives associated with poverty. Can we do this without spoiling the global commons?

A world of 8+ billion that consumes at the average pace of the United States is one with tremendous environmental liabilities. An example of resource constraints leading to the collapse of a civilization on a small scale is the case of the remote, isolated civilization on Easter Island (Reddy et al., 2019). The opportunity is real for the less developed world—with aid from the developed world—to learn from the mistakes made along the way by the affluent countries. This allows the developing countries to leapfrog the catastrophes experienced from the mistakes of those who climbed the ladder first and is the best hope for a global future. Civilization regression or collapse due to resource constraints or climatic conditions is a palpable concern.

Environmental degradation ranging from local soils and groundwater contamination, air quality degradation, mining, and finite resources such as fossil fuels, increasing construction waste, and climate change, are all exacerbated by population growth. These acute problems and their relations to the built environment will be discussed in later chapters after a big-picture look at human well-being, population, technology, and economic activity.

3.3 Historical United States and global energy use

Human society has developed coincidently with increasing energy use, fire being among the most primeval of sources and still the foundation of our energy systems. Human history with fire in the Earth system has been complex, involving both suppression of natural landscape fire and utilization of fire for our energy needs, beginning on a grand scale with the Industrial Revolution. Our first taming of fire involves biofuels such as wood, and biofuels are still a small component of our energy mix (Fig. 3.4) and significant in developing countries.

3.3 Historical United States and global energy use

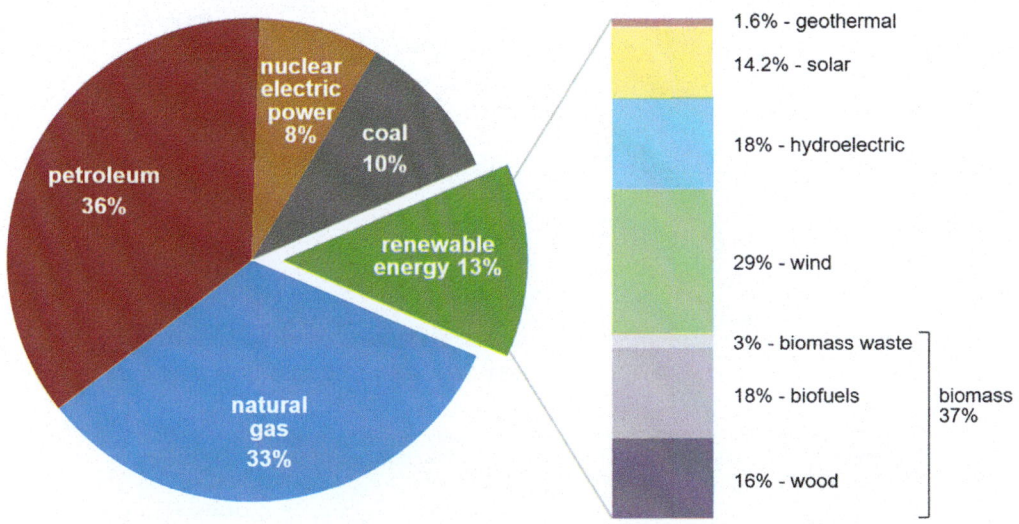

FIG. 3.4

US primary energy consumption by source in 2022.

Public domain image from US EIA.

Energy use only began to scale into megawatt and gigawatt scales with the Industrial Revolution, which came into full bloom in the late 19th century (Fig. 3.5). Fossil fuels became important beginning in the latter 19th century. Other sources had their peak contributions to what was a comparably tiny global energy pie centuries ago, including animal power, largely for agricultural production and transportation, and whale oil for illumination. Another cautionary tale: whale oil declined due to the near hunt-to-extinction of whales as well as the ascendance of alternatives, including those from the first petroleum production wells near Titusville, PA, in 1858. Since then, we entered and exited a peak in the coal age in the early 20th century and now have a global energy pie dominated by the dominant contributions from oil and natural gas plus a declining contribution from coal (Figs. 3.4 and 3.5). Though still a smaller contribution, the growth in renewable energy sources, particularly solar and wind, has been rapid over the past couple of decades.

The 2022 US primary energy source distribution also shows the dominance of fossil fuels (petroleum, natural gas, and coal in that order) in our energy system (Figs. 3.4 and 3.5). According to the US

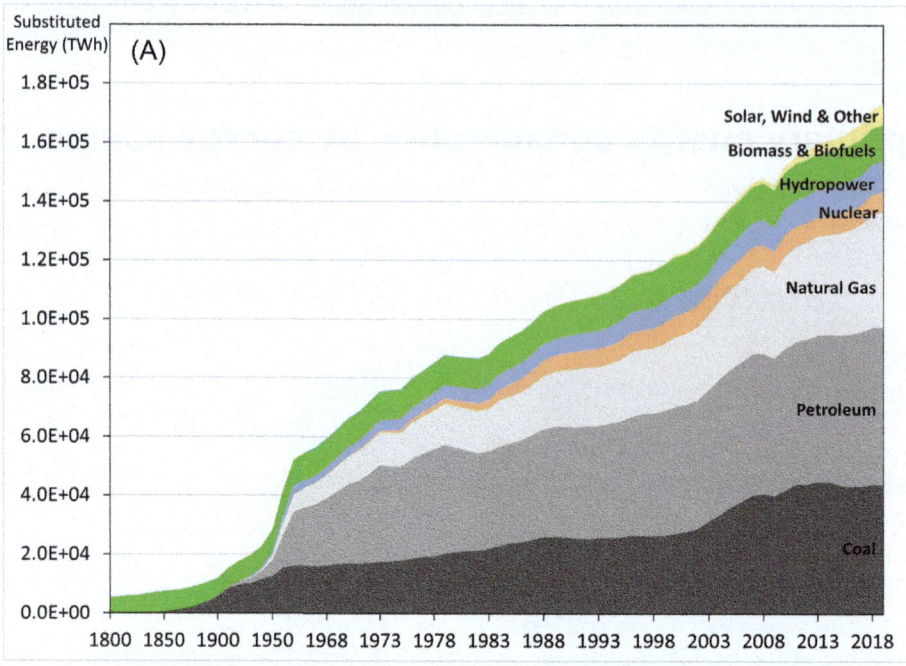
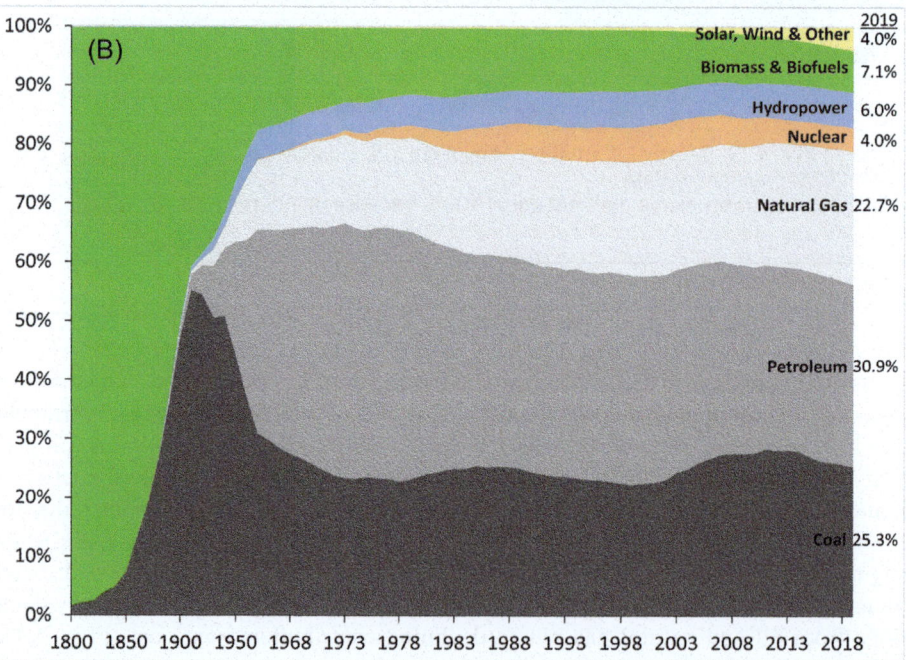

FIG. 3.5

Global annual substituted energy use (A) by major fuel type in equivalent terawatt-hours (TWh) and (B) fractional contributions from major sources. The rapid expansion of petroleum use began in the mid-20th century. The fraction of energy from fossil fuels has remained stubbornly in the range of 80%–85% globally since ~1950. Substituted energy denotes that the output energy for electrical energy generation (nuclear, solar, wind, and hydro) is converted to input energy needed with a typical fossil generator efficiency of 40%.

Plots by the author and data from Our World in Data with primary sources including the BP Statistical Review of World Energy and references therein.

Energy Information Administration (EIA), around 80% of US energy across all sectors is fossil. The US EIA projects that without concerted efforts, this proportion will stay roughly constant through the middle of the century. Contrastingly, science says we must reduce our greenhouse gas emissions by approaching 100% by mid-century to avoid the worst impacts of climate change by limiting the global average temperature rise to $1.5 < \Delta T < 2°C$, as discussed later. This is a huge task that only becomes more monumental with further delays.

Among other energy sources in the United States, nuclear fission reactors (for electricity generation) and a wide basket of renewable energy sources comprise 8% and 11% of the overall mix (Fig. 3.4). For the last few decades, the nuclear contribution has been ~20% of US electrical generation from approximately one hundred 1000 MW fission power stations. Renewable energy sources—solar, wind, biofuels, hydro, geothermal, wave and tidal, and assorted other small contributors—have shown robust growth over the last 20 years, particularly wind and solar. They now comprise over 10% of the energy pie in the United States.

Several trends in the last two decades are worth discussing (Fig. 3.6). First, the decline of coal-fired electricity has been dramatic and rapid in the United States. Since 2019, there have been short-term periods where summed renewable energy contributions now exceed those from coal in the United States, and the curves are even now close to intersecting on an annual basis (Fig. 3.6).

Secondly, the transportation end-use sector has become the largest energy consumer in the United States, exceeding the industrial, commercial, and residential. It is primarily dependent upon petroleum-based fuels, with small contributions from biofuels, natural gas, and increasingly electrified vehicles with battery storage. And, despite the progress with electrical vehicles, the transportation sector is considerably harder to decarbonize. This is particularly the case with long-haul trucking and the growing aviation emissions.

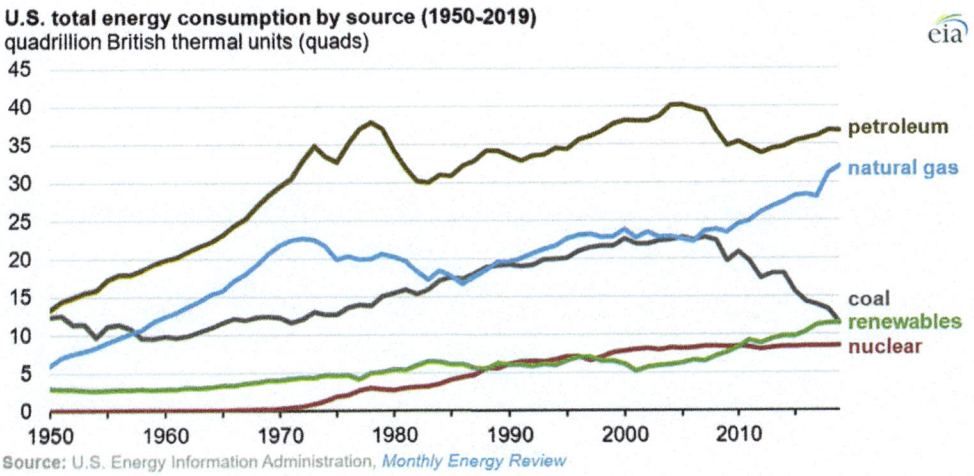

FIG. 3.6

Historical (1950–2019) US energy consumption by source.

Public domain images from US EIA.

The United States is one of the major global energy consumers (Fig. 3.7, only exceeded by China) and among the highest on a per capita basis. The rate of change in the last 30 years shows the growing importance of newly industrializing nations such as China and India, however (Fig. 3.7).

Even within the United States, the per capita use varies considerably by state (Fig. 3.8), as shown for electricity consumption, which varies from ~7 to 27 MWh/person. The drivers for this include climate differences, population density, industrial density and type, urban vs rural, energy-producing states, energy costs in the given state, state programs to reduce consumption, and socioeconomic/political differences. Take the state with the highest per capita usage, rural Wyoming has a four-season climate, very low population density and thus large transport costs, a large industrial consumption, particularly with fossil fuel production, and among the least expensive energy costs. California, on the other hand, where most of the population lives on the mild climate coast, has a more technological industry and less heavy industry, higher density, more urban, and higher energy costs. It has also been focusing on energy efficiency efforts through myriads of programs since the 1970s.

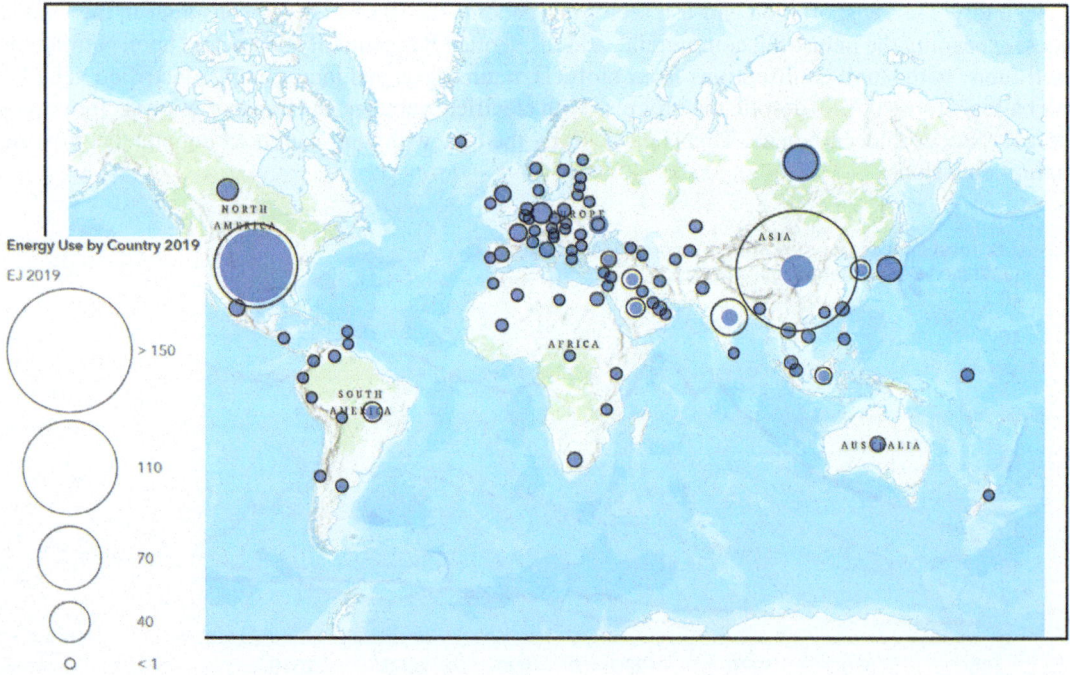

FIG. 3.7

Global energy use by region comparing 1990 *(blue-filled circles)* and 2019 *(open circles)*. Map lines delineate study areas and do not necessarily depict accepted national boundaries.

3.3 Historical United States and global energy use

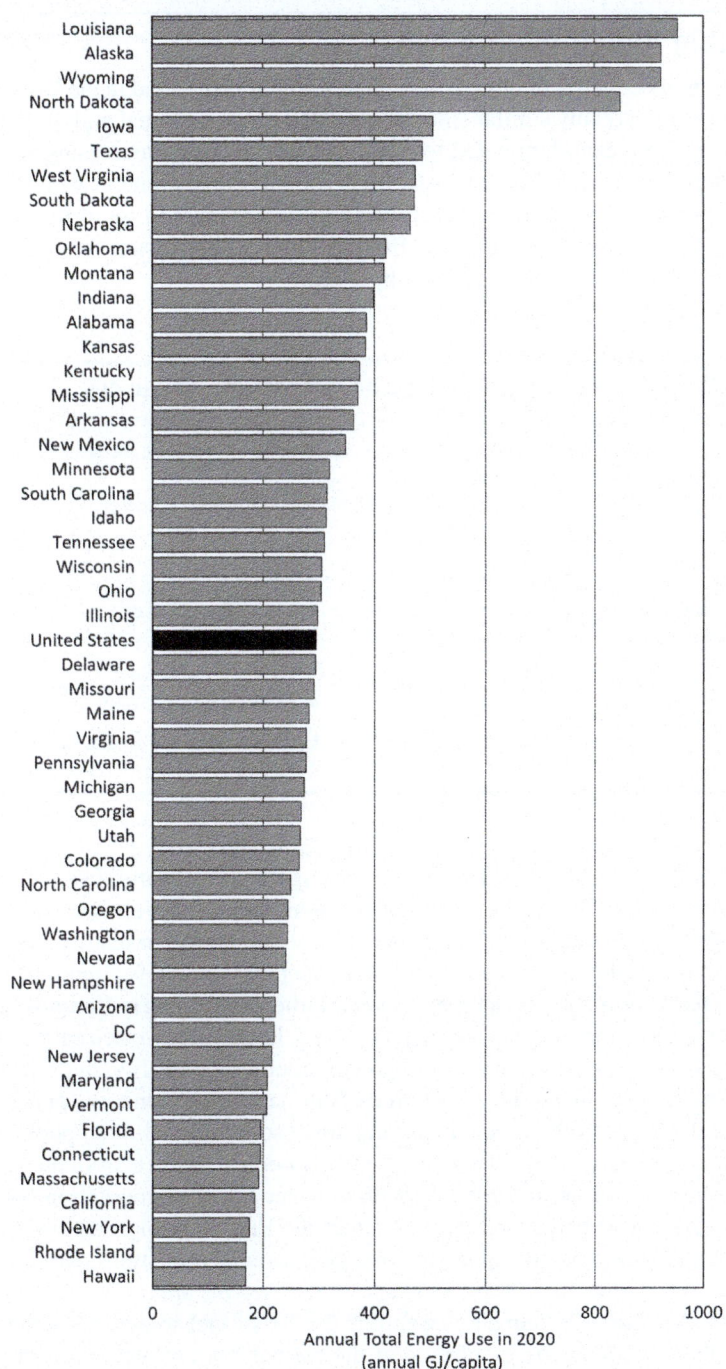

FIG. 3.8

Per capita annual energy use in 2020 across all end uses by US state.

Author image with data from US EIA.

3.4 Sustainable global energy use

Finite fossil fuels comprise ~80% of our energy sources both globally and in the United States, and this contribution has been remarkably similar since about 1950 on the national and global scale (Fig. 3.5). The mix has changed from coal-dominated to oil-dominated to a now increasing share of natural gas during the last century. The rate of use of fossil fuels far exceeds the rate at which they are forming, considering the millions of years of formation time. This finite nature of fossil fuels makes this rate of use unsustainable. An example illustrating the finite lifetime of petroleum is given below using data from the BP Statistical Review of Energy (Example 3.2).

EXAMPLE 3.2 Estimated lifetime of finite petroleum resources

Problem: The BP Statistical Review of Energy compiles by nation the production and consumption of all major energy resources. Using the global consumption of oil and the proven reserves of oil, estimate the remaining lifetime of our petroleum resource, assuming consumption remains constant.

Given: Consumption rates and resources from a recent BP Statistical Review of Energy.
Find: $T_{exhaustion}$.
Assume: consumption rate constant.
Solution:
BP Statistical Review of Energy from 2022 gives 2020 Proved Petroleum Resources as 1732 thousand million bbl of oil and a daily consumption rate of 94,088 thousand bbl daily.

$$\dot{V} = \frac{\text{Volume}}{\text{Time}} \text{ or Time} = \frac{\text{Volume}}{\dot{V}} = 1.732\text{E}12\,\text{bbl} \times \frac{\text{Day}}{94,088,000\,\text{bbl}} \times \frac{\text{Year}}{365\,\text{days}} = 50.4\,\text{years}$$

The proved resources are the economically producible estimated resource. Even the oil industry gives a limited lifetime of oil resources.

A global look at this energy history shows the increasing fraction of natural gas and renewables in the electrical generation mix supplanting the declining coal generation in the electric sector. Renewable energy is defined as that which is either inexhaustible (e.g., solar energy) or readily replaceable in a short amount of time (e.g., biofuels from growing annual crops) (Dunlap, 2019). Renewable energy sources can be replaced in a reasonable amount of time (<1 year) rather than taking millions of years such as fossil fuels or are strictly depleted without renewal such as U-235. Renewables, often called "natural" sources, include solar thermal, solar photovoltaics, wind power, geothermal energy, hydropower, biomass energy, and various ocean thermal and kinetic sources. Many renewable energy sources (not to mention fossil fuels) ultimately derive from the sun, including direct solar thermal, solar photovoltaics, wind energy from the differential heating of land masses and water bodies, biofuels from the growth of biomass, and hydropower from evaporation and condensation of water at higher gravitation potentials. The only sources that are not solar are nuclear, geothermal (from the Earth's primordial energy in its core from radioactive decay), and tidal power (from the moon's gravitational attraction). Although generally considered low impact, renewables are not zero impact, of course. We are trading a fuel-intensive energy sector for a materials-intensive resource by swapping out fossil fuel sources with renewable sources.

The growth in the renewable energy portion of the pie has been robust in recent decades, particularly wind and more recently solar (factors of ~3 and ~10, respectively, in the last decade). This has been driven by increasing awareness of the environmental liabilities with conventional energy sources, climate change concerns, governmental policies and tax incentives favoring renewables, and private sector interest due to the scaling up of production coupled with rapid cost decreases. Despite rapid growth, it should be noted that wind and solar still account for small fractions ~2.5% and 1%, respectively, of the overall energy mix in the United States (Fig. 3.4).

Globally, in terms of the total fraction of electrical generation that is renewable, the European continent has been the most aggressive adopter of renewable energy sources (20+% renewable fraction of electrical generation), while the slowest to adopt are the Middle Eastern and Commonwealth of Independent States (<3% of electrical generation).

A continuing challenge remains how to divorce human development from continued growth in energy and materials use, particularly nonsustainable sources. There are models which may serve as examples of how to accomplish this. For example, since 1973 the state of California flat-lined state electric usage and kept per capita total energy consumption stable while growing its economy, population, and vehicle miles traveled. California has also made a concerted statewide effort to emphasize efficiency, conservation, and renewables since the mid-1970s energy crisis to reduce its energy consumption. Thus, the last energy source is that of demand-side management. Reductions in our energy use are often the most economical and environmentally sensitive of all sources. These result from both curtailment of nonessential energy use and end-use efficiency gains. These will be discussed in chapters focused on the building thermal envelope and more efficient engineered systems in buildings (HVAC, lighting, among others).

3.5 Anatomy of centralized electrical generating units

Heat engines discussed previously have a major application in the electrical power generating industry. A traditional electrical generating unit (EGU), if powered by fossil fuels, takes a large mass flow rate of fuel, extracts the stored chemical energy from it, and exhausts an equally large mass flow rate mostly into the atmosphere (Fig. 3.9). A steam cycle power plant works based on the Rankine cycle, and the latent heat of vaporization plays a critical role in its operation. Just as the cyclic heat engine requires the cold reservoir to exhaust waste heat, the power plant rejects heat into the atmosphere via its cooling towers. Other plants that have cooling ponds or older plants may use natural lake, river, or ocean water as a cooling water source.

Based on the Carnot relationship, the higher the temperature difference between the hot and cold reservoirs (e.g., flame temperature vs ambient temperature), the higher the Carnot efficiency. Typical Rankine cycle EGU thermal efficiencies range from 30% to 40%, mostly a result of the thermodynamic constraints imposed by Carnot. This means something on the order of two-thirds of the thermal energy must be immediately dumped into the cold reservoir at the plant. The boiler only requires a significant source of heat, including nuclear fission, concentrating solar thermal technologies such as power towers and parabolic trough collectors, a geothermal heat source, or other heat source. A distinction with natural gas turbine EGUs that run on a Brayton cycle or a combined Brayton-Rankine cycle is notable. Such turbines, along with renewables, have displaced much of the coal-fired Rankine cycle generators and can exceed 60% thermal efficiency for the most efficient combined cycle varieties.

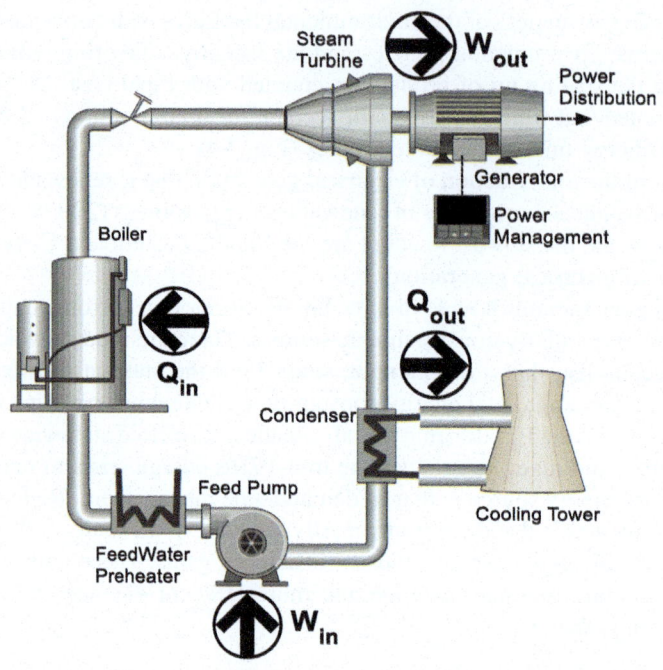

FIG. 3.9

Typical Rankine (steam) cycle power plant showing major components and energy flows.

The two chemical reactions of most relevance for the combustion of natural gas and coal (approximated as pure carbon) are summarized below (Eqs. 3.3 and 3.4).

$$CH_4 + 2O_2 \rightarrow CO_2 + 2H_2O + 213 \, kcal/mol \, (or \, 890 \, kJ/mol) \quad (3.3)$$

$$C + O_2 \rightarrow CO_2 + 94 \, kcal/mol \, (or \, 393 \, kJ/mol) \quad (3.4)$$

The following example shows the calculations of relevance for a typical coal-fired power generation facility illustrated in Fig. 3.10. The scale of mass flow rates of fuel, flue gas containing a high concentration of carbon dioxide and water vapor, the fuel feed rate, and cooling water used in such facilities is staggering, as shown in the figure.

Electrical generation resources are characterized as being baseload (always producing), load-following (dispatchable and following the demand curve), and peaking resources (those called upon only occasionally to meet periods of peak demand). Most grid-tied EGUs (numbered in the hundreds in the United States) serve as baseload generators or always on sources with capacity factors ranging from 50 to 90+% for such facilities. Many load-following plants are natural gas turbines that are used on demand and can be started up and modulated rapidly. The dispatchable nature of such resources makes them useful for electrical "load-following" and creates challenges with the large integration of renewable sources. As will be discussed later in the section on microgrids, a more flexible, two-way electrical grid is evolving to help integrate distributed generation from renewable sources that are inherently more variable. Newer taxonomy includes "fuel-saving" variable renewable energy (VRE) resources, "fast-burst" balancing resources such as battery storage systems and demand response approaches, which are

FIG. 3.10

Coal-fired power plant showing the scale of air pollution control devices, flue gas stacks, and the high flow rate of emissions.

available for timescales of hours. Other "firm" decarbonized resources include nuclear, natural gas with carbon capture and sequestration (CCS), and bioenergy that may also use CCS (Sepulveda et al., 2018). CCS is currently only done on pilot and demonstration scales, as will be discussed later.

3.6 Historical and current materials use

Global material use and energy use are intertwined by the large share of material flows related to the movement of fuels. The use of materials by humans was minuscule in paleolithic times owing to the human population in millions rather than billions, the hunter-gatherer nature of humans, and the limited needs for shelter and clothing amounting to kilograms of material per capita annually (Krausmann, 2012).

Historical material use shows an increase coincident with energy use beginning in the mid-20th century following World War II (Fig. 3.11). The largest contribution is construction materials. The effects of economic downturns and wars both temporarily depressed materials usage.

Materials use at some level also scales with GDP. Wiedmann et al. (2015) found a 6% increase in material footprint for every 10% increase in GDP when all leakage of materials is considered. Studies differ as to whether, as economies develop, the materials use and GDP relationship can be decoupled (Wiedmann et al., 2015). The 20th century witnessed an eightfold increase in the mass of materials produced, while one of the most important transitions was from renewable sources (i.e., mostly biomass) to finite resources (i.e., fossil fuels, metals, and minerals) (Krausmann, 2012).

The agricultural era increased the per capita materials used primarily as biomass. The Industrial Age added in what came to be the dominant contributions from fossil fuels plus industrial minerals, metals, and plastics. Overall, the most consumed materials are water and concrete, which will be discussed in later chapters dedicated to these important materials used extensively in the built environment. The mass of fossil fuels that we are removing from the Earth and depositing into the atmosphere is staggering as shown in Example 3.3.

EXAMPLE 3.3 Coal-fired power plant inputs and outputs

Problem: How much coal does a typical large-scale coal-fired power plant consume? Most people would guess a train car full a week or maybe one per day for a large power plant. Let us take a look at a 700 MW$_e$ coal-fired power plant that has a typical thermal efficiency of 35% and burns coal with a heating value of 22,000 kJ/kg, ash content of 5.5%, and sulfur content of 3.8%.

1. What is the daily input rate of coal (kg/day)?
2. If the efficiency of removal of SO_2 is 95%, how much SO_2 is emitted per day?
3. If 30% of ash drops out as bottom ash and the PM control is 98.5% efficient, how much fly ash is emitted in kg/day?
4. How much water is required for cooling if it is heated from 25°C to 40°C?
5. The above heat is transferred to the atmosphere via vaporization. How much water must be vaporized (and replaced) per day?

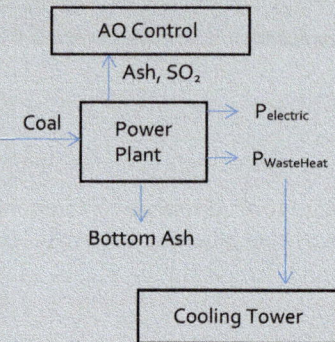

Given: 700 MW$_e$; Coal E_c = 22,000 kJ/kg; ash = 0.055; S = 0.038; 30% bottom ash; 98.5% PM & 95% SO_2 removal; ΔT = 15°C.
Find: P_{th}; kg/day SO_2 and fly ash; BTU/min to H_2O; kg/day H_2O vaporized.
Assume: SS operation.
Solution:

$$\eta = \frac{P_{electric}}{P_{thermal}}; P_{th} = \frac{P_{elec}}{\eta} = \frac{700 \text{ MW}}{0.35} = 2000 \text{ MW thermal input}$$

$$\dot{M}_c = \text{Coal In} = \frac{P_{th}}{\text{Coal } E_c} = \frac{2000\text{E}3 \frac{\text{kJ}}{\text{s}}}{22,000 \frac{\text{kJ}}{\text{kg}}} = 90.9 \frac{\text{kg}}{\text{s}} = 7.855\text{E}6 \frac{\text{kg}}{\text{day}}$$

To find the rate of emissions of fly ash into the atmosphere, we can use the coal mass flow rate, the fraction of ash left after combusting, the fraction that is airborne (vs the bottom ash), and the fraction not removed in the air pollution control device.

Continued

EXAMPLE 3.3 Coal-fired power plant inputs and outputs—cont'd

Fly Ash Emission Rate = Coal Mass Flow Rate × Coal Ash Fraction × Airborne Fraction × $(1-\eta_{\text{air pollution control}})$

$$= \dot{M}_c(f_{\text{ash}})(1-f_{\text{Bottom Ash}})(1-\eta_{\text{APC,PM}})$$

$$= 7.855\text{E}6 \frac{\text{kg}}{\text{day}} \times 0.055(1-0.3)(1-0.985) = 4535 \frac{\text{kg}}{\text{day}} \text{ of PM emissions}$$

Similarly for sulfur dioxide emissions (note the ratio of the molecular weight of SO_2 and S):

$$SO_2 \text{ Out} = \dot{M}_c(f_{\text{sulfur}})(1-\eta_{\text{APC,SO2}})\left(\frac{M SO_2}{M \text{sulfur}}\right) = 7.855\text{E}6 \frac{\text{kg}}{\text{day}} \times 0.038(1-0.95)\left(\frac{64}{32}\right) = 29,847 \frac{\text{kg}}{\text{day}} \text{ of } SO_2$$

Now to find the water needed for this power plant, we note that the waste heat, found from the difference in thermal power in and the electric power out, must be taken away by heating or vaporizing cooling water.

$$\Delta \dot{h} = \dot{m}c\Delta T = (P_{\text{th}} - P_{\text{elec}}) = 1300\text{E}6 \frac{\text{J}}{\text{s}}$$

$$\dot{m} = \frac{\Delta \dot{h}}{c\Delta T} = \frac{1300\text{E}6 \frac{\text{J}}{\text{s}}}{\frac{4.18\text{J}}{\text{g}^\circ\text{C}}(15^\circ\text{C})} = 20.7\text{E}6 \frac{\text{g}}{\text{s}} = 20.7\text{E}3 \frac{\text{kg}}{\text{s}} \text{ water heated}$$

$$\dot{m}_v = \frac{\Delta \dot{h}}{(\text{heat}VAP)} = \frac{1300\text{E}6 \frac{\text{J}}{\text{s}}}{2250\text{E}3 \frac{\text{J}}{\text{kg}}} = 578 \frac{\text{kg}}{\text{s}} \text{ of water vaporized}$$

The vast flows of energy and mass associated with a power plant are clear.

Materials use involves impacts from the point of extraction, through their use and maintenance, and at their disposal. The materials that humans use include both those for energy generation as well as those comprising our built environment such as metals, minerals, plastics, ceramics, and organic materials. Concrete and steel as discussed later are two major construction materials that are contributors to greenhouse gas emissions by the carbon embodied in their production.

Wood and timber resources have dominated buildings for thousands of years. Wood is key for residential buildings and low-rise buildings for many centuries (Mehta et al., 2018). Wood's favorable strength-to-weight ratio, its capacity to handle compressive, tensile, and shearing loads, and its general availability and low cost (compared to concrete and steel) have led to its utility. There are limits for tall structures, but timber is becoming more prevalent for mid-rise structures. Wood or wood-like materials renew rapidly, especially for bamboo and similar grasses, and the growth is driven by the sun. The downsides of wood include its land and water use, habitat, and other ecosystem uses, and soil resources required. Also, as an organic material, it can be subject to fire, insects, and rot.

The annual global average material use per capita has risen from 6 tons/year in 1970 to over 10 tons/year presently (UNEP, 2016). According to the University of Michigan Center for Sustainable Systems, the lifetime materials used by the average US citizen amount to 3.19 million pounds or an astounding ~1600 short tons per person. Several chapter problems and examples allow the students to calculate for themselves these quantities.

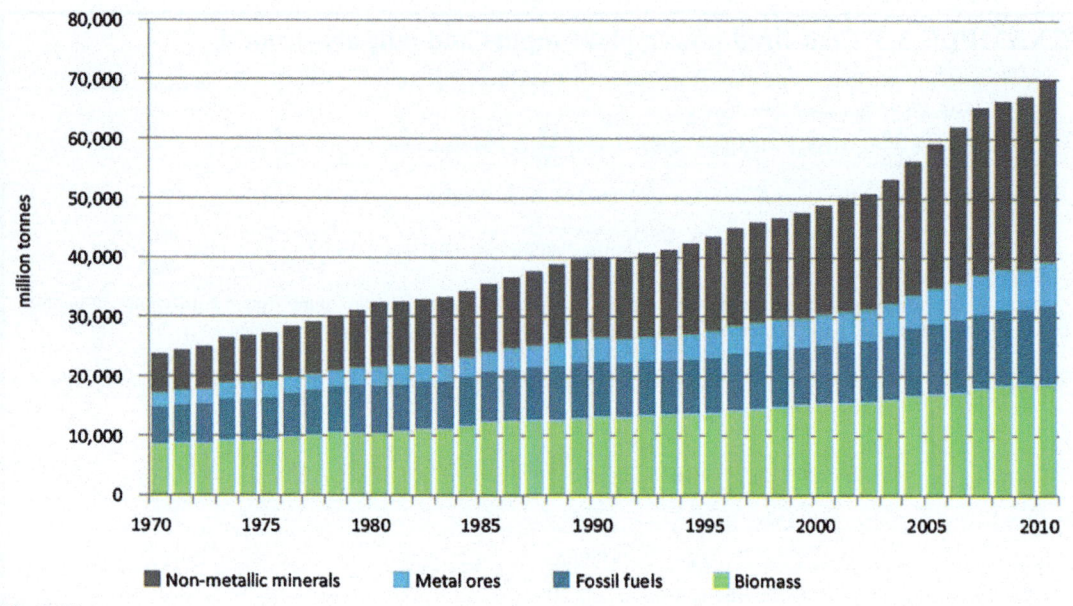

FIG. 3.11

Historical global materials usage and type in megatons from 1970 to 2010 (UNEP, 2016).

United Nations public domain figure.

3.7 Energy, materials, and sustainability

A few of the major concepts applied to energy, materials, and sustainability are reviewed here. The Brundage Commission summarized elegantly the meaning of sustainability in terms of meeting present needs without compromising the needs of future generations. Since that time, the field of sustainability has developed and taken on new concepts and approaches. Sustainability has become intertwined with an expanded view of human well-being, as discussed earlier in the chapter concerning the Human Development Index. A new approach, the triple-bottom-line approach, integrates three separate facets: environmental protection, economic viability, and social equity concerns—into our organizational aims.

One of the main approaches for evaluating the impacts of a process is termed life-cycle analysis (LCA). Life-cycle analysis is a holistic method for calculating the cradle-to-grave (or preferably to new cradle) impacts or costs associated with a particular activity, product, or device. It is a useful framework to compare alternative solutions to a given environmental problem or production issue (Azapagic, 2018). The approach considers the effects of extraction, processing, production, delivery, use, maintenance, and end-of-life disposal, including potential recycling and reuse of materials. Its four components include the (1) goal and scope definition, (2) inventory analysis to identify materials and energy use, (3) analysis of impacts such as climate, air quality, and water pollution, and (4) interpretation of the results. LCA is governed by International Standards Organization (ISO) standards and is navigable by numerous tools and databases; it has also been expanded to social LCA to consider social impacts on workers, consumers, society, and other stakeholders (Azapagic, 2018).

Sustainability has become endemic to our energy and material use as we are still so reliant on finite resources with significant climate and environmental liabilities. The example of fossil fuel lifetime earlier in the chapter underscored this finite nature. This pursuit of sustainability is integral to the built environment as well, considering the energy used in constructing and operating buildings plus the materials contributions to greenhouse gas emissions, particularly those from concrete and steel.

The materials and energy worlds are tightly linked in numerous ways such as energy use for mining and material processing. Historically, the operational energy use of a building dominated the life-cycle energy use and carbon emissions, as well as the conversations about green building. The embodied energy used in producing raw materials through to finished, delivered products is increasingly recognized as significant. Embodied energy will become more important as the low-hanging fruit in reducing operational building energy use is addressed.

The production of construction materials is associated with ~25% of global greenhouse gas emissions, and material efficiency offers a "third pillar" for deep decarbonization of the global economy (Pauliuk et al., 2021). Improved accounting of embodied energy and carbon of construction materials will be required, as these figures can vary in time and space and will likely prove to be harder to track than real-time building energy use.

As will be detailed in the materials chapter, the energy that goes into building materials per unit mass has a vast range. Among the most energy-intensive (energy per unit mass) are metals, particularly aluminum, and petroleum-derived synthetics, including carpeting, PVC, plastics, and foam-based insulation. Otherwise, wood and mineral-based products generally are low on the scale of energy intensity per unit mass. Though its energy per unit mass is modest, the large mass of concrete used in a structure (depending on foundation type) ensures that it comprises a large fraction of the embodied energy of a building. Much of this comes from the Portland cement content of the concrete. This will be discussed in detail later in the chapter on building materials.

3.8 Chapter summary and conclusions

Global industrialization has advanced humanity while also encountering energy and material resource constraints and environmental damages. Overall, expanding population plus its energy and resource use has become detrimental to our climate stability and the future of humans on this planet. The Industrial Age has been underpinned by the "taming of fire," i.e., using combustion to provide the vast energy needed for industry. Fossil fuels provide 80% of our energy for half a century in both the United States and globally. Materials use has increased by almost an order of magnitude in the last century. This is not a sustainable approach due to their finite lifetime, environmental liabilities, and climate change impacts, all calling for a new industrial approach.

Vast changes are emerging in the energy sector as well as the built environment in the United States and abroad. This is also seen in the declining role of coal and the increasing importance of natural gas and renewables in the US energy mix, beckoning a lower carbon future. The electrical grid is morphing from one anchored by centralized power generation facilities to more renewable energy sources and more widely distributed generation.

A new energy economy requires a vast scale-up of production to provide for the material intensity of new energy sources. Materials and energy are intricately linked by embodied energy, the energy required in producing and delivering materials. Our society is becoming more efficient with ongoing end-use energy in buildings, and overall, the economy produces a unit of output with less energy. In the future, the embodied energy in materials will become relatively more important.

58 Chapter 3 Energy and building history

The question now and into the future is whether we can sustain a vibrant construction industry while minimizing materials and nonrenewable energy use. Newer, more sustainable approaches and materials are focusing on reducing impacts or even becoming carbon negative. There are no easy answers, or they would have been implemented already. In the end, the emerging solutions are multipronged, evolving, and subject to further revision as we learn along the way.

3.9 End of chapter exercises

(1) **Concepts:** Do some research. What is the current floor area of all buildings globally? Find a country with a comparable land area.
(2) **Concepts:** Diagram the major components of a Rankine cycle electrical power generation facility. Don't forget the air pollution control, cooling towers, and generator/transmission lines!
(3) **Concepts:** The figure below shows the diagram of a steam cycle power plant. Is this configuration possible, and why or why not?

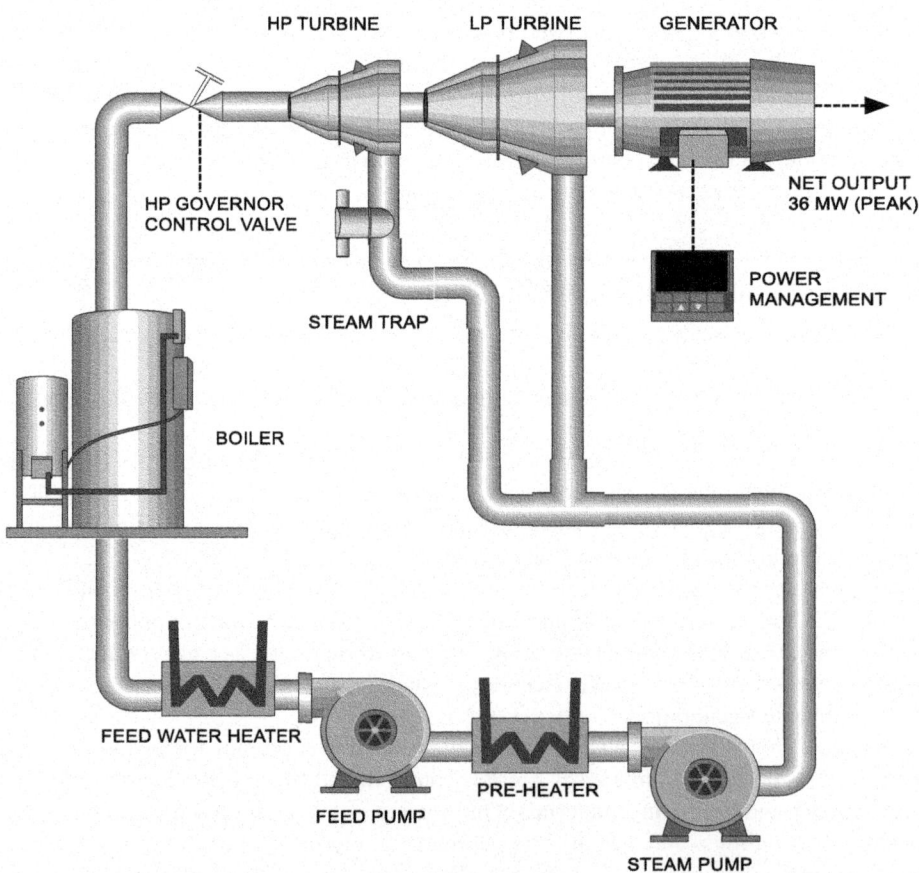

(4) **Concepts:** An electrical generating power station involves multiple conversions of energy from one form to another. List, in order, the sequence of energy forms starting from how the energy in a lump of coal was formed and ending with the energy of a hair dryer in the home. Indicate where the energy conversion happens. Hint: There are at least seven major forms to think about!

(5) **Concepts:** Historically, it has taken 50–100 years or more to transition from one dominant energy source to another strictly via market choices. Several excellent books detail the history and scale of energy use and transitions. Discuss some ways that the transition can be accelerated.

(6) **Problem:** I'm starting an energy hedge fund, and I need to know where to invest. The current World Proven Reserves of oil, natural gas, and coal resources are shown in the table below from BP. Help me out and fill out the last column of the table. I need you to calculate an estimate of how long these resources will last at current production rates. Watch your units!!

Note: "Proved reserves of oil—Generally taken to be those quantities that geological and engineering information indicates with reasonable certainty can be recovered in the future from known reservoirs under existing economic and operating conditions."

Estimating organization	Oil recoverable resources (thousand million bbl)	Consumption rate (billions liters/day)	Years until fully consumed
BP Statistical Review 2015	1700	14.542	?
	Natural gas recoverable resources (trillion cubic feet)	Consumption rate (billion cubic meters/day)	
	6606.4	9.296	?
	Coal recoverable resources (millions short of US tons)	Consumption rate (MTOE/year)	
	982,743	3881.8	?

Note: MTOE is a million tons of oil equivalent. For our purposes, 1 ton of oil is equivalent to approximately 2.25 tons of coal (half and half mix of hard and soft coal).

Revisit the calculation of the lifetime of current natural gas supplies. Rather than constant use, this time assumes a growth rate of 2.5% annually in terms of consumption rate. How long will the supply last under these conditions? You will have to use a little calculus.

(7) **Problem:** Think about the refueling of a gasoline car vs recharging an EV.
   (a) Compare to the average US home electrical service panel and the power it can deliver.
   (b) Now compare to the gasoline refueling rate. What is the rate of energy delivery for refueling the gasoline car? Assume 3 min and 20 gal to refuel.

(8) **Problem:** The Population Bomb written by Stanford biologist Dr. Paul Ehrlich was published in 1968. Assume that the human population growth rate at that time, 2% annually, continued unabated since 1968.
   (a) Calculate what the current human population would have been in the year 2022. There were 3.5 billion people in 1968 (now there are 8 billion).
   (b) What was the actual effective exponential growth rate over time (%/year)?

Chapter 3 Energy and building history

(9) **Problem:** Let's look at human population growth which has been exponential albeit with a varying rate over time (now declining). Let's consider our current growth rate of 1% annually and the maximum growth rate of about 2% some decades ago. Make any appropriate assumptions. Comment on the numbers.
   (a) How long before we have a human occupying every square meter of the land surface of Earth?
   (b) How long before the human population mass exceeds the mass of the Earth under each scenario?

(10) **Problem:** Assume that the human population growth rate at 2% annually continues unabated. Calculate in what year the mass of humans would equal Earth's mass. Repeat the calculation assuming we get really reproductive and our growth rate is 5%/year.

(11) **Problem:** It's been said that the entire world population could fit in Texas. How feasible does this seem? Think about the land requirements for food and other material needs.

(12) **Problem:** Estimate the mass flow rate of exhaust gas (in kg/s and m^3/s) from a 1200 MW_e power plant (34% efficient) burning coal (assume pure carbon) according to $C + O_2 \rightarrow CO_2$ with a heating value of 25,000 kJ/kg. You can assume STP298. What is the rate of CO_2 emissions in kg/s and m^3/s? Does it seem like it will be easy to capture and sequester this?

(13) **Problem:** A boiler produces steam at 520°C and this steam is used to run a heat engine to produce mechanical energy. It is desired to use a river as the cold reservoir and to have a Carnot efficiency of at least 45%. (a) What is the maximum allowable temperature for the cold reservoir? (b) Is this feasible with river water?

(14) **Problem:** It's estimated that a person requires a minimum of approximately 1.2 acres of arable land to support subsistence (basic food, clothing, shelter) for that person. Compare the needs of the current global population to the land area of the United States and the globe, and then compare to the arable land area in the United States and the globe.

(15) **Problem:** Let's look at human population growth which has been exponential growth albeit with a varying rate over time (now declining). Let us consider our current growth rate of ~1% annually and the maximum growth rate of about 2% in the mid-20th century. Make any appropriate assumptions. Comment on the numbers. How long before we have a human occupying every square meter of the land surface of Earth under each scenario?

(16) **Problem:** Assume we could use all MSW in the United States for electrical power generation. How many 1 GW power plants would it fuel? Use reasonable assumptions for a Rankine cycle power plant and MSW energy content.

(17) **Problem:** The 2000-Watt Society envisions average per-capita use of all forms of energy as 2000 watts continuous. Energy use ranges widely by region, however; while Africans and Bangladeshis use less than 500 watts per capita, Americans use on average 12,000 watts apiece! Calculate the global per capita energy use currently. In 2018, 598 Quads were used across the entire globe in all sectors.

(18) **Problem:** Estimate the mass flow rate of CO_2 from a typical 1000 MW electric coal-fired power plant with typical thermal efficiency. Assume coal has an energy content of 10,000 BTU/lb and can be represented as C_1H_2.

(19) **Problem:** Find the average annual gas and electric use for a house in the United States. Convert into total GJ per year. Compared to the 600 GJ of embodied energy to build a home, how many years before the operational energy use surpasses the embodied energy of construction?

CHAPTER 4

Environmental problems from the local to global scale: Motivation for why we need to build better

Learning objectives

(1) Comprehend the breadth, scale, and diversity of threats to the environment that human activities drive.
(2) Recognize the paramount threat from global climate change and how building better is a key facet of a global response.
(3) Link the human threats to the environment to the built sector through the critical role of energy and materials used in building construction and operations.

After taking a big-picture look at the scale and issues related to population and consumption of finite resources in Chapter 3, it is useful to understand interconnected environmental problems whether by land, air, or sea. The following chapter reviews some of the major themes associated with the environmental threats and degradation of the natural world, and in particular, the numerous points where it links to the built environment. This chapter summarizes the main topics that would be discussed in more detail in an introductory class in environmental science or engineering as well as illustrates a subset of some of the key calculations. This is intended for the students outside these disciplines rather than those students with an extensive background in environmental science and engineering for which this chapter may be a review.

Crafting solutions to these problems in the building sector is predicated on understanding the problems and avoiding the pitfalls of solving one problem while not creating another. The list of topics here is broad, though the reader is referred to additional textbooks that address these problems in more detail than can be done here (Cooper, 2015; Davis and Cornwell, 2022).

The same ingenuity we used to have mastery over nature is now crucial to preserving it and the biosphere on which we depend. Never has the need been more critical for enhanced sustainability throughout our society—especially the built environment—as one key approach to the mitigation of environmental problems.

4.1 Infrastructure and environment

Environmental stresses felt globally are a subset of a larger issue with a broader deterioration of infrastructure and larger sustainable development challenges. The American Association of Civil Engineers periodically publishes an evaluation of the US infrastructure health once every 4 years (America's Infrastructure Report Card 2021 | GPA: C−). The evaluations have not been particularly complimentary with a current overall rating of C− and a range from D− (transit) to B (rail) (Table 4.1). Compared with other countries, the United States invests a small fraction of its GDP in upgrading the nation's infrastructure. Small improvements were noted from the previous report due to the recent focus on infrastructure by the federal government, an area where the two US political parties may be able to agree. Recovery funding from the COVID-19 pandemic has also created opportunities for improvements, though the general status of infrastructure in the United States remains dismal.

Deficiencies in infrastructure include the subcategories bolded directly related to environment and green building: drinking water, energy, hazardous waste, solid waste, stormwater, and wastewater (Table 4.1). The needed funding for upgrading the deficiencies in infrastructure and the funding gap are shown, amounting to $2.6 trillion in the most recent report card. Clearly, the need for infrastructure improvement exists, and the capacity to fund and undertake these upgrades is unidentified. Although the United States is considered among the most deficient in maintaining its infrastructure, other nations also face similar challenges. The most pressing environmental-related problems that link to infrastructure in many ways are discussed in the rest of this chapter.

Table 4.1 American Society of Civil Engineers gradebook for US infrastructure (2013–21).

ASCE category	2013 Report card	2017 Report card	2021 Report card
Aviation	D	D	D+
Bridges	C+	C+	C
Drinking water	D	D	D
Energy	D+	D+	C−
Hazardous waste	D	D+	D+
Inland waterways	D+	D+	D+
Levees	D−	D	D
Ports	C	C+	B−
Public parks	C−	D+	D+
Rail	C+	B	B
Roads	D	D	D
Schools	D	D+	D+
Solid waste	B−	C+	C+
Stormwater	–	–	D
Transit	D	D−	D−
Wastewater	D	D+	D+
Overall infrastructure grade	D+	D+	C−
Total upgrade funding required (decadal)	$3.64 T	$4.59 T	$5.94 T
Funding gap (decadal)	$1.61 T	$2.06 T	$2.59 T

4.2 Hydrological resources on planet earth

Water use in buildings and landscapes is a key environmental concern and merits understanding of the large-scale water cycle. The hydrological cycle, along with the carbon cycle discussed later, are among the most critical to the functioning of the Earth system and human needs. Though there is a tiny exchange with space and formation/destruction into constituent hydrogen and oxygen, mass balance dictates that the quantity of water on the planet is essentially constant. It does, however, change phase, location, and shifts from freshwater and saltwater reservoirs.

4.2 Hydrological resources on planet earth

Fig. 4.1 shows major reservoirs and interactions between reservoirs of water and their interactions in the hydrological cycle. Most of the world's water is saltwater, and most of the freshwater is in a frozen state (Fig. 4.2). A small fraction of liquid freshwater is the source of our potable drinking water and a sink for our wastewater. Stocks of freshwater include surface water such as riparian (river) and limnic (lake) resources as well as groundwater (Fig. 4.2). Reservoirs of freshwater also include the atmosphere, the biosphere as part of all plants, animals, and microbes, as all life forms depend on water. The interfaces between land and water including freshwater wetlands and coastal estuaries are where much of the biodiversity on the planet exists.

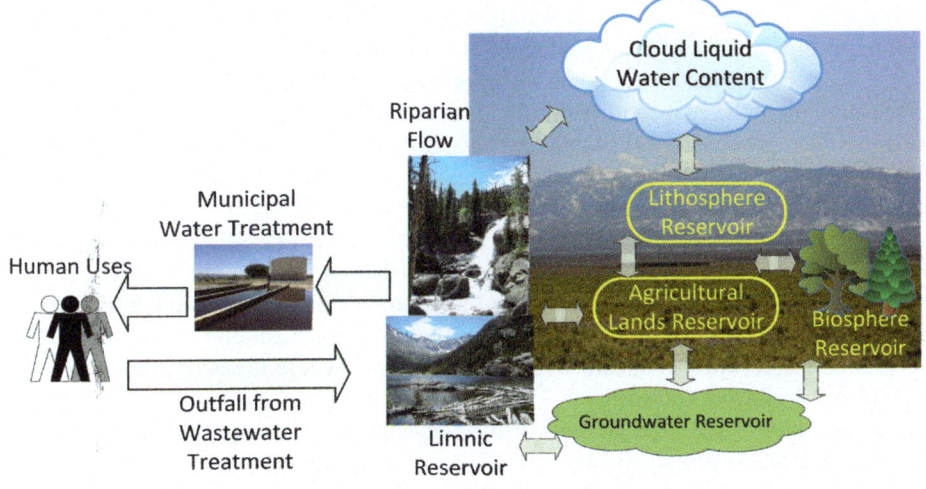

FIG. 4.1

Schematic showing major freshwater stocks and flows that play a role in Earth's hydrological cycle.

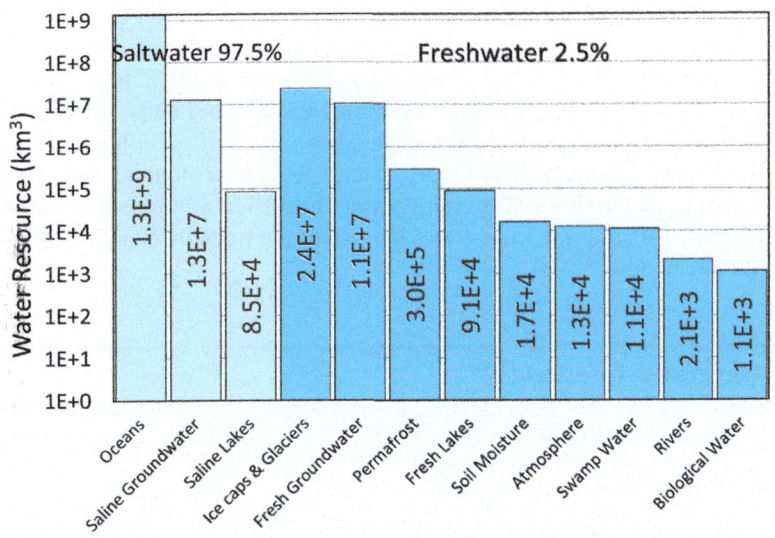

FIG. 4.2

Global water resources by category.

Data from USGS and Shiklomanov, 1993.

Freshwater resources that are easily and economically retrievable are under increasing threats to their quantity, quality, and timing. As a long-standing global environmental problem and an ongoing geopolitical issue, freshwater access is essential due to water's role in the biosphere and civilization. Arid environments such as the Western United States and northern Mexico, the Middle East and northern Africa (MENA), much of the Australian subcontinent, and the Gobi Desert region of Asia are each case studies in water constraints limiting development in such regions. Many of these regions rely upon the slow release of water from mountain snowpack in nearby mountain ranges in spring. The runoff washes down the drainages and passes through freshwater flows and reservoirs beginning in spring providing an extended period of water availability. The Rio Grande and Colorado Rivers in the desert southwest US are dominated by snowmelt and its runoff. Surface water resources have become a critical issue in reservoirs such as Lakes Mead and Powell on the Colorado River in the Western United States. Increasing demands have combined with decreasing snowpack and exacerbated by the southwestern mega-drought of the 21st century (Williams et al., 2020). As more of the precipitation is falling as rain in the shoulder seasons, runoff is both early and fast, and water management becomes even more challenging. Such water availability issues are critical to the evolution of the building industry in impacted regions as they can be a hard constraint on any new development in this region.

One of the most important but largely hidden resources is groundwater. The subsurface has a sub-saturated vadose zone near the surface and an underlying water-saturated zone or aquifer defined by the "water table." This can be thought of as an underground mobile reservoir of water contained in the void spaces of the soil matrix. Aquifers have an effective flow rate much smaller than rivers. Groundwater resource depletion has resulted in subsidence in select regions where it has been extensively pumped such as the Central Valley of California.

Despite the ubiquity of saltwater in the world's oceans, the status of some major salt lake ecosystems is also stressed due to human consumptive issues and climate change (Wurtsbaugh et al., 2017). For example, the role of declining snowpack in the surrounding Rocky Mountains is also a key contributor to the decline of the extent and volume of the Great Salt Lake in Utah, USA (Hall et al., 2021). Roughly one-third of its extent in 1980, the declining Great Salt Lake (among others) has equally dire consequences for habitats and humanity. Conversely, due to rising sea levels, saltwater intrusion on freshwater resources is a growing concern, especially for coastal regions.

With freshwater resources, we have both a quality and quantity problem. Its quality can be degraded due to contamination from physical, chemical, and biological contaminants. The importance of water treatment is the subject of a subsequent chapter of this book focused specifically on water issues and their critical relationship to green building. Hydrosystems require substantial energy and material use and are an integral part of all buildings. The energy-water nexus is a relationship of greater importance in the last several decades. Example 4.1 illustrates the size of the frozen water portion of our freshwater resources—and the threats to it.

> **EXAMPLE 4.1 Water resources**
>
> **Problem:** In 2017, the iceberg A-68 calved from the Antarctic ice sheet. The iceberg is roughly 55 km × 150 km and 650 ft thick. The specific gravity of ice is 0.934. What is the mass of this iceberg? How many Empire State Buildings (331,000 t) does it equate to? How long could it provide potable water to a city of 5 million?
> **Given:** A-68 (55 km × 150 km × 650 ft thick); sg = 0.934
> **Find:** M_{ice} ESBs
> **Assume:** uniform density

Continued

> **EXAMPLE 4.1 Water resources—cont'd**
>
> **Solution**
> Mass = volume × density = 55,000 m × 150,000 m × 650 ft × (m/3.3 ft) × 934 kg/m^3 = 1.52 × 10^{15} kg   or   1.52 × 10^{12} Tons = 1.52 Tt (which would occupy ~1.52 × 10^{12} m^3 as liquid water)
> Empire State Buildings = 1.52 × 10^{12} t/(331,000 t/Empire State Building) = 4.6 × 10^6 Empire State Buildings
> If this could be used to provide fresh water for a city of 5 million, how long would it last? Use reasonable assumptions. A standard assumption for water and wastewater treatment is roughly 100 gal per capita per day.
> Time = Vol/volumetric flow rate = 1.52 × 10^{12} m^3/(5 × 10^6 people × 100 gal/person/day × 1 m^3/264 gal) = 802,560 days or 2199 years!
> The declining quantities of ice mass are removing the "air conditioning" of planet.

4.3 Land use, soils, and materials extraction

Like water use, the land use and effects on the solid Earth associated with human civilization are a crucial area of environmental damage and intimately tied to the built environment. Our civilization has expanded with increased agricultural use, industrial activities from extraction to production to disposal, and the increasing fraction of land developed for human habitation (Foley et al., 2005).

A global map product from the European Space Agency (ESA) of the existing land-use cover shows the extent of developed land and agricultural usage in particular (Fig. 4.3). It is estimated nearly 70% of the ice-free land surface has been developed, urbanized, mined, farmed, or otherwise disturbed by human use in some manner with roughly a third of that being degraded (IPCC, 2019). Not only do these altered landscapes fragment the habitats of other native species, but also threaten the very functioning of an interconnected biosphere and the value of the ecosystem services that it provides to humans.

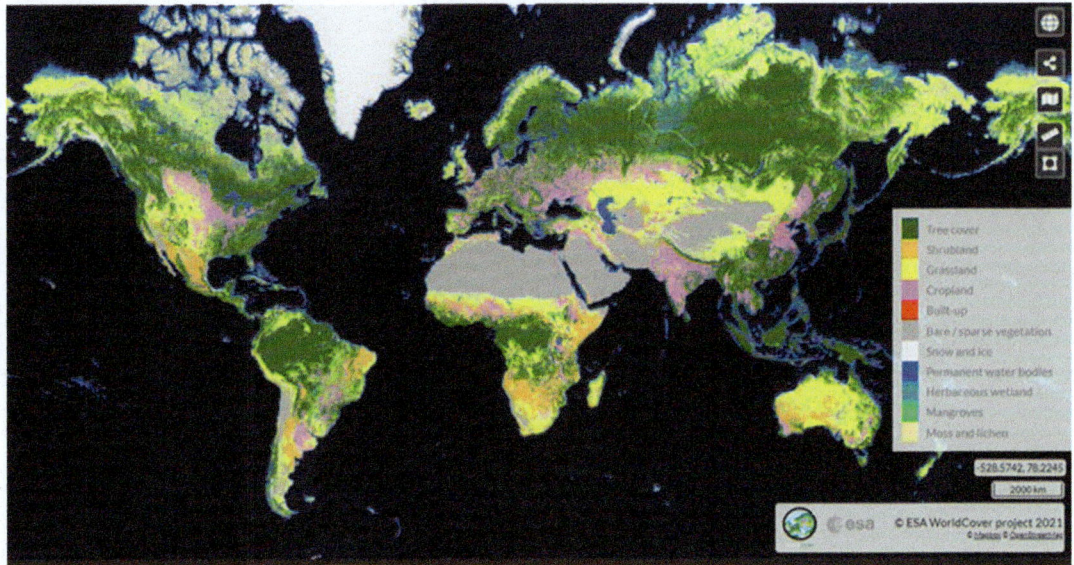

FIG. 4.3

Global land-use cover (© ESA WorldCover project 2020 / Contains modified Copernicus Sentinel data (2020) processed by ESA WorldCover consortium, used under Creative Commons license) (Zanaga et al., 2021).

Soil quality is a bit harder to define and regulate than air or water quality. The Soils Society of America defines it as the capacity to support biological activity, foster environmental quality, and sustain the health of plants and animals dependent upon the soils (Karlen et al., 1997). This includes multiple uses including supporting human civilization and health, agriculture, forestry, and wildland ecosystems. Soils represent one of the largest stocks of carbon, nitrogen, phosphorus, and many other key terrestrial elements (Batjes, 2014). Soil sequestration of carbon is one of the key carbon stocks and also offers a cost-effective mitigation tool for the climate crisis which will be discussed later (Griscom et al., 2017; Lal, 2004).

When discussing soils, one is also invariably discussing groundwater properties as well. The primary characteristics are the porosity and permeability of the soil through which the groundwater flows. The porosity is a measure of the void space fraction (% in the soil matrix while the permeability is the ability of fluids to flow through this solid matrix so it depends on porosity, pore size distribution, and other physical properties (Jackson, 2019). The permeability (m^2) ultimately relates to the hydraulic conductivity and the velocity (m/s) of the groundwater flow through soil (Jackson, 2019). Darcy's law describes the specific discharge q (m/s) which is the volumetric flow rate over the cross-sectional area as a function of the hydraulic conductivity (K, m/s) and the hydraulic gradient (dh/dL, m/m) or gradient in pressure head (m/m) (Example 4.2).

$$q = \frac{Q}{A} = -K\frac{dh}{dL} \qquad (4.1)$$

EXAMPLE 4.2 Darcy's law flow rate calculation

Problem: A groundwater flow is found to have a hydraulic head decrease (decrease in pressure equivalent to a column of water of a given height) of 10 m of water over a 100 m flow path length. It is a typical US limestone. Find the volumetric flow rate through a cross-sectional area of $10 m^2$.
Given: 10 mH_2O/100 m in limestone
Find: Q in xsa = $10 m^2$
Assume: Uniform properties

Solution:
For a typical US limestone, the hydraulic conductivity is $K = 2 \times 10^{-4}$ m/s ± factor of 10

$$q = \frac{Q}{A} = -K\frac{dh}{dL}$$

$$Q = -K\frac{dh}{dL}A = -\left(2 \times 10^{-4}\frac{m}{s}\right)\left(\frac{-10\,m}{100\,m}\right)(10\,m^2) = 2 \times 10^{-4}\frac{m^3}{s} = \frac{200\,cm^3}{s}$$

Ground water resources are finite and require careful management.

4.4 Materials extraction and processing

Concerning material sourcing, it is said that if it is not grown it is mined. Mining by its nature is a material, energy, and water-intensive industry that impacts land, sea, and air. Much of the waste products and hence environmental consequences owe to the small ratio of mined metals or minerals to the total amount of Earth mined and processed (Dudka and Adriano, 1997). The mining and mill tailings are a long-standing problem in the mining industry. The other key factor is that the mining process allows the mobilization of many naturally occurring substances including heavy metals, radionuclides, and other compounds that when made bioavailable become hazardous.

The primary impacts from mining and materials processing include (a) materials, water, and fuels use; (b) wildlife and habitat destruction or disruption; (c) direct atmospheric (e.g., heavy metals, SO_2) or emissions to surface water bodies; (d) tailings piles and other solid wastes; (e) soils and groundwater contamination with chemicals use (e.g., acids) and their potential for runoff and contamination; and (f) land disturbance including erosion and inadequate remediation of disturbed lands. A key determinant of these impacts is the nature of surface vs underground mining.

Mining has inherent risks for the workers in the profession. Health effects for miners include both acute (e.g., accidents, fires, over-exertion) and chronic (lung cancer) morbidity and mortality. Mining impacts are not limited to the active production period of the mine. Acid-mine drainage is a case in point where disturbed lands, often under the water table for underground mines, result in the pooling of water in abandoned mine shafts. These waters can easily become contaminated with the naturally occurring soluble or suspended materials left over as well as any remaining processing chemicals.

The interconnectedness of materials and energy use is also undeniable. The movement of fossil fuels is one of the most significant mined materials directly related to energy use. It is also a significant constraint for transitioning to a low-carbon energy infrastructure as new energy technologies such as solar PV and wind require significant inputs of metals and minerals. This quantity, however, pales in comparison to the scale of fossil fuel extraction to provide fuels for our existing energy system.

4.5 Solid, hazardous, and construction and demolition waste

Solid waste (i.e., garbage) disposal issue came into particular relief in 1987 with the MOBRO 4000 "Garbage Barge" that sailed from New York City to the Caribbean and back unsuccessfully looking for a landfill to accept its solid waste. In developed countries, solid waste generation has grown roughly in proportion to the scale of the economy. A time history of waste generation in the United States is shown in Fig. 4.4. The average solid waste generation in the United States is $\sim$4.5 lbs/capita/day, meaning the average American generates their weight in solid waste in little more than a month. The good news is that this per capita generation rate of solid waste has more or less stabilized over the past several decades after steadily increasing up until $\sim$1990.

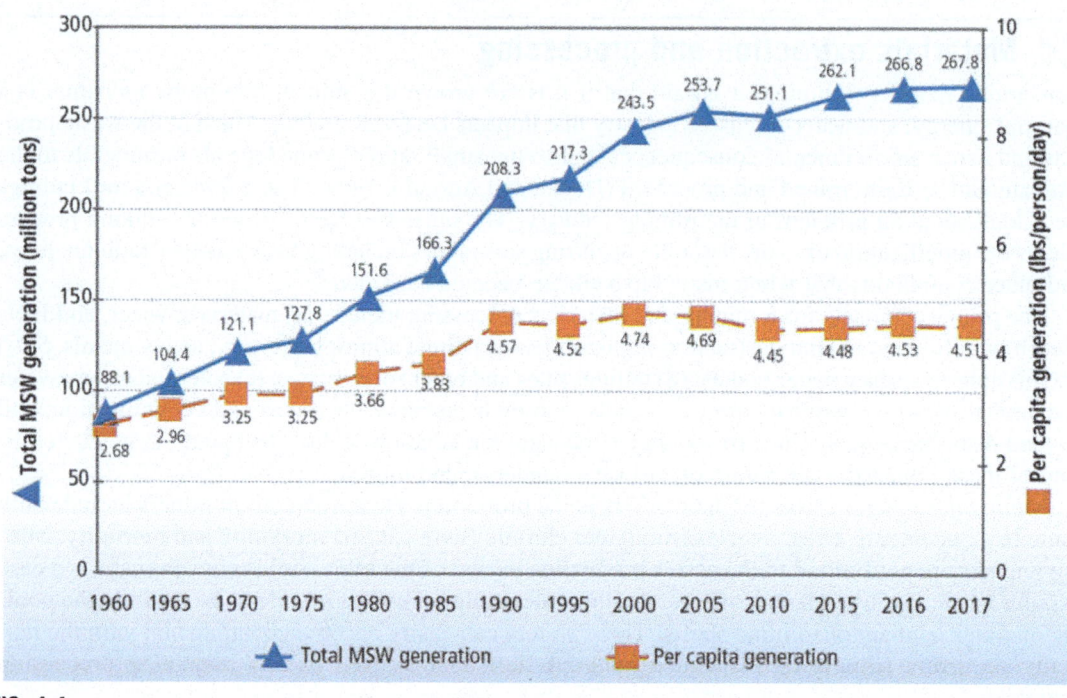

FIG. 4.4

Time history (1960–2017) of US total and per capita municipal solid waste (MSW) generation (USEPA, 2019).

The old approach of digging a hole and burying our garbage has evolved into a much more sophisticated system to manage solid waste from the household to the city scale. The options and appropriate solutions for solid waste now encompass reducing generation, minimizing waste in both design and production, diverting waste back into the production process, reusing materials, recycling, composting, incineration, or proper landfilling of waste at the backend. Recycling operations include elements as low tech as manual sorting and high tech as automated sorting using density, size, and optical segregation methods. The efforts to reclaim solid waste have "flattened the curve" in terms of total waste generated and per capita waste in the United States (Fig. 4.4). In part, waste generation has followed the increase in wealth in our society (USEPA, 2019), although those curves have diverged in some nations in recent decades. The economy continued to grow over that period, and thus the solid waste per unit of GDP has declined. The average cost for landfilling ("tipping" fees charged to haulers to landfills) more than doubled from the mid-1980s to the mid-1990s, giving an economic incentive for finding alternatives to landfilling.

Increased diversion has been hard-won gains with a broad range of scales of local community recycling, composting, and waste reduction efforts from major cities to small localities, involving public, private, commercial, industrial, and residential sources. The US diversion rate (fraction not landfilled) is now approaching 50% with ~25% recycled ~10% composted, and the other 13% combusted with energy recovery. This of course varies considerably regionally concerning the material type, its recyclability, the economic value, regional recycling options, and market conditions. Nearly 75% of paper products are recycled with a low of 13% of plastics; 69% of yard waste while only 6% of food waste is composted (USEPA, 2019). Other nations with higher land and landfilling costs such as Japan and select European nations have much higher diversion rates.

Construction and demolition (C&D) waste is the fraction of solid waste related to buildings. It includes the scraps of construction materials and the debris from demolishing existing built structures undergoing renovation or replacement. Typically, it also includes road and bridge debris though excludes land-clearing biomass debris. It will be discussed in more detail in Chapter 7 as it is so important to building construction impacts.

The waste materials we produce vary considerably in terms of their hazard level to both humans and the environment. Hazardous materials are multiphase: solid, liquid, gas, or a mixture of phases. The most hazardous materials include those that are reactive, strong acids or bases, flammable, corrosive, or have known toxic health effects and are categorized as "hazardous waste" rather than solid waste. It requires special handling, treatment, and disposal to prevent posing environmental or human health hazards. The National Fire Protection Association (NFPA) diamond, shown here for gasoline, describes the hazards associated with a given chemical. The hazards are rated from 0 (nonhazardous) to 4 (highest hazard) for human health (blue), flammability (red), reactivity (yellow), and any special hazards listed in the white box (radioactive, corrosive, water reactive, or other hazards).

The US EPA estimates there are over 50,000 hazardous waste sites in the United States alone (Watts et al., 2022). Though sites have been or are currently under remediation, many of these are legacy sites from many decades ago when the standard approach was disposing of such waste in unlined pits or ponds (Watts et al., 2022).

Although many other federal and state statutes come into play with hazardous waste, the Resource Conservation and Recovery Act of 1976 and the Hazardous and Solid Waste Act of 1984 dictate the proper disposal of hazardous wastes. The Comprehensive Environmental Response, Compensation, and Liability Act (CERCLA) of 1980, also called "Superfund," and the subsequent Superfund Amendments and Reauthorization Act (SARA) in 1986 established the regulatory framework for remediating the abandoned US hazardous waste sites (Watts et al., 2022).

4.6 Microplastics and marine debris

Societal use of plastics, derived from petroleum, has grown much faster than petroleum fuel use since the 1960s. Over the last half-century, the fraction of plastics in the solid waste stream has increased by an order of magnitude, now comprising over 10% of the mass of the solid waste. The largest market use for plastics has become single-use packaging material, meaning it all becomes solid waste almost immediately (Jambeck et al., 2015).

The 2018 global annual production of plastics reached 359 MT/year, according to PlasticsEurope (https://www.plasticseurope.org/en/resources/market-data). The recycling rates for plastic have remained below 10% in the United States. Economical recycling and reuse are still a challenge for this material, and its persistence in the environment is often decades or longer. This means a significant fraction remains in landfills, the ocean, and as uncontained refuse scattered throughout our society. It becomes a more significant issue when it does start to degrade with long chain bonds breaking and forming smaller pieces of plastic (1 mm or smaller) that are both more harmful and more difficult to remove for disposal. A study of the land flux of plastics into the ocean estimated 275 MT/year generated in coastal nations with a mass flow rate of 4.8–12.7 MT/year reaching the oceans (Jambeck et al., 2015). It is projected to continue to increase over the next decades unless robust solid waste handling systems are enacted to reduce this waste issue (Borrelle et al., 2020).

Plastics in macro-, micro-, and nanosizes have accumulated in soils, aquatic, and atmospheric reservoirs due to the scale up in use and poor disposal techniques (Bolan et al., 2020). Microplastics are those <5mm in dimension and are a new concern with the scale and extent of their presence, particularly in the oceans (Rochman and Hoellein, 2020). Many synthetic microplastics are fibers derived from synthetic textiles used for clothing. Although large fractions of the fibers are from natural sources which can confound the analysis (Suaria et al., 2020). Its environmental ubiquity includes finding microplastics in rainfall in remote regions (Brahney et al., 2020), terrestrial ecosystems including soils (Rillig and Lehmann, 2020), and also found in human tissues (Vethaak and Legler, 2021). The microplastic particle can enter the body through ingestion of food or water or via inhalation from atmospheric transport (Evangeliou et al., 2020). The threat is submicrometer particles that can cross boundaries in human organs and tissues such as the epithelial boundary in the lungs or intestines (Vethaak and Legler, 2021). The smallest particles are able to cross cellular boundaries and potentially lead to inflammation responses.

From ship-based measurements on transects of all the major convergence zones and many other ocean regions, the accumulation of plastic in the oceans is estimated as exceeding 5.25 trillion pieces weighing 269,000 t (Eriksen et al., 2014). Clearly, the lifetime is not unlimited as the stock is smaller than the annual addition rate to the oceans. The marine debris problem is particularly acute in the ocean convergence zones, such as the aptly named "Great Pacific Garbage Patch" in the northern Pacific Ocean. The region is defined as a broad region between California and Hawaii; it is one of the five gyre zones where waters converge in the world oceans. The current data show that plastic debris is rapidly accumulating in this region and is largely a function of ocean currents combined with surrounding nations' population, development level, and robustness of solid waste management systems (Lebreton et al., 2018). The distribution by mass is dominated by macroplastics 5cm and larger (92%) while microplastics dominate the number count (94%) estimated at 1.8 trillion pieces (Lebreton et al., 2018). The microplastics that can exceed 100,000 particles per m^3 of water are most damaging to marine life, particularly the fibrous particles (Wright et al., 2013) (Example 4.3).

EXAMPLE 4.3 Ocean plastics

Problem: The North Pacific convergence zone is approximately 8×10^6 km^2. Assume a well-mixed 100-ft surface layer and a plastics concentration of five pieces for every m^3 of water and a spherical particle with a diameter of 5mm. The bulk density of plastic is 0.9 g/cm^3. Calculate the mass of plastics in tons in this zone.

Given: N Pacific 8×10^6 km^2; 5/m^3; $\Delta z = 100$ ft; $D = 5$ mm; $\rho = 0.9$ g/cm^3

Find: $M_{plastic}$

Assume: consumption rate constant

Solution:
Garbage patch water Vol = area × depth = 8×10^6 km^2 $(1000 m/km)^2$ × 100 ft (m/3.28 ft) = 2.439×10^{14} m^3
Mass particle = volume × density = $\pi/6 \times D^3 \times \rho$ = 3.14159/6 × $(0.005 m)^3$ × 900 kg/m^3 = 5.89×10^{-5} kg
Mass Conc = Num Conc × mass particle = 5/m^3 × 5.89×10^{-5} kg = 2.945×10^{-4} kg/m^3
Mass = MassConc × WaterVol = 2.945×10^{-4} kg/m^3 × 2.439×10^{14} m^3 = 7.183×10^{10} kg = 7.183×10^7 t of plastic in the garbage patch… ~72 million tonnes of plastic!

4.7 Biodiversity, habitat loss, and species extinction

Biodiversity describes the variations and variability of plant, animal, fungal, and microbial living organisms (i.e., the biosphere) from the genetic scale of individual organisms to species scale to ecosystem scale to the global scale. Human influences reducing biodiversity include land use and overdevelopment as dominant ones, over-harvesting of natural resources, facilitating the spread of invasive species, altering climatic conditions, and contamination of air, land, and water resources (Vitousek et al., 1997). This impacts the richness, evenness, and species prevalence in different ecosystems as well as potential extinctions in the extreme case (Chapin et al., 2000). The flipside is possible too where introduced, exotic species dominate ecosystems, excluding native species and causing ecological changes in productivity, soil and water salinity, and vegetative litter. This has been the case, for example, with the tamarisk species (salt cedar) in Southwestern United States desert regions. Other invasive species such as cheatgrass and kudzu have caused similar disturbances in other ecosystems (Chapin et al., 2000).

Early examples of humans' impacts on species prevalence are the effects of overfishing of specific marine species over the past centuries (Jackson et al., 2001). One of the most persistent and harmful effects of human development on the planet that sustains it is the loss of habitat and the resulting loss in global biodiversity with up to one-half of land areas transformed by human activities and rates of biodiversity loss increasing rapidly in the last century (Butchart et al., 2010; Vitousek et al., 1997). Via increasing human land disturbance, our efforts to enhance the built environment in part have resulted in significant losses in species diversity due to human encroachment on natural habitats (Foley et al., 2005). This has accelerated the rate of species loss via extinction by an estimated factor of 100–1000 (most species are still not well-documented leading to such a wide estimate range). Habitat fragmentation has been as well linked as a key player in the land-use category affecting species biodiversity, though less distinctly than outright loss (Fahrig, 2003). Other increasing land-use shifts, most notably agricultural lands displacing natural habitats, have similarly impacted biodiversity.

Despite our ingenuity, humans are critically dependent on the natural world. Ecosystem services, the provision of food, water, and raw materials for human use, the mitigation of environmental damage, and other benefits that nature provides us are at risk from decreased biodiversity. Ecosystem services include water, air, and climate regulation; mitigation of storm impacts such as with estuaries, pollination, organism, and species propagation. In the end, this impacts the productivity of resources that humans depend upon such as forests, aquatic resources, or marine life. Ecosystem changes impact the services provided by the natural world. Humans benefit from ecosystem services with no other cost than maintaining a healthy ecosystem both on land and in the marine environment (Chapin et al., 2000; Hooper et al., 2005; Worm et al., 2006). One prime example related is that of the decline of pollinators (colony collapse) and its profound implications for agriculture that are still under extensive study (Potts et al., 2010). No single driver has been implicated in pollinator declines with at times multiple contributing factors including habitat loss and fragmentation, pathogens, climatic changes, invasive species, and agrochemicals. Another important area of emerging infectious diseases and the linkage to causing and resulting from biodiversity changes (Daszak et al., 2000).

With biodiversity, the role of climate change is paramount and has to this date been a strong driver of species distribution and movement plus increasingly a concern for species extinction (Thomas et al., 2004). The rapid warming of the last century has caused species dislocation higher in altitude at a rate of 11 m per decade on average and northward at a rate of 16.9 km per decade (Chen et al., 2011). In terms of species loss, one of the sensitive species of most concern is ocean coral reefs where the combination of warming ocean waters and acidification both linked to increasing atmospheric CO_2 is resulting in large-scale coral bleaching and declines in some global regions (Hoegh-Guldberg et al., 2007; Hughes et al., 2003).

With a recognition of the problem and efforts to address it, there have been local to regional improvements in biodiversity, while overall indicators show continuing global declines in biodiversity (Butchart et al., 2010). The pressure on biodiversity has continued to increase and relates to human population growth, both urban-industrial and agricultural growth, and the conversion of wildlands to developed lands. The picture is clear that humans have come to dominate the planet, and with population and development trends only our concerted action can minimize the harm to the planet that sustains us.

The rate of biodiversity loss has raised sufficient alarm that this period may be the beginning of the world's sixth "great mass extinction event" where the majority of Earth's species disappear over the period of 1 million years (Ceballos et al., 2015). The rates of extinction currently exceed those of the periods leading to the five previous events (Barnosky et al., 2011). The human contribution to these trends has enabled a likely state shift in the biological system and a transition from the geological epoch from the Holocene when humans emerged to the "Anthropocene" (not an official designation at this point) when humans have come to dominate the Earth (Barnosky et al., 2012; Waters et al., 2016). The United Nations Environment Programme tracks the issue of biodiversity and notes that efforts to address the problem have increased in recent decades with the conservation efforts through protected lands in countries around the world.

4.8 Ambient and indoor air pollution

Historically, and particularly during the 20th century, the effects of growing industrialization in the United States resulted in severely degraded ambient air quality in many industrialized regions. Mid-20th century smog episodes resulted in hundreds or even thousands of acute deaths and illnesses, primarily cardiopulmonary episodes in cities such as London, Pittsburgh, and St. Louis.

Pollution events in such cities led to the development of formation of regulatory agencies like the US Environmental Protection Agency (EPA), stricter emission limits, and other regulatory approaches. The EPA and the major Clean Air Act provisions were implemented in 1970 (with amendments approximately every 7 years following). Human health impacts are the primary driver of air quality regulations. Secondary standards address air quality-related values such as visibility, ecosystem impacts, material damage, and others. In the United States, the EPA has promulgated National Ambient Air Quality Standards (NAAQS) that limit the concentrations of seven criteria air pollutants that have been shown to impact human health (Table 4.2). Depending on the hazards of each pollutant, enforceable ambient limits are set at an appropriate concentration and specified averaging time for each pollutant species. The standards are based on the best available science examining the toxicological and epidemiological evidence of thresholds for effects. As shown in the table and footnotes in Table 4.2, there are many additional details shown such as specified rules regarding multiple-year averaging and which percentile of the highest concentrations to compare against the NAAQS. Lead and PM are in the aerosol phase (solid or liquid particles suspended in air), whereas CO, NO_2, SO_2, and O_3 are gas-phase species, typically measured as a mixing ratio in parts per million (ppm) or billion (ppb). Notably, CO_2 is not a criteria pollutant as it is not thought to be a human health hazard at ambient concentrations.

4.8 Ambient and indoor air pollution

Table 4.2 Summary table of National Ambient Air Quality Standards (NAAQS) for seven criteria pollutants regulated by the US EPA.

Pollutant		Primary/ secondary	Averaging time	Level	Form
Carbon monoxide (CO)		Primary	8 h	9 ppm	Not to be exceeded more than once per year
			1 h	35 ppm	
Lead (Pb)		Primary and secondary	Rolling 3-month average	0.15 µg/m^3 [a]	Not to be exceeded
Nitrogen dioxide (NO$_2$)		Primary	1 h	100 ppb	98th percentile of 1-h daily maximum concentrations averaged over 3 years
		Primary and secondary	1 year	53 ppb[b]	Annual mean
Ozone (O$_3$)		Primary and secondary	8 h	0.070 ppm[c]	Annual fourth-highest daily maximum 8-h concentration averaged over 3 years
Particle pollution (PM)	PM$_{2.5}$	Primary	1 year	9.0 µg/m^3	Annual mean averaged over 3 years
		Secondary	1 year	15.0 µg/m^3	Annual mean averaged over 3 years
		Primary and secondary	24 h	35 µg/m^3	98th percentile averaged over 3 years
	PM$_{10}$	Primary and secondary	24 h	150 µg/m^3	Not to be exceeded more than once per year on average over 3 years
Sulfur dioxide (SO$_2$)		Primary	1 h	75 ppb[d]	99th percentile of 1-h daily maximum concentrations averaged over 3 years
		Secondary	3 h	0.5 ppm	Not to be exceeded more than once per year

[a]In areas designated nonattainment for the Pb standards prior to the promulgation of the current (2008) standards, and for which implementation plans to attain or maintain the current (2008) standards have not been submitted and approved, the previous standards (1.5 µg/m^3 as a calendar quarter average) also remain in effect.
[b]The level of the annual NO$_2$ standard is 0.053 ppm. It is shown here in terms of ppb for the purposes of clearer comparison to the 1-h standard level.
[c]Final rule signed October 1, 2015, and effective December 28, 2015. The previous (2008) O$_3$ standards are not revoked and remain in effect for designated areas. Additionally, some areas may have certain continuing implementation obligations under the prior revoked 1-h (1979) and 8-h (1997) O$_3$ standards.
[d]The previous SO$_2$ standards (0.14 ppm 24-h and 0.03 ppm annual) will additionally remain in effect in certain areas: (1) any area for which it is not yet 1 year since the effective date of designation under the current (2010) standards, and (2) any area for which an implementation plan providing for attainment of the current (2010) standard has not been submitted and approved and which is designated nonattainment under the previous SO$_2$ standards or is not meeting the requirements of a SIP call under the previous SO$_2$ standards (40 CFR 50.4(3)). A SIP call is an EPA action requiring a state to resubmit all or part of its State Implementation Plan to demonstrate attainment of the required NAAQS.

Suspended aerosol particles or particulate matter (PM) are regulated by the US EPA and other regulatory bodies based on mass concentrations of particles (typically measured in µg/m^3). The PM is quantified for defined sizes, e.g., for PM$_{2.5}$ and PM$_{10}$ for sizes less than 2.5 and 10 µm, respectively. PM$_{2.5}$ may be the best single indicator of the severity of air pollution problems and is one component of the unitless Air Quality Index, a composite indicator commonly reported in public media sources and weather reports (Fig. 4.5). A separate but related issue is what is termed photochemical smog.

Los Angeles, its industrial density, generally sunny conditions, geography, and its dependence on the automobile all contributed to making it notorious for "bad air." In LA and other sunbelt cities, nitrogen oxides (NO_x), volatile organic carbon (VOC) compounds, and sunlight cause photochemical reactions that form the power oxidant ozone (O_3).

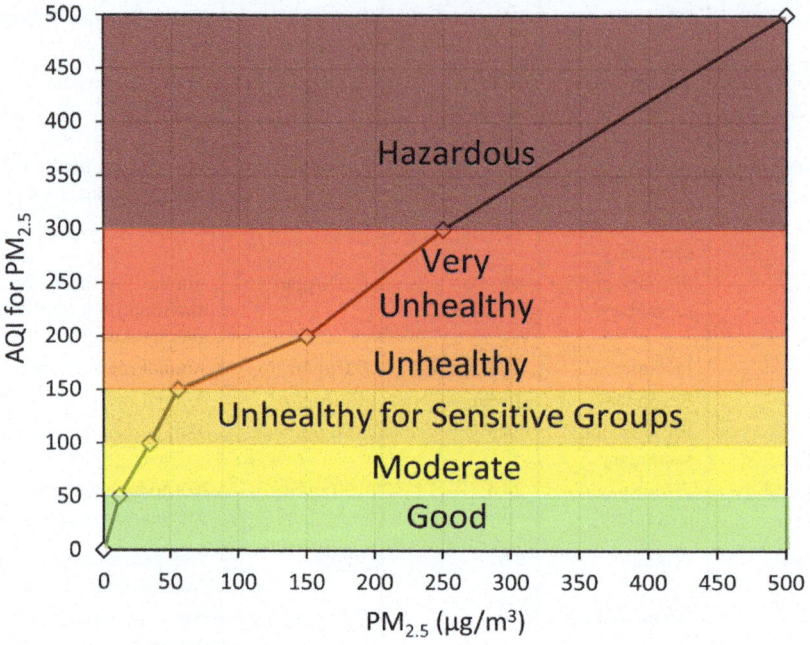

FIG. 4.5

Relationship between $PM_{2.5}$ mass concentration and the composite indicator "air quality index" (AQI) that is often reported to the public.

Contamination of the air with solid or liquid particles (aerosols or particulate matter) and trace gases causes human health effects, visibility degradation, particularly an issue in scenic areas, ecosystem impacts, materials damage, forest and crop damage, and contributes to global climate change (Seinfeld and Pandis, 2016). Historically, large population density, heavy industrialization, and the uncontrolled combustion of solid and liquid fuels degrade air quality. The noted London Smog episode of 1952 (among others of that era) resulted in 12,000 excess deaths, with effects persisting for 2 months following the ~1-week pollution episode (Bell and Davis, 2001; Bell et al., 2004). In most acute air pollution events, the meteorological context of the event plays a substantial role in the occurrence or at least the severity of the event (e.g., vertical atmospheric stability, low winds during inversion events).

The visible haze in polluted environments, also contributed by natural sources such as wildland fire and dust storms, is largely due to fine-mode particulate matter, or $PM_{2.5}$ (particles with aerodynamic diameters less than 2.5 μm). These particle sizes are individually invisible to humans (Fig. 4.6), but their combined effects are clearly visible as a hazy atmosphere (Fig. 4.7). Typical aerosol particles are of a size that attenuates atmospheric radiation via scattering and absorbing visible light. The sources of $PM_{2.5}$ include both human (power generation, industry, vehicles) as well as natural sources (biomass burning, dust storms); $PM_{2.5}$ is directly emitted (primary) and forms in the atmosphere (secondary) from precursor emissions.

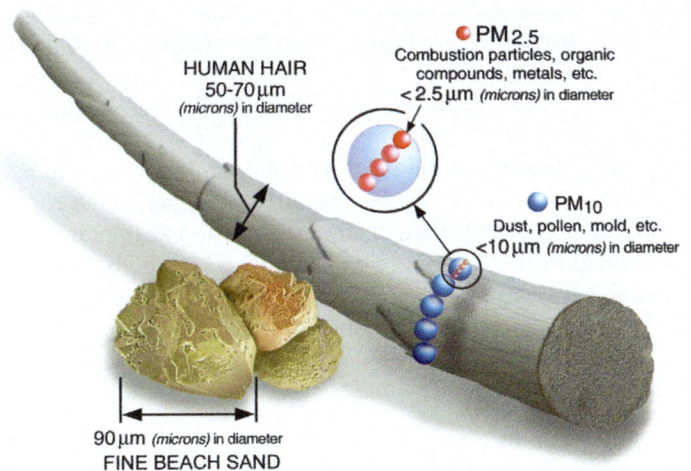

FIG. 4.6
The size scale of PM$_{2.5}$ and PM10, the size ranges of particulate matter regulated by the National Ambient Air Quality Standards (NAAQS) (public domain image US EPA).

FIG. 4.7
Impact of long-range transported smoke from Oregon fires on visibility in Yosemite National Park in the summer of 2002.

Fine-mode aerosols ($PM_{2.5}$) are also the most damaging to human health as they are capable of penetrating deep into the lungs (Seinfeld and Pandis, 2016), causing damage to the cardiopulmonary system (Pope et al., 2009). The Global Burden of Disease study found air pollution-related deaths as the third largest cause of death globally (Fig. 4.8). Despite progress, air pollution costs are still staggering in terms of health impacts and resulting economic costs. A recent study estimate of the health and climate costs of burning fossil fuels in the United States attributed nearly 95% of the $850 billion per year in costs to the premature deaths of 107,000 people per year due to elevated $PM_{2.5}$ concentrations (De Alwis and Limaye, 2021).

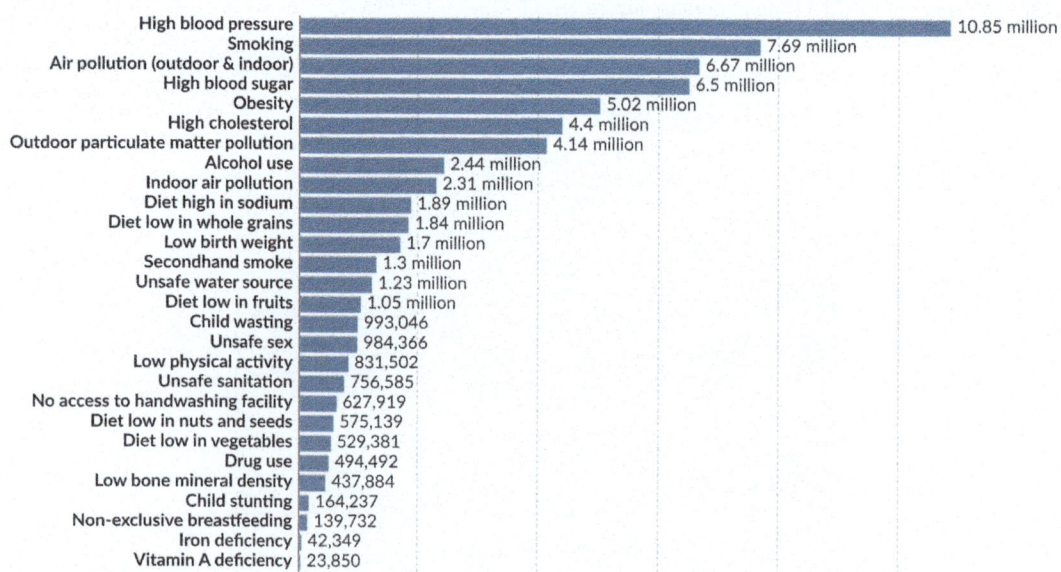

Data source: IHME, Global Burden of Disease (2019)   OurWorldInData.org/causes-of-death | CC BY

Note: Risk factors[1] are not mutually exclusive. The sum of deaths attributed to each risk factor can exceed the total number of deaths.

1. **Risk factor:** A risk factor is a condition or behavior that increases the likelihood of developing a given disease or injury, or an outcome such as death. The impact of a risk factor is estimated in different ways. For example, a common approach is to estimate the number of deaths that would occur if the risk factor was absent. Risk factors are not mutually exclusive: people can be exposed to multiple risk factors, which contribute to their disease or death. Because of this, the number of deaths caused by each risk factor is typically estimated separately. Read more: How do researchers estimate the death toll caused by each risk factor, whether it's smoking, obesity or air pollution? Read more: Why isn't it possible to sum up the death toll from different risk factors?

2. **Civil Registration and Vital Statistics system:** A Civil Registration and Vital Statistics system (CRVS) is an administrative system in a country that manages information on births, marriages, deaths and divorces. It generates and stores 'vital records' and legal documents such as birth certificates and death certificates. You can read more about how deaths are registered around the world in our article: How are causes of death registered around the world?

FIG. 4.8

Global causes of mortality in 2019 (Abbafati et al., 2020; Roser and Ritchie, 2021).

Creative Commons image from OurWorldInData.org.

4.8 Ambient and indoor air pollution

Air quality monitoring is conducted by the USEPA, other federal regulatory bodies, and the state environmental compliance agencies at hundreds of monitoring stations in the United States and by equivalent agencies worldwide. Though more advanced real-time measurements have been developed, the measurement of mass concentration is simply and accurately quantified by filter collection and analysis. The PM mass concentration is the sampled mass found before and after weighings divided by the volume of air sampled found from the volumetric flow rate and sample time (Eq. 10.7).

$$[PM] = \frac{(m_f - m_i)}{Q \times t} \tag{4.2}$$

The interagency monitoring for visually protected environments (IMPROVE) network monitors regional PM chemical composition at rural and remote sites such as national parks and wilderness areas (IMPROVE (colostate.edu)). IMPROVE is a collaborative association of state, tribal, and federal agencies, plus international partners. The US Environmental Protection Agency is the primary funding source, with contracting and research support from the National Park Service. The Air Quality Group at the University of California, Davis, is the central analytical laboratory, with ionic analysis provided by the Research Triangle Institute and carbon analysis provided by Desert Research Institute. Trends in $PM_{2.5}$ in the United States show that air quality in a regional sense has improved, particularly reduced ammonium sulfate in the eastern United States due to Clean Air Act reductions in SO_2 emissions (Fig. 4.9). The Western United States, which started cleaner, has seen fewer improvements and in some cases backtracking due to carbonaceous aerosols from increased wildland fire (Fig. 4.9). An example of the data collected is given in Fig. 4.10 for the site at Bandolier National Monument in northern New Mexico showing the importance of seasonal biomass smoke.

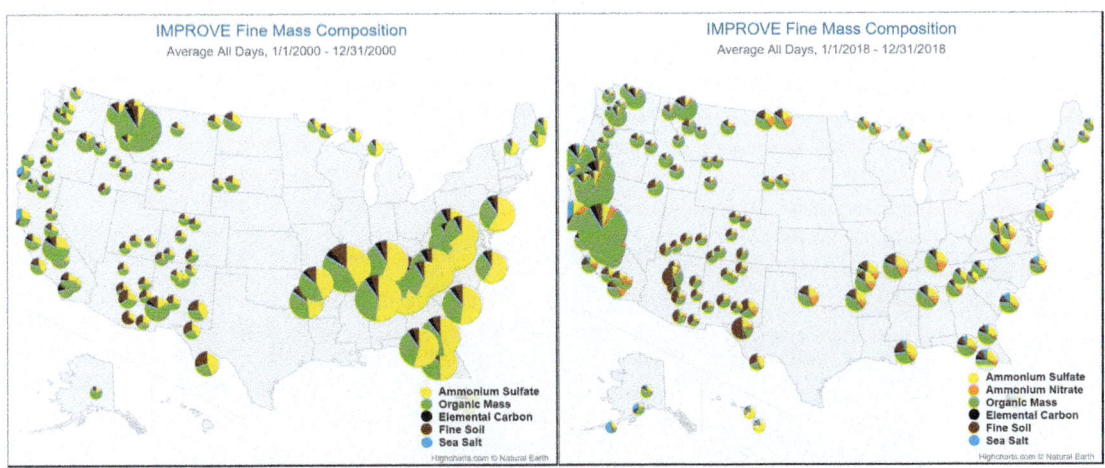

FIG. 4.9

Comparison of 2000 vs 2018 air quality in background US locations represented by "Class I visibility" sites such as national parks and monuments. The largest pies are approximately 13 µg/m³ as compared to the annual National Ambient Air Quality Standard of 9 µg/m³.

Data for 2000–18 is from the WRAP Technical Support System (TSS) website (http://views.cira.colostate.edu/tss/) on November 15, 2020.

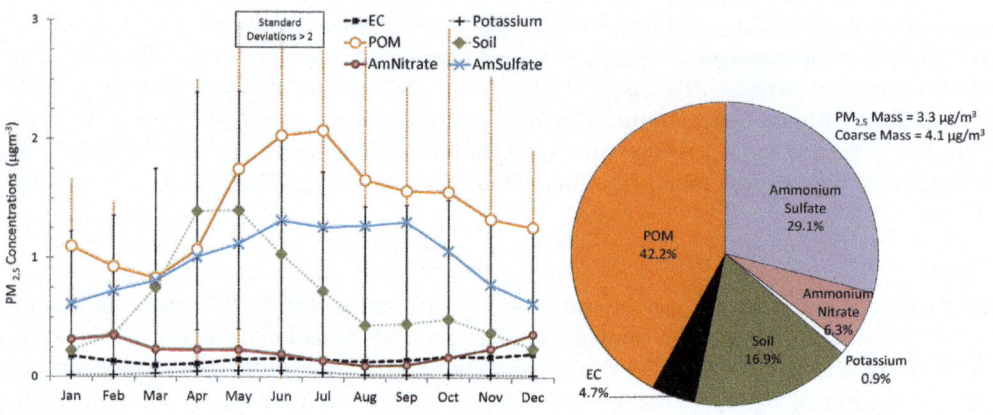

FIG. 4.10

IMPROVE network measurements (1988–2014) showing $PM_{2.5}$ composition and seasonality at Bandolier National Monument in northern New Mexico. The large contribution from organic carbon in summer and fall is contributed to by biomass burning smoke (data and protocols as described in Malm et al., 1994).

Overall, air quality in major urban areas such as Los Angeles, New York, Chicago, and many industrialized countries has improved over the last 50 years (Fig. 4.11). Another similarly successful response has been to acid rain which peaked in the 1980s but improved with Clean Air Act Amendments addressing the source of the problem, mainly SO_2 and NO_x emissions from coal-fired power plants and other combustion sources (Fig. 4.12). The response was aided by the issuance of tradable permits or

FIG. 4.11

Visible haze is still common though less severe over the Los Angeles, CA, airshed in 2011. Before 1970, it is said that some grew up in Los Angeles, not knowing it was surrounded by mountains due to poor visibility. Air quality in many urban areas has improved dramatically over the past half-century due to the Clean Air Act and other legislation.

"emissions trading." The improvements in air quality were much quicker and cheaper than anyone predicted. Similar approaches for climate change, though nascent in regions like California and several northeastern states and other nations, have not been universally adopted to internalize the cost of greenhouse gas emissions on a global scale.

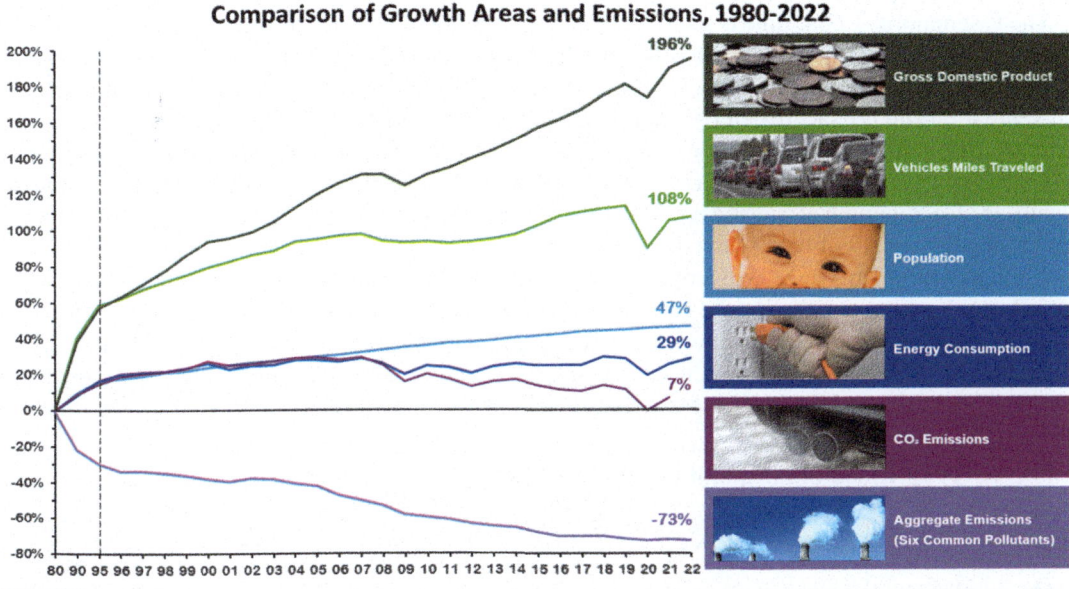

FIG. 4.12

Historical (1980–2022) US emission changes and demographic data including annual changes in US population, vehicle miles traveled, nationwide energy consumption, CO_2 emissions, and criteria air pollutant emissions (public domain US EPA plot).

Another example of a success story is the case of stratospheric ozone depletion which was uncovered in the 1970s. The culprit behind the problem turned out to be chlorofluorocarbons, the compounds increasingly used as refrigerants and for other applications such as propellants. Originally, CFCs were considered "wonder" chemicals in the early 20th century because of their refrigerant properties, nonflammable nature, nontoxic nature, as well as being chemically inert. The lack of reactivity assured a long atmospheric lifetime that allowed the accumulation and migration of these compounds to the stratosphere. Here they were photolyzed and formed halogen radicals, most notably chlorine radicals, where a single radical would lead to the catalyzed destruction of hundreds or more ozone molecules causing the ozone "hole" over Antarctica. Scientists Mario Molina, Sherwood Rowland, and Paul Crutzen developed a chemical mechanism and model showing that CFCs were reaching the stratosphere, leading to the depletion of stratospheric ozone which protects humans from excessive UV exposure. The Nobel Prize in Chemistry was subsequently awarded for this work in the 1990s. The hypothesis was confirmed by measurements made by Susan Solomon and colleagues in the early 1980s showing that indeed CFCs were responsible for the developing ozone "hole" over Antarctica. This set the stage for the phaseout of CFCs via the Montreal Protocol in 1987 and provides a model of international cooperation to solve a global environmental problem (Seinfeld and Pandis, 1998). The world responded by putting into place a gradual phaseout of production and substitution of less harmful compounds with non-ozone-depleting alternatives. Though setbacks have occurred, the Montreal Protocol and subsequent revisions have been an unmitigated success at addressing stratospheric ozone depletion achieved on an international scale.

The vast majority of our time is spent indoors, on the order of 90% for the average person, although any regulation on indoor air quality is voluntary. In roughly a half-hour, the outdoor air becomes indoor air. Indoor air quality is ameliorated by any indoor removal mechanisms including air cleaning devices and ventilation and is exacerbated by unique indoor sources, both natural and anthropogenic. The recognition of indoor air quality, disease transmission, and human health has evolved dramatically since the COVID-19 pandemic (Morawska et al., 2020). Later in Chapter 12, approaches such as filtration (as quantified by a minimum efficiency reporting value or MERV rating) and other techniques for reducing indoor air pollutants will be discussed.

Air quality problems have evolved from local- to global-scale issues alongside the development of our industrial society, paralleling the growth and scale of human activities (Table 4.3). What started as issues in select urban-industrial areas progressed from regional-scale issues such as acid rain and regional haze, to global-scale issues such as stratospheric ozone depletion and global climate change. The latter has been the most vexing atmospheric problem of the last 50 years.

Table 4.3 Progression of air quality issues over the past half-century in the United States.

Problem	Spatial scale	Temporal scale	Status
Urban air pollution	Localized-urban airshed	Hours to days	Substantial progress in developed world though challenges remain
Acid deposition and visibility impairment	Regional (Northeast United States) to transnational	Days to months	Largely addressed through Cap & Trade Program in the United States
Stratospheric ozone depletion	Hemispherical (ozone hole over the Antarctic)	Seasonal to multidecadal	Largely addressed through Montreal Protocol & recovery in progress
Global climate change	Global scale	Decadal to millennial	Continuing acceleration of emissions and impacts; reaching a tipping point toward action?

Aerosols are also an important component of the climate system affecting the radiative balance of the planet in a way analogous to their visibility effects described above (Fig. 4.11) (Seinfeld and Pandis, 2016). Historically, following major volcanic eruptions such as Mt. Tambora in 1815 and Mt. Pinatubo in 1992, the Earth cooled for a couple of years by about 0.5–1°C due to the stratospheric injection of the sulfate aerosol precursor SO_2 and dust. The haze caused by aerosols that limit visibility in the horizontal direction also affects the vertical radiation balance due to light scattering and absorption by particles. Though aerosols cause net cooling, the magnitude and even sign of this radiative forcing term is a function of the particle size distribution, aerosol refractive index, concentration, and extent of the aerosol layer. This is magnified by the indirect effects aerosols have on cloud formation, properties, and lifetimes (IPCC, 2021b).

The climate crisis is the next topic of discussion and is among the most important driving green building. Climate change and air quality issues are distinct problems that have many crossovers. First, emissions of greenhouse gases are roughly 75% related to combustion of fuels, and reductions of these activities and emissions will have co-benefits on air quality. A changing climate also has important feedback on air quality via changes in temperature, absolute humidity, cloud cover, precipitation, ventilation rates, and other atmospheric characteristics. The most impacted parameters are ozone and $PM_{2.5}$ which are themselves climate forcers, though many air pollutants are intertwined with climate change (Jacob and Winner, 2009). For example, higher temperatures favor O_3 formation while increased water vapor favors lower background O_3. $PM_{2.5}$ is more difficult to model due to its diversity of sources and complex atmospheric chemistry. However, it is accepted that reduced mid-latitude cyclonic activity, increased wildland fire, increasing biogenic emissions, and increasing atmospheric stagnation events with reduced ventilation will all lead to increasing $PM_{2.5}$.

4.9 Anthropogenic climate change and the climate crisis

The preceding discussion leads to arguably the most profound planetary impact of human activities: human disturbance of the carbon cycle and increasing anthropogenic climate disruption (IPCC, 2013, 2021b). The primary means by which humans are upsetting the carbon balance and altering the planetary energy balance is the release of greenhouse gases, most notably CO_2 and CH_4, as well as aerosol particles, as previously discussed.

The climate crisis has to this point been a slow-moving disaster that has only reached crisis status due to our continued inability to address it. Some consider it perplexing that CO_2 would be considered a pollutant as it is non-reactive and generally benign at current atmospheric concentrations. It is the goal of any good combustion engineer to convert as much of a hydrocarbon fuel under complete combustion to CO_2. Otherwise, efficiency suffers, and the traditional air pollutants like CO, VOCs, and particulate matter increase. After all, CO_2 is just plant food, one of the essential atmospheric gases involved in the biosphere, right?

As with other air pollutants, it is the quantity or the dose that makes the poison. A pollutant is often dictated by its concentration and impacts, not its presence. As a frame of reference, a carbon monoxide concentration matching the current global average CO_2 concentration of ~420 ppm would have fatal consequences for humans. Likewise, fatal infections of COVID are at an infection to human body percentage far less than 0.042%. A small number does not negate its significance.

4.9.1 Overview of fundamental climate system physics

Detailed climate physics is beyond the scope here though can be found in many other works (Dessler, 2015; Jacobson, 2012). The physics of how greenhouse gases (GHG) keep the planet habitable has long been well understood since the 19th century. Scientists of note contributing to the greenhouse theory include Fourier, Foote, Tyndall, Ångström, and Arrhenius, the latter who determined that higher GHG concentrations lead to warming and devised the first climate model predicting a ~4–5°C increase in global average temperature with a doubling of CO_2. This sensitivity, likely and hopefully on the warm side, is reasonably in line with current climate models. The climate change problem relates once again to an energy problem as a radiative forcing function on the climate system.

82 **Chapter 4** Environmental problems from the local to global scale

Climate determines the habitability of a location for plant and animal species and even drives evolutionary changes in those species. Climate zones are a key variable in terms of green building solutions as well as what works in the cloudy, humid eastern half of the US differs from that in the high and dry western half. A cooling strategy based on ventilation, thermal mass, insulation, and evaporative cooling will work far better in dry regions compared to humid regions.

The US climate zones based on temperature, rainfall, and (driven by) elevation are shown in Fig. 4.13. The west–east gradient of arid to wet and the north-to-south gradient of cold to hot is apparent. The Rocky Mountains serve as a major barrier to the typical west-to-east atmospheric motion in the mid-latitude Northern Hemisphere. The atmospheric moisture is condensed and rained out more often on the upwind, western side of these mountain ranges, whereas descending that warms adiabatically and is drier affects the climate on the downwind side of the mountain ranges. The influence of elevation is also clear where a given map location becomes colder and wetter in more mountainous locations.

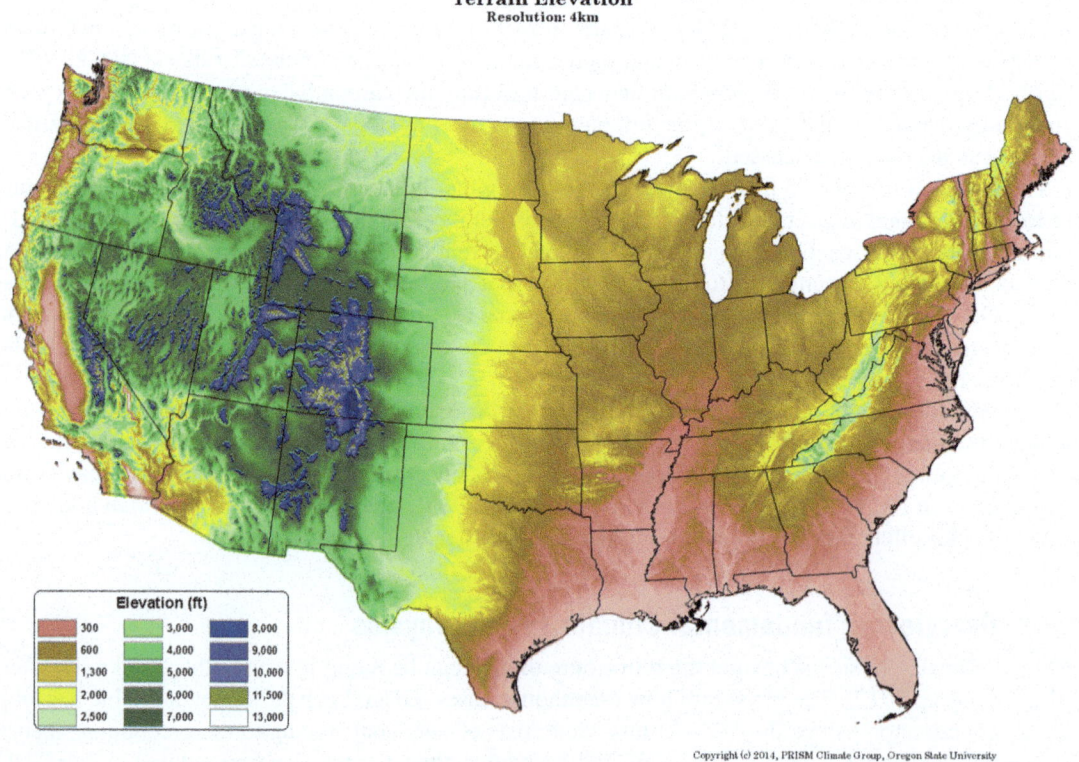

FIG. 4.13

Continental US surface elevations, mean temperature, and mean annual precipitation averaged over 30 years.
Copyright © 2015 PRISM Climate Group, Oregon State University, maps created by author 9 August 2024.

continued

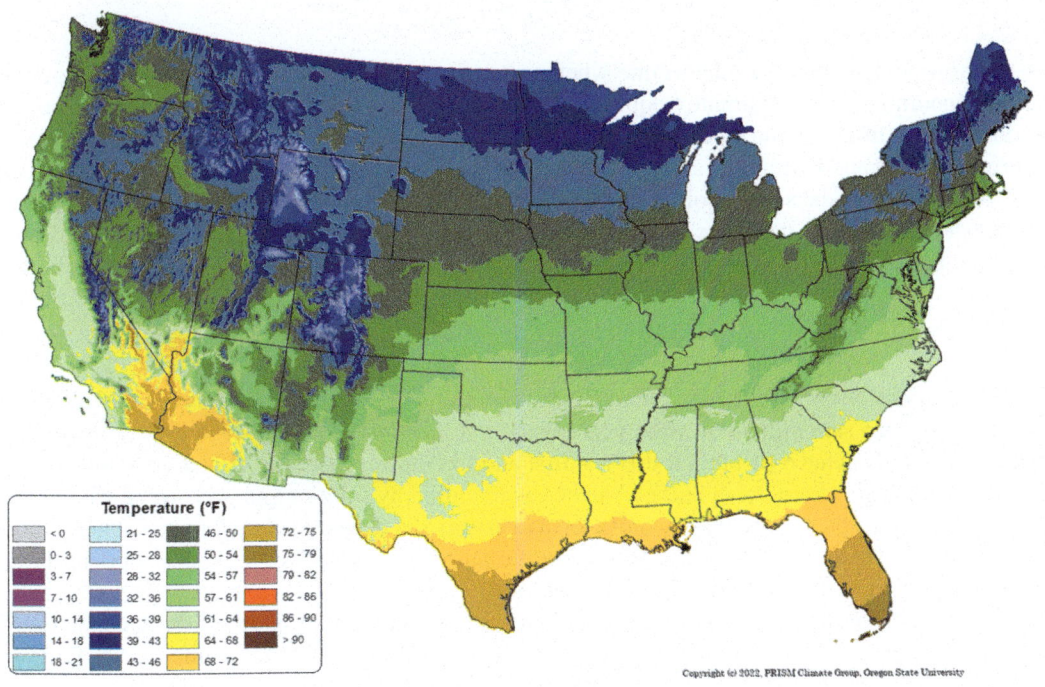

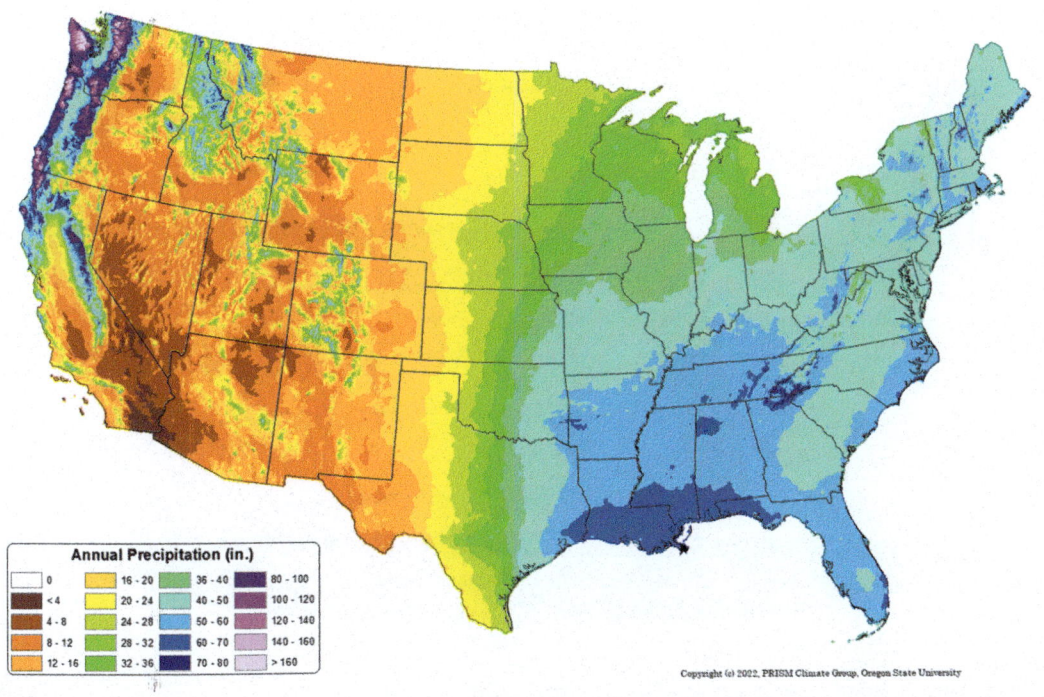

FIG. 4.13—cont'd

A region's climate zone will determine whether it is more dominantly a warm climate (requiring more space cooling) or a cool climate (requiring more space heating). Some locations, primarily coastal locations, require little heating or cooling due to thermal inertia of the water body. Near the coast in southern California and Hawaii, it rarely gets below freezing nor much above 30°C (~90°F). The concept of heating-degree days and cooling-degree days quantifies this. The HDD and CDD are given in Eq. (4.3, where T_{avg} is the daily average temperature (usually taken as the mean of the min and max) and $T_{baseline}$ is a baseline "comfortable" temperature often taken as 65°F.

$$\text{HDD} = \sum_{i=1}^{365} \left(T_{avg} - T_{baseline} \right) \tag{4.3}$$

A distinction between weather and climate is the timeframe as climate considers decadal or greater changes. It has been said climate is what you expect at a location and weather is what you get. The climate is of course dictated by natural drivers, first and foremost, before the effects of human perturbations. In the end, it reduces to an energy balance as modified by solar output and orbital parameters, and atmospheric transparency to visible radiation and energy out determined by the atmospheric transparency to infrared radiation (Fig. 4.14). Natural perturbations are caused by volcanic eruptions, large-scale fires, meteorite impacts, plate tectonics, and other similar events.

The Earth seeks an energy balance equilibrium. Sitting in the vacuum of space, Earth can only receive energy in as incoming solar radiation and out of the system via infrared radiation emitted by the Earth for which equilibrium is sought (Fig. 4.15). Although some of the sun's energy is converted into work to move winds and seas, thermal expansion, and numerous others, all of this stays within the Earth-atmosphere boundary, not leaving the system or perturbing the first law energy balance in Eq. (2.16) applied at the

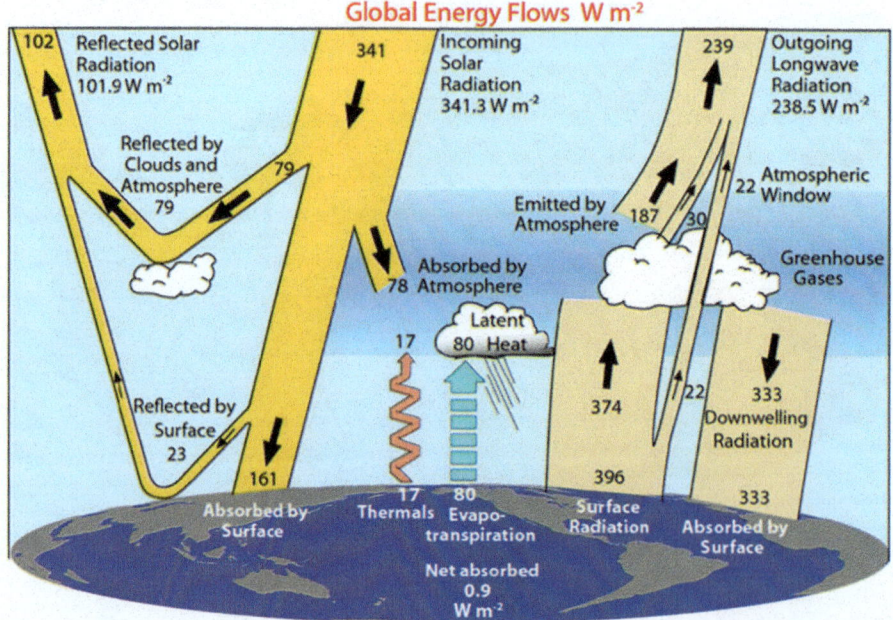

FIG. 4.14

Earth's radiation balance shows incoming and outgoing radiative fluxes.

Public domain image from UCAR Center for Science Education.

top of the atmosphere. Without a work term, when an energy imbalance exists, the only options are storage of this energy (i.e., increased internal energy, U, following Eq. 2.12), phase changes, or increased radiative heat rejection due to a higher effective Earth temperature (Eq. 2.11). The equilibrium temperature of Earth devoid of greenhouse gases would be about 33°C colder than its current equilibrium temperature of 288 K, making this planet likely uninhabitable (or at least a very different biosphere!).

4.9.2 Radiative forcing of climate

Small atmospheric concentrations of greenhouse gases—most notably water vapor, carbon dioxide, methane, nitrous oxide, and CFCs, among others—are responsible for regulating the outgoing infrared radiation. Absorption bands of these greenhouse gases have been known (and progressively more accurately measured) since the 19th century (Fig. 4.15). This IR absorption process, a function of the concentration of these gases and their specific absorption cross-section, prevents some fraction of the Earth's radiated energy from leaving the system. Due to their increasing concentrations, including the water vapor feedback from its increased concentration as the atmosphere warms, the Earth system is accumulating excess energy, primarily in ocean thermal storage.

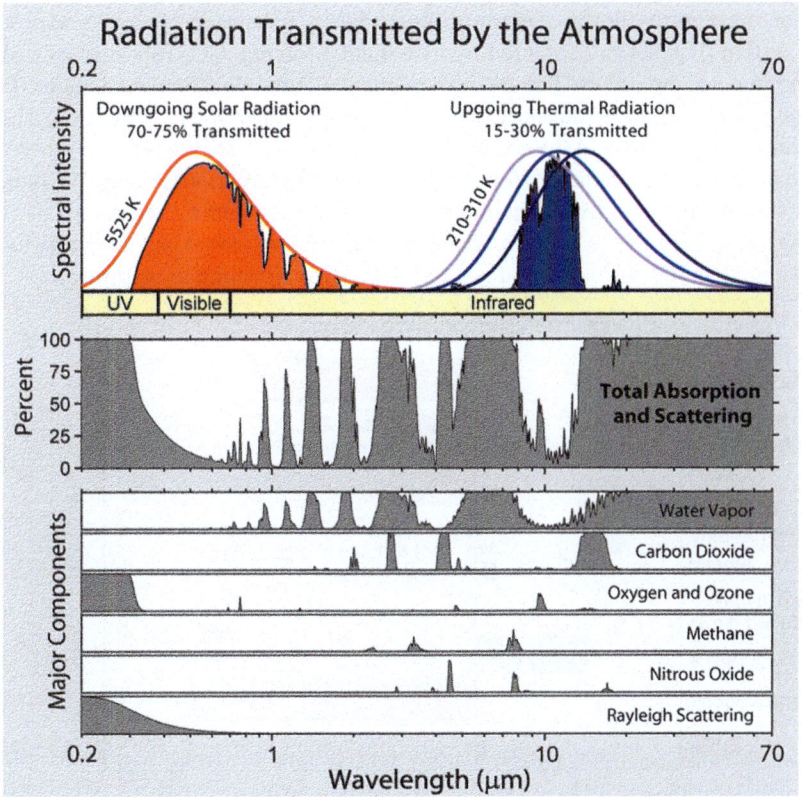

FIG. 4.15

Absorption bands of important greenhouse gases.

This figure was prepared by Robert A. Rohde as part of the Global Warming Art Project, and it is licensed under the Creative Commons Attribution-Share Alike 3.0 Unported license.

The key anthropogenic perturbation of the climate system is the release of vast quantities of carbon dioxide (CO_2), methane (CH_4), nitrous oxide (N_2O), sulfur hexafluoride (SF_6), chlorofluorocarbons (CFCs) and their derivatives, and several other greenhouse gases that absorb in IR radiation bands in regions nonoverlapping with water vapor (Fig. 4.15).

There are other important tweaks to the system as well including particulate matter or aerosols emitted by humans as well as changes in surface reflectivity overdeveloped lands. Aerosol particles discussed in the air quality section are shown to impart a net cooling effect globally. Aerosols and their cloud effects are more complex and uncertain as some aerosols such as soot carbon cause a warming effect aloft in the atmosphere. The net aerosol effect is cooling though, which will disappear quickly as we clean up our emissions, unmasking more of the long-lived greenhouse forcing. The aerosol climate effect is much more complex than greenhouse warming in that it depends on the particle size distribution, refractive index, particle shape and mixing, and ambient relative humidity. A subset of aerosols, termed soot, black carbon, or light-absorbing carbon depending on the technique, have a warming effect due to absorption. Reduction of light-absorbing carbon would be one of the quickest ways to reduce atmospheric warming in short order. Due to their heterogeneity in time and space, the aerosol effect has one of the largest uncertainties, particularly when considering the cloud impacts including lifetime, reflectivity, droplet size distribution, and precipitation probability.

Radiative forcing perturbs the top of the atmosphere energy balance through several processes relevant to the climate system (Fig. 4.16). Measured by its radiative forcing, CO_2 followed by CH_4 are the most important anthropogenic emissions. The Intergovernmental Panel on Climate Change (IPCC) publishes periodic updates to its exhaustive summary of climate change research with Volume I: The Physical Science Basis summarizing the state of the science. As shown in Fig. 4.16, the IPCC synthesizes the best estimate from the literature of the radiative forcing due to these perturbations, mostly the human influences. This comparison shows the long-lived greenhouse gases on the top with CO_2 and CH_4 as contributing the largest warming influences of $+1$ to $+2$ W/m^2. Smaller contributions from long- and short-lived greenhouse gases are shown next with contributions of -0.5 to $+0.5$ W/m^2 (cooling to warming).

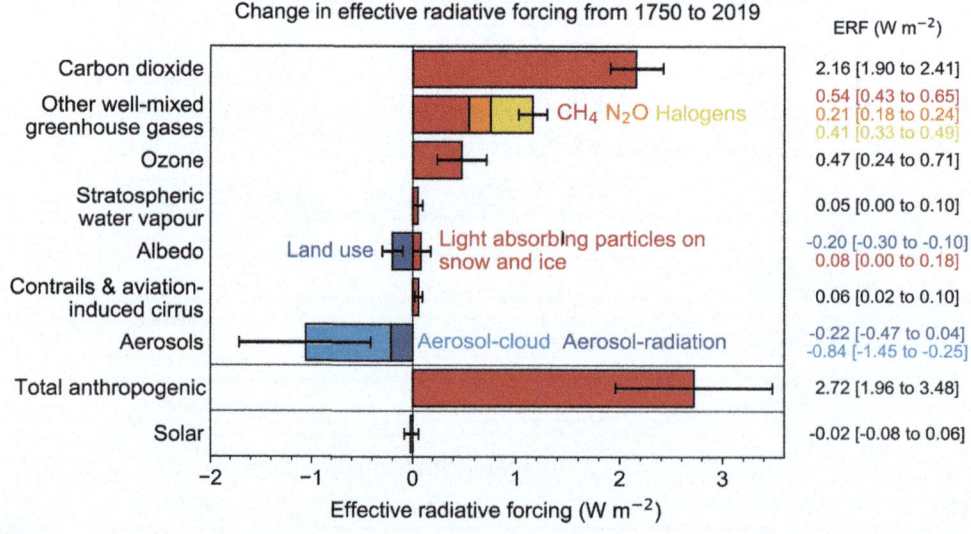

FIG. 4.16

4.9 Anthropogenic climate change and the climate crisis

FIG. 4.16—cont'd

Public domain figure from IPCC AR6 Fig. 7.6 | Change in effective radiative forcing (ERF) from 1750 to 2019 by contributing forcing agents (carbon dioxide, other well-mixed greenhouse gases (WMGHGs), ozone, stratospheric water vapor, surface albedo, contrails and aviation-induced cirrus, aerosols, anthropogenic total, and solar). Solid bars represent best estimates, and very likely (5%–95%) ranges are given by error bars. Non-CO_2 WMGHGs are further broken down into contributions from methane (CH_4), nitrous oxide (N_2O), and halogenated compounds. Surface albedo is broken down into land-use changes and light-absorbing particles on snow and ice. Aerosols are broken down into contributions from aerosol–cloud interactions (ERFaci) and aerosol–radiation interactions (ERFari). For aerosols and solar, the 2019 single-year values are given (Table 7.8 in IPCC), which differ from the headline assessments in both cases. Volcanic forcing is not shown due to the episodic nature of volcanic eruptions. Further details on data sources and processing are available in the chapter data table (Table 7.SM.14 in IPCC) (IPCC, 2021b).

The major greenhouse gases have different global concentrations as well as the efficiency at which they function as greenhouse gases (Table 4.4). CO_2 is a weak greenhouse gas, but because of its relatively large atmospheric concentration of about 420 ppm as of 2024, it is of great importance. The Global Warming Potential (GWP) of a given gas ratios its efficiency to that of CO_2 which has a GWP of 1 (Example 4.4). The emissions of greenhouse gas i can be converted to the equivalent CO_2 emissions (CO_2,e) by multiplying the gas emissions by the GWP.

$$[CO_2e] = [i](GWP_i)$$

Table 4.4 Radiative forcing efficiency of important greenhouse gases (Pearson and Derwent, 2022).

Gas species	Radiative forcing efficiency (W/m²/ppm)
CO_2	0.0137
CH_4	0.363
N_2O	0.30
O_3	2.57
CFC-12	320
CFC-11	260
HCFC-22	210

> **EXAMPLE 4.4 Greenhouse gas radiative forcing**
> **Problem:** The approximate top of the atmosphere radiative forcing due to the human enhancement in the greenhouse gas CO_2 is $\sim +1.5\,W/m^2$. Compare this with the annual average energy added to the climate system to the total global energy consumption. Compare this to the energy released by the Hiroshima atomic bomb.
> **Given:** $\Delta F = 1.5\,W/m^2$
> **Find:** Power; Compare to the atom bomb and human global energy consumption
> **Assume:** steady state power consumption
>
> **Solution:**
> $$\text{Annual energy added from } CO_2 = \frac{\text{Power}}{\text{Area}} \times \text{Earth surface area} \times \text{time}$$
>
> $$E_{CO_2} = 1.5\frac{W}{m^2} \times 4\pi \times R_e^2 \times \text{Time} = 1.5\frac{W}{m^2} \times 4\pi \times (6371 \times 10^3\,m)^2 \times \frac{365\,\text{days}}{\text{year}} \times \frac{24(3600)s}{\text{day}} = \frac{2.41 \times 10^{22}\,J}{\text{year}}$$
>
> $$\text{Equivalent power} = 7.65 \times 10^{14}\,\frac{J}{s} = 765\,TW$$
>
> Total 2010 global energy use $= 524$ Quads/year $= (524 \times 10^{15}\,BTU/\text{year}) \times (1055\,J/BTU) = 5.528 \times 10^{20}\,J/\text{year} = 17.5\,TW$
> So enhanced CO_2 is adding heat at the rate of 50 times global human energy use!
> The Hiroshima nuclear blast released 67 TJ (67×10^{12} J) of energy. How many blasts is this equivalent to?
> Rate $= (7.65 \times 10^{14}\,J/s) / (67 \times 10^{12}\,J/\text{Hiroshima}) = 11$ blasts per second of energy added to the Earth system via the CO_2 radiative forcing. The actual energy added to the system is somewhat lower as the temperature of the Earth has increased to reject some of the added energy.

4.9.3 The carbon cycle and a human fingerprint

As the biosphere is populated by carbon-based life forms, the carbon cycle (along with the aforementioned water cycle) is arguably the most important cycle to life on Earth (Fig. 4.17). One of the fundamental global elemental cycles, the carbon cycle, represents how the element carbon (C), cycles through global reservoirs via fluxes among them (Fig. 4.17). The carbon cycle includes both biogenic carbon in numerous molecular forms (e.g., biomass contained in forests) as well as inorganic forms of carbon such as carbonates in the lithosphere.

Important reservoirs include the atmosphere, the biosphere, the lithosphere containing soil carbon, carbonates, fossil carbon forms in coal, oil, and natural gas, and dissolved and sedimented species such as in the hydrosphere. A balance of these fluxes gives equilibrium. The point is often raised that certain natural fluxes of carbon exceed by an order of magnitude or more the human flux of ~ 10 GTC/year of CO_2 emissions that is added from the combustion of fossil fuels and cement production. However, the human tweak to the cycle is sufficient to represent a mass imbalance that is leading to a measurable increase in both the atmosphere (CO_2 and CH_4 most notably) and ocean (carbonic acid, bicarbonate ion) concentrations of carbonaceous species, already proving to be highly damaging to ecosystems and economies.

4.9 Anthropogenic climate change and the climate crisis

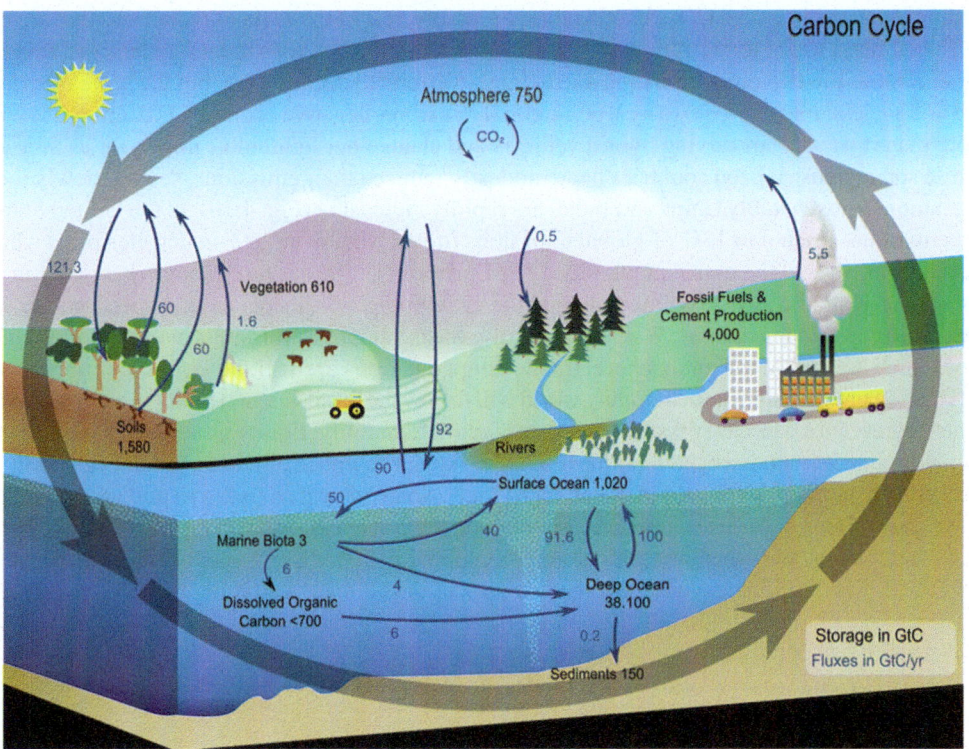

FIG. 4.17

The global carbon cycle shows fluxes and stocks from natural and anthropogenic activities (NASA public domain image via Wikimedia Commons).

Anthropogenic climate change is ~75% a CO_2 problem and ~75% associated with energy production. However, other tweaks to the system are also of relevance—offering rapid and cost-effective mitigation measures in some cases—most notably land use, cement, steel, and agriculture. The global emissions as a function of time are shown in Fig. 4.18, and the geographic distribution and 30-year change in emissions of anthropogenic CO_2 are represented in Fig. 4.19, where the filled circles represent emissions in 1990 and the open circles represent the 2019 emissions in annual gigatons of CO_2 emissions (~36 GT/year in 2021). This is a staggeringly large mass of material in the form of fuels.

Examining data from the Global Carbon Project, the current decadal scale trend in CO_2 emissions is still increasing but flattening globally (Fig. 4.18). The growth of greenhouse gas emissions, generally a given since the middle of the last century, has begun to show moderation following the Paris Agreement (Le Quéré et al., 2020). The shifts to lower carbon fuels, the increases in efficiency, and other improvements have been largely offset by the increasing global demand for air travel, electricity, and natural gas consumption, particularly though not exclusively in developing world economies like China and India (Jackson et al., 2019).

Significant short-term negative perturbations in global GHG emissions have occurred with major global economic events, for example, in 2009 with the Great Recession and in 2020 with the global COVID-19 pandemic. 2020 witnessed a large temporary plunge with rapid recovery due to the COVID-19 global pandemic (Le Quéré et al., 2020). This pandemic-related emissions reduction in 2020 is the largest since World War II though proved to be short-lived and of little consequence to continuing increases in global CO_2 concentrations. Notably, it is of the scale of reductions that are needed on an annual basis to get to net zero emissions by mid-century. The recent data underscore several key points: emissions are beginning to flatten due to human efforts to decarbonize, but the scale of reductions needed and societal impacts to go to a carbon-neutral or carbon-negative economy is daunting. It will take far more than if we all start driving hybrid vehicles and change our lightbulbs, important as they are.

While some industrialized countries have modestly reduced their emissions, several newly industrializing nations, most notably China and India, are rapidly increasing (Fig. 4.18). In the United States, the annual emissions amount to 13% of global emissions (down from over 20% in decades past) which are generated from ~4% of the global population. The European Union has fared somewhat better in reducing emissions. It should also be pointed out that the developing world's per capita emissions are still far less than more developed nations. When considering the cumulative emissions over the Industrial Age, most of which are still in the atmosphere, the developed nations are still the largest emitters as well.

The default prospect that the developing world will follow the fossil-intensive pathway of the developed countries represents a daunting prospect in terms of climate stability. Equity issues also weigh heavily as developing countries witnessed the wealthier countries having "climbed the ladder" of development during the last century, in part with the aid of cheap fossil fuels. Some view it as trying to pull up the ladder preventing the developing countries from progressing under climate concerns. Who will reduce their emissions, by how much, and when are critical areas of global deliberations? Such disagreements between the developed and developing world have been a critical impediment to solving the climate crisis.

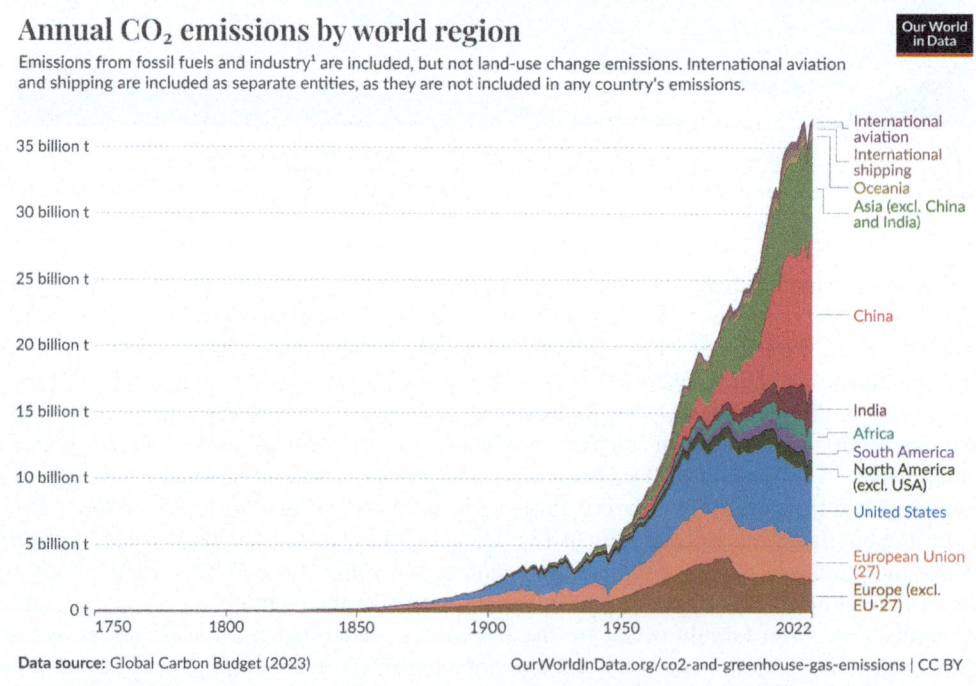

FIG. 4.18

Historical (1750–2022) global CO_2 emissions timeline by nation (Ritchie et al., 2023).

Creative Commons BY license from Our World in Data.

4.9 Anthropogenic climate change and the climate crisis

FIG. 4.19

Geographic distribution of the magnitude and changes from 1990 to 2019 in CO_2 emissions from the largest emissions sources.

Author image with data from the US EPA FLIGHT database.

A map of CO_2 emissions from major stationary sources is shown in Fig. 4.20. As such, smaller sources such as buildings, vehicles, and small engines are all excluded (they would roughly track with population density). Larger point sources are in some respects easier to control, replace, or phase out than a billion or more mobile, distributed sources. The small to moderate sources will be a more difficult category to eliminate. The open circles represent major (>100,000 TPY) greenhouse gas emission sources in 1990, while the filled circles represent the same in 2019. The largest sources are mostly centralized coal-fired or gas-turbine power generation facilities concentrated in the industrial Ohio River and southeastern United States but spread throughout the interior of the nation. Other important though less significant sources include ethanol production facilities, gas processing facilities, refineries, mining operations, and large landfills.

Greenhouse gas emissions do show gradual though clear changes in the United States in the last decade as many of the larger sources (generally coal-fired power plants) have either decreased in magnitude or retired entirely. The overall emissions from the power generation sector have declined from 2.33 GTC/year in 2010 to 1.67 GTC/year in 2019. Of note, the US economy-wide (including all major and minor) total greenhouse gas emissions peaked in 2005 at 7.39 GTC/year (with 6.13 GTC associated with CO_2), dropping in 2010 to 6.98 (5.70) and dropping further to 6.68 GT/year (5.42) in 2019. The shift from coal-fired electricity generation to a wider mix including more renewables and natural gas was a significant driving force. Conversely, GHG emissions from the domestic oil and gas production sector (OGA) have shown a marked increase (excluding refineries which have remained relatively constant). Emissions from the OGA sector have increased from 0.079 in 2010 to 0.341 GT/year in 2019, coincident with the increase in domestic OGA production. The appearance of small to midsize sources is particularly evident in areas of intensive production including the Bakken formation in North Dakota and the Permian Basin in west Texas and southeast New Mexico (Fig. 4.20).

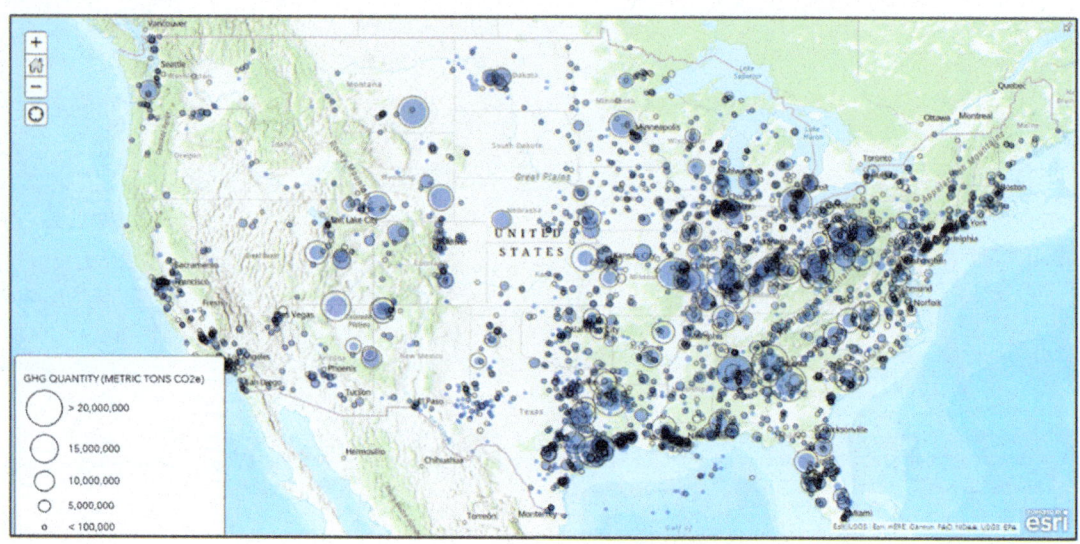

FIG. 4.20

Continental US major (>100,000 TPY) greenhouse gas emission point sources comparing 2010–19 (filled circles).

Data is from the US EPA FLIGHT database.

4.9.4 Overwhelming evidence for anthropogenic climate change

There is broad general agreement among scientists that the planet is warming, and human activities are increasingly driving climatic changes. Anthropogenic (deriving from human influence) climate change has been the term used in scientific circles from the early days as the changes involved are more than just the vague sounding "global warming." It is more complex than a uniform warming effect. For example, a weakening Gulf Stream ocean circulation risks dropping the temperatures in Europe by 10°C or more based on paleoclimatic evidence (Caesar et al., 2018). The latest IPCC AR6 states in the Synthesis Report Summary (IPCC, 2023):

> Human activities, principally through emissions of greenhouse gases, have unequivocally caused global warming, with global surface temperature reaching 1.1°C above 1850–1900 in 2011–20. Global greenhouse gas emissions have continued to increase, with unequal historical and ongoing contributions arising from unsustainable energy use, land use, and land-use change, lifestyles, and patterns of consumption and production across regions, between and within countries, and among individuals (high confidence).

Global temperature history is shown in Fig. 4.21, and the temperature anomaly since the mid-20th century is approximately 1°C. The 10 hottest years in the instrumental record are the last 10, 2014–23. The ocean's temperature has increased less rapidly due to its thermal inertia and long equilibration time for deep mixing (Fig. 4.22). Nonetheless, most of the energy added to the system has gone into the ocean assuring a semipermanent atmospheric warming. A change in average temperature of a few degrees or even 1°C should not be underestimated as it means a significant shift in climate (Table 4.5).

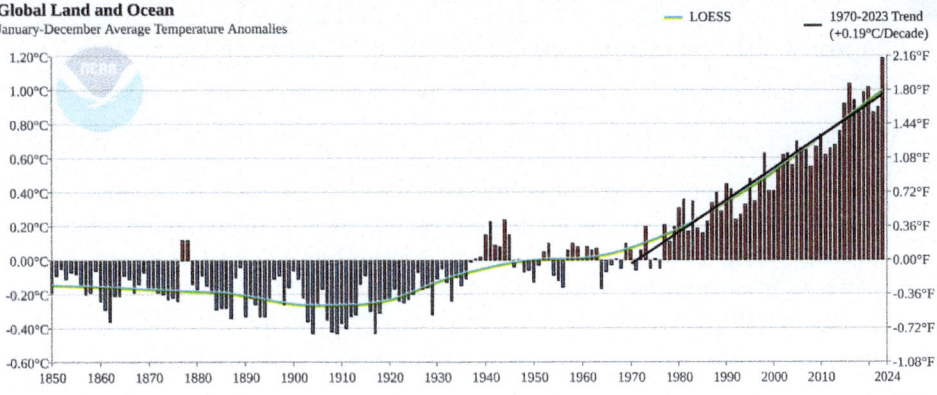

FIG. 4.21

Global annual average surface air temperature over land and ocean during the instrumental record (1880–2023). The 10 hottest years on record occurred in the last 10. The decadal trend over the last half-century is +0.19°C/decade.

Data from NOAA-NCEI available at Climate at a Glance | National Centers for Environmental Information (NCEI) (noaa.gov).

Data sources:

- CSIRO (Commonwealth Scientific and Industrial Research Organization) (2024). Update to data originally published in Domingues, C. M., Church, J. A., White, N. J., Gleckler, P. J., Wijffels, S.E., Barker, P.M., & Dunn, J. R. (2008). Improved estimates of upper-ocean warming and multi-decadal sea-level rise. *Nature, 453*(7198), 1090–1093. https://doi.org/10.1038/nature07080
- IAP (Institute of Atmospheric Physics). (2024). Update to data originally published in Cheng, L., Trenberth, K.E., Fasullo, J., Boyer, T., Abraham, J., & Zhu, J. (2017). Improved estimates of ocean heat content from 1960 to 2015. *Science Advances, 3*(3), e1601545. https://doi.org/10.1126/sciadv.1601545
- MRI/JMA (Meteorological Research Institute/Japan Meteorological Agency). (2024). *Global ocean heat content.* www.data.jma.go.jp/gmd/kaiyou/english/ohc/ohc_global_en.html
- NOAA (National Oceanic and Atmospheric Administration). (2024). *Global ocean heat and salt content: Seasonal, yearly, and pentadal fields.* www.nodc.noaa.gov/OC5/3M_HEAT_CONTENT

For more information, visit U.S. EPA's "Climate Change Indicators in the United States" at www.epa.gov/climate-indicators.

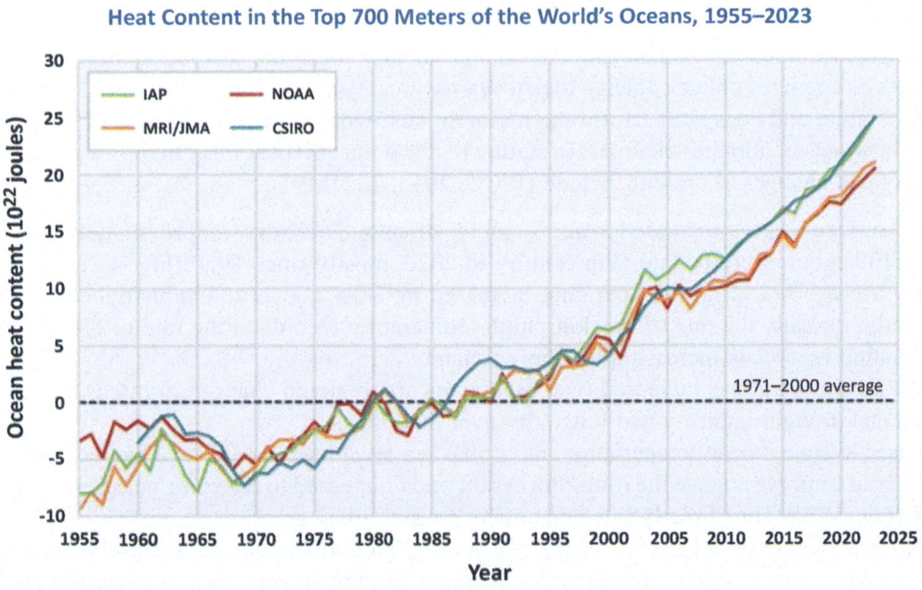

FIG. 4.22

Four datasets show the upper 700 m Ocean Heat Anomaly, i.e., the energy accumulation in the world's oceans is due to the energy imbalance imparted by greenhouse gases and other perturbations to the climate system

Public domain image from US EPA.

Table 4.5 Examples of US cities illustrate the significance of the seemingly small difference in average temperatures. $\Delta T \sim 1°C$ is the current global average change to date over the Industrial Age, while $\Delta T \sim 4°C$ is within the range of prediction for a doubling of atmospheric CO_2.

Cities with average $\Delta T \sim 1°C$ (current warming)	Cities with average $\Delta T \sim 4°C$ (upper end of the range predicted with a doubling of GHG)
Seattle to Portland	Albuquerque to El Paso
Minneapolis to Milwaukee	San Francisco to San Diego
Milwaukee to Chicago	Pittsburgh to Nashville
Philadelphia to Louisville	Chicago to St. Louis
Washington DC to Norfolk	New York to Atlanta
Raleigh to Atlanta	Austin to Phoenix

The Intergovernmental Panel on Climate Change (IPCC), under the auspices of the United Nations, produces assessments of climate change to date approximately every 6 years. The IPCC synthesizes the existing literature and does not perform the research itself working on a consensus of its member nations and hundreds of climate scientists. In stating that it is unequivocal that humans warmed the Earth system, notable changes at present include (IPCC, 2013, 2021b):

- Global surface T_{avg} has increased in almost all locations and overall by approximately 1.07°C (1.1 K or 1.8°F) beginning in the late 19th century to 2020, mostly since 1950 (Fig. 4.21).
- Global average sea surface temperature increased by $\sim 0.8°C$ over the industrial era.
- Currently, globally the rate of breaking high-temperature records to the rate of breaking low-temperature records is increasing by approximately 2:1.
- Global precipitation has increased over land areas and regional changes in intense precipitation events and drought intensity and longevity over land areas.
- Increased average ocean temperature and global sea level rise of ~ 0.2 m. This has continued annually at 2 mm/year since the mid-20th century and increased to 3.73 mm/year between 2006 and 2018 (Fig. 4.23). This directly threatens many coastal urban areas.
- Global average ocean pH has increased $\sim 30\%$ as related to Henry's Law dissolution of CO_2.
- Over the last 30 years, large increases in wildland fire-burned areas have occurred in the Western United States and other global regions.
- Loss of ice mass from the majority of land ice, Northern Hemisphere snow cover, and a drastic reduction in Arctic sea-ice extent (Antarctic changes are inconclusive).
- Biosphere changes are consistent with anthropogenic warming, and the growing season has lengthened by 2 days per decade since the 1950s.
- Damage to coral reefs and other particularly vulnerable species is accelerating.

4.9 Anthropogenic climate change and the climate crisis

FIG. 4.23

The global mean sea level continues to rise, and overall ocean acidity continues to increase both being driven by anthropogenic emissions of greenhouse gases. Erosion of coastlines is a considerable threat to coastal infrastructure from rising sea levels as well as more intense storm surges.

One of the underappreciated, parallel problems of enhanced concentration of atmospheric CO_2 is the effects on global ocean acidity. A reduction in average pH of ~0.1 pH units has already occurred. This change, along with ocean heat waves, has contributed to episodic large-scale diebacks of ocean coral reefs in regions such as the Great Barrier Reef in Australia (Anthony et al., 2008; Hughes et al., 2017). Altering the global ocean pH is a profound change that will have many ramifications on marine health as well as the global food chain. It also affects the coastal built environment (Example 4.5).

EXAMPLE 4.5 Ocean acidification

Problem: The approximate change in global average ocean pH due to Henry's law dissolution of atmospheric CO_2 is 0.1. This sounds like a negligible change. How much has the approximate ocean acidity changed?
Given: $\Delta pH\ -0.1$
Find: % change in dissolved CO_2
Assume: equilibrium; no secondary carbonate chemistry

Solution:

$$pH = 8.2 = -\log(H^+) \quad [H^+] = 10^{-8.2} = 6.310(10^{-9}) \quad \text{(initial conditions)}$$
$$pH = 8.1 = -\log(H^+) \quad [H^+] = 10^{-8.1} = 7.943(10^{-9}) \quad \text{(current conditions)}$$

$$\Delta[H^+] = 7.943(10^{-9}) - 6.310(10^{-9}) = 1.634(10^{-9}) \quad \text{(change)}$$

$$\text{fractional increase in } [H^+] = \frac{1.634(10^{-9})}{6.310(10^{-9})} = 0.26 \text{ or } 26\% \text{ increase}$$

Continued

> **EXAMPLE 4.5 Ocean acidification—cont'd**
> This makes sense that is of the same order of magnitude as the change in atmospheric CO_2 but lags it somewhat as the mixing to the deep ocean is a long process requiring long equilibration times. Some of this carbonic acid will dissociate to form H+ and bicarbonate and carbonate ions drawing more into solution.

Correlation of course does not prove causation. However, when a physical mechanism (trapping of outgoing infrared energy by greenhouse gases) clearly explains the relationship, and no other natural process can do so, it is an extraordinarily compelling case and actionable. An anthropogenic "fingerprint" or profile of greenhouse gas-driven climate change confirms the anthropogenic driver of the warming over the last century. More specifically, attribution studies seek to find distinguishing features of a "human fingerprint" on climate or specific events, and this is more difficult. A few of the notable facets related to a human fingerprint include (IPCC, 2013):

- The mass balance from the stoichiometry of the combustion of fossil fuels shows humans are collectively emitting ~40 GT/year of CO_2, about half of which comprises the atmospheric annual increase while the remainder is taken up by land and sea.
- The isotopic ratio of carbon species shows this is a fossil carbon signature (sequestered for millions of years) rather than biogenic carbon.
- Nights have warmed more than days as would be expected from enhanced greenhouse gases.
- Winters have warmed more than summers, again as expected with enhanced greenhouse warming.
- The troposphere has warmed while the stratosphere has cooled, consistent with greenhouse gases trapping outgoing IR radiation in the troposphere and preventing it from reaching the stratosphere.

4.9.5 Modeling current and future climate disruption and relevance to green building

Climate disruption has altered the natural world and humans' relationship with it on a global scale. What does the future look like? The adage often attributed to Yogi Berra that "it's hard to make predictions, especially about the future" and the unattributed adage that "all models are wrong, but some are still useful" are both of relevance here. Climate models, or general circulation models (GCMs), are numerical models that track over the long term the fundamental physics of the atmosphere (conservation of mass, energy, and momentum). To picture a climate model, think of putting a grid of boxes across the planet's surface and stacked to the top of the atmosphere and tracking over years to centuries many parameters such as temperature, pressure, humidity, and many others in brief time steps over long time intervals. Even though they are complex and require supercomputers, they are a simplification of true physics and chemistry. Nonetheless, they have become tremendously more sophisticated in recent decades, particularly in their representation of the interactions between atmosphere, land, ocean, cloud formation, and atmospheric chemistry. Their ability to predict past climatic changes and their range of agreement make them an instructive tool to forecast scenarios. The several dozen models predict a warming of ~2 to 4°C with a doubling of atmospheric CO_2. All do agree the planet will continue to warm with continued greenhouse gas emissions; it will accelerate with the worst-case scenarios of emission pathways, and the resulting impacts already palpable will become profound in upcoming decades.

The spread of predictions of global average temperature with a doubling of CO_2 is approximately 2–4°C among several dozen independent GCMs from around the world. This is not a bad agreement for models of this complexity, and they generally track the actual temperature changes (Fig. 4.24). In the late 19th century, Arrhenius, with his simple mathematical model predicted ~5°C of warming with doubled CO_2, so not terribly different from today's models. Moreover, many of the qualitative observations mentioned in Section 4.9 are observed in these models. Models get right much more than what they get wrong: the general magnitude and velocity of global warming, the polar and particularly the Arctic amplified warming, storm tracks moving toward the poles, a rising tropopause, greater warming at night and winter, and numerous other key features. Naysayers like to focus on the uncertainties, but climate models get many features right and can "hindcast" the changes to date in global climate reasonably accurately. Certainly, many improvements are happening, but it is hard to argue these are so fundamentally flawed to be useless. In fact, the greatest uncertainty in predicting future climate is the emissions pathway humankind will choose rather than the physics or modeling of the Earth system.

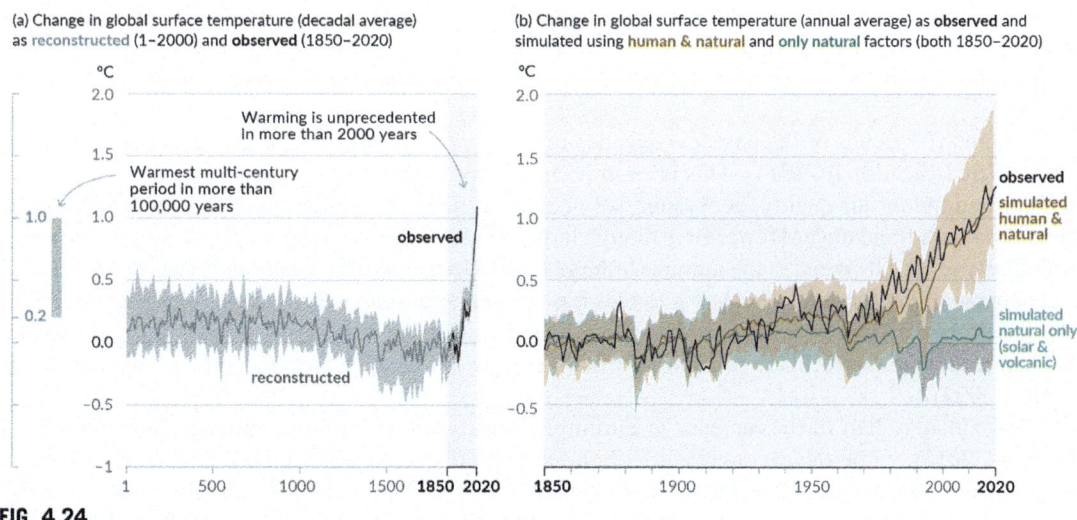

FIG. 4.24

The temperature history of the planet for 2000 years vs model simulations including both natural and human forcings Figure SPM-1 in IPCC Summary for Policymakers (IPCC, 2021b).

Emission scenarios include the RCP8.5 emissions scenario which invokes $8.5\,W/m^2$ radiative forcing by 2100 which is the most plausible worst-case scenario. Among the four plausible emissions scenarios, it is the highest emission scenario considered with no efforts to control emissions before 2100 (sometimes called "business as usual") which would project the most severe risks. This scenario entails a global average CO_2 concentration of approximately 1200 ppm and a global temperature change of 4°C, a profoundly different planet than we have become accustomed to.

Climatic changes have accompanied all of the major species extinction events, and one of the hardest things to predict is the follow-on effects, whether it be disease emergence, financial strains, international conflicts, or another destabilization. Among future impacts of most concern are the collapse of all or part of the Greenland or Antarctic ice sheets and accompanied sea level rise, large-scale bleaching and die-back of coral reefs, fire-related loss of a major forest area such as the Amazon, slowing or cessation of the Atlantic Meridional Ocean Circulation (AMOC, including the "Gulf Stream"), large-scale ecosystem shifts in the tropical rainforests or boreal forests, and Arctic loss of permafrost and accompanied methane release (Lenton et al., 2019). Such changes do not have an easy "undo" option by reducing or even removing our excess greenhouse gases from the atmosphere. Under a significantly warmer climate, the restoration of such critical natural phenomena is essentially impossible on relevant time scales, and thus the change is "irreversible." These become much more likely as the planet exceeds 1.5–2°C of warming, considered the threshold of avoiding the most severe climate consequences. Discussion of the impacts and the need to prepare for these changes in the built environment are key topics ahead.

4.9.6 Case study: Wildland fire and the built environment

Here we will focus on one impact of climate change: the connection of a warming climate to US wildfire trends. The resilience of the built environment in high-fire hazard regions is a topic that will be discussed in the closing chapter among many interconnections between climate change and extreme weather.

Wildland fire trends of the past few decades underscore a changing relationship between humans and wildfire (Rocca et al., 2014). This is an important convergence between land use, climate change, ambient and indoor air quality, ecosystem services, and the built environment. Recent wildland fire statistics show a trend toward fewer fires though larger total area burned (Fig. 4.25). A notable increase, particularly in the US West, in the number of large (>1000 acre) wildfires has occurred. The increased total acreage burned annually links to a longer fire season (approximately 78 days) plus increased fuel aridity with hotter, drier conditions for longer periods (Westerling et al., 2011, 2006). Many fire-prone landscapes around the world are suffering similar increases in intensity, duration, and spread of wildfire (UNEP, 2022).

Approximately half of the increase in burning is linked to our warming climate (Abatzoglou and Williams, 2016), while factors such as fuel buildup and poor management are added contributors. A decadal scale "mega-drought" in the Western United States combining increasing aridity and naturally occurring dry periods ("dry" droughts) with climate change-driven heat ("hot" drought) amplifies wildfire impacts (Williams et al., 2020). A projection of the impacts of climate change on biomass burning in the Western United States predicts a 54% increase in annual burned area by mid-century (Spracklen et al., 2009).

4.9 Anthropogenic climate change and the climate crisis

US trends in wildland fires (Fig. 4.25) are at once a response (due to heat and drought) and driver (greenhouse gas and aerosol emissions) of climate change. Thus the "natural" designation of wildfire is more accurately called "perturbed natural" with a growing human fingerprint. The climate-fire-air quality feedback cycle is highly uncertain and may depend on ecosystem; e.g., studies in the boreal forest found that the forest regeneration caused a net carbon uptake following fire disturbance (Mack et al., 2021). One variable is the contribution of stored soil carbon or "legacy carbon" that is released in high-intensity fires in young forests making this a net source of atmospheric CO_2 (Walker et al., 2019).

Wildfire season in select US regions has expanded to a year-round threat with behavior often described as "unprecedented" in terms of speed, size, and intensity. As illustration, the Marshall Fire in late December 2021 in Boulder County, Colorado, is a fire in the Wildland-Urban Interface (WUI) that destroyed over 1000 homes and caused 2 human and 1000 pet fatalities (Juliano et al., 2023; Silberstein et al., 2023). This fire sustained by 100 mph winds occurred in winter in a cold climate, not in a forested region but in the grasslands area east of Front Range of the Rocky Mountains (Irvine and Andre, 2023). Other devastating recent events such as the 2023 Lahaina Fire in Hawaii and the 2018 Camp Fire in California shared common features: wet winters, proliferation of grasses, followed by parched summers, and intense winds creating "blowtorch" like fires. Elsewhere in the west, New Mexico suffered its two largest wildfire events in the modern record in 2022, and many other locations have experienced their largest fires on record in the last 20 years. Consistent with anecdotal evidence from fire managers on the ground, data for the summed fire radiative power measured by satellites show a clearly increasing trend in severity of fires across the United States (Cunningham et al., 2024).

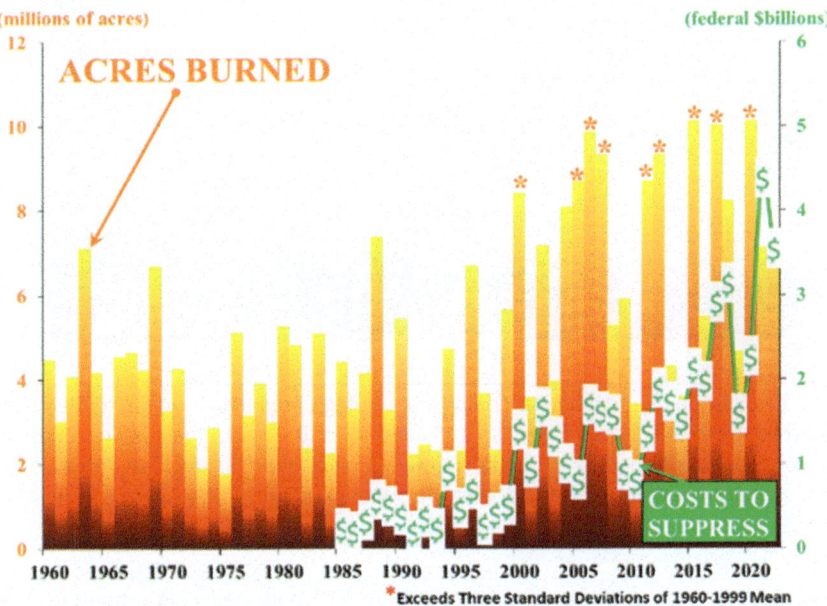

FIG. 4.25

US Wildland fire total acres burned and federal fire suppression costs 1960–2022. Starred markers indicate values exceeding the 1960–99 respective averages

Data from the National Interagency Fire Center, NIFC.

Other global regions such as the Mediterranean, sub-Saharan Africa, the Amazon rain forest, Southeast Asia, and Australia are facing increased fire losses as well. Australia is another nation that has faced a vastly degraded situation related to the extent and severity of bushfires. This culminated in the disastrous fire season of 2019–20 burning 17 million hectares. These bushfires destroyed over 3000 homes while causing 33 human deaths, a (conservatively) estimated 1 billion animal deaths, and 100 times that including insects (Richards et al., 2020).

The United States Forest Service (USFS) has been operating more recently effectively as the "US Fire Service." The USFS now spends into the billions per year or over half of its budget on fire suppression compared to 16% in 1995 (USFS, 2015) (Fig. 4.25). This reduces land management effectiveness, preparedness, recovery, prescribed fires, recreation, and other core activities. The failed policy of fire suppression has been reoriented toward the reintroduction of fire both as a feature of natural landscapes and a management tool allowing wildfires to burn and using prescribed fire (Fig. 4.26), an important tool not without hazards to both the natural and built environments.

FIG. 4.26

Prescribed fire at the Bosque del Apache National Wildlife Refuge. Reintroducing fire to fire-dependent landscapes is crucial to climate change response and adaptation. It is a tool that restores ecosystem health including control of invasive species such as the salt cedar (Tamarix) shown in the inset. Introduced species such as salt cedar are common invasives that displace native cottonwood and willow species in riparian areas of the Western United States.

Increases in biomass smoke emissions, a major source of both particulate matter (PM) and greenhouse gases, are emerging concerns for indoor air quality, atmospheric visibility, climate and cloud feedback, human health, and regulatory compliance. Health outcomes are primarily associated with the respiratory system and cardiovascular impacts (Reid et al., 2016). Research shows that $PM_{2.5}$ from

4.9 Anthropogenic climate change and the climate crisis

wildland fires is likely more toxic than general background $PM_{2.5}$ (Aguilera et al., 2021). Fire has increased particulate organic material annual $PM_{2.5}$ by up to $5\,\mu g/m^3$ in some western locations (Burke and Childs, 2023; Burke et al., 2023; Childs et al., 2022). In the United States, this has included a 27-fold increase in a decade in the population exposed to $PM_{2.5} > 100\,\mu g/m^3$ for 24 h (Childs et al., 2022). $PM_{2.5}$ from ambient smoke is transported 100s to 1000s of km, affecting air quality on a regional to continental scale as well as affecting indoor air quality (Carrico et al., 2016). Hard-won gains in air quality due to the Clean Air Act are beginning to be eroded by smoke contributions (McClure and Jaffe, 2018).

The fires in 2023 spanning Canada were described as uncontrollable, arrived early, and persisted for months, affecting over 16.5 million hectares (40 million acres) (Tracking Canada's Extreme 2023 Fire Season (nasa.gov)). The fires were unique in their scale (doubling the record year), wide swath of impacts, and early timing and duration. The impacts were acute in Canada and millions were affected by the smoke plume from these fires impacting much of Canada, the northern United States (Fig. 4.27), and across the Atlantic in western Europe. As with many hazards, though climate change does not cause such discrete events, it increases the probability and severity of these events. Similar events should be expected going forward and will influence greatly where and how we build our civilization.

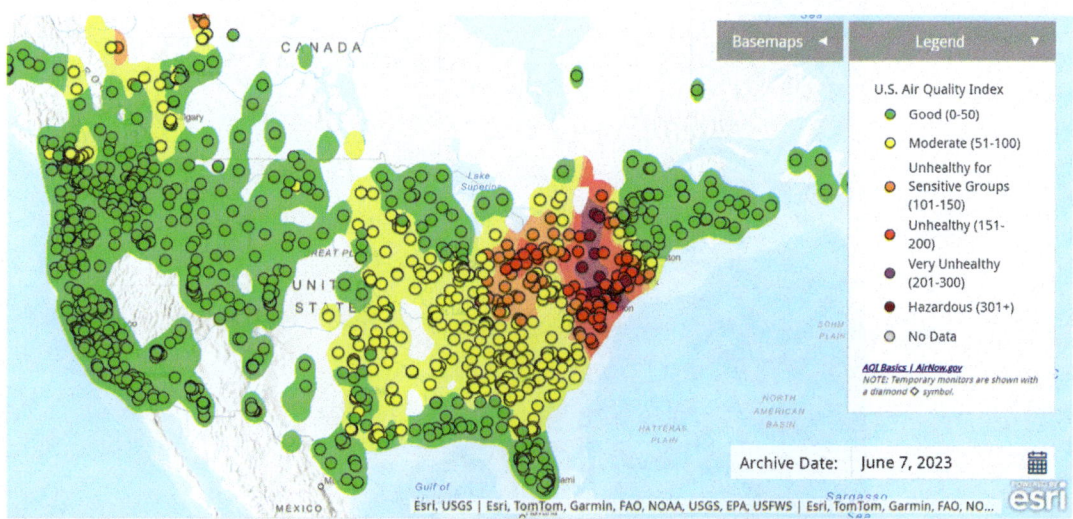

FIG. 4.27

Map of North America from the EPA AQS air quality network showing the $PM_{2.5}$ concentration on 9 June 2023 during which time much of the northeastern United States was impacted by a massive wildfire outbreak in Quebec, Canada.

4.10 Environmental policy and environmental justice

Global climate change is the quintessential environmental problem requiring a role of national to international agreement and policy. Governmental policies, international agreements, regulations, and statutes—at the international, federal, tribal, state, and local levels—have evolved over many years to address aspects of many of the environmental problems discussed above. Particularly following the National Environmental Policy Act (NEPA) in 1970 in the United States, the regulatory framework has been developed to minimize these problems (Milford, 2018). NEPA established the US EPA at this time (Fig. 4.28). NEPA requires federal agencies to consider the environmental impacts in a comprehensive manner of any major public infrastructure project, requires consideration of alternative design options, and mandates a means for public participation. The environmental statutes also feature regulatory enforcement mechanisms, penalty stipulations for violations of the regulations, and mechanisms for civil suits by the public for harm caused by violation of the standards (Milford, 2018). This experience provides models for addressing climate change.

FIG. 4.28

The USEPA facility in Research Triangle Park, North Carolina. The research into environmental quality problems and the development and enforcement of prudent regulations have resulted in great improvements in many air, land, and water quality problems.

All US federally funded significant infrastructure projects must go through an environmental review process called an Environmental Impact Statement (EIS) (similar processes in other countries). The EIS must consider the major project alternatives that are available, and the options selected. It then provides a comprehensive written review of all potential project impacts in terms of water quality, soils and land use, stormwater runoff and flooding potential, hazardous materials, construction impacts minimization, air quality and greenhouse gas emissions, endangered species, scenic vistas, archeology, historical resources, and socioeconomic impacts including issues of environmental justice (EJ). Outreach events to stakeholders are required, and public comments given or submitted in writing must be considered and given response particularly in regard to how the project will seek to minimize and measure the impacts listed above. These processes are more carefully considering the climate impacts as well.

Policies to address our climate crisis are nascent and at the local to global scales. Despite a level of scientific certainty calling for action far earlier, progress on addressing anthropogenic climate change has been difficult on a global scale. Efforts are accelerating, though the Paris Climate Agreement and predecessors are still only a first step rather than the final word in emissions reductions needed. The primary goal of the Paris Agreement is to limit global warming temperature anomaly to less than 2°C (as a global average) and preferably less than 1.5°C. The ramifications of 1.5°C vs 2°C of warming are drastic in terms of the impacts on human civilization and infrastructure. With the inertia in our political systems, not to mention the climate system, the 1.5°C threshold is unlikely attainable without some overshoot. Though the Paris Agreement had 196 signatories, binding emission limits and an enforcement mechanism are lacking, instead relying on commitments of voluntary efforts. With successive US presidential administration changes, US participation much less the all-important leadership role needed from the US has been elusive.

4.11 Chapter summary and conclusions

Global-scale issues with water, air, and waste became urgent in the 20th century underscored by mid-century smog events, rivers catching on fire, and Superfund scale contamination issues. Government oversight responded in kind, and the environmental engineering field developed. The interconnected nature of these challenges is vital to grasp as we have to address multiple problems simultaneously.

Water is abundant on the surface of the planet, although the vast majority is saline seawater, and the largest component of freshwater is in the solid state. Water issues span both quantity and quality. Water treatment focuses on minimizing the risks of disease transmission. The primary reservoirs from which humans extract domestic municipally treated water are surface water resources such as lakes and rivers and groundwater aquifers. The shallow depth, water-unsaturated zone is termed the vadose zone, while the saturated zone of the solid Earth where there is flow through a permeable media is an aquifer, an underground slow-moving "river." Many water-constrained regions including much of the Western United States are particularly dependent upon mountain snowpack. Our lakes, rivers, and wells are no longer sufficient, and humans are restricting usage and using unconventional resources such as wastewater reuse and desalination. The dwindling water levels in reservoirs along the Colorado River and its reservoirs are a case study in the real-time water shortage.

Air pollutants include solids, liquids, and gases. $PM_{2.5}$, or fine-mode particles, are important to health effects, haze, and climate. Twentieth-century air quality problems led to efforts to mitigate these problems. Some of the issues of regional air quality like acid rain have improved recently, and stratospheric ozone depletion has stabilized as humans implemented solutions. Part of the success of the Clean Air Act used market-based mechanisms such as tradable permits. Our efforts to address environmental problems related to the health of air, water, and land, though not perfect or complete, offer hopeful models that we will rise to the climate challenge.

After decades of increasing solid waste generation and disposal issues, the good news is that the previously increasing upward trend in both total and per capita solid waste generated have both flattened starting around the year 1990. Solid waste as well as hazardous waste contain a significant fraction of construction and demolition waste associated with the building industry.

Over the past several decades, the impacts of climate change have been emerging quickly compared to geological scale changes. Humans' climate impacts are complex including the warming effects of increasing concentrations of long-lived greenhouse gases. The net cooling of aerosol particles has offset or "masked" some of the global warming from greenhouse gases. These and other changes such as surface reflectivity have altered the planetary energy balance. The continuing increase in greenhouse gas emissions has become the dominant driver resulting in the net warming of the atmosphere and oceans. The importance of changing climate zones to green building cannot be understated as it determines locally applicable solutions. The science of anthropogenic climate change, long compelling, has only become more certain over time. Though science indicates we have to decarbonize our economy quickly, progress has been slow.

The challenge is turning our greenhouse gas emissions curve down and then to zero by the mid-century to avoid the worst impacts. The global pandemic disruption reduced CO_2 emissions by ~7% temporarily, a scale needed from now to mid-century to decarbonize our society. The good news is tools and techniques are largely available, but the political will and the investments in scaling an appropriate response are still lagging. The side benefits in many cases are improvements in air, water, soil, and materials use. Many of the solutions directly involve the built sector as will be discussed ahead.

4.12 End of chapter exercises

(1) **Concepts**: Use the following tool to calculate your carbon footprint. Discuss in a paragraph. Any surprises? How would you go about reducing your carbon footprint?
Carbon Footprint Calculator | Climate Change | US EPA

(2) **Concepts**: Although they are indispensable tools, do we need global-scale climate models to show us that enhanced greenhouse gas forcing will be disruptive to the planet?

(3) **Concepts**: Consider the greenhouse gas warming and aerosol cooling effects. Could these simply cancel one another and lead to no net impact? Explain.

(4) **Concepts:** Municipalities are required to meet enforceable drinking water limits prescribed in the Safe Drinking Water Standards. Locate the required Annual Drinking Water Quality Report for your community. Give a one-page synopsis of drinking water in your community. Where is it sourced? Do your community violate any water quality standards? How does it compare to the standards? Which are pollutants of most concern? Take a glass of municipal water and drink it. What is your assessment of the taste, odor, and appearance of your municipal water?

(5) Concepts: Which of the following is most accurate:
The Earth primarily gains energy via radiation and loses it via convection
The Earth primarily gains energy via conduction and loses it via convection
The Earth primarily gains energy via convection and loses it via convection
The Earth primarily gains energy via radiation and loses it via radiation

(6) Problem: Air pollution is often characterized by high concentrations of very small particles ($D_p < 10\,\mu m$). What are the ratios of the (a) surface areas and (b) masses between a $10\,\mu m$ and a $10\,nm$ spherical particle that have the same density? For the same mass concentration, which population of particles would be more harmful to health?

(7) Problem: A 1000 MWe 33% efficient steam electric power plant is burning coal with an energy content of 11,500 BTU/lb. What is the coal input per day (lb/day) and the rate of CO_2 generation (lb/year)? Assume coal is pure carbon.

(8) Problem: Approximate the global use of fossil fuels ($\sim$85% of total energy use) as entirely coal combustion which can be approximated as pure carbon (specific gravity $=0.8$). Calculate the size of a cube of material we are taking annually from the ground and putting it into the atmosphere (dimension of perfect cube in km)? Alternatively, what surface area (km^2) would it occupy if instead, it was a mine that went 10 m deep that is pure coal? Use reasonable assumptions.

(9) Problem: You reduce your power plant's emission by 1 t of SO_2. SO_2 forms sulfate ion (-2) in solution (assume 100% conversion) while ammonia gas forms ammonium ion ($+1$) in solution. Assume excess ammonia is available. These react to form ammonium sulfate solid particles. Show a balanced reaction to produce ammonium sulfate. Assume this ammonium sulfate is in the form of dry 300 nm diameter particles. You reduce 1 t of SO_2 emissions; how many ammonium sulfate 300 nm particles have you removed from the atmosphere assuming complete conversion? You can take the density to be $1.13\,g/cm^3$.

(10) Problem: Given the global increase of sulfur hexafluoride of approximately 3 ppt from 2009 to 2020, find the carbon dioxide equivalent mass $CO_{2,e}$ added to the atmosphere.

(11) Problem: The N. Pacific convergence zone (great Pacific Garbage Patch) is approximately $8 \times 10^6\,km^2$. Assume a well-mixed 100-ft surface layer and a plastics concentration of five pieces for every m^3 of water and a spherical particle with a diameter of 5 mm. The density of plastics is $0.9\,g/cm^3$. Calculate the mass of plastics in this zone.

(12) Problem: The human respiration rate is $\sim$15 breaths/min at 0.75 L and 21% by volume O_2 content. Find the mass flow rate of air ingested and find the mass flow rate of oxygen ingested by someone breathing. Pretend the human developed gills and needed to get this O_2 from dissolved oxygen (DO) in water. The DO concentration in water is approximately 10 mg/L at average Earth temperature. If the person were forced to get this same O_2 from DO, find the mass flow rate of water that would need to be ingested. You can assume STP conditions.

(13) Problem: Estimate the total number of particles inhaled by a human over their lifetime at the $PM_{2.5}$ annual standard. You can use a 300 nm diameter spherical particle as the characteristic particle with the density of water.

(14) Problem: The leaking well in Aliso Canyon in California leaked methane at a rate of up to 110,000 pounds per hour for several months. The leak episode began on October 23, 2015, and was finally plugged on February 11, 2016. Assume the average leak flow rate during the entire episode is $0.7 \times$ max flow rate. Assume natural gas is pure methane. Calculate the released mass of methane in metric tons of CH_4 and metric tons of equivalent CO_2 (on a 100-year basis for CO_2 equivalent). If this is mixed into the atmosphere, how much did it add to the global atmospheric mass of CH_4? You can use 1.8 ppm by volume and find the mass of the atmosphere from surface pressure and knowing the Earth is ~6370 km in radius.

(15) **Problem:** Atmospheric rivers have been in the news of late and the rain they can deposit especially in the West Coast states. According to an NYTimes article, "This vapor plume will be enormous, hundreds of miles wide and more than 1,200 miles long and seething with ferocious winds. It will be carrying so much water that if you converted it all to liquid, its flow would be about 26 times what the Mississippi River discharges into the Gulf of Mexico at any given moment." Test this assertion to see if a typical atmospheric river is an order of magnitude or so greater flow than the mighty Mississippi. Assume a system with a 2 km thick boundary layer and a 400 km wide river, saturated air at 75F, homogenous airmass at 1 atm. You may find the Antoine Equation discussed earlier in the book of use.

(16) **Problem:** You decide to run your house off-grid using a standalone diesel generator. How many gallons per year would it require to generate the average US household electricity consumption? What are the resulting CO_2 emissions in tons per year? Make reasonable assumptions for the chemical formula for diesel and its energy content and efficiency of a diesel generator.

(17) **Problem:** A large coal-fired power plant is outputting 1333 MW of electrical power. Its thermal efficiency is 32%. It is burning coal with an energy content of 11,000 BTU/lb. A train car carries approximately 100 t of coal.
   (a) How many coal cars per day does this power plant require?
   (b) A train car occupies approximately 4000 cu ft of volume. What is the volume of coal burned per day at the above power plant?
   (c) The typical house has a volume of approximately 20,000 cu ft. How many houses full of coal would this volume of coal fill?
   (d) If the coal is pure carbon, how much CO_2 in tons/year would this power plant produce? At 298 K and 1 atm, how many Pentagon-sized office buildings would this occupy?

(18) **Problem:** Calculate and compare the direct CO_2 footprint of using propane vs methane for cooking (compare in kg CO_2 per kJ and in lbCO_2/megaBTU). Assume complete combustion and balanced stoichiometry for the combustion of each.

(19) **Problem:** Global average atmospheric temperature has increased by approximately ~1 K while ocean temperature has increased ~0.5 K over the Industrial Age, primarily due to enhanced anthropogenic greenhouse gases in the atmosphere. Assume these are uniform throughout (they are not exactly) and assume a constant pressure process and you can assume reasonable, constant specific heats. Find the change in enthalpy (energy added to system) of each associated with the temperature change (assume constant pressure-specific heat for air). Assume we return to a preindustrial atmosphere. How many times could one "reheat" the atmosphere by 1 K via the ocean thermal storage? What does that say about the longevity of the warming of the atmosphere?

(20) **Problem:** An iceberg the size of Delaware broke off of the Larsen C ice shelf in Antarctica. It is 5000 km^2 in area with a thickness of 350 m. Assume the specific gravity of ice is 0.93.
   (a) What is the mass of this iceberg?
   (b) There are 1×15 J/s of heat continuously added to the Earth system due to enhanced greenhouse gases in the atmosphere. If this were completely input into melting this iceberg, how long would it take?

4.12 End of chapter exercises

(21) **Problem:** The globe combusts ~10 Gt/year of fossil carbon fuels. Assume density = 0.8 g/cm^3. If one could stack this in a 1 m diameter cylindrical column, how many times could it reach the moon?

(22) **Problem:** You know the annual global CO_2 emissions are approximately 35 Gt/year.
  (a) Calculate how much this increases the CO_2 concentration in ppmv assuming it all ends up in the atmosphere. You can find the mass of the atmosphere knowing the surface pressure on the surface area of a 6371 km radius of Earth.
  (b) Compare this to an estimated global flow of treated municipal water. Assume the 10 States Standard average water use applies to municipal water treatment and extrapolate globally.

(23) **Problem:** Find the primary annual standard for $PM_{2.5}$ in μg/m^3. Assume this concentration and that all particles are spheres, water, and are 200 nm in diameter. Calculate the number of particles per m^3. With the same mass concentration, if the particles coagulate to form a population of 0.5 μm diameter spheres, what is the new number concentration? Assume 30°C and 0.85 atm and water droplets of unit density.

(24) **Problem:** A concentration of ozone (O_3) in the air considered to have negative effects on health is 70 ppb (as an 8-h average). This is a volume mixing ratio and is the same as saying 70 molecules of ozone for every 1 billion molecules of air. How many molecules of ozone does this represent in units of molecules per cm^3? The pressure is 0.86 atm and the temperature is 30°C.

(25) **Problem:** Estimate the average emission factor per mile of the US fleet. (a) Calculate the annual emissions in gigatons of CO_2 from the US fleet of passenger vehicles. The fleet is 200 million vehicles, the average mileage is 24 MPG, and the average driving distance is 12,000 miles/vehicle/year. Assume STP, fuel density is 0.7 kg/L, complete stoichiometric combustion, and that gasoline is pure octane. (b) Calculate the CO_2 emissions in lb/mil for the average vehicle.

(26) **Problem:** (a) Compare the CO_2 emissions (in US tons) from the following two vehicles on an annual basis (assume the typical 15,000 miles driven per year in each vehicle). (b) Find the volume occupied by this emission if it is pure CO_2 at 1 atm and 20°C. (c) If this mass of CO_2 were diluted into the air until its concentration was 400 ppmv (the present atmospheric concentration), how much volume would the air+CO_2 mixture occupy? (d) For the mixture, how many Pentagons of volume does this translate into knowing there are 2.4×10^6 m^3 in the Pentagon?

(27) **Problem:** Global fossil fuel combustion is 7.2 Pg C year^{-1}. How much does this add to the atmospheric burden of CO_2 in ppmv if it all stays in the atmosphere? Assume the generic fossil fuel form is CH_2 plus a well-mixed atmosphere with a MW of air of 28.97 g/mol.

(28) **Problem:** Assume natural gas is pure methane. Find the required volume (STP298) of CH_4 needed to produce 1 lb of CO_2.

(29) **Problem:** The global emission of CO_2 in 2020 is approximately 35 Gt annually. How many Empire State Buildings is this mass equivalent to? At 1 atm, how many Empire State Buildings would this fill with pure CO_2?
   Referencing the Empire State Building Fact Sheet (https://www.esbnyc.com/sites/default/files/esb_fact_sheet_4_9_14_4.pdf)

(30) **Problem:** Compare the energy scales of anthropogenic climate change to date and the last mass extinction event due to the 14 km diameter asteroid impact 66×10^6 years ago which likely lead to the downfall of the dinosaurs. Use an asteroid mass of 1600 Gt. Presume the asteroid impacts the

Earth while traveling at 35 km/s and coming to a dead stop. You can compare the ocean heat anomaly to this energy scale or calculate it from the 2 W/m² forcing at the top of the atmosphere averaged over the surface area of the Earth ($r = 6371$ km) for 50 years.

(31) Problem: Estimate the mass flow rate of coal and the volumetric flow rate of exhaust gas (in m³/s) from a 1200 MWe power plant (34% efficient) burning coal (assume pure carbon) according to $C + O_2 \rightarrow CO_2$ with a heating value of 25,000 kJ/kg. You can assume STP298.

(32) Problem: The accepted emission factor for gasoline-powered vehicle emissions is 8.887 kg CO_2/gal gasoline. Starting from the combustion of pure octane (C_8H_{18}, 0.75 g/cm³, 124,000 BTU/gal; 25 mpg), see how close you get to this value. Find the emission factor for CO_2 per km traveled.

(33) Problem: If we had a planet devoid of ocean, how much would the atmosphere have warmed with the input of energy shown by the ocean heat anomaly (you may need to do some research)? You can assume the ocean volume is 1.36×18 m³. You can estimate the mass of the atmosphere knowing it exerts a pressure of 101,325 Pa over the surface area of the sphere defined by the Earth with a radius of 6371 km.

CHAPTER 5

Global solutions to the climate crisis: Our friends with cobenefits

Learning objectives

(1) Evaluate the breadth and necessity of solutions to the climate crisis.
(2) Recognize that no single solution is sufficient to the scale of the problem.
(3) Understand the role, importance, and scale of green building solutions.

The urgency of anthropogenic climate change has emerged as a driving force for green building. Connections to the built world will be highlighted for many solutions, but familiarity with all the major approaches is useful.

Have no illusions, addressing climate change on a scale that is meaningful is a global "heavy lift" with our ~80+% reliance we have on fossil fuels for energy. Nearly every sector of our global economy emits greenhouse gases by some pathway, and every individual contributes to emissions.

The consensus among the scientific community is that progress in decarbonizing our civilization is happening, but the pace needs acceleration to meet net-zero carbon by mid-century to avoid the worst of climate impacts (Bistline et al., 2022). Past energy transitions occurred over 50–100 years, and we do not have time to allow a natural progression without concerted assistance. We have a scant couple of decades to reengineer to decarbonize our energy system, a system that took a few centuries to implement!

There are positive side effects in implementing solutions as well. Many of the ills discussed in the chapter on environmental issues will benefit from reducing our climate impacts. For example, the same soot emissions we reduce for climate warming impacts also help with ambient air quality issues related to human health, visibility, and ecosystem health.

Finally, the technical solutions are numerous and all imperfect. There is no silver bullet and only tarnished buckshot, although each is important incrementally. The solutions are also inextricably linked to myriad economic, social, political, and organizational factors as discussed in a later chapter.

5.1 Climate change mitigation approaches: Enhancing sustainability

A former presidential science adviser, Prof. John Holdren, stated that we have three general responses to climate change: mitigation, adaptation, and last human suffering. The latter increases if the others are ignored. Mitigation is reducing the source of the problem, e.g., reducing or eliminating greenhouse gas emissions, while adaptation is altering our lifestyle and infrastructure to accommodate the changes occurring. We cannot solely reforest, solarize, or geoengineer are way out of the problem on their own, and even a large deployment of large-scale solutions is needed just to flatline the growth of emissions that would happen otherwise (Pacala and Socolow, 2004).

The slow nature of climate responses means anthropogenic changes are without a quick "undo." As humans continue to push the climate system warmer, the chances of an abrupt, irreversible climatic change or "tipping points" increase as well (Solomon et al., 2009). The response rate of humans combined with the lag-time of the climate system to perturbations makes the possibility of triggering such tipping points real. An example of feedback is increased methane released from the permafrost which risks the situation where temperatures will continue to rise even as humans reduce emissions (Steffen et al., 2018). Among other known and currently monitored tipping points include mass dieback of coral reefs or the Amazon rainforest, collapse of Greenland or West Antarctic ice sheets, disturbance of monsoonal flows, and slowdown of the Atlantic Meridional Overturning Circulation or ocean conveyor belt.

An emergency is defined when both the risk level and consequences and thus urgency to respond are high. If the reaction time for human institutions to take action is longer than the intervention time left, system control is surrendered (Lenton et al., 2019). The evidence to date tells us our current trajectory in terms of the human ecological footprint is not one that can be sustained long-term (Hoekstra and Wiedmann, 2014). However, it is not too late to dramatically change the trajectory of human civilization with respect to climate stability.

Traditional approaches to solving environmental problems first relied upon "build a taller smokestack" or "dilution is the solution to pollution" approaches. As a general philosophy, EPA's pyramid of choices offers a more robust approach (Fig. 5.1). A problem with approaching greenhouse gases traditionally from a control technology standpoint is that CO_2 is not a by-product of combustion as is SO_2, NOx, CO, and particulate matter. CO_2 is THE product (along with water vapor and heat), and the more efficient the combustion process the higher the emissions. In the exhaust stream, CO_2 concentration is orders of magnitude higher than the regulated criteria air pollutants. Essentially, we are using the entire atmosphere and ocean to dilute our emissions, and it is not feasible any longer without risking catastrophic damages.

The EPA Office of Pollution Prevention has developed a hierarchy of waste management choices ordered in preferability and applicable to problems of solid waste and food waste (https://www.epa.gov/p2) (Fig. 5.1). First, if one can prevent pollution or minimize waste, this forms the base of the (inverted) pyramid. Next, if recycling of material is possible—or better yet reuse—this should be pursued. Next is the treatment of the waste and this can be thought of as tailpipe controls (e.g., catalytic converters on vehicles). Finally, safe disposal is appropriate for wastes that cannot be otherwise addressed.

The expanded "'10 R's" of circular economy build on the familiar reduce, reuse, and recycle mantra with the following list: refuse, rethink, reduce, reuse, repair, refurbish, remanufacture, repurpose, recycle, and recover in decreasing order for the circular economy (Morseletto, 2020; Potting et al., 2017; Saidani et al., 2019). This also extends to the construction materials industries that are thinking about circular economy principles (Orsini and Marrone, 2019; Potting et al., 2017; Tayebi-Khorami et al., 2019).

A conceptualization of applying this hierarchy to climate responses is shown in Fig. 5.2 ranking solutions from most preferable to least preferable. Major solutions to the climate crisis will be introduced in this chapter; those linked to the built environment are emphasized and will be discussed in more detail in later chapters. A key factor is transforming our present linear, once-through economy into a circular economy where waste is recycled back to the front of the process.

Often the quickest way to get peoples' attention is through their wallets by putting a price on carbon emissions. Quoting Steven Chu, former presidential science advisor: "Most economists agree that the

5.1 Climate change mitigation approaches: Enhancing sustainability

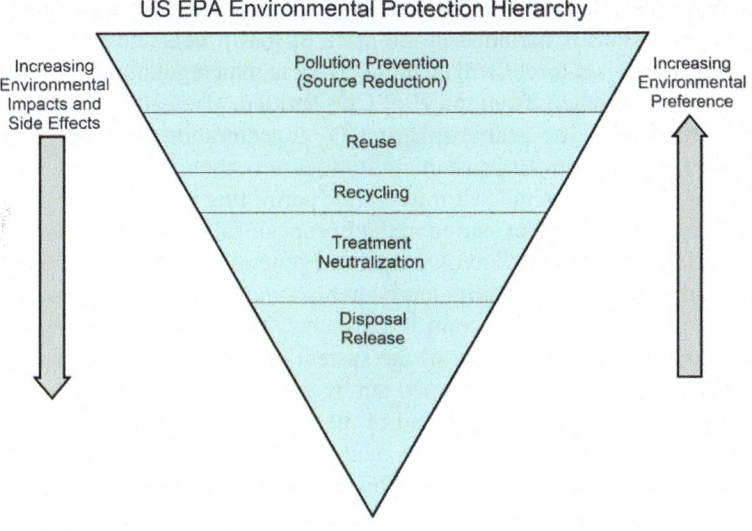

FIG. 5.1

Hierarchy of choices from the US EPA Office of Pollution Prevention for addressing environmental problems where the top widest part of the pyramid represents the most desirable or first approaches (public domain US EPA image from https://www.epa.gov/p2).

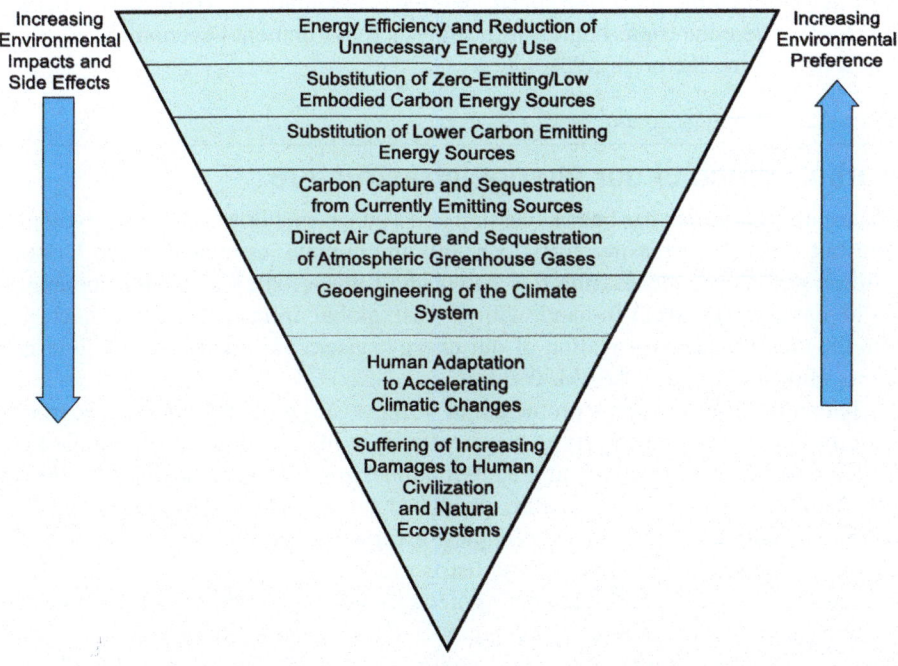

FIG. 5.2

Hierarchy of choices for addressing climate change where the top widest part of the pyramid represents the most desirable or first approaches.

most efficient mechanism to stimulate the shift to a low-carbon economy is a meaningful carbon price. If the cost of carbon emissions is included in the price of fossil fuels and primary energy-intensive industries, they argue that market forces will be more efficient than regulations." Greenhouse gas emissions impart an extra cost to society from ton X of CO_2 emitted, also called the social cost of carbon. The social cost of carbon increases as atmospheric CO_2 concentration increases, as effects amplify.

A transition from a growth focus to a sustainability focused economy is a huge but necessary transformation. Market-based systems include various carbon permitting and trading systems, the earliest of which is the Emissions Trading System started in the European Union in 2005. The upsides of market-based approaches are less governmental involvement and generally a more cost-effective solution. The downsides have included that the availability has been high and the price low on permits in some trading systems leading to little incentive to control emissions. There are debates on appropriate carbon offsets, and some approaches allow gaming of the system leading to less effective outcomes.

It should be emphatically reiterated as pointed out by the IPCC that none of the solutions to climate change, all likely necessary, are sufficient in and of themselves to fix the damage and prevent future catastrophes. Life Cycle Analysis (LCA) of costs mentioned previously is an approach that quantifies the costs from cradle to grave and can help guide which solutions are most appropriate. One of the most comprehensive and specific treatments of solutions examined on a life-cycle basis is the book Drawdown by Paul Hawken (2017). The ongoing Project Drawdown (https://drawdown.org/) assesses life-cycle CO_2 emissions reductions from an exhaustive and broad list of mitigation options. The top three were found to be related to refrigeration management, deployment of onshore wind turbines, and a significant global effort to reduce food waste (Hawken, 2017). The role of green building was integral to ~20 of the solutions with the most potential. The scalability and practical rollout of all solutions also become a key question. Furthermore, the social, political, economic, and other factors become part of the picture. This is where the triple-bottom-line analysis (environment+economic+equity) becomes helpful as well in defining the preferred solutions.

5.2 Decarbonization of our energy infrastructure

Societal reliance on combustion has been a blessing and a curse, enabling economic development in the industrial era but exacting increasing costs to health and climate, estimated in the United States at nearly $1 trillion per year and increasing (De Alwis and Limaye, 2021). The development of our industrial world has evolved hand-in-hand with a vast global infrastructure for energy resources (Fig. 5.3). In the end, the decarbonization of our energy system and society as a whole is the surest way to address climate change (NASEM, 2023).

Climate change is ~75% an energy problem and ~75% a CO_2 problem, and the decarbonization of the energy sector is a vital goal. Such efforts are necessary at the individual level, municipal level, via the private sector, and most importantly at the international level. The data are clear in that the energy sources with the highest risk to human health, quantified as the death rate from accidents and air pollution, are also generally those that have the highest greenhouse gas emissions (Fig. 5.4). Clearly, an energy transition to those sources lower on the list is needed.

Attention has been focused on implementing net-zero emissions sources of energy (Davis et al., 2018). By some measures, the emissions have moderated or even decreased in some industrialized countries, at least in comparison to projections under "business as usual." For example, since 2005, the CO_2 emissions from the electrical generation sector in the United States are approximately half

5.2 Decarbonization of our energy infrastructure

FIG. 5.3
Petroleum refinery (located at the coast near sea level) for turning crude oil into finished fuels and other products. The global infrastructure built around fossil fuel production and use is massive and much of it is susceptible to climate change.

of what the models projected (Wiser et al., 2021). This has largely occurred due to the combination of natural gas and renewable sources displacing coal-generated electricity over recent decades. This also improved air quality including emissions of $PM_{2.5}$ and other air pollutants. However, natural gas electrical generation is at best an interim solution. Recent research is showing the leakiness of natural gas extraction and delivery infrastructure is larger than bottom-up estimates negating at least some of the climate benefits when compared to coal (Chen et al., 2022).

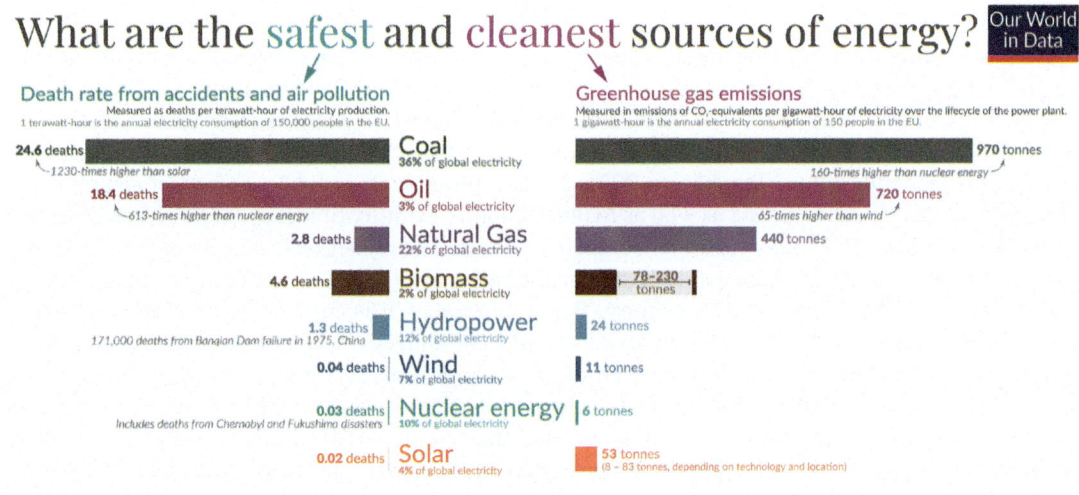

FIG. 5.4
Rank ordering of major global energy technologies including both the human health effects on the left and greenhouse gas emissions on the right per produced MWh of electricity (Creative Commons, Our World in Data (Ritchie, 2020)).

The concept of a "carbon footprint" helps to quantify from a life-cycle perspective the summed emissions of greenhouse gases associated with the actions of an individual, a facility or process, a nation, or the globe (Hertwich and Peters, 2009). For the individual, it incorporates the lifecycle (operational as well as embodied emissions) from an individual's transportation, heating and cooling, electricity use, industrial and commercial consumption, food and agriculture, and other emission sources. It incorporates the other greenhouse gases besides CO_2 and converts them to a CO_2-equivalent based on their individual global warming potential (GWP in Table 4.4) (Wuebbles, 2013). Thus, a kg of CH_4 has the same climate impact of ~30 kg of CO_2 over the timeframe of 100 years. The CO_2 emission factors (mass of CO_2 per unit of fuel or activity) are given in Table 5.1. The student is encouraged to use one of the many carbon calculators to estimate their own "carbon footprint." (e.g., the US EPA has the following: https://www3.epa.gov/carbon-footprint-calculator/).

Table 5.1 CO_2 emission factors for energy-related quantities US EIA and (CFSS, 2020 Collection).

Activity	CO_2,e emission factor (SI)	CO_2,e emission factor (US)
1 kWh of electricity from the grid	0.433 kg/kWh	0.953 lb/kWh
1 kWh of electricity from coal	1 kg/kWh	2.2 lb/kWh
1 therm combusted of CH_4	2.75 kg CO_2/kg fuel	11.7 lb/therm
1 gal of gasoline	2.3 kg/L	19.6 lb/gallon
1 gal of diesel	2.7 kg/L	22.4 lb/gallon
1 lb of anthracite	2.8 kg/kg coal	2.8 lb/lb coal
Mile of average passenger vehicle travel	0.22 kg/km	0.78 lb/mi
Aviation emissions	0.11 kg/passenger-km	0.39 lb/passenger-mile

Despite halting progress at the international level over the past 30 years, discussion of other municipal and industry efforts is important as well. Several states including California and Washington have mandated an end date in the 2030s for the sale of internal combustion engine passenger vehicles. Likewise, publicly traded automakers have planned dates for ending production of combustion vehicles, most notably General Motors, Inc. has set 2035 for their production end date. Efforts of shareholders to bring to bear the risks to climate as well as to the economic viability of legacy, high-carbon footprint companies have become another fulcrum to climate progress. The transition to a lower carbon economy has begun though will take decades of effort; historically, the transition between dominant energy sources has lasted the better part of a century. Beginning today, it is entirely feasible that a properly incentivized and regulated transition can take us to a decarbonized energy system by mid-century.

The scale, rate, completeness, and required mix of energy sources needed to accomplish decarbonization is an area of continued research and scientific debate. Recent analyses have claimed that a global energy infrastructure based on wind, water, and solar (with massive improvements in efficiency and suitable energy storage technologies) is feasible by mid-century (Jacobson, 2020) while counterpoints have cast doubt on the feasibility and cost-effectiveness of a "100% renewable electricity" approach (Clack et al., 2017). Certainly, major obstacles begin to emerge as intermittent renewables become a larger fraction of the grid due to their materials needs and mining, generally low energy density and land use concerns, site specificity, intermittent nature, and the consequent need for energy storage. Geographic and source type diversity helps, and energy storage technologies including battery farms, pumped hydro storage, other gravity-based systems, compressed air in geological formations, and numerous others are in various states of commercialization. Although it remains unclear the cost-effectiveness, scalability, and other technical challenges of each storage technology, integrating these with renewable distributed generation is critical.

5.3 Energy conservation, efficiency, and curtailment: A building-centric decarbonization

Energy efficiency has an inherent advantage as one of the lowest economic and environmental cost of energy resources. We defined the efficiency of a process earlier as it relates to thermodynamics as what you get out divided by required process inputs and limited to a maximum of 1 (or 100%). For a given output, fewer input resources mean lower associated emissions.

Efficiency is also intimately tied to building design, operations, and onsite energy use. Efficiency of resources, energy, dollars, time, and materials are all familiar concepts to engineers. The lowest cost megawatt is often the one not produced, or a "negawatt," as described by energy efficiency pioneer Dr. Amory Lovins. Various governmental programs (e.g., Energy Star from US DOE) as well as private nonprofit organizations (e.g., Rocky Mountain Institute) have promoted a focus on efficiency approaches as a top-tier choice resource.

Emphasis on energy efficiency emerged in the 1970s with the global oil price shocks. Savings in energy use for the United States since those efforts began are estimated to be almost 50% reduction in current annual energy use compared to what was projected at the time. US total energy use approached 100 Quads in 2020 rather than 200 Quads extrapolating growth trends in the 1970s (Sweeney, 2016). In the United States and many countries, the energy intensity—energy use per unit of economic output—has declined in absolute terms. This has happened through countless small changes sustained over decades and across all major sectors—industrial, residential, commercial, municipal, and transportation.

Improved energy efficiency can resonate with consumers for economic, environmental, and national security reasons. Policies and technologies to harness these efficiencies ideally take these incentives into account. A sector with demonstrable efficiency improvements driven by governmental standards in the last 50 years has been passenger vehicles. The passenger fleet average fuel efficiency has mirrored the CAFE fuel efficiency standards mandated by the US federal government during both times when they have increased (e.g., late 1970s–80s) and times when they have stagnated (1990s–2000s). An example of the potential with transportation emissions reduction is given here (Example 5.1). More specific technologies and approaches regarding efficiency as applied to the building sector are discussed in later chapters.

Another related aspect of energy conservation is the curtailment of nonuseful energy consumption, i.e., elimination of "wasted energy." Contributors include phantom loads due to plug-in devices that are not actively in use. This is also the approach related to turning off the lights when leaving an empty room or turning off other appliances not in active use. Parents around the world are familiar with the futility of attempts to condition the outside world by children who leave doors open during the heating or cooling season or leave televisions on in empty rooms. The elimination of these unnecessary loads is sometimes referred to as curtailment to differentiate it from efficiency.

> **EXAMPLE 5.1 Fleet replacement and CO_2 emissions**
>
> **Problem:** Estimate the average total CO_2 emissions and the emission factor in kg of CO_2 per km traveled (and lb per mile) of the US fleet average automobile. Use appropriate assumptions.
> **Given:** US fleet average
> **Find:** Total CO_2 annual emissions and CO_2 emission factor (kg/km and lb/mi)
> **Assume:**
> 25 mpg average vehicle
> 1 gal of gasoline emits 20 lb CO_2
> 270×10^6 passenger vehicles in the United States
> Average vehicle usage is 15,000 mi/year
> **Solution:**
> Emission Factor:
> CO_2 EF = 20 lbs/25 mi = 0.8 lb/mi = 0.225 kg/km
> Fleet Emissions:
> Mass CO_2 = 270×10^6 vehicles (15,000 mi/year) (0.8 lb CO_2/mi) = 3.24×12 lbs of CO_2
> $M = 1.5 \times 10^9$ TPY or 1.5 GT/year
> This is a significant fraction of the 35–40 GT/year of CO_2 emitted globally. Replacing all these with hybrid vehicles would roughly cut this in half, and replacing all with EVs charged via renewable energy would eliminate all but the embodied carbon of producing EVs.

Improving energy end-use is called demand side management by utilities. The effectiveness of demand reduction in some cases may be tempered by a rebound effect of the decreased marginal cost of each unit of energy and thus increased total usage due to increased economic activity. This is something akin to the thought process that it is OK if I drive more because I'm driving a hybrid vehicle. This has been studied in economics and was first described as the Jevons paradox as it applied to coal combustion in the United Kingdom with increasing steam engine efficiency. As coal combustion processes became more efficient, overall consumption increased as the entire industry expanded with its increased economic efficiency. Efficiencies of scale, enabled by more efficient consumption per use, can increase overall consumption in cases with price elastic demand. Similar concepts have been applied to traffic congestion on roadways as well as computer network traffic. The key to preventing a rebound effect is to ensure the cost of use reflects the actual cost (including environmental) and to pursue a balanced approach to both low-impact energy supply and demand management.

Despite the economic efficiency associated with efficiency upgrades, many barriers and market failures can prevent fuller implementation, even when the payback is short or immediate. These factors include lack of knowledge, limited time frame for ownership or rentals, high up-front costs, transaction costs, lock-in with technology and infrastructure, limited feedback information on effectiveness, and general apathy (Sweeney, 2016). One of the more interesting problems to address is the split incentives for rental housing units. Owners lack the incentive to minimize tenants' ongoing energy costs, while tenants have little incentive for investing in energy upgrades to a property they do not own. The energy efficiency aspect of climate stabilization is intimately tied to building systems, and these linkages will be explored in more detail in subsequent chapters (Example 5.1).

5.4 Additional built environment solutions to the climate crisis

The transformation of the built environment to a more sustainable approach is critical for climate change mitigation with the cobenefits of improving overall environmental health. The urgency of building contributions to emissions reductions is underscored by a recent update to the "wedges" GHG emissions stabilization paper of the early 2000s (Pacala and Socolow, 2004). The original paper mapped pathways to flattening the global emissions curve by implementing a series of 1 GtC carbon emissions reductions by mid-century (Johnson et al., 2021; Pacala and Socolow, 2004). In the update paper that assessed where we stand, the backtracking regarding the "buildings" wedge was noted and attributed to accelerated improvements in living conditions in the developing world (a good thing overall) and a lack of coordinated policies in many developed nations (not a good thing).

An "electrify everything" or "beneficial electrification" approach has taken hold of all sectors including utilities, transportation, industry, and the building sector. Berkeley, CA became the first municipality to ban new natural gas hookups in residential new construction and others have followed rapidly though it remains contentious. Electrification offers the opportunity to replace fuel burning with electrical consumption that is continuing to become cleaner and have lower carbon alternatives down the road even if today's electrical grid is still evolving toward lower emissions. Recent efforts focus on eliminating combustion in buildings and electrifying all energy-consuming processes.

The effort toward electrification has been given detailed information for the homeowner to approach this comprehensively on the residential scale across all end uses (Armstrong et al., 2021). Key appliances to accomplish this include heat pump space and water heating, induction cooktops, heat, and energy recovery ventilators, combination/condensing clothes washers/dryers with heat pumps, and incorporating electric vehicles. These have the added benefit of lower fire risk than combustion sources and not producing combustion gases with their health and safety implications.

Two other approaches are deep energy retrofits which involve retrofitting the infrastructure in a building (adding 240 V circuits, solar photovoltaics, and other building envelope upgrades), and "box swapping" which is substituting electric appliances for gas-consuming or other fuel-burning appliances (Armstrong et al., 2021). The former is what a building owner might embark upon while a renter of a building would likely only pursue the latter and typically in a supplemental manner. For example, a renter may use a small portable induction cook plate to reduce the use of a gas range.

There are also considerations for new construction and remodeling including a complete choice of systems as well as materials. Materials and their embodied energy and emissions such as concrete and steel are discussed elsewhere in the book, as they are carbon-intensive materials due to the processing from mining to delivery as finished construction products. Carbon-sequestering materials, including novel concrete mixes as well as mass timber, offer hope for negative emissions technology that can take excess atmospheric CO_2 and sequester it in building materials.

5.5 Agriculture, forestry, and land use changes: Reinventing our food system

The agricultural, forestry and land use (AFOLU) are key contributors to climate change though subordinate to the primacy of our energy systems. Lands management offers ways to both diminish or accelerate greenhouse gas emissions (IPCC, 2014). Displacing forested lands with over-tilled, irrigated lands requiring large inputs of fertilizer, pesticides, and herbicides, all serving a ruminant-focused animal production system, are ways to exacerbate climate change as compared with lower impact agriculture. A more sustainable agricultural system has benefits beyond climate including habitat preservation, land use efficiency, reductions of herbicide, pesticide, and fertilizer runoff and eutrophication, and a more resilient food production system.

Other land use changes include reforestation of previously forested areas, afforestation of new lands, prevention of over-grazing, and bioenergy applications in connection with insect-damaged forests. Soils management and sequestration of carbon offer a key tool for reducing our global carbon footprint (IPCC, 2019). Methods that have the potential to sequester additional carbon in soil include biochar and other techniques that retain soil health such as no-till planting. These are often used as carbon offsets by industries that have difficulties decarbonizing their operations. Concerns arise regarding the permanence and "additionality" of some of these offsets. In other words, do they result in increased carbon sequestration beyond what would have happened otherwise? Or are we just counting uncertain carbon reductions that were in the pipeline? This is not an uncommon question across the board. Permanence relates to whether a forest planted for carbon sequestration stays in place or is potentially harvested or burned in wildfire as will be discussed in a later chapter on resiliency.

The forestry aspect relates directly to green building. The responsible management of forested regions such as rainforest areas has become important globally over the last decades. Several certification and verification programs have evolved to better manage forestry-related emissions and other impacts. The Forestry Stewardship Council (FSC) has one of the most recognized programs for certifying sustainable wood products.

Otherwise, at first glance, this sector seems to be only tangentially related to buildings. However, a recent direction for agricultural production has been indoor growing operations. These are often designed as "vertical indoor farms," also called Controlled Environment Agriculture or Z-farming due to its negligible land use (Eigenbrod and Gruda, 2015). Vertical farms have the advantages of being very land area efficient (two to three orders of magnitude more production per acre) as well as locations near markets reducing transport costs. These will become more attractive in increasingly urbanized areas with expensive and constrained arable land resources. The indoor farms are also water, soil, and nutrient-efficient, and typically free from pesticides and herbicides. These are often some variety of hydroponic or aeroponic systems. Current limitations are that the energy use and carbon footprint are typically higher than outdoor farms as are the costs due to energy inputs. The lighting systems, as well as the heating system (most often natural gas), make this the case, although the use of ever-improving LED lighting has helped to moderate electric use (Yeh and Chung, 2009). The weather extremes driven by climate change will likely drive more agriculture indoors in addition to the aforementioned benefits. Likewise, efforts at green roofs as discussed later can also contribute to food production in urban areas. The development of a truly sustainable indoor farm will be a boon to reducing the impacts of agriculture.

5.6 Carbon capture, use, and sequestration

Many climate analyses find that carbon capture, use, and sequestration (CCUS or CCS) will be mandatory to meet the challenging goals of keeping the global average $\Delta T < 2°C$ by mid-century (IPCC, 2005). CCUS began as a technique in the early 20th century as a means of enhanced oil recovery

(EOR), where CO_2 is injected to enhance production in declining reservoirs. Later conceptions sequester the CO_2 in saline reservoirs (deep groundwater layers that are too saline for other uses) or other suitable geological formations such as basalts (Fig. 5.5). It has even been proposed down the road that CO_2 will be delivered rather than taken for EOR from natural reservoirs such as Bravo Dome, Sheep Mountain, and McElmo Dome in the Rocky Mountains, reversing the flow of CO_2 through those pipelines.

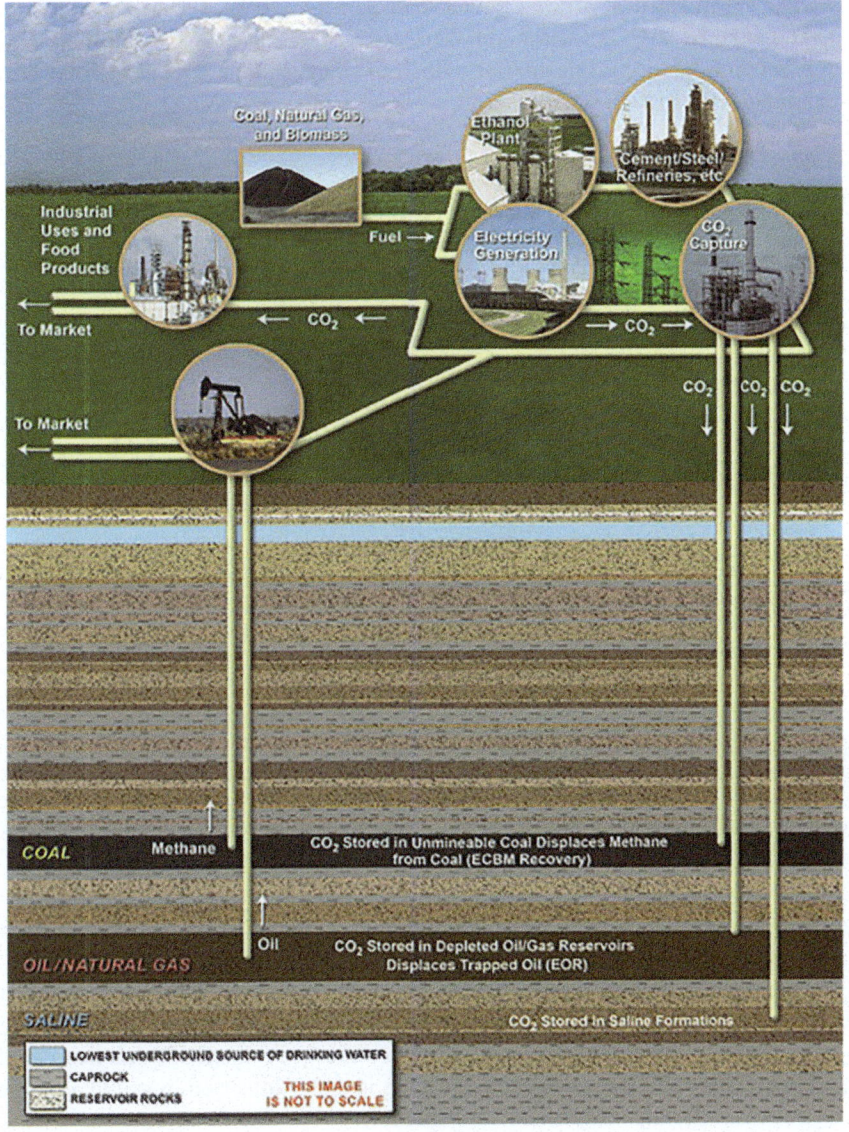

FIG. 5.5

An overview of carbon capture sources and sinks (public domain image from DOE National Energy Technology Lab Carbon Storage FAQs | netl.doe.gov).

The number of significant sources and large sinks in the US is not considered presently a limiting factor (Fig. 5.6). Various major industries have CO_2 effluent streams that are 5% to ~100% (Bains et al., 2017). Globally, a handful of demonstration power plants with CCUS (Petra Nova, Kemper, and others) have intermittently shown that stack-gas capture is technically feasible on a utility scale though economically challenging. Removal of CO_2 from point sources such as power plants or ethanol facilities is most cost-effective due to its high concentration. Decreasing separation costs and energy are required with higher concentration capture as specified by Sherwood's Rule dictating separation efficiency. Capturing CO_2 in-stack at a point source can be highly efficient at 85%–95% capture rates (IPCC, 2005). The capture of CO_2 is best suited to large point sources with simple effluent streams such as power generation facilities, but smaller industrial sources that are <0.1 Mt/year are also potentially viable candidates (Psarras et al., 2017).

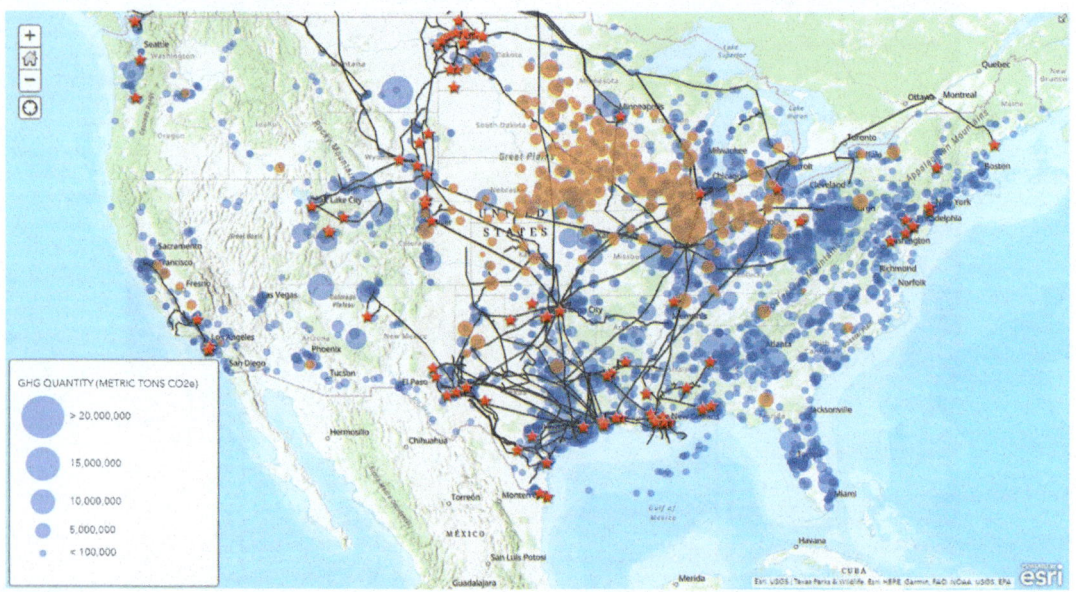

FIG. 5.6

ArcGIS map of major CO_2 point sources (*blue dots*) as well as ethanol refineries (*brown dots*, scaled by ethanol production) as of 2020. Also shown are major pipeline routes (*black lines*) and rail terminals for crude delivery (*stars*). Data are from the US EPA FLIGHT database (https://ghgdata.epa.gov/).

CCUS relies upon chemical or physical capture techniques including absorption by amines (ammonia like organic compounds), metal–organic frameworks (MOFs), porous materials, and membrane separation. The primary method pursued for capture is absorption tower systems using typically an amine capture solution (Leung et al., 2014). The most viable transport method is pipelines, although liquefied transport by rail is possible. Although enhanced oil recovery (EOR) has been applied for decades, the most sustainable and carbon-friendly storage is either injection in saline aquifers or basalts where the CO_2 becomes permanently chemically bound.

The mass flow rates of CO_2 into the 100's of kg/s for a typical steam cycle coal-fired power plant make it daunting to consider capturing, separating, compressing, delivering, and sequestering at scale.

As such, CCUS technologies, though progressing, have not been scaled or proven economic (Bui et al., 2018). Many approaches involve a substantial energy cost, so it is best if renewable energy sources can be harnessed to run these, as it is otherwise introducing a large parasitic load. For the power sector, current cost estimates are an increase in conventional electrical generation costs by 30%–50% (Chu et al., 2017). The costs, depending on the industry, are typically in the range of $10–100 per ton for current CCS technologies on stack capture from point sources (Abramson et al., 2020).

The federal government and other municipal entities have been increasingly incentivizing CCUS at the industrial scale. In the United States these fall under the "45Q" tax credits in the IRS tax code that began in 2008 with a $10/metric tonne (t) tax credit for enhanced oil recovery and $20/t for saline storage. The current 2024 45Q tax credits are summarized in Table 5.2, although it has many additional details and nuances. Tax credits under 45Q in the US have accelerated interest in CCUS as it provides a federal tax credit of $60/t for carbon reuse such as enhanced oil recovery (EOR) and $85/t for saline storage of CO_2. Other concerns beyond cost range from leakage to induced seismicity to water use that can be limiting in water-constrained regions (Rosa et al., 2020). The production of coproducts with CO_2 removal and storage is a benefit that may help offset some of the costs, including carbon-sequestering construction materials such as mass timber (Hepburn et al., 2019). SimCCS, an optimization model to link CO_2 sources, transport pathways, and sinks, is a tool to find cost-effective solutions for capture, transport, and CO_2 storage (Middleton et al., 2020) (Example 5.2).

Table 5.2 US federal tax credits for CCUS contained the 45Q provisions of the US IRS Tax Code as of 2024 ($/metric tonne, t).

	Geologic storage in saline or other formations	Carbon reuse (fuels, chemicals, products)	Geologic storage in oil and gas fields
Industry capture	$85/t	$60/t	$60/t
Direct air capture	$180/t	$130/t	$130/t

EXAMPLE 5.2 CCUS viability

Problem: You are eligible for 45Q tax credits for power plant carbon dioxide emissions of 18,750 TPY or greater. Estimate the $MW_{thermal}$ and $MW_{electric}$ size of such a power plant for coal and gas. Use the standard reactions assuming pure carbon and methane. Assume 35% thermal efficiency for the steam cycle coal-fired power plant and 42% for the simple turbine natural gas plant.

Given: 45Q tax credits for >18,750 TPY
Find: MW of a typical coal or gas power plant
Assume: Steady state, steady flow; 35% and 42% efficiency, respectively
Solution:
CO_2 output = 18,750(10³) kg/year × 1 year/(24 × 365 × 3600 s/year) = 0.595 kg/s of CO_2 generation
CO_2 mol/s = 0.595 kg/s × (mol/0.044 kg) = 13.5 mol/s

Continued

> **EXAMPLE 5.2 CCUS viability—cont'd**
>
> $C + O_2 \rightarrow CO_2 + 94\,kcal/mol$ (or $393\,kJ/mol$)
> Carbon combusted $= 13.5\,mol/s \times (12\,g/mol) \times (1/0.35) = 464\,g/s$ of C burned
> Thermal power generated $= 13.5\,mol/s \times 393\,kJ/mol = 5306\,kJ/s = 5.306\,MW_{thermal}$
> Or approximately $1.857\,MW_{electric}$ output at 35% plant thermal efficiency
> This would be considered a small power generator, and most utility-scale generators would be far above this threshold.
> $CO_2\,mol/s = 0.595\,kg/s \times (mol/0.044\,kg) = 13.5\,mol/s$
> $CH_4 + 2O_2 \rightarrow CO_2 + 2H_2O + 213\,kcal/mol$ (or $890\,kJ/mol$)
> Carbon consumed $= 13.5\,mol/s \times (12\,g/mol) \times (1/0.42) = 386\,g/s$ of C burned
> Thermal power generated $= 13.5\,mol/s \times 890\,kJ/mol = 12,015\,kJ/s = 12.02\,MW_{thermal}$
> Or approximately $5\,MW_{electric}$ output at 42% plant thermal efficiency.
> Power plants of this output are quite small meaning many power generation facilities would qualify.

To meet ambitious climate goals, humans will eventually be faced with the need to remove greenhouse gases from the atmosphere, and "negative emissions" technologies are a key tool to assist in this effort (Creutzig et al., 2019). Bioenergy with CCS (BECCS) and direct air capture (DAC) are two viable methods for doing so. BECCS offers a means where nature's CO_2 sequesterers, plants, can be used to produce a biomass fuel that can then be combusted for power generation with CCS to capture and sequester the generated CO_2. The technology offers promise with the drawback that biomass fuels have a low energy density, and thus the requirements for land use are large to make a significant impact. The challenge with DAC is that the concentration in air is much lower at 0.04% vs the 10%–15% typical of the stack gas in a combustion source. Current costs for atmospheric DAC are cost-prohibitive at ~$100s to $1000/t of CO_2, although this cost will decline as it is optimized and scaled. Operations such as the Climeworks Orca plant in Iceland are designed to remove 4000 TPY from the atmosphere and operate using geothermal energy. A second-generation plant, Mammoth, also in Iceland, upped this by an order of magnitude to 36,000 TPY. This still needs scaling up by orders of magnitude to displace the annual emissions of CO_2 globally which are about 1 million times larger. The target is $100/t removal costs, and the current space needed is estimated as 400 times less than a typical regrowing forest for the same CO_2 removal.

As it now stands the energy and economic costs of CCS, particularly the direct air capture due to the lower ambient concentration, make it cost-prohibitive and possibly even a net detriment (Jacobson, 2019). When the carbon intensity of our grid and society has been drastically reduced it may make sense to look closer at DAC. Atmospheric carbon dioxide removal is in its infancy and the field is active with research and startup companies pursuing various avenues for DAC. The counterarguments to CCUS are that it perpetuates our addiction to fossil fuels when cheaper solutions in the form of renewable energy are a better route. It does behoove us to research cost-effective CCUS as even after our emissions have been zeroed out, we will need to pursue DAC to take the excess CO_2 out of the atmosphere (Example 5.3).

EXAMPLE 5.3 Current costs and scaling for direct air capture

Problem: Orca, the world's largest direct air capture facility as of 2023 removes 4000 metric TPY of CO_2. It runs on geothermal energy and will soon be succeeded by an order of magnitude larger plant "Mammoth." Find the annual dollar and energy cost for atmospheric removal equivalent to our current annual global CO_2 emissions. How much scale-up would the current demonstration plant in Iceland need to meet this need?
a. How many Orcas would be required to negate the global annual CO_2 emissions (approaching 40 GT/year)?
b. The annual capture quantity is equivalent to what time period of global emissions (e.g., 1 month's emissions)?
c. At 10–20 kg/year for the average tree's CO_2 uptake, how many trees does it represent?
d. The energy (provided in this case by geothermal energy) cost per tonne of CO_2 captured is approximately 2650 kWh. How much energy would be needed? Compare this to the global steady state power use that totals ~18 TW for all end uses.
e. At $10–15 million per unit, what would the present cost be?

Given: Orca 4000 TPY of CO_2 removal, $10 million cost; 2650 kWh/t
    Find: Orca plants needed and dollar/energy cost to displace global emissions
    Assume: 10–20 kg CO_2/year uptake for a tree
    Solution:
CO_2 uptake rate $= 4000$ TPY $= 4 \times 10^3$ t/year (1 year/365(24)3600 s) $= 0.000127$ t/s
40 GT/year $= 4 \times 10^{10}$ t/year (1 year/365(24)3600 s) $= 1268$ t/s
It would require $1268/0.000127 = 1 \times 10^7$ or 10,000,000 Orca facilities to offset the total human emissions
The Orca plant is 4000 t/(1268 t/s) $= 3.15$ s of our emissions
4000 TPY $= 4 \times 10^6$ kg/year or the equivalent of at least 200,000 trees as on average trees uptake 10 to 20 kg of CO_2 per year
The annual energy use would be 40×10^9 t/year (0.002650 GWh/t) $= 1.06 \times 10^8$ GWh
The total annual energy use of our global energy system is 18,000 GW $\times$ (365) $\times$ (24 h) $= 1.576 \times 10^8$ GWh
Thus, this is of the same order of magnitude as our entire energy use
This would cost $1 \times 10^7 (\$10 \times 10^6) = \$1 \times 10^{14}$ or $100 trillion
This compares to the global GDP in 2019 (total "worth" of the economic goods and services) of ~$88 trillion so roughly the same scale as the global economy to direct air capture an amount equal to our global annual CO_2 emissions (ignoring the other GHG like CH_4). Obviously, this has a long way to go.

5.7 Development of a hydrogen economy

A likely even longer-term process revolves around the concept of using hydrogen (compressed to a liquid typically) as a transportation-friendly energy carrier. There is a lot of recent momentum around hydrogen as a solution for hard-to-decarbonize sources requiring high energy density fuels producing elevated temperatures. Hydrogen may be indeed the ultimate solution to a ready replacement for natural gas and a way to reduce aviation emissions, heavy transport emissions such as container ships, on-road emissions from heavy trucking, and other long-haul transport. Other hard-to-decarbonize sectors such as concrete and steel may be able to use hydrogen combustion in their production processes. The development of a hydrogen infrastructure, though in the research phase for several decades, will undoubtedly require decades more of development to reach anything similar to our existing petroleum-based infrastructure. The relevant hydrogen combustion reaction is given as follows (Eq. 5.1).

$$H_2 + \tfrac{1}{2} O_2 \rightarrow H_2O \, (+286 \, kJ/mol) \tag{5.1}$$

Other applications include semi-trailer trucks, aircraft, cement manufacturing, and power generation turbines. For example, the Intermountain Power Agency operated by the Los Angeles Department of Water and Power, the main purchaser of the power, will shutter 1800 MW of coal power and replace it with 840 MW of natural gas that can be in part substituted with hydrogen including renewable production of hydrogen (https://www.ipautah.com/ipp-renewed/) (Example 5.4).

EXAMPLE 5.4 Hydrogen energy density

Problem: Compare the energy density (J/kg) of hydrogen, natural gas (assume CH_4), and bituminous coal.
  Given: H_2, CH_4, bituminous coal
  Find: J/kg of H_2 vs CH_4 vs Bituminous coal
  Assume: Bituminous coal is ~30 MJ/kg
  Solution:
  Bituminous coal: $E_c = 30$ MJ/kg
  Methane: $E_c = 890$ MJ/kmol × (kmol/16 kg) = 55.63 MJ/kg
  Hydrogen: $E_c = 286$ MJ/kmol × (kmol/2 kg) = 143 MJ/kg
  Methane is about twice as energy dense as coal. Hydrogen is the clear winner in terms of energy per unit mass. The downside is that hydrogen is in the gas phase at atmospheric conditions (and thus has a low energy per unit volume). Thus, it must be compressed and stored effectively at high pressure or liquified.

Hydrogen is also the basis for many fuel cell technologies. Fuel cells provide an alternative to combustion engines, realizing a chemical reaction in a cell not unlike a battery. They also emit only water vapor and heat, making them available for combined heat and power. They require the replenishment of the hydrogen fuel, and thus the question becomes the source of the hydrogen (e.g., fossil fuels vs electrolysis using solar PV). Fuel cells are scalable down to the residential scale.

Hydrogen has a high energy content per unit mass, but low energy content per unit volume in its normal gaseous state. Thus, it is normally compressed and liquified which requires special tanks and handling adding costs. The details on its environmental friendliness come down to how the H_2 is produced, liquified, transported, and stored, all determining its total carbon intensity. The bookends in terms of that are hydrogen produced via renewable energy sources through electrolysis (green hydrogen) and hydrogen produced from natural gas without any carbon capture and sequestration (gray hydrogen).

5.8 Adaptation to an altered climate

The speed, quasi-permanency, and lag time of anthropogenic climate change necessitate that one of the key tools is adaptation to an altered climate (Wuebbles, 2013). This is another area where the building industry will play a vital role as it will have to respond to greater extremes of heat, water, wind, and other extreme weather. In some sense, adaptation feels akin to surrendering to the problems we created. At the same time, construction practices, codes, and materials will have to evolve in response to changes that are "baked in" at this point, and thus adaptation is a key factor. At the most extreme end, it will mean abandoning some of the highest risk areas. Adaptation must be pursued in concert with mitigation as it will become exponentially more expensive to adapt the further the climate system

5.9 Geoengineering approaches: Purposely altering the radiative balance

is perturbed. The closing chapter of the book will discuss in much greater detail the approaches to adaptation and resiliency in the built sector as a response to climate change.

5.9 Geoengineering approaches: Purposely altering the radiative balance

In decades past, geoengineering of the climate seemed like a radical idea with intensive side effects. It still is, but our lack of climate progress has necessitated further study as a "last ditch" option. Due to the slow pace of humans to respond to our current negligent greenhouse gas geoengineering experiment driving climate changes, an effort to "reverse geoengineer" the climate system to compensate for the ongoing warming is under active research. This policy approach is fraught with technical and logistical complexities, some of which are likely "unknown unknowns." However, these mitigation efforts, when pursued as a complementary and incremental approach may offer some measure of reduced catastrophic warming effects. It may offer some additional time to reduce our emissions, achieving net zero by mid-century.

The most likely geoengineering scheme involves some form of planetary albedo management to reflect incoming solar radiation or enhance outgoing infrared radiation (Fig. 5.7). One of the most likely scenarios is to mimic the effects of a volcanic eruption though on an ongoing basis. For example,

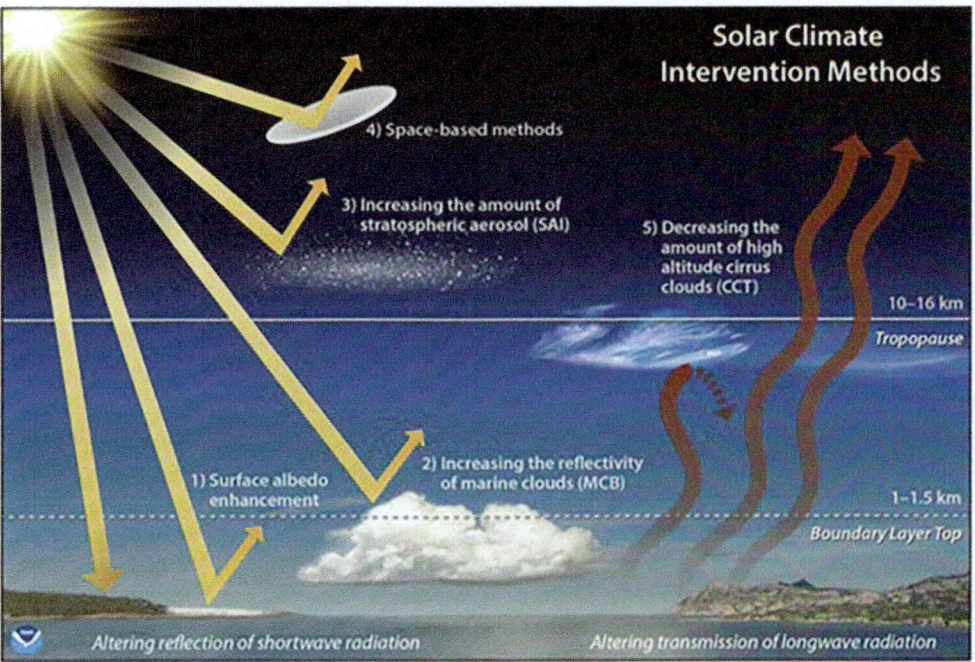

FIG. 5.7

Geoengineering schemes (public domain image from NOAA and US National Academy of Science, Credit: Chelsea Thompson, University of Colorado CIRES and NOAA Chemical Sciences Laboratory).

the stratospheric aerosol injection (SAI) using sulfate aerosols is considered one of the most viable and cost-effective approaches available. It however is rife with uncertainties, unintended effects, and general downsides (Robock et al., 2009). One should note that this is not the "inverse" of the greenhouse gas effect of trapping outgoing infrared radiation. It will offset the increased global average temperature much as a major volcanic eruption does but does nothing related to ocean acidification. The approach also likely will impact the hydrological cycle, regional weather patterns, the status of ozone layer recovery, along with other unforeseen consequences. It requires constant stratospheric SO_2 inputs or greenhouse warming will reassert itself within months of ceasing SAI. It also does nothing to address ocean acidification due to increased atmospheric CO_2 nor does it help divert us off a path of increasing use of unsustainable resources.

Less invasive smaller-scale geoengineering approaches are also possible, at least one of which applies to the building sector. Cool roofs and more reflective pavement surfaces are examples in the built sector that are discussed in other sections. A regional-scale effort to increase marine cloud cover by injecting sea salt aerosols may work on a scale that could help save the major reef systems. Other efforts include trying to prevent ice sheet disintegration by stabilizing the basal layer or brightening and/or enhancing freezing at the surface.

GEOMIP and CDRMIP are two scientific associations working on modeling issues related to geoengineering schemes involving aerosols and clouds for the former and atmospheric removal of carbon dioxide for the latter. While the fundamental capabilities and structures exist, computer modeling efforts from the microphysics to global scale climate models must be optimized and modified from current existing general circulation models (GCMs) to model the effects of geoengineering approaches. The bottom line with geoengineering is that there are large uncertainties, known and unknown, as well as many downsides (Robock et al., 2013; Robock et al., 2009). It is certainly not a quick-fix and at best should be viewed as a means of possible partial moderation of only the warming effects of global climate changes (Irvine et al., 2019), buying us time while we go about the hard work of achieving net-zero greenhouse gas emissions.

5.10 Chapter summary and conclusions

Climate change requires the reinvention of our energy, materials, and agricultural systems. The general approaches are mitigation, adaptation, and the least desirable, human suffering. All three will be part of our future, although we have choices as to the extent of each. In terms of mitigation, a wide range of solutions exist and are becoming more cost-effective, centered on low-carbon energy sources. Analysis in the compilation of solutions in "Drawdown" is a good first stop in learning about those with the most promise and scalability (Hawken, 2017).

First, a net-zero carbon emissions energy system is required by mid-century to avoid the most severe climatic disruptions. Energy alternatives are becoming cost-competitive and more sustainable. Radically improved efficiency is a key contributor to decarbonization. Zero carbon emissions sources include renewables such as solar, wind, geothermal, wave, and tidal, as well as nuclear power. Buildings are central to both renewable energy integration and efficiency improvements. Despite their utility, real challenges await their increasing integration. These are being addressed by improved engineering and operational solutions. Hard-to-decarbonize energy uses such as concrete and steel may benefit from electrification or a future hydrogen economy. Other land use solutions including soil carbon sequestration can contribute.

IPCC projections for staying below 2°C of warming require carbon capture, use, and sequestration (CCUS). CCUS is evolving though still costly and unproven at scale. The standard approach for CO_2 capture is stack gas capture at the power plant. After eliminating our emissions, we will later need negative emissions technologies. Direct air capture (DAC) removes the accumulated CO_2 from the atmosphere, although it is costly at ~$100s to $1000 per ton. Direct air capture is much more space efficient and permanent compared with forest uptake and certainly has a role in our carbon future. Its scale-up to a meaningful contribution will take decades and is presently very costly. Orca in Iceland is a demonstration-scale plant, but it is a useful case study. Another option is bioenergy with CCS where biomass fuels are grown and used for electricity generation while capturing and sequestering the CO_2 emissions.

Adaptation, or changing our infrastructure to accommodate an altered climate, will be a part of our response. It should also be underscored again that technical solutions are only viable in the real world of social, political, equity, and economic factors, among others. A later chapter looks more closely at these intertwined issues and how triple-bottom-line solutions incorporating environmental, social, and economic factors are the most likely to succeed.

Geoengineering is where we purposely modify the climate to counteract some of the worst effects of global warming. All of these geoengineering approaches have known (and likely unknown) significant downsides but should not be ignored. Mitigation is the best approach (control at the source), but other general approaches including adaptation and managed retreat in some areas that are not salvageable will become increasingly more important.

5.11 End of chapter exercises

(1) **Concepts:** Consider the expected continuing trends toward climate extremes (temperature, precipitation, etc.). Discuss some aspects of how this will impact the construction industry.

(2) **Concepts:** Human adaptation to climate change includes all of the following EXCEPT:
   a. Raising infrastructure in coastal areas
   b. Building homes and buildings to green standards as well as more resilient to storms and extremes
   c. Building infrastructure more resilient to storms and extremes
   d. Actively modifying the planetary radiation balance through geoengineering
   e. Reducing urban heat island effects by tree planting

(3) **Concepts:** The general definition of sustainability is:
   a. Keeping the electric system online
   b. Meeting the needs of the current generation without compromising those of future generations
   c. Assuring the transportation infrastructure does not become overloaded
   d. Preventing breakdowns in the economic system
   e. Recycling more materials

(4) **Problem:** One can compare the climate forcing humans are exerting on the climate to a nuclear war. How often are we launching a full nuclear exchange against the climate system? A typical nuclear weapon is 500 kt (TNT equivalent). Let's imagine a full nuclear exchange between the

United States and Russia (10,000 nukes). Compare this energy release to global climate forcing. How often are we waging a full nuclear assault on the climate system?

(5) Problem: According to the California carbon offset program, the forested areas in Northern California have the capacity to store on the order of 100 tons of CO_2 per acre (the rainforest regions up to 200). Assume a mid-range value optimistically. If the United States were to annually reforest to offset its entire CO_2 emissions, how much land area would be needed?

(6) Problem: Find the emission factor for CO_2 (lb/mi) for the following vehicle. You have a Subaru Crosstrek that has been averaging 35 miles per gallon overall. Find its CO_2 emission factor in lb CO_2/mi traveled assuming combustion of pure octane (0.75 g/L density of gasoline can be assumed). First, find a balanced reaction for octane combustion.

(7) Problem: Wind power developer Pattern Energy is completing the largest single-stage wind project in east central New Mexico that, once fully operational in December 2021. The development includes four wind farms totaling 1050 MW and a transmission line from New Mexico's eastern plains to western markets. Use an appropriate capacity factor (available uptime) for wind generation and the average US household electrical consumption. How many households will this windfarm service (ignore storage issues)?

(8) Problem: Find the carbon footprint of the average US household: The average US home uses 10,632 kW-h of electricity, 68 BTU of natural gas, and 620 gal of gasoline annually. Use typical national average CO_2 emission factors per unit of use for each contribution.

(9) Problem: The Intergovernmental Panel on Climate Change was established in 1988. Find data for global carbon emissions. How much of our emissions have happened before and after the IPCC?

(10) Problem: Assume we get serious about climate change and decide that the complete energy infrastructure of the globe needs to be replaced in the next 50 years by nuclear power plants (all end uses). Also, assume the developing world (6 billion now plus 3 billion more to be added) wants to obtain the standard of living of Spain assume 150 GJ/year per person) and also obtain its energy use all from this fleet of power plants. The proposition is that our entire energy infrastructure will be replaced by electricity from a noncarbonaceous source, and 9 billion more people need to attain the energy consumption per capita of Spain (assume they start from ~0 energy use now). Using reasonable assumptions, estimate the number of large power plants (1 GWe) per year that the world will need to build in the next 50 years and the total price tag....it is a large number.

(11) Problem: The currently accepted estimate for the social cost of carbon is $50/t. What are the costs associated with the lifetime emissions of a vehicle with fleet-average efficiency (assume 200,000-mile lifetime)?

(12) Problem: Enchant Energy is a private development firm working to install a CCS system on one of the units at the San Juan Generating Station. The estimated goal is that it will emit just 249 pounds of CO_2 per Megawatt Hour (MWh) of power produced. Compare this to the direct emissions from a combined cycle natural gas generator (assume pure CH_4) and from burning coal uncontrolled (assume pure C). Assume 35% thermal efficiency for coal, 50% for combined cycle CH_4, methane energy content of 213 kcal/mol, and pure carbon at 94 kcal/mol. Ignore fugitive methane emissions and consider only direct CO_2 emissions.

(13) Problem: How much further out (km) would the earth's orbital have to increase to reduce the incoming solar flux by 4 W/m^2 to compensate for GHG forcing? Look at the current ocean heat anomaly below (joules that have been added to the ocean). If we could somehow turn the ocean heat anomaly (400 ZJ) energy from the current excess GHG energy stored in the ocean to move the earth to a farther orbital (assume 100% efficiency), how much of the heat anomaly could we "consume"?

(14) **Problem:** Volcanic eruptions provide one of the best confirmations that aerosol particles (of the right size and composition) cause a net cooling of the atmosphere. For example, the Pinatubo eruption in 1992 caused a global cooling of 0.5 K for about 2 years. Mt Tambora released a solids volume of 12 mi^3 of aerosols into the atmosphere!
  a. Calculate the number of particles generated assuming a uniform size of 1 μm.
  b. What fraction of the fraction of sun's radiation would it block? Think about the earth's sunlit half and that it blocks a fraction of the solar flux intercepted by the cross-sectional area of the disk. Assume a uniform distribution around the globe and that the particles scatter all the light they intercept but no multiple scattering. This is all a vast simplification but instructive nonetheless!!

CHAPTER 6

Green design, delivery, commissioning, and auditing: Starting the project right

Learning objectives

(1) Evaluate the goals and importance of the up-front integrated design and planning to deliver green buildings.
(2) Distinguish the different choices in new construction versus renovations.
(3) Understand the importance and elements of the verification and auditing processes.

The best time for integrating sustainability into a construction project is early in the planning phase as a myriad of environmental impacts can be avoided or minimized. The second-best time is as soon as the project has started as possible. The planning process is connected from the societal level down to the individual site scale to the subsystems level as will be discussed here.

6.1 Site sustainable development

Drilling down from the regional planning scale, the next level of early focus before a construction or renovation project breaks ground is the sustainability of the site. Site selection is essentially not an option with existing buildings but is critical for new construction or additions. The US Green Building Council's (USGBC) Leadership in Energy and Environmental Design (LEED) v4 program focuses intently on sustainable sites and acknowledges that the building and site do not exist in isolation. Site concerns are crucial to the overall environmental integrity of the development process and often represent the most cost-effective efforts to make a high-performing building.

The first site issue for new construction in an urban area is the concept of infill development vs outlying areas that often can contribute to urban sprawl. Infill projects, particularly if they are live-work-play-shop mixed-use developments with high density and nearby amenities, dramatically reduce transportation-related emissions, giving residents the possibility to live car-free. They also save the development costs of bringing utilities and infrastructure to far-flung greenfield sites. Often, a redevelopment site is more centrally located as well, giving it transportation advantages in terms of transit, cycling, and shorter work commutes for employees. Avoiding the need for commuting and enabling alternate modes, particularly walking and cycling, links to the Location and Transportation category of LEED v4. Such central locations often have complexities and costs that outlying sites do not, including soil remediation. Barriers to success include potential local opposition as well as the issues associated with gentrification and pricing out long-time residents that may not have the means to continue living in a more costly area. To be successful and scalable, such projects require careful coordination between neighborhoods, developers, and municipalities.

Brownfield sites are typically urban-industrial sites that may suffer from a contamination problem (e.g., an EPA Superfund site), and their redevelopment may require remediation of the site for healthy human habitation. Grayfields are previously developed through uncontaminated sites, while greenfields are sites that are currently undeveloped and resemble most closely an undisturbed ecosystem. In general, brownfields > grayfields > greenfields when it comes to the sustainability of choosing sites for new construction. This, of course, is dependent on effective site cleanup. Taking a contaminated urban-industrial site, remediating the site, and returning it to functionally developed land is much preferred vs using an undeveloped "greenfield" site. For example, the remediation and redevelopment of a USEPA Superfund site is one of the most impactful (and often expensive) efforts that can be undertaken (Kruger and Seville, 2012).

The sustainable site plan should minimize impacts by considering the water, winds, sun, topography, and soils of a given site. Elements for sustainable sites include rainwater management, local habitat preservation, minimization of earth moving required for the site, hardscape design and placement, promoting onsite biodiversity, retaining open spaces, and enhancing natural water bodies. Hardscapes (without rainwater harvesting) result in enhanced site runoff, which can overwhelm systems used to manage municipal stormwater and degrade natural infiltration resources. This in turn can carry pollutants (e.g., leaked vehicle fluids, tire and brake materials, spilled chemicals) to receiving waterbodies. The planning and site development stages are when runoff, heat island contributions, light and sound pollution and many other problems can be minimized.

The use of natural features of the landscape to serve their intended function is emphasized with site planning. Or, as a reasonable alternative in some cases, the use of engineered systems to mimic these functions is the next best option. For example, the ideal is to preserve existing wetland areas and the ecosystem services they provide (Fig. 6.1). A second alternative may be to use constructed wetlands that are designed to manage stormwater flow, provide habitat, and use nature to treat stormwater flow, flood mitigation, or other purposes. When the functionality of wetlands is compromised, the results can be flooding as well as degraded water quality. Urban areas such as Houston, Texas, are often cited for the fraction of impermeable pavement covering previous wetlands, and this contributes to flooding issues in such regions. Appropriate zoning and floodplain restrictions are important to the planning efforts as well.

A certification program called Sustainable SITES Initiative (Table 6.1) was developed by the Green Building Certification Incorporated (GBCI) specifically to address site sustainability, irrespective of buildings that may be present (https://sustainablesites.org/). SITES focuses specifically on landscape attributes including how to transform landscapes into carbon sinks versus carbon sources. Though independent of LEED, the two programs work similarly and in a complementary fashion. Sustainable SITES applies to a wide range of developed and undeveloped sites including open spaces, local, state, and national parks, botanic gardens, arboretums, streetscapes, plazas, commercial sites, retail and office areas, corporate campuses, neighborhoods, individual yards, educational campuses, museums, and hospitals. It begins from the point of land development (or redevelopment).

The developed footprint on the site should be minimized as measured by the floor area ratio that takes the building footprint divided by the total buildable land on the site. This is more about properly sizing the building to the site to maintain natural areas than trying to maximize lot size and thus decrease development density. Building up rather than building out can help in this effort. Other aspects

6.1 Site sustainable development

FIG. 6.1
Management of stormwater runoff is an important site consideration that will be discussed in greater detail in Chapters 10 and 11 (public domain image from US EPA).

Table 6.1 Sustainable SITES credit categories and examples.

Section	Sustainable sites credit category	Example nonmandatory point subcategory	Example effort(s)
1	Site context	Context C1.5: Redevelop degraded sites	Perform site remediation before reusing formerly contaminated industrial sites
2	Predesign assessment and planning	Predesign C2.4: Engage users and stakeholders	Conduct a town-hall-style meeting to solicit input from site users and impacted stakeholders
3	Site design-water	Water C3.4: Reduce outdoor water use	Eliminate landscape use of potable, surface, or groundwater with substitution of collected rainwater, graywater, onsite treated water, condensate water, and other unconventional sources

Continued

Table 6.1 Sustainable SITES credit categories and examples—cont'd

Section	Sustainable sites credit category	Example nonmandatory point subcategory	Example effort(s)
4	Site design-soil + vegetation	Soil + vegetation C4.9: Reduce urban heat island effects	Use highly reflective roof and paving materials, onsite plantings and shade structures, and vegetated roof areas to reduce urban heat island
5	Site design-materials selection	Materials C5.4: Use salvaged materials and plants	Retain or reuse existing materials and plants already on-site
6	Site design-human health and well-being	HHWB C6.5: Support physical activity	Establish site trails and infrastructure such as bike racks to encourage site users to be physically active
7	Construction	Construction C7.6: Divert reusable vegetation, rocks, and soil from disposal	Reuse 100% of land materials disturbed during construction either onsite or within 50 miles, including noninvasive vegetation
8	Operations + maintenance	O+M C8.7: Protect air quality during landscape maintenance	Create low-maintenance landscapes; use non- or electric-powered landscape tools
9	Education + performance monitoring	Education C9.1: Promote sustainability awareness and education	Provide onsite interpretive and educational materials relevant to site sustainability efforts
10	Innovation or exemplary performance	Innovation C10.1: Innovation or exemplary performance	Exceed the site credit thresholds substantially or provide a new category of credit for SITES

that are important in the design and planning phase include the site lighting plan, where artificial lighting can be thoughtfully used to prevent "light pollution" from stray light escaping the site to maintain "dark skies" in areas outside of cities.

As discussed in Section 6.3, using and renovating existing buildings is usually preferable over new construction with all the waste generated, materials use, embodied energy and carbon, and other costs associated with new construction.

6.2 Global scale concerns to integrated design process

In 2015, the 193 members of the United Nations agreed to 17 Sustainable Development Goals (SDGs) that aim to end poverty, promote prosperity, ensure equality, and protect the environment (https://www.undp.org/sustainable-development-goals) (Table 6.2). These are a good framework to view a truly sustainable approach to construction, as all 17 SDGs arguably are indirectly related to green building. Over half of these are directly related to sustainable buildings including 6 Clean water and sanitation; 7 Alternative and clean energy; 9 Industry, innovation, and infrastructure; 11 Sustainable cities and communities; 12 Responsible consumption and production; 13 Climate action; 14 Life below water; and 15 Life on land.

Table 6.2 UN sustainable development goals and their relationship to green building.

Goal number	UN development goal description	Relationship to green building
1	No poverty	Access to quality housing for all incomes
2	Zero hunger	Urban gardening and food production
3	Good health and well-being	Healthy indoor spaces
4	Quality education	Educational facilities and infrastructure
5	Gender equality	Building design for gender equality
6	Clean water and sanitation	Onsite and offsite systems for water treatment
7	Affordable and clean energy	Onsite and offsite renewable energy systems
8	Decent work and economic growth	Buildings that meet the triple bottom line goals of environment + economics + equity
9	Industry, innovation, and infrastructure	Infrastructure designed for innovative green building systems and techniques
10	Reduced inequalities	Community infrastructure designed to meet the needs of all community members
11	Sustainable cities and communities	Development and planning on the community scale
12	Responsible consumption and production	Materials production and use for buildings
13	Climate action	Reducing built-sector emissions
14	Life below water	Onsite wetlands management
15	Life on land	Sustainable site development
16	Peace, justice, and strong institutions	Professional and accrediting organizations with robust systems
17	Partnerships for the goals	Teamwork focused on sustainable construction among architects, engineers, construction, trades, and other project participants

Efforts at green building have become more holistic and now consider the community planning aspects of building our infrastructure as a first-order concern. These are the first steps toward the planning, construction, and commissioning of green-built buildings. Urban density is desirable to avoid suburban sprawl and the traffic and other ills that it creates, ranging from air pollution, congestion, lost productivity, traffic accidents, and others. However, at the same time, a site should not be overbuilt with buildings and parking areas displacing all-natural features. A "concrete jungle" in an urban area is not the goal either.

The term low impact development (LID) denotes an approach that focuses on stormwater management. To protect aquatic ecosystems, LID tries to mimic natural hydrological processes to reuse stormwater where practical and to enhance infiltration and evapotranspiration elsewhere.

The smart growth movement embodies many of the paradigms for community-level planning (www.smartgrowthamerica.org). Its 10 core principles are as follows:

(1) Build mixed-use (e.g., residential, commercial, workplaces, recreation, outdoor spaces) neighborhoods for transportation efficiency.

(2) Promote compact urban areas favoring building up rather than out to reduce sprawl.
(3) Create a range of housing options for all lifestyles and economic means.
(4) Foster walkable neighborhoods to reduce congestion, environmental impacts, and improve fitness.
(5) Support unique community features including history, architecture, art, and open spaces.
(6) Natural areas preservation and accessibility in neighborhoods.
(7) Develop existing sites and cities rather than greenfields.
(8) Provide multimodal transportation options (e.g., walk, cycle, mass transit, passenger vehicle carpooling).
(9) Make the development process fair and transparent for developers to make the right choices.
(10) Involve all stakeholders and residents in community decisions.

The community-scale aspects of green buildings are integrated with zoning and codes. Historically, underserved communities suffered the brunt of the environmental impacts of global development. The fact that marginalized communities are typically on the south and east sides of urban areas is related to the impacts of contaminated wastewater (rivers in the United States run mostly north to south) and polluted air (in the US winds generally blow from west to east). The use of codes to exclude and marginalize groups, known as redlining practices, are abuses that must be eliminated due to their corrosive effects on those excluded. The lack of economic, social, and political power in marginalized communities enabled such abuses. It was always easier to locate an interstate highway next to a community of color, and some of them even bisected right through the heart of these neighborhoods (Lewis, 2013).

Several categories that merit points in green building certification programs require a level of community planning in addition to the site and building development. The "15-minute city" is the live-work-play concept where shopping, workplaces, recreation, and most of the necessities of life are available within a 15-min walk or bike ride. Though it does not eschew the option of private vehicles, it gives residents the option of forgoing the costs of vehicle ownership and maintenance. An example of a live-work-play urban development in Calgary, Alberta, is shown in Fig. 6.2. The residential towers are serviced by street-level shopping and park-like indoor–outdoor green space above this, which features the amenities of this development.

These include encouraging physical activity and enabling alternative transit beyond commuting in the single-passenger vehicle. Community development of shared assets such as walking trails, bike paths, and other outdoor recreation is part of this effort as well (Fig. 6.3).

An integrative design approach marries the designers, constructors, and users of a building early in the process for brainstorming, planning, and design decision-making. This includes architects designing the building and grounds, engineers specifying systems, the construction crew and trades, the commissioning agent, the operations management, and the end users. Often these players perform their subtasks in isolation, leading to a less optimal holistic design as subsystems do not integrate effectively.

The integrated design process (IDP) is often implemented in green building "whole building" design. The essence of IDP is that rather than subcontractors fulfilling their specific roles, the building design and delivery feature much more interaction and communication to creatively identify solutions that benefit multiple goals (Kibert, 2016). There are numerous systems in a building that benefit from IDP, and one of the key areas that transcends multiple disciplines is the design of the building's thermal envelope that separates the conditioned space from the outdoors. The National Institute of Building Sciences offers a Whole Building Design Guide in a web portal to guide this effort for the construction of federal facilities and other uses (https://www.wbdg.org/).

FIG. 6.2
Live-work-play residential development in Calgary, Alberta, featuring street-level shopping and services plus a mezzanine rooftop amenity level with workout space and shared greenspace.

FIG. 6.3
Outdoor amenities including green space such as this park near the Flatirons in Boulder, Colorado, can enhance sustainability on a community level by providing recreation plus walking and biking alternatives. They also entail hazards including wildland fire in the region termed the wildland–urban interface.

As discussed previously, the American Society of Civil Engineers frequently updates its "report card" on the status of large-scale infrastructure in the United States. The latest in 2021 shows modest improvements yet still an overall poor condition given a grade of C−. The deficits in maintaining and improving US infrastructure are a real concern and a ripe area for public and private investments.

6.3 Building renovation projects

The shiny new green buildings make the photo ops, but it is the hard-fought, messy renovation projects that win the war. Building renovation and reuse is vital to green building (Fig. 6.4). Arguably, the greenest building in the world is the one that already exists, particularly from a materials and embodied carbon perspective. The New London Architecture (NLA) group has a program called "Don't Move, Improve!" that focuses on the renovation of existing buildings. While some buildings merit major reconstruction or even replacement, the majority of the buildings that will exist in 2050 are already built. Moreover, our climate crisis requires decarbonization of our world in a short few decades, elevating the importance of the embodied carbon that is expended during construction or renovation. Almost invariably, the reuse of existing buildings has far less carbon footprint than a teardown and rebuild (King, 2017). It also allows energy upgrades that will reduce operating emissions as well. Thus, existing building renovations that enhance environmental integrity and reduce energy use using low-embodied energy materials are imperative. Just like any renovation, a green renovation can range from simple upgrades to deep energy retrofits.

Adaptive reuse is repurposing an existing building, often obsolete in purpose or functionality, for a new use. It obviates the waste associated with building demolition as well as much of the energy and materials associated with new construction. Historical buildings are prime candidates, and adaptive reuse often integrates both exterior and interior historical architectural features into the repurposed building.

A current repurpose solution in the commercial areas of urban areas has been repurposing unleased office space in major urban areas into new uses, including residential units. This has accelerated following the global pandemic and decreasing office space needs with increasing work-from-home options, though this may be relaxing with more "return to work" mandates. Urban areas such as San Francisco, New York, and Chicago face office vacancy rates of ~20% or greater after the pandemic. They also face affordable housing challenges, creating the opportunity to convert unused office spaces to residential. Projects that have been undertaken include the LaSalle Reimagined Initiative in Chicago.

Major challenges with renovating urban office towers into housing or other uses are real. These include access to the outside, operable windows, natural light, and the much more extensive distributed plumbing needed for residential units as compared with office spaces. Every point in a residential space should be within 30 ft of an exterior window, which was often the case with early 20th-century office buildings but less so with newer office buildings that have large floor plans, extensive indoor lighting, and nonoperable windows. This can be further constrained by differing local codes for residential and office space, in particular for seismically active areas. Many office buildings lack the plumbing and kitchen space needed in residential spaces while having a plethora of elevators that are not necessary in residential spaces. Creative solutions have carved out inner atriums that connect with the light and air of the outside world for interior spaces. These projects will continue to develop and will face significant challenges.

One key to thoughtful repurposing is to work reasonably within the constraints of the existing building, its structure, and the existing site. The more invasive a rehabilitation becomes, the more it will have cascading effects on other parts of the building, increasing both costs and impacts. This inevitably means compromises in terms of the optimal versus the practical design of the space that is being repurposed. Deep energy retrofits are the ultimate in energy efficiency efforts but are often prohibitive in terms of scale and cost.

FIG. 6.4

Renovation of a 1950s schoolhouse building reclaimed into a residence. "Outsulation" is reinsulating the exterior surface.

6.4 Case study: Ghirardelli Square, San Francisco, California

One of the early documented cases of an adaptive reuse project is Ghirardelli Square in San Francisco, California. Adaptive reuse repurposes a building or site that has outlived its original purpose, while its condition still offers a new life and purpose (Brand, 1994; Merlino, 2018). Ideally, it retains some of the original character and historic nature of a building while adapting it to a new use. Ghirardelli Square (Fig. 6.5) was redeveloped in the 1960s preceding the National Historical Preservation Act of 1966. This project opened in 1964 and was completed in 1968 when all the chocolate production moved off-site. It took a former chocolate processing facility and renovated this into restaurants and tourist attractions including a nearby cable car stop. It is a series of interconnected buildings connected by patios and staircases and following the topography of the hill sloping toward the shore. It also connects the waterfront to the surrounding Russian Hill neighborhood. It is widely considered the first successful adaptive reuse project in the United States, taking a former industrial site and converting it into commercial buildings. Ghirardelli Square was added to the National Historic Register in 1982 and is one of the most touristed and iconic sites in San Francisco.

FIG. 6.5
Ghirardelli Square in San Francisco, California, is one of the first documented adaptive reuse projects.

6.5 Building sustainable design: Appropriate sizing

A structure in its basic form needs to shelter the occupants from the elements and serve its intended purpose of a residence, business, factory, etc. This is the minimum, and good design needs to consider human needs, functional uses of the space, desires, and aspirations. The first order of business is right-sizing the structure, as generally all costs, materials use, and environmental impacts all scale with the building size. The elements to consider are, of course, the building uses, the site and scaling to it, the surrounding neighborhood and its scaling, and any planned future adaptations. Simple structures minimize materials use and facilitate low energy use and possible PV integration.

The world of architecture is focused on the design of the built environment although it also naturally links to engineering. Post-World War II prosperity for some decades allowed the design of buildings meant to oppose the local environment due to plentiful and cheap materials and energy resources. More recently, a better building approach is to work with the local climate and environment. The sustainable architectural design approach takes account of the effect of the environment on the building and vice versa (Kibert, 2016).

Over the last half century, buildings have increased in floor area (in square footage) (Fig. 6.6), the median price for new construction, and the associated environmental impacts. From 1970 to the present day, residential square footage increased from ~139 m^2 (~1500 ft^2) to approaching 230 m^2 (2500 ft^2) in 2020, while the average number of occupants has decreased (Mehta et al., 2018).

6.5 Building sustainable design: Appropriate sizing

However, homeowners are realizing that quantity does not equate to quality, and a countervailing trend of "tiny houses" and other smaller structures has taken hold in recent years as well. Per unit floor space, the cost, maintenance, environmental impacts, carbon footprints, and energy use for small dwellings can be significant. However, the overall magnitudes of the impacts as well as construction and operation costs are less for smaller buildings. Example 6.1 illustrates some of these trends.

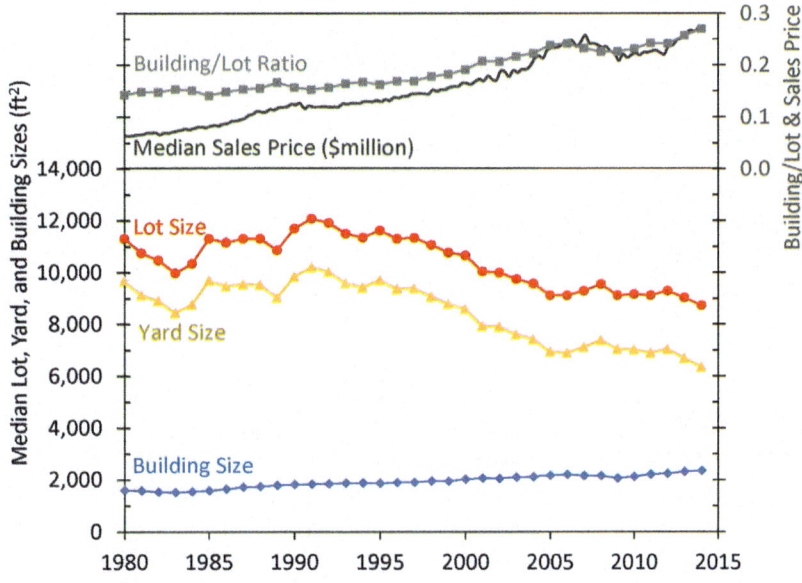

FIG. 6.6

US median residential lot, building, and yard sizes as well as building/lot ratio over time (plot by author and data from US Federal Reserve (Bowen and Li, 2017)).

EXAMPLE 6.1 HOUSING COSTS

Problem: What are the costs ($/ft^2) of US residential housing in 1980, 2000, 2007, and 2014? Assume constant exponential growth of floor area, cost, and cost/ft^2, and find the annual rate of increase r. Comment on the data.

Given: Data on residential costs
Find: $/ft^2 (1980, 2000, 2007, and 2014) and annual r for ft^2, $, $/ft^2
Assume: First-order growth
Solution:
Residential median square footage from Fig. 6.6 (source: US Federal Reserve Economic Data):
In 1980, the median square footage: 1624 ft^2
In 2000, the median square footage: 2050 ft^2
In 2007, the median square footage: 2170 ft^2
In 2014, the median square footage: 2366 ft^2

$$C = C_o \exp(rt)$$

$$r = \frac{\ln\left(\frac{C}{C_o}\right)}{t} = \frac{\ln\left(\frac{2366}{1624}\right)}{34 \text{ years}} = \frac{0.0107}{\text{year}}$$

Residential median sales price from Fig. 6.6 (source: US Federal Reserve Economic Data):

In 1980, the US median sales price: $64,750
In 2000, the US median sales price: $167,550
In 2007, the US median sales price: $244,950
In 2014, the US median sales price: $285,775

$$r = \frac{\ln\left(\frac{C}{C_o}\right)}{t} = \frac{\ln\left(\frac{285,777}{64,750}\right)}{34 \text{ years}} = \frac{0.0437}{\text{year}}$$

Residential cost per ft² of floor space (unadjusted for inflation):

1980 cost/ft² = $64,750/1624 ft² = $39.87/ft²
2000 cost/ft² = $167,550/2050 ft² = $81.83/ft²
2007 cost/ft² = $244,950/2170 ft² = $112.88/ft²
2014 cost/ft² = $285,775/2366 ft² = $120.78/ft²

$$r = \frac{\ln\left(\frac{C}{C_o}\right)}{t} = \frac{\ln\left(\frac{120.78}{39.87}\right)}{34 \text{ years}} = \frac{0.0326}{\text{year}}$$

Comments: The size of homes has increased (while the number of occupants has declined). However, the cost and cost per square foot have increased more rapidly than home size. For reference, the inflation rate was approximately 3.27% annually during 1980–2014 (according to the US Bureau of Labor Statistics data), matching the cost per square foot increase. The median home price has increased more rapidly than inflation, contributing to the home affordability crisis. As an update, the median home price in 2024 was $420,800, which follows a rapid increase during the pandemic and stagnation for 2 years, or ~3.87% annually since 2014.

6.6 Building as a system and net zero

One of the most vital concepts in building green and effective buildings is that a building operates as a complex system through which air, water, electricity, light, heat, and other material flows occur. The field of building science or building physics investigates these complex interdependencies. As discussed ahead, the many subsystems (HVAC, electrical, lighting, building structure, thermal envelope, plumbing, and control systems) all work in concert, making the building itself a complex integrated system or "organism" (Kruger and Seville, 2012). "Whole house" thinking recognizes the interdependency of these systems and consideration of the effects holistically. Changes in one realm are likely to have impacts in other realms. For example, retrofitting electrical service or the plumbing system may likely disturb the building's thermal envelope if not done carefully. As another, the improvement in air sealing of a structure may introduce indoor air quality problems, including radon, particulate matter, or mold and mildew.

Simplicity of design is a construction and operational benefit that can also lend itself well to building aesthetics. Although overemphasized, particularly in the real estate world, the aesthetics of a building are important to sustainable design. Visually striking, audacious, innovative designs have their place in new buildings if they serve a purpose. Good design provides form and function, as any architect will profess. A sustainable building may be a masterpiece of sustainability and include all the facets here, but if not designed to be attractive to buyers, it will not be as impactful (nor marketable). Predominantly, form will follow the function. Beyond that, the architect and engineers have a broad palette of choices in both the interior and exterior. A clunky building that is unattractive, looks too much like a Rube Goldberg contraption, or is too unique or specialized will have a limited audience.

A net-zero energy house produces as much energy as it consumes. The design phase is where a net-zero energy house is challenging enough for new construction and even more challenging for a renovation application. It is also possible to design a new or renovated structure that facilitates the future adoption of onsite renewable energy systems. California requires that at a minimum, new residential buildings are prewired to add a rooftop solar photovoltaic system in the future. A small fraction of "temples to sustainable design," though laudable, is far less important than the more massive adoption of greener-built starter homes or greener renovations. The "pretty good house" philosophy espoused on the Green Building Alliance's website applies to the renovation approach (Kolbert et al., 2022). Do not let the perfect be the enemy of major improvements. Work with the reality you have and move the efficiency ball forward but accept imperfections and non-idealities in renovation situations.

6.7 Building construction and manufacturing

The planning process is also the best time to consider and minimize the environmental impacts of construction practices as well. As discussed previously, the quantity of construction and demolition (C&D) waste is twice our municipal solid waste generation rate, according to the USEPA, and growing. Planning for a durable and adaptable building is probably the most consequential in terms of materials use. A building lasting centuries rather than decades will be subject to far less C&D waste generation as well as lower embodied energy and carbon from new materials. Its functionality must be capable of transformations over its lifetime.

Construction scheduling is a planning activity that can help minimize the footprint of the construction site (Pearce et al., 2018). Also, an appropriate staging area for construction materials helps minimize the soil erosion and runoff problems associated with construction sites. A thoughtful plan to minimize erosion and stormwater runoff with both practices and barriers is important. This can be very problematic for water treatment facilities, as even natural stormwater treatment can be overstressed by the runoff from construction sites. Less reliance on invasive "yellow metal" (i.e., heavy construction equipment) and instituting penalties for damaging vegetation are all worth considering at this stage (Pearce et al., 2018).

A recent focus has been advanced techniques for manufacturing components of buildings that can be rapidly assembled onsite. Manufactured components can follow the systems thinking approach by integrating the subsystems into a whole, resulting in less waste, better air sealing, and lower costs. Prefabrication of many elements of a building is a way to minimize waste rather than onsite "stick builds" which are custom onsite builds. This is becoming an area of significant startup company activity with prefabricated components including wall sections, plumbing, and electrical as well as 3D-printed components or entire structures. The final assembly of these components occurs quickly onsite. Where onsite construction occurs, the waste generation can be minimized by designing the building in "equal units" to avoid cutoffs, and onsite reuse and recycling of any materials possible. The repeatability and automation possible with manufactured homes provide not only the opportunity for reduced environmental costs but also significant cost reductions, which may help alleviate problems such as affordable housing and homelessness in urban areas. One constraint for buildings manufactured offsite is that the prescriptive building codes vary substantially by municipality. Part of this is unavoidable, as California has to build to earthquake codes, Florida has to build to hurricane codes, etc. For manufactured housing, though, this is a barrier equivalent to automakers having to custom-build to each county or city's highly variable regulations. As a result, other countries that have national-level building codes and/or performance-based codes are leading in manufactured housing.

As a further benefit, the scale of manufactured housing offers the scalability of green building solutions. For example, the builder Clayton announced in June 2024 that all its ~40,000/year manufactured homes, all energy star and zero energy ready homes, would use heat pump water heaters, representing a 30% boost in the market for those units. This also helps dispel the myth that green homes are either ultra-high-end or minimalist tiny homes.

6.8 Design for sustainable transportation options

Though the focus in this book is the built environment, this is vitally linked to other spheres, including the transportation sector. Much of our economy is associated with moving people and goods from one part of the built environment to another. Thus, the transportation sector is intimately connected to the built environment through public thoroughfares and transit modes and governed by transportation, building, and zoning codes. Transportation of goods, services, and people is important to consider in the planning phase whether in new developments or renovations. This was the philosophy behind the redevelopment of the Salesforce headquarters complex in San Francisco, which integrates a transit hub and elevated park-like greenspace (Fig. 6.7). It makes little sense to build a state-of-the-art eco-palace located such that each of its employees is incentivized to commute in a single occupancy vehicle 50 miles per day to work there. Here, a few of the key intersections of transportation with buildings are highlighted.

A deep dive into sustainable transportation is beyond the scope of this book. The most important aspects related to the built sector are to help transcend the past model of the single-passenger commuter in an internal combustion vehicle. Buildings can help facilitate the increasingly multimodal nature of commuting options including mass transit of all varieties, including electric vehicles ridesharing, van pools, walking, and cycling (Fig. 6.8). Other essential elements include electric vehicle charging, sheltered transit stops, showering and secure bike storage, and incentives for using these alternatives. These are featured prominently in programs like the Leadership in Energy and Environmental Design (LEED) program and the Sustainable SITES program, both of which will be discussed in more detail in Chapter 14 on the socio-economic context of green building.

Beyond commuting, the movement of freight is also important. Emissions associated with increased international trade and supply chain logistics are of growing importance though challenging to apportion to the producer vs. transporter vs. end-user (Davis et al., 2018). An emerging area of research involved "last-mile logistics" and how this impacts the environmental and broader sustainability of

6.8 Design for sustainable transportation options

FIG. 6.7

Photo taken from the LEED Platinum Salesforce Tower in San Francisco, California, showing an example of infill redevelopment. It includes a multimodal transportation hub with pedestrian/cyclist facilities, a high-speed rail station, and a bus level served by the cable-stayed bridge in the background. The green roof 4-acre park is elevated above street level and contains a walking/jogging path, yoga classroom, and extensive gardens (Fullmetal2887, CC BY-SA 4.0 (https://creativecommons.org/licenses/by-sa/4.0), via Wikimedia Commons).

the delivery of materials (de Oliveira et al., 2017). For the retail sector, the advent and growth of e-commerce over the last 25 years, along with delivery options including crowdsourcing, cargo bikes, drones, and other options, has made the sustainability of materials delivery a disrupted market and a subject of developing research (Correa et al., 2012; de Oliveira et al., 2017; Olsson et al., 2019).

Telecommuting work options have accelerated with the COVID-19 pandemic, and the proliferation of remote work options offers meaningful reducing commuting energy and time. The office spaces are often still available, open, and operational using energy, though in some cases office space can be decreased. With work-from-home efforts, the jury is out on how much this can reduce commercial energy use, albeit with an increase in residential use. A counter-trend of back-to-the office (BTO) orders has also emerged in 2024, potentially reducing the significance of remote work.

FIG. 6.8

In some municipalities, a significant fraction of commuting is via cycling (Bike to Work Day in Fort Collins, Colorado). Planning for alternative transportation modes is essential for green buildings. The Leadership in Energy and Environmental Design (LEED) program awards points in its green building rating system for making provisions for alternative transportation at the workplace including cycling and walking.

6.9 Case study: Fuel & Iron redevelopment in Pueblo, Colorado

An example of a commercial building renovation is the Fuel & Iron Food Hall redevelopment in the historic steel city of Pueblo, Colorado, USA (Fig. 6.9). Ostensibly, the green building features are not the focus of the project. The project, however, represents a model of adaptive reuse in a building that stood empty for half a century. The project is a classic infill project located in the downtown area of Pueblo, Colorado, which formerly was the steel production hub of the western United States. A 33,000-square-foot brick-and-mortar building, it was built in 1915 for the Holmes Hardware building after the original wood structure was lost to fire. The project focus is redeveloping incubator space for local restauranteurs to develop their businesses in a low-capital-cost space, much more cost-effective than a stand-alone restaurant building. Restauranteurs (approximately five kitchens) are leased commercial kitchen space for up to 3 years, sharing dining space, a central bar, and a coffee/desert shop. The Fuel and Iron project also includes low-income housing targeted toward restaurant workers as a live-work-and play development. The upper floors of the building will be affordable housing with ~28 units targeting the employees of the food hall. The income range of residents will be 30%–60% of the median local income. The facility opened in spring 2023 with all restaurants and most of the apartments leased. Locally grown produce will be a future emphasis, with a plot of land set aside for urban farming.

FIG. 6.9
The Fuel & Iron adaptive reuse of the 1915 Holmes Hardware Building in Pueblo, Colorado, while under construction. It is now used as a food hall for 5 restaurants on the ground level and 28 income-restricted apartments on the upper two levels.

6.10 Building orientation and passive solar considerations

Depending on the climate, a significant fraction of a building's heating needs can be supplied via proper passive solar design. Passive solar is not appropriate in all climates or locations, as locations closer to the poles or typically overcast locations can be far less reliant upon passive solar. The initial stages are the best time to incorporate passive solar features in the building, particularly the orientation and location of the building on its site, as well as the surface areas covered by glass.

An east–west orientation of the long axis of the house, significant window area on the equatorial (south side in the Northern Hemisphere) side of the building with deep overhangs, judicious window areas on the north, east, and west sides, and an open sloped roof area on the southside for potential solar collection are all desirable features (Fig. 6.10). This will be discussed in more detail in the section on passive solar.

Trees that will be preserved on the site and impact the shading of the building should be considered both presently and with future growth and maturity. Trees or other shade structures are often useful on the east and west sides of the structure to block the direct low-angle summer sun in the morning and afternoon. Deciduous trees allow more solar gain during the winter although the morning and afternoon gains on the east and west sides are small due to the arc of the sun in winter. Depending on the latitude of the site, an optimal range exists for the depth of overhangs that will allow direct winter sun while excluding the high summer sun. In the end, a balance has to be struck between the benefits of onsite, mature landscaping, and passive solar access.

Considering the site wind profile is a good idea, even if wind power generation is not feasible on the site. This is usually the case for most urban or suburban locations. Siting the building to avoid or shed the winter winds while allowing summer breezes during evenings to ventilate the home can reduce building space conditioning loads.

FIG. 6.10

"Earthship" house near Taos, New Mexico. Notice the east–west orientation with a large amount of window area for passive solar gain. The high summer sun is not excluded by overhangs to allow direct sunlight into the greenhouse integrated with the house.

6.11 Building codes, commissioning, and auditing of buildings

The first order is for buildings to meet all building codes, which can vary significantly depending on locality. The International Building Code (IBC) specifies many of these, of which local codes often derive. Nonetheless, building codes, including those focused on energy efficiency and green building, are locally driven, and it is beyond the scope of this book to consider this high variability. The codes are slowly evolving statutes that govern the minimum requirements a building must meet and focus on safety and operability. The prescriptions for energy efficiency (e.g., minimum insulation levels) have become more stringent over time though this varies based on locality and the climate zone.

All human systems and electromechanical systems are subject to flaws. Too often, installed equipment does not meet up to its specifications due to construction errors, inadequate quality control, or a range of other non-idealities. Commissioning involves a "pretest" of installed systems to verify their performance, usually at the closeout of building construction (Dykstra, 2016). Commissioning can be described as "passing off of the keys" and, more importantly, knowledge from the builder to the user. The process can help identify performance issues, "train" the owner and users on building operations, and allow them to make minor change requests. Increasingly, building codes as well as green building certification programs require commissioning as well as extensive documentation of the process and materials. Engineers use a variety of tools to aid in the commissioning process; these tools are also part of the typical energy audit process on existing buildings and systems.

Just as building commissioning can occur at any time both for new construction and retrofits, audits for energy use are becoming common throughout the building's lifetime (Krarti, 2011). The energy audit of a building not only helps with commissioning a building but is also a prerequisite for performing a green renovation project. Establishing a baseline of performance is important for verifying initial performance and for measuring improvements or degradation in performance over time.

The tools for conducting energy audits have evolved over the last decades to give diagnostic indicators on building performance. Some audit commonalities include utility analysis, a building walk-through, and cataloging of systems to determine the current status of building energy use and existing systems. Much can be learned from examining the building energy and water usage and how these vary over a year. A range of other both diagnostic tools (blower door and thermal imaging) and modeling tools (Manual J calculations) are available for the auditing process and are discussed later with respect to the building thermal envelope.

6.12 Design for future adaptability

The life cycle and adaptation of buildings over time are examined at length in Stewart Brand's book "How Buildings Learn" (Brand, 1994). Well-constructed buildings, with lifetimes extending to centuries, adapt to their climate, changes in culture and social norms, economic conditions, surrounding neighborhood characteristics, and a myriad of other factors as human uses of the building evolve.

At some level, this can be planned from the start by making the design as flexible as possible (Kruger and Seville, 2012). An example is the design of a residential structure that could be easily adapted to the stages of occupants' lives from a new relationship to retired empty nesters. Flexibility in room use is one key. Adapting a children's bedroom to a guest room or office is not an uncommon transition.

It is unfortunate and somewhat avoidable to be forced to abandon a family home for changing lifestyles and needs. Adaptable floor plans with handicap accessibility features include sufficiently wide hallways and doors, minimizing stairs, and other planned details. A single-level residence is one example of a more easily adaptable building, but this can be used in multistory buildings as well. If designed appropriately, adaptable buildings can be designed to both accommodate and change over time to allow aging in place where occupants can remain living in the same home with minor modifications. These same principles apply to commercial buildings, which may change needs more frequently.

Universal design principles are a good starting point to make the building accessible to humans of all ages and physical and mental characteristics (Andreae et al., 1998). The Americans with Disabilities Act prescribes ADA design principles that incorporate accessibility in the broadest sense. Accessible fixtures in bathrooms and kitchens and the use of safety bars and grab handles are important. This includes things such as minimizing trip hazards and making switches, handles, and appliances accessible. Open, larger spaces, doorways, and hallways are more easily navigated than numerous closed-in smaller rooms. One of the early proponents of universal design for accessibility by everybody was Marc Harrison who pioneered such design in manufactured houses in the 1970s.

6.13 Chapter summary and conclusions

The design phase of the project is a critical first step to facilitating a green design and implementation. This occurs via implementing universal design principles at scales ranging from regional/urban scale to an individual structure. An ideal integrated design process incorporates all the trades—architects, engineers, general contractors, installers, interior designers, landscapers, and end users—early in the process to optimize the many subsystems that comprise the whole-house system. Considering the site's properties and how a building integrates with the site is a vital early step. This includes proper building sizing to its task, building orientation, integration with the site and landscape, and consideration of how the building connects to the outside world, such as transportation. Despite recent interest in "tiny houses," the trends have been toward larger houses and smaller lots with increasing urbanization.

Sustainability is not enhanced by design that is awkward, short-lived, flimsy, unaesthetic, high maintenance, nondurable, uncomfortable, inflexible, unhealthy, poorly functioning, or unconnected to the surrounding environment or community. An ideally sustainable design will be appropriately sized and oriented properly onsite, in a reasonably high urban density redevelopment (or infill) site, near transportation and utilities, and have natural amenities onsite or nearby that can be integrated into the design. It should avoid culturally, archeologically, or naturally a protected area, be outside the 100-year (and preferably even the 500-year) flood plains, have access to renewable energy sources, and be designed to be efficient, comfortable, and aesthetic. Sustainable developments are more often mixed-use where residents can work, live, shop, and play within a single developed area to avoid the need for commuting and enable alternate modes, particularly walking and cycling. Renovation projects also invoke many of the same considerations though many of the decisions regarding the building and its site are locked in. A range of certifications, including LEED and Sustainable SITES programs, focus on these site issues, and will be discussed in Chapter 14 on socio-economic aspects of green building.

6.14 End of chapter exercises

(1) **Concepts:** Describe the groups that would participate in a design charrette for new building construction. What are some of the questions that would be discussed?

(2) **Concepts:** The former Stapleton Airport site in the urban center of Denver, Colorado was redeveloped into a residential community after Denver International Airport opened. Devise a two-page green site plan for a residential lot in this redevelopment project

(3) **Concepts:** Discuss several of the challenges with converting all of the now unused office space in major commercial centers to residential housing units.

(4) **Concepts:** Earthship architecture, popular in Taos, New Mexico, among other places, is an off-grid residential building approach. Discuss three unique green features of the Earthship construction.

(5) **Concepts:** Identify an adaptive reuse project, preferably in your community. Discuss its original and repurposed uses, including the adaptations needed to make this happen.

(6) **Problem:** You have a hemispheric-shaped building (radius 20′) vs a square footprint building with the same floor area with 10′ wall heights. Compute the surface area exposed to the atmosphere and the SA/Volume ratios. Considering strictly heat loss and considering all else equal, which would you prefer? Also, find the floor area to surface area (FA/SA) ratio.

CHAPTER 7

Materials use in buildings: Choosing the best options

Learning objectives

(1) Comprehend the immense use of materials in the built environment and the factors that drive their choices.
(2) Recognize the vital role of concrete in our world, its environmental impacts, and the ways to reduce them.
(3) Examine ways to minimize embodied energy, embodied carbon, and construction use related to material choices.

7.1 Overview of materials

Essentially, the materials' mass on Earth exists and will simply be transferred from one form to another due to chemistry and phase changes. There are minor exchanges with the rest of the universe, but otherwise, the transformations are simply changes in molecular form and/or phase. Mass balance may be applied from the microscopic to the building scale to the planetary scale. It is from this existing mass that the materials that we construct our buildings with come. In the big picture, these can be divided into materials that are grown (organics) or materials that are mined (minerals, metals, inorganic compounds).

As with an energy footprint, a given activity has a "materials footprint" as well. Buildings are responsible for about half of the total material use in society. Much of the remainder of the mass could be described as the stuff we put in and around buildings and move between them. Embodied carbon emissions follow suit, with the fraction of embodied CO_2 emissions associated with the built sector material uses at ~100% (concrete), ~59% (iron and steel), and ~41% (aluminum). There are currently extensive efforts across all materials industries to enhance the sustainability and ultimately the long-term viability of construction materials.

Overall, the global annual materials footprint totals an estimated 70 Gtonnes (Gt) annually in 2008, or approximately 10 t/capita/year (Wiedmann et al., 2015). As a benchmark, the current emission of CO_2 is approximately 35 Gt/year (or ~10 Gt/year of carbon when the oxygen is subtracted out). These are exceptionally large mass flow rates! Although not exactly linear, for every 10% increase in GDP, a nation uses 6% more materials on average (Wiedmann et al., 2015).

Much like the energy use per capita, the material use per capita also has a direct relationship with the Human Development Index (HDI), which combines economic, social, and environmental factors. However, the variation in material usage per capita in countries with high HDI is quite large.

Despite some nations reducing the material intensity of their economies, the question of whether overall material use and the global economy can be decoupled (i.e., grow economies without growing material use) is still largely unanswered (Wiedmann et al., 2015). The decoupling, or a relative decoupling (fewer materials per unit GDP) in select countries has occurred but is difficult to verify, often related to the role of changing international trade in goods and raw materials (Wiedmann et al., 2015).

The mass in constructing the average home is strikingly large, particularly when including the foundation mass, which is often concrete. The National Association of Home Builders (NAHB) and the US Department of Energy (DOE) estimate that construction of an average 2000 ft^2 home uses 200 tons of materials and produces 8000 lbs of construction waste. As a construction rule of thumb, the weight of materials is 200 pounds per square foot for a single-level home, 275 for two levels, and 350 for three levels. This includes the foundation but not especially heavy features such as tile roofing or extensive masonry work (Kruger and Seville, 2012).

Material properties profoundly impact the structural characteristics, durability, embodied carbon and energy, ongoing energy use, and overall impacts and sustainability of a building. Detailed material characteristics (strength, density, degradation, etc.) are an immense field, and this chapter will only provide an overview with a focus on sustainability. The use and properties of materials as they relate to the building's thermal envelope will be covered in Chapter 8.

7.2 Materials attributes and sustainability

Sustainability will, of course, form a complex matrix of attributes to consider along with other material features such as tensile and compressive strength, strength-to-weight ratio, elasticity, durability, flammability, acoustics, and of course material cost and availability. The discussion focus here is on the sustainability of various materials rather than the typical material attributes the engineer would consider. Sustainable material choices involve minimizing external impacts—environmental, social, and economic. It is often stated that when building bikes, choose two out of three of the following: lightweight, strong, and cheap. For buildings, the green aspect adds another dimension to these three, driving the imperative for materials engineering.

In terms of green attributes, the following are preferable: (a) low-embodied energy; (b) high recycled content; (c) highly recyclable; (d) locally to regionally produced; (e) "healthy" and thus low emissions in terms of off-gassing; (f) highly durable and maintainable; (g) contributing to low operational energy use in the final product (high R-value, thermal mass, reflectivity, or some other property depending on the application); and (h) adaptable to future reuses as well as a changing climate. Some of the other key attributes of sustainable material use include:

- The use of local materials (and contractors) is a priority and typically enhances the sustainability of buildings by minimizing transportation costs and emissions.
- The longevity and robustness of materials and their low-maintenance functionality are key attributes.
- Materials can help or hinder indoor climate control. Thermal mass and hygroscopic mass materials absorb and emit enormous quantities of heat and moisture, respectively, and thus modulate the indoor temperature and humidity within a given comfortable range. For example, a large thermal

mass is provided by rammed earth waste tires and other waste materials. An even greater exchange of energy involves phase-change materials, which are becoming more common in building applications.
- The heat transfer properties of materials, including conductivity, density, permeability, void space, and thermal storage, are a key consideration and will be discussed in more detail in the chapter on the building thermal envelope.
- The use of materials can enhance or reduce the production of solid waste. As a rule of thumb, approximately 4 lbs of waste material is produced per square foot of floor area during construction (Magwood, 2017). Conversely, the use of waste materials can be a sink for such materials rather than a source such as "Earthship" homes. Zero waste usually translates into ~90% diversion of waste from landfills in practice.
- The energy and carbon footprint of materials can be quantified and minimized via life cycle analyses. For construction materials not associated with mechanical, electrical, and plumbing (MEP) systems, the footprint is almost always embodied energy and carbon rather than operational.
- Carbon-sequestering materials are preferred rather than ones that contribute embodied carbon emissions associated with their production and transport. Several of these will be discussed here, such as timber-frame buildings, cross-laminated timber, and CO_2-sequestering concrete.

7.3 Embodied energy and embodied carbon

Building embodied energy is a key metric for material use and gauging the impacts of a building. The energy use of a building is divided into operational and embodied energy. For mechanical, electrical, and plumbing (MEP) systems, often the operational dominates, while for general materials, the embodied energy is paramount. Embodied energy is the energy associated with the extraction, fabrication, transport, construction, and disposal of the materials that go into a structure (Table 7.1). The energy use (and carbon emissions for embodied carbon) occurs upfront rather than the operation over the building's lifetime. Life cycle assessment (LCA) is the method for evaluating the embodied energy and/or carbon of a product from cradle to grave (Kruger and Seville, 2012).

The operational energy of the building has been the past focus, and for good reasons, as the average building approaches 90% of its energy consumption via operations. In a world where climate change demands we reach net zero CO_2 emissions over the next several decades, embodied energy and carbon take on new importance. Much of the impact of materials comes from the extraction and processing of the materials. Mining operations typically use energy-intensive machinery, high-temperature processes, or aggressive chemical use for the processing of mined minerals. As operational emissions are further reduced, the embodied emissions associated with building materials become more significant, albeit harder to measure (Dixit et al., 2012). This also underscores the importance of the durability of buildings to avoid the upfront embodied energy and carbon of new construction (Fig. 7.1).

The recent Inflation Reduction Act of 2022 has a specific "buy clean" focus on reducing emissions associated with building materials at their manufacture, particularly concerning the top two embodied carbon materials of concrete and steel. It set into place provisions for federal procurement and details on product declarations to help consumers choose materials with lower impacts. Concrete and steel will be discussed in more detail as follows.

156 Chapter 7 Materials use in buildings: Choosing the best options

FIG. 7.1

The longevity of a structure (built in 1717) and the use of local materials that are low-embodied energy and carbon, such as adobe, are desirable properties associated with the mission at the Pecos National Monument near Santa Fe, NM.

A recent graphic from the IPCC shows the carbon footprint of various building materials (Fig. 7.2). Notably, wood products represent significant embodied energy. Depending on the details of their growth and harvesting, they can be net carbon sequesters or emitters. Steel and cement are carbon-intensive building materials that are a part of the "difficult to eliminate" greenhouse gas emissions (Davis et al., 2018). Concrete requires both significant fossil energy in its production and CO_2 emissions in the concrete curing process, i.e., the chemistry of turning limestone into quick lime, which releases CO_2 mostly related to Portland cement.

Manufactured masonry products such as concrete, brick, and blocks have significant embodied energy and carbon. Adding to this tally is the scale of use for these materials, which makes them a significant emissions source globally (Fig. 4.4). Metals such as steel and iron have both the highest per kg embodied energy and carbon. Concrete and steel are each responsible for approximately 5%–8% of global greenhouse gas emissions. The significant ranges of values for a given material represent where and how the raw materials and products are produced and delivered. A compilation of approximate embodied energy values for selected materials is given in Table 7.1 from data taken from Kibert (2016).

7.3 Embodied energy and embodied carbon

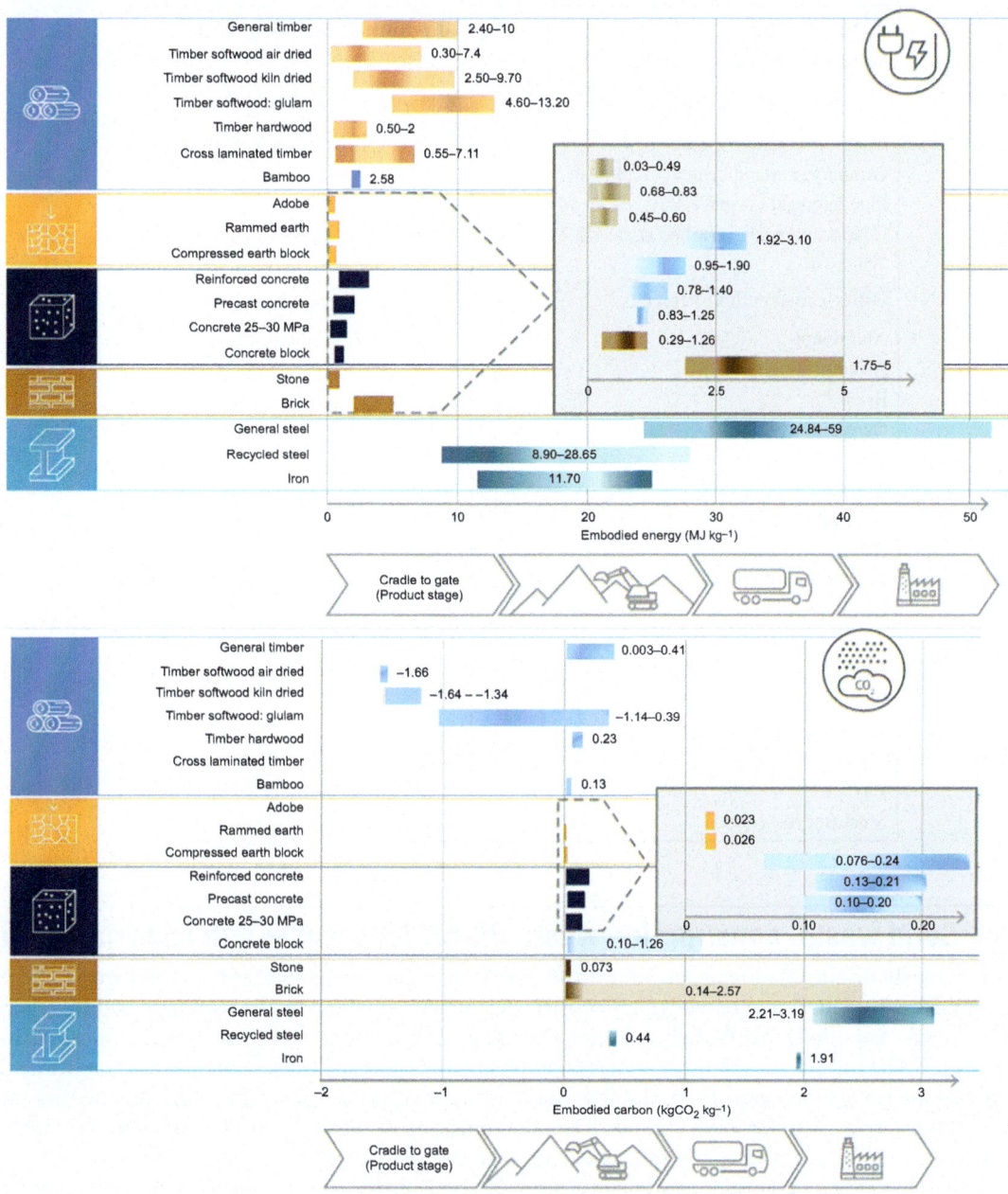

FIG. 7.2

Embodied energy and embodied carbon for several common building materials (used with permission originally Fig. 9.9 from the IPCC AR6 (Cabeza et al., 2022)).

Table 7.1 Data on the embodied energy of common building materials (Kibert, 2016).

Material (mass fraction of concrete)	Embodied energy in MJ/kg
Concrete (average mix)	0.95 (equals 817,000 BTU/t)
Ordinary portland cement (12%)	7.322
Fine aggregate (sand, 34%)	0.0488
Coarse aggregate (crushed stone, 48%)	0.116
Water (6%)	0
Other construction materials	**Embodied energy in MJ/kg**
Aluminum	227 (8.1 for recycled aluminum)
Asphalt shingles	9
Brick	2.5
Carpet	148
Concrete	1.3
Copper	70.6
Drywall (gypsum)	6.1
Fiberglass insulation	30.3
Linoleum	116
Lumber (general)	2.5
Mineral wool	14.6
Paint	90.3
Particle board	8
Polystyrene insulation	117
Plywood	10.4
PVC	70
Steel (recycled steel)	32 (8.9)

7.4 Solid waste, construction waste, hazardous waste and material reuse

As previously discussed, municipal solid waste (MSW) includes the nonhazardous solid waste generated by the residential, commercial, and industrial sectors as governed by the Resource and Recovery Act (RCRA) Subtitle D. MSW includes products and packaging, food waste, and yard waste as major categories. The old-school solution was a regular pickup by a garbage truck, which hauled the waste to a dump (unlined in the early days). Solid waste management in the last half century has become far more sophisticated with far more options. Also, the transition occurring is from a once-through, linear process to a circular economy where waste materials are cycled back into a process or building.

Construction and demolition waste and debris (C&D) is a special category of solid waste that has taken on a new importance to solid waste management and green building. C&D waste falls mostly under the nonhazardous waste category, similar to municipal solid waste but differing in composition (Fig. 7.3). It includes waste from buildings (32% of the total), infrastructure like roads and bridges (44%), and other construction sources (23%). The US generation rate in 2015 was 548 million tons, roughly twice the MSW generation. The overall mass is dominated by concrete (70%), asphalt paving (15%), wood (7%), and other materials (8%) such as drywall, plastics, metals, ceramics, etc. Roughly 95% of the C&D waste is from demolition rather than scrap material from new construction (USEPA, 2019) (Fig. 7.3).

C&D waste has difficulty with material reuse and recycling in part because USEPA data shows that ~85% of C&D waste is concrete in various forms (Fig. 7.3). This is logical considering ~30 GT of concrete is produced annually, and it is only exceeded by water in its global use. However, more sophisticated techniques for managing C&D waste are emerging, and approximately 75% is diverted from landfills. Reuse options for concrete include materials for erosion control and base materials for roads. Recycled aggregate concrete has durable reuse possibilities as the coarse aggregate in the production of new concrete, particularly for nonstructural applications (Akhtar and Sarmah, 2018; Guo et al., 2018).

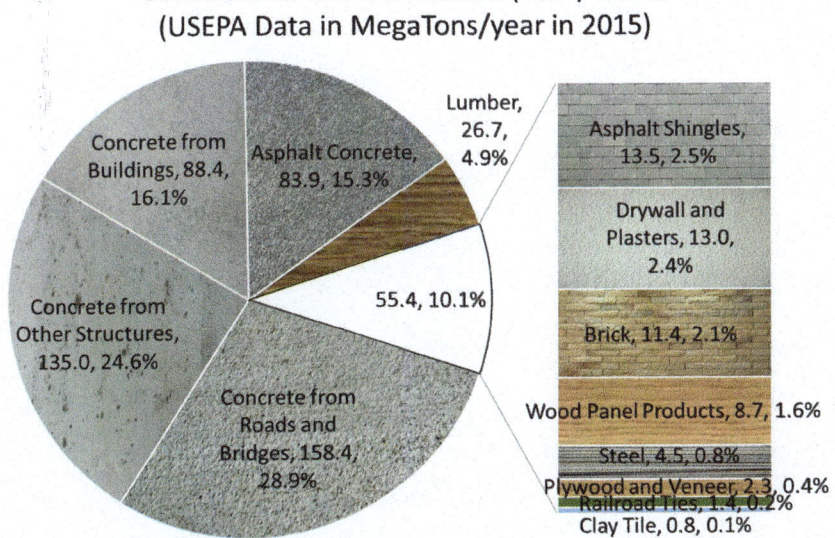

FIG. 7.3

US Construction and Demolition (C&D) Waste Composition Breakdown in Tons/year (Total = 0.548 Mt/year, data from US EPA).

Another effort to minimize C&D waste at the end of the life cycle is to focus on the deconstruction of buildings rather than demolition. Although it can increase the time commitment, a fraction of the building materials in a building under deconstruction can be repurposed. An example (Fig. 7.4) is a Works Public Administration (WPA) building that was built in 1936 and used for more than six decades as an elementary school in Socorro, NM. The building and two other associated buildings had been vacant for 25 years and had deteriorated beyond use. Deconstruction efforts can be lengthy, painstaking, and expensive compared to simple demolition. These issues are compounded when dealing with hazardous materials such as asbestos, lead-based paint, and other hazards such as still-active extensive gas and electric utilities. The complications with this project meant it cost ~$1 million for three buildings to be removed and the site renovated. Materials that were saved included large-span ceiling joists from the gymnasium as well as assorted scrap metals. After remediation and demolition, the site must be regraded and drainage provided before it can be potentially reused. The wall structure of the original building shows the use of adobe bricks in the wall structures (Fig. 7.4).

FIG. 7.4

Deconstruction in 2023 of a WPA-era elementary school Adobe block building from 1936 in Socorro, NM. The building had deteriorated and was beyond repair and the costs to upgrade.

Most construction waste material can, in theory, be diverted from landfills. Depending on the material, the diversion of waste materials into new products by recycling often includes lower energy use and GHG emissions as compared to processing virgin materials. The energy reductions and GHG benefits of diverting waste are calculated in the EPA WARM model (www.epa.gov/warm). Rather than creating solid waste that needs to be transported and disposed of off-site, a range of possibilities exists for reuse, repurposing, recycling, and other diversions of this waste stream (Fig. 7.5). Some items, such as doors, windows, appliances, and hardware fixtures, can be reused elsewhere. Several companies are pursuing material reuse, with Habitat Restore associated with Habitat for Humanity as a most notable example. Much of the paper, cardboard, aluminum, glass, and some plastics can be recycled through municipal recycling programs. Scrap wood that cannot be reused can be chipped to provide onsite mulch for landscaping needs. Concrete and masonry waste can be used to some extent as an alternative coarse aggregate in new concrete, as rubble foundation for fences and walls, drainage material, or fill material. Drywall scrap that cannot be repurposed can be depapered as a soil amendment.

FIG. 7.5

Earthship architecture near Taos, New Mexico, includes some of the earliest and most effective sustainable buildings constructed using waste materials (note the rammed earth scrap tire foundation and glass bottle wall), passive solar architecture (long access oriented east–west with large window area), solar PV on the roof, automatic ventilation, and integrated greenhouses for growing food.

A number of high-profile hazardous materials discoveries have been associated with materials and buildings, including asbestos, radon, and lead-based paint. Treated wood products, particularly from decades past, are another hazardous material. Creosote, a liquid containing over 100 identified organic carbon compounds produced from coal tar, is used to soak railroad ties, and pressure-treated wood using pentachlorophenol or chromium copper arsenate are two examples that are far less common. Other hazardous materials, such as lead-based paint and asbestos (e.g., in insulation or flooring tiles), were eliminated decades ago but are still found in historic buildings. Less hazardous compounds are still present, including volatile organic carbon compounds (VOCs) from carpets, paints, and adhesives, which are discussed more in the indoor air quality section later in the book. Exposures to hazardous substances in buildings have multiple pathways, including volatilization from solid surfaces or from soil vapor penetration to a building, ingestion or skin exposure to contaminants in the water supply in a building, and direct contact with hazardous building materials (Fig. 7.6). Exposure pathways include contact with contaminated materials and soils, leaching of the waste material into surface or groundwater, followed by ingestion, and respiration of volatilized or aerosolized hazardous material.

The International Living Futures Institute maintains a Living Building Challenge "Red List" of chemical compounds to avoid in building materials (https://living-future.org/red-list/). This list of over 800 individual compounds includes numerous chlorinated compounds, *per-* and *poly-*fluorinated alkyl substances (PFAS), asbestos, certain VOCs such as benzene and formaldehyde, heavy metals such as lead and mercury, and select organic carbon compounds. The building sector has sought to minimize use and reduce the hazards associated with building materials used. This included low VOC coatings, less hazardous wood preservatives, and eliminating benzene and urea formaldehyde from materials, among others. These efforts are important to continue to promote healthy indoor environments and will be discussed later.

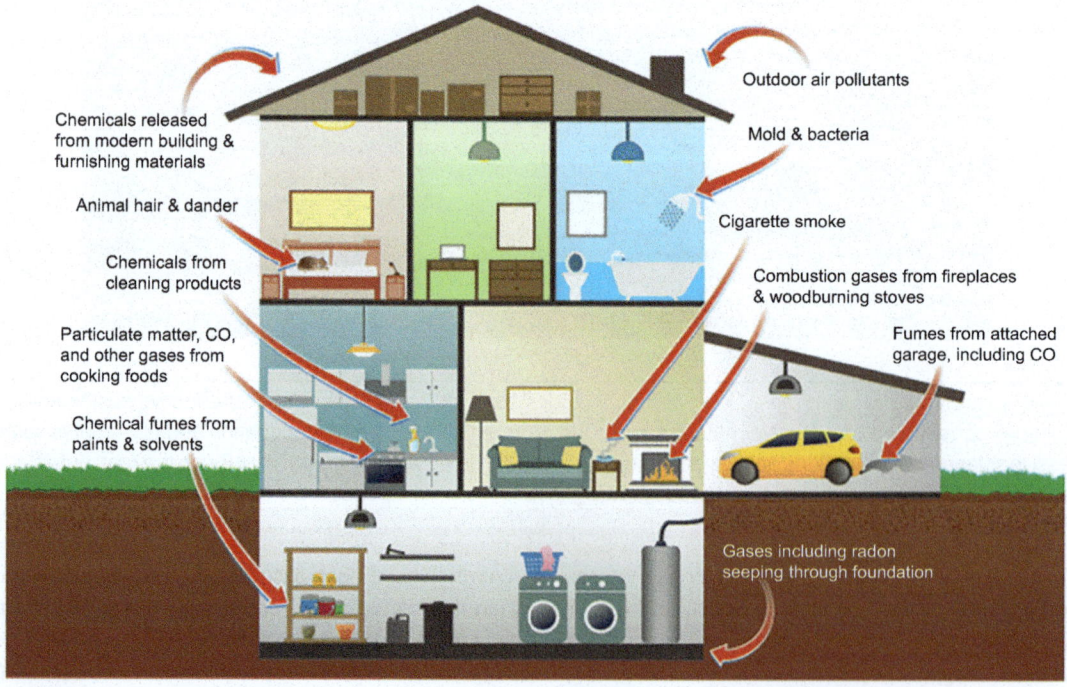

FIG. 7.6

Exposure sources and routes to hazardous substances in an indoor building environment from the EPA Exposure Tool.

US EPA public domain image available at: https://www.epa.gov/expobox/exposure-assessment-tools-approaches-indirect-estimation-scenario-evaluation.

7.5 Concrete and cement: Scale, composition, and impacts

It is hard to overstate the importance of concrete to construction and the climate impacts due to its vast global use (~30 Gt/year). The massive use is illustrated by some of the largest public works projects, e.g., the Hoover Dam alone required 3.3×10^6 m^3 of concrete, then a world record. China's Three Gorges Dam on the Yangtze River was an order of magnitude larger with 27.2×10^6 cubic meters of concrete. Concrete has distinct strength and formability advantages as a construction material; it is a durable material that can be both structural and a finish material (e.g., polished concrete floors).

Sand (fine aggregate), water, Portland cement, and coarse aggregate such as gravel are the main ingredients of concrete (Table 7.1). Hence, water is itself a construction material. However, since water is more of a "consumable" in buildings and is important in its own right, it is discussed in a later chapter. By way of its use in concrete, sand is a material with its own concerns, as discussed later in this section.

In terms of greenhouse gas emissions, CO_2 emissions from concrete are the third largest contributor after fossil fuels and land use change (Andrew, 2018). Moreover, approximately two-thirds of concrete-related CO_2 emissions have occurred since 1990. Ordinary Portland Cement (OPC) predominates its embodied energy and carbon. OPC has two major contributors: (a) the high-temperature kiln that requires fossil fuels and (b) the chemical processing of calcining that releases CO_2 when $CaCO_3$ is converted to lime CaO. Cement use in concrete currently exceeds 4 Gt/year globally (Andrew, 2018, 2019), or roughly 10% of the concrete mass. Thus, at an average of 0.75 kg CO_2/kg cement produced, it contributes approximately 3 Gt/year of the 40 Gt/year total CO_2 emissions, or roughly 8%. This would rank cement as third, behind only China and the US if it were a nation (Lehne and Preston, 2018).

An accelerating global effort has been multifaceted in making concrete more sustainable in recent decades (Damtoft et al., 2008; Meyer, 2009). The first US low-carbon concrete code was enacted by Marin County CA in 2019 (Andrew, 2019) (https://www.marincounty.gov/departments/cda/sustainability). Research into the use of waste materials and low-carbon concrete has proliferated in the last decade to reduce emissions associated with concrete (Siddique and Cachim, 2018) and includes:

- Reduction in overall concrete use by increased efficiency onsite and in production
- Reduction in the concrete quantity needed in a project and in particular the more carbon-intensive high-strength concrete
- Process changes to lower the temperature and batch time with OPC to improve the efficiency of production processes
- Substitution of larger fractions of hydrated lime or Supplementary Cementitious Materials (SCMs) in place of Portland cement as dictated by the strength needs of the concrete mix
- The reduction of water use for concrete and cleanup activities
- Injection of CO_2 into the concrete versus the typical air entrainment
- Use of waste materials and recycled concrete materials for some fraction or even all of the coarse aggregate
- Use of prefabricated components (wall, roof, and foundation systems) to allow more rapid assembly onsite as well as tighter buildings and lower embodied carbon.

Researchers have multiple constraints and focus on the quantity, strength, durability, workability, as well as environmental attributes of the replacements for Portland cement. Many universities and companies, both startup and established, are working on better concrete mixes to minimize emissions and water use, all while maintaining concrete strength and workability. Much of the focus is on replacing Portland cement with supplementary cementitious materials (SCMs). SCMs include fly ash, blast furnace slag, pulverized calcium carbonate, natural pozzolans from volcanic rock, and silica fume, involving two industries that are subject to climate concerns themselves (coal-fired power plants and steel production). Appropriately specifying concrete strength and quantity is also key to not over- or under-designing the structural components. Process modifications can substitute lower carbon fuels and/or reduce the temperature in the rotary kiln.

Concrete, in addition to its carbon footprint, also uptakes atmospheric CO_2 over its lifetime. This happens as the lime component reacts to form $CaCO_3$, offsetting some of its carbon footprint and offering hope for exploiting this process for carbon capture (Xi et al., 2016). Many of the innovators pursuing low-carbon concrete seek to minimize the release of CO_2 in production and maximize its uptake over time (Table 7.2), although the research is in its infancy. Laboratory research has sought to replace the majority of Portland cement with coal fly ash, given the right chemistry tweaks to maintain its strength and durability. Still, others are working on recycling concrete to produce new Portland cement using electric arc furnaces powered by renewable energy sources. The landscape of innovation in the area of reimagined concrete is vast.

Table 7.2 An incomplete list of companies pursuing novel concrete mixtures to help reduce greenhouse gas emissions and environmental impacts.

Company	Concrete-related product
Osto	Recycled plastic waste as aggregate to make carbon-negative concrete
Pozzitive	Post-consumer glass as a replacement for some fraction of the Portland cement
Hempitecture	Hempcrete uses hemp fiber as a bio-based additive to concrete
Sublime systems	Electrochemically driven cement production process rather than a kiln process heated by fuel combustion
Carbon cure	Carbon-sequestering concretes using various chemistries including using CO_2 injection into new concrete
LC^3	Limestone calcined clay cement, where SCMs include limestone, clay, gypsum, and fly ash. These SCMs minimize the downsides of an individual SCM.
Blue planet CCS	Produces geomass $CaCO_3$ using nonpurified CO_2 to grow synthetic limestone for use as aggregate at the rate of 0.5 ton of CO_2 sequestered per ton of concrete
Acrete PTE	Uses much larger fractions of fly ash for the cementitious material
Brimstone	Silicate-based cement that is carbon negative

Reimagined concrete also takes the form of prefabricated units (Fig. 7.7) that (a) minimize the amount of Portland cement to reduce the carbon footprint, (b) use recycled polystyrene or other lightweight materials in the mix to increase their insulating value, and (c) are often lighter and sometimes cheaper than poured-in-place concrete. The blocks that can be built into structural walls can often be easily carried by one to two construction workers.

Other interesting low-carbon materials now available are non-kiln dried bricks, K-Briqs, produced by Kenoteq in Scotland that have less than 10% of the carbon footprint of conventional bricks and are produced from construction waste. Others abound, produced by small startup companies, and it remains to be seen which of the emerging concrete alternatives will take hold in the market based on structural, insulating, cost-effectiveness, code compliance, and scale-up potential.

Many other alternative materials are emerging, including "hempcrete" which is a mixture of hemp plant core with lime to make a material that is concrete-like in appearance (US Hemp Building | Building a Sustainable World (https://www.ushba.org/)). Although currently less strong (~3 MPa) than traditional structural concrete, it can be used in light or non-structural uses (e.g., cladding, partition walls, and others) and has a much higher R-value than pure concrete. The hempcrete material also is a carbon-sequestering material and thus reverses some of the GHG impacts of traditional concrete.

Permeable concretes are another area of innovation that simultaneously decreases roadway surface water runoff and flooding potential while improving groundwater recharge and water quality (Martin et al., 2018). Permeability is increased by decreasing the fine aggregate used in the mix (e.g., less sand). They are more costly pavement surfaces and are mostly suitable for low loads (minor roads and parking lots). There is potential for project costs to decrease if stormwater infrastructure is reduced or avoided. The permeable concrete can filter sediment, reduce runoff erosion, capture some chemical contaminants, and reduce the size of stormwater runoff infrastructure needed. Additional benefits include less hydroplaning and splash and thus greater safety, less road noise, and reduced heat island effect (Martin et al., 2018) (Example 7.1).

FIG. 7.7

Prefabricated concrete blocks can be produced that have high strength to weight, a higher R-value, and a lower carbon footprint. Some are lightweight enough that they can be manually carried by one or two workers vs machinery.

EXAMPLE 7.1 Renovation of deteriorating concrete staircases

Problem: Several case studies and problems use the concrete block building shown in Fig. 7.8. The first is a concrete staircase replacement for the four staircases in this former commercial building. Though concrete is a durable material, the left image in Fig. 7.9 shows that it is still prone to weathering, cracking, and spalling due to water, freeze–thaw cycles, temperature cycles, and sun exposure.

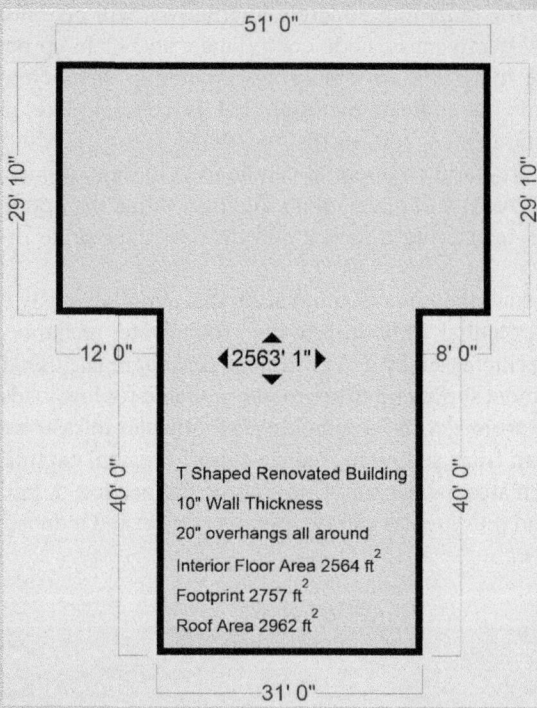

FIG. 7.8

The footprint of the building in this and in other calculations in this book.

Four of these concrete steps/landings were in similar condition—still structurally stable but in similar states of disintegration. What are the options when you need to replace these? What are some of the plusses and minuses, physical design parameters, and limitations? The following table assesses the major options (Table 7.3).

Solution: There were four of these porches of similar age and condition (Fig. 7.9). The homeowner chose the tile-over renovation project repairing existing concrete steps, including concrete cleaning and patchwork, crack isolation membrane, and overlay of exterior porcelain tile, a dense material that resists water and freeze damage, using thin-set mortar. The following compares the energy use and carbon footprint vs the most impactful option, full replacement with concrete.

Porch/Staircases Concrete Volume = for four porches, the sum of stairs and porch volumes for each. The average size porch was 6 ft W × 5 ft D × 2 ft T with a staircase that is on average 5 ft W × 2 ft D × 1 ft T.

Volume = $4[(6\,\text{ft} \times 5\,\text{ft} \times 2\,\text{ft}) + (5\,\text{ft} \times 2\,\text{ft} \times 1\,\text{ft})]$.

$V = 4(60\,\text{ft}^3 + 10\,\text{ft}^3) = 280\,\text{ft}^3$ (this is about $10\,\text{yd}^3$, which is more than a typical concrete mixer truck).

Concrete Mass & Embodied Energy
$M = \rho V = 280\,\text{ft}^3 \times 145\,\text{lbs/ft}^3$ (using an average density for concrete of 145 lbs/ft^3)
$M = 40{,}600$ lbs, or ~18.5 t of concrete!
$E = 18{,}454$ kg of concrete (1.3 MJ/kg) = 24 GJ of embodied energy

It also results in a release of ~2 t of CO_2 (0.1 t CO_2/ton of concrete). This is about half a year of the average automobile's emissions!!

Continued

7.5 Concrete and cement: Scale, composition, and impacts

EXAMPLE 7.1 Renovation of deteriorating concrete staircases—cont'd

Table 7.3 The main options for addressing four deteriorating concrete porches.

Option	Description	Cost ($ = $1000)	Longevity	Aesthetics	Labor	Eco-liabilities	Other concerns
Concrete repair & resurfacing	Clean, patch, repair, resurface, or paint/stain	$$	Medium	Marginally improved	Labor-intensive (homeowner)	Moderate concrete use	Would it adhere or cause crack transmission?
Replace	Remove and repour	$$$$$$$$$$$$	Long	Largely improved	Labor-intensive (professional)	Large concrete use and concrete waste	Disruptive to landscape
Tile over	Clean, patch, crack isolation, porcelain tile	$	Long (?)	Largely improved	Labor intensive (homeowner)	Small-tile and moderate mortar use	Largest time investment
Do nothing	Allow continued deterioration through their natural life	~0	Short	Continued Declining	None immediately	None immediately	Shortest timeframe for eventual repairs needed

Tile Embodied Energy (300 ft^2 with $\rho = 4 \text{ lbs/ft}^2$ were purchased to tile the exposed surfaces).
Weight of tile $= (4 \text{ lbs/ft}^2)(300 \text{ ft}^2) = 1200$ lbs or 0.5 t.
Embodied energy tile (using brick's embodied energy) $= (545 \text{ kg})(2.5 \text{ MJ/kg}) = 1.36 \text{ GJ}$ for tile.
Embodied energy 400 kg mortar (using concrete embodied energy) $= (400 \text{ kg})(1.3 \text{ MJ/kg}) = 0.5 \text{ GJ}$ for mortar.
Total $E = 1.86$ GJ of embodied energy for tile and mortar, an order of magnitude smaller than concrete.

CO_2 generation (using brick as a surrogate for tile and concrete as a surrogate for mortar):
$M = (0.5 \text{ t})(0.75 \text{ t } CO_2/\text{t material}) = 0.375 \text{ t}$ of CO_2 emissions (using brick as a surrogate for tile).
$M = (0.4 \text{ t})(0.1 \text{ t } CO_2/\text{t material}) = 0.04 \text{ t}$ of CO_2 emissions (using concrete as a surrogate for mortar).
$M = 0.42 \text{ t}$ of CO_2 emissions, which is one-fifth of the concrete option.

Finally, the choice for tile came down to two options—one manufactured in neighboring Texas and the other overseas in Asia. The added transport costs for such a mass and the use of local materials dictated the first option. The additional energy and carbon associated with this are shown below using standard costs and emission factors (Rodrigue, 2020).
Tile Transport Energy Difference (Overseas vs US tile manufacturer, ~12,000 km difference).
$E = 12,000 \text{ km} (0.5 \text{ t})(0.1 \text{ MJ/(km t)}) = 600 \text{ MJ}$ or 0.6 GJ.
Transport Emissions Difference $= (12,000 \text{ km})(30 \text{ g } CO_2/(\text{km·t}))(0.5 \text{ t}) = 180 \text{ kg } CO_2$ or 400 lb of CO_2.

FIG. 7.9
Renovation project repairing existing concrete steps, including concrete cleaning and patchwork, crack isolation membrane, and overlay of exterior porcelain tile, a dense material that resists water and freeze damage.

According to the UN Environment Programme, sand use is 50 GT/year, much of it for concrete production. As a component of concrete and glass, sand is the largest material consumed resource after water and air (UNEP, 2014, 2019). This is increasing rapidly, particularly in the developing world, including the two largest nations, China and India. Unlike water and air, sand is less easily treated and regenerated, although options for the reuse and recycling of glass as fine aggregate are expanding. At the same time, constraints and impacts are increasingly associated with the mining of virgin sand. As an indicator of its importance, organized crime has even emerged around illegal production sites globally (UNEP, 2014, 2019).

Sand makes not only the building block for concrete but also glass used in windows and photovoltaic panels and is a key mineral for the global energy transition. Its role in remediating disturbed landscapes and mitigating coastal erosion is also vital. Locations like the desert regions of the southwest US have plentiful sand (and yet much more in the world's great deserts and beaches), so the constraints are surprising. Much of desert sand is too smooth due to wind erosion effects to effectively be used for concrete or is in protected areas such as the stunningly beautiful White Sands National Monument (Fig. 7.10). Though sand seems plentiful, it is finite, and the increasing demands and depleting reservoirs for construction-grade sand have led to concerns about an emerging shortage of usable sand. The grade needed for concrete is mostly mined from river sand which is sharper and coarser than most beach sand. Sand in such riverine ecosystems is particularly susceptible to ecosystem damage with unsustainable extraction. Sands in many oceanside locations are depleted and/or under threat from pressures from tourism, sea-level rise, and resource demand. The constraints on sand demand a better approach to sustainably using this resource in an increasingly urbanized world. This can be facilitated by pursuing a circular economy that recovers waste materials like glass for new uses (Example 7.2).

FIG. 7.10

White Sands National Park in southern New Mexico is a protected land composed of highly reflective gypsum-rich dunes. Despite seemingly endless deserts and beaches, the available sand faces constraints and sustainability concerns (UNEP, 2019) (Example 7.2).

> **EXAMPLE 7.2 Sand constraints**
> **Problem:** White Sands National Park (Fig. 7.10) covers 145,762 acres. Assume an average dune depth of 40 ft and dunes cover half the park area. Assuming a sand particle size of $D = 50\,\mu m$ and spherical sand particles, how many grains of sand are in the dunes? Gypsum (the main component) has a density of $1.4\,g/cm^3$, what is the mass of sand in the dunes?
> **Given:** White Sands NP 145,762 acres; sphere $D = 50\,\mu m$; $\rho = 1.4\,g/cm^3$.
> **Find:** Number of particles.
> **Assume:** uniform horizontal layer of 40 ft; half the area is dunes.
> **Solution:**
>
> $$\text{Volume} = 0.5 \times 147{,}762\,\text{acres} \times 43{,}560\frac{ft^2}{acre} \times 40\,ft \times \left(\frac{1m}{3.28ft}\right)^3 = 3.65(10^9)\,m^3$$
>
> $$\text{Mass of Sand} = \rho V = 1400\frac{kg}{m^3} \times 3.65(10^9)\,m^3 = 0.5 \times 10^{13}\,kg = 5\,\text{billion tonnes} = 5\,GT$$
>
> $$\text{Single Particle Volume} = \frac{\pi}{6}D^3 = \frac{\pi}{6} \times (5 \times 10^{-5}\,m)^3 = 6.55(10^{-14})\,m^3$$
>
> $$\text{Number of Particles} = \frac{\text{Total Volume}}{\text{Single Particle Volume}} = \frac{3.65(10^9)\,m^3}{6.55(10^{-14})\,m^3} = 5.6(10^{22})\,\text{particles!}$$
>
> This easily exceeds the number of stars estimated in the universe. Also noteworthy, the human use of materials is roughly 15 White Sands NPs every year.

Another important related material is asphalt, which is used in roofing products, roadways, parking lots, driveways, pathways, and the like. It is made from the heavier semisolid material that comes out of the fractionating tower at a petroleum refinery plus the addition of aggregate material. Though it's cost per unit area is lower, asphalt has less strength and durability than concrete. Asphalt surfaces are also a material that contributes more to urban heat island effects due to it heating more effectively than concrete (Fig. 7.11).

7.6 Steel and other metals

The process of making steel is both energy-intensive and a prodigious greenhouse gas emitter due to the intrinsic process of coking. Coking steel uses coal as the carbon input for reducing iron oxide and thus has CO_2 as an inherent byproduct. Every metric ton of steel produced in 2020 emitted almost twice that much carbon dioxide into the atmosphere, according to the World Steel Association, with 1 t of steel emitting roughly equivalent to ~1.8 t of CO_2 emissions. In the overall magnitude of greenhouse gas emissions (2.6 Gt/year in 2020), steel is roughly comparable to overall emissions from concrete.

Alternatives include using renewably produced hydrogen as the energy source (green H_2) along with renewably produced electricity for use in electric arc furnaces (EAFs) to replace the carbon-intensive coking process. Roughly 70% of steel produced in the US is now from EAFs due to the large fraction of recycled steel. Other advanced technology methods for steelmaking are in commercialization, such as the electrolysis process used by Boston Metals, a spinoff company from the Massachusetts Institute of Technology.

Steel framing is a viable choice for commercial new construction and offers advantages over timber framing regarding recyclability and recycled metal content plus more rot and fire resistance (Kruger and Seville, 2012). The key drawback is the conductivity of metal and thermal bridging through structural members. As discussed in Chapter 8 on building thermal envelopes, some of this can be mitigated with advanced wall structures.

FIG. 7.11

Comparison of asphalt (lower part of the picture) and concrete (upper part of the picture) surface temperatures on a hot, sunny July day (97°F).

7.7 Structural wood products: Mass timber and cross laminated timber (CLT)

Considering the large impacts of traditional materials, including their high-embodied energy and embodied emissions, new construction products focus on carbon-sequestering materials. It is important to distinguish this stored carbon (a good thing) from the embodied carbon that represents atmospheric emissions from the production of building materials. Wood can result in considerable carbon storage if managed sustainably, and the net carbon storage can outweigh the embodied emissions.

7.7 Structural wood products: Mass timber and cross laminated timber (CLT)

Negative emission buildings, or "drawdown buildings," offer a future with carbon-negative buildings where they serve as sinks of carbon. Any new building does involve stored carbon in the form of materials carbon sequestration (quasi-permanent removal of CO_2 from the atmosphere and conversion to solid or liquid phase), which is on the order of 5–12 t of stored CO_2 in an average residential dwelling. According to the US Census Bureau, approximately 1.5 million residential structures are built annually in the US. Efforts are underway to enhance this stored carbon.

Many details as to how wood is grown and harvested determine its sustainability. The Forest Stewardship Council oversees the sustainability of wood products, their growth and harvesting, and the production of wood products. Urban wood is scrap timber increasingly harvested to reduce shredding, chipping, or burning of the wood while adding a distinctive character to the wood. This is still an evolving area of research and effort toward making construction more sustainable.

Another growing application is cross-laminated timber (CLT) or mass timber construction for structural members in both high-rise and low-rise applications. CLT builds up large structural beams, posts, and panels in a layered fashion. They use perpendicular orientations of the layers to maximize strength and minimize warping. These large, dense timbers also have advantages with respect to rot, decay, infestation, and fire resistance. It can offer reduced weight and improved earthquake performance as compared to traditional concrete and steel typically called for in larger structures. The problems of pests, rot, fire, and other hazards of an organic building material are designed to be minimized with its density. Variations of related products include I-beams, laminated veneer lumber, oriented-strand lumber, parallel strand lumber, and glue-laminated timber. All involve engineered wood products built up from smaller dimension layers.

To date, the largest mass timber structures are several tens of stories. Two examples include the 18-story building, the Haut, in the Netherlands and a 25-story building, the Ascent, in Milwaukee, Wisconsin, made of mass timber. Currently planned for expected completion in 2026 are a 32-story apartment building also in Milwaukee, "The Edison," and the 42-story building "Atlassian Central" in Sydney, Australia. A building completed in 2023 at Nanyang Technological University in Singapore is a 6-story, nearly 500,000 ft^2 mass timber business school facility.

Many variables exist regarding the comparison to conventional building construction (harvest, transport, waste generation, and disposal), but the promise of reducing embodied energy and emissions with these advanced timber products is substantial. This serves also to make it better able to handle structural, wind, and earthquake loads. Rather than the (0.5–2 t CO_2e per t of material) emissions associated with concrete and steel, the sequestration of carbon on the order of 1 t CO_2e per m^3 by timber buildings is possible. It also has the potential to reduce construction waste, as much of the process can take place in factories offsite rather than onsite stick-building. It can avoid some of the finish materials as its aesthetics mean it can be left exposed with even a raw finish rather than the drywall or other finish materials used over metal, concrete, and typical lumber. If CLT can lead to sustainable harvesting of wood products rather than clear-cutting common in the past, it can be a significant contributor to climate mitigation and sustainability. CLT has other advantages and will continue to grow as an active area of research and building applications.

Increased wood use is still a topic that is debated as to the ecological benefit and thus is currently under intensive research. In its life cycle analysis are many variables that even make the carbon footprint uncertain, much less the wider impacts of best practices in forest management. Carbon neutrality is strongly a function of species, the managed ecosystem properties, age and harvest frequency, kiln-drying of the wood, its end-use product, and its lifetime, and in particular the management of the forest land, including harvest details and soil management. The bottom line for structural wood products is an open area of research. The evaluation of CLT and mass timber performance will continue over the years and decades ahead.

Another novel application is a wood product that, by processing involving cell wall engineering, is moldable, foldable, and modular and thus amenable to form factors such as hexagonal honeycombs. Though still under intensive research, when evaluated on environmental impacts ranging from global warming potential to smog formation, moldable wood is superior to traditional aluminum alloy honeycomb structures (Xiao et al., 2021). Another example is TRIQBRIQ, a German company making wood bricks for structural purposes that have strength properties comparable to concrete while sequestering carbon in the building structure.

7.8 Bamboo, hemp straw bale, and other structural plant fibers

Carbon-negative building materials are emerging that allow the building itself to sequester carbon that otherwise would be released into the atmosphere, causing additional warming. Many of the products are returning to materials directly available from nature versus those requiring intensive production from raw materials (e.g., concrete) or those from petroleum (e.g., carpeting, or other synthetics). Many non- or lightly-processed earthen materials are gaining more prevalence in the construction market: adobe blocks, strawbale, cob, which mixes sand, clay, and straw, rammed earth, hemp/lime, and mass timber, among others (Halliday, 2008) (Fig. 7.12). Adobe, cob, strawbale hemp, bamboo, and other similar earthen materials are embraced as alternative construction materials, albeit many are ancient building materials (Halliday, 2008).

FIG. 7.12

An exterior wall revealing some of the materials used in the Black Range Lodge in Kingston, NM, including reused glass bottles, cordwood, strawbale, adobe blocks, rammed earth, and locally harvested stone.

7.8 Bamboo, hemp straw bale, and other structural plant fibers

Strawbale uses high-density bales of straw that can be used for both interior and exterior walls that are then covered in a stucco-like finish. It can form load-bearing walls without framing, and its advantages include structural stability, a large thermal mass combined with high R-value walls, and the use of a renewable resource for construction. Using it in wet locations requires additional construction details to ensure it remains dry to avoid rot. A series of straw bale books provides a comprehensive coverage of straw bale and a range of other related materials for the practitioner's guide to designing and building with these (Magwood et al., 2005). Strawbale building styles are highly diverse, from primitive to modern, with one-room huts to yurts to tiny homes to luxe high-end homes.

Bamboo, though a material of ancient use in Asian countries, has emerged as a viable alternative to wood-based products that sequester carbon in its tissues via its rapid growth. As the largest of grass species, its attributes include its fast-growing nature (in cases, daily growth reported in feet per day), its short growing cycle of months to a few years, regeneration from the same stems, which aids in soil conservation, and little need for input of fertilizer or water. Bamboo, as a grass-like species, produces a material with many of the attributes of wood and its fibrous nature with hardness exceeding that of the best hardwoods. It has performed well in high winds, earthquakes, and other extreme events where more rigid, heavier materials like steel and concrete might fail. A review gives the diverse applications of bamboo both indoors and outdoors, including furniture, fencing, flooring, and many other applications (Stangler, 2008). A builder, BLDUS, has created a code-compliant "grass house" in Washington, DC, comprised of mostly bamboo building products.

Structurally, bamboo expands and contracts more with atmospheric RH changes than wood, and therefore the engineered laminated products are more suited to extreme or highly variable humidity. Though the benefits outweigh its drawbacks, improved management is key, as large plantations harvesting monocultures will exacerbate its land use issues.

Bamboo is increasingly accepted as a structural material as it possesses similar strength-to-weight properties while having much quicker regeneration and a much higher forest carbon density (Braham and Casillas, 2021). Bamboo's growth rate plateaus at around 10 years. When continually harvested, bamboo provides an annual carbon sink of ~200 t/hectare of CO_2 sequestration, exceeding the typical treed forest by about an order of magnitude. Alternately, this means about 10% of the land area is required for a given mass of material. The rhizome and root structure can provide ecosystem benefits by enhancing wetlands and preventing soil erosion.

On the downside, most, if not all, must be shipped across the Pacific from bamboo regions of southeast Asia. Its processing is energy-intensive, and it uses glues or binders that can off-gas as well as make it susceptible to water damage. This may be mitigated in some cases where traditional timber produced in North America is shipped, processed, and produced into end products often after roundtrip transport overseas to less expensive production options in Asia. Bamboo can be damagingly invasive if planted in a hospitable ecosystem where it is nonnative.

Hemp is another bio-based material that can be used to make insulation, flooring, a concrete-like replacement, and even a wood like lumber product (Fig. 7.13). A manufacturer in the US, Hempitecture, is leading the way in using hemp in construction materials for strength and sustainability. Hemp is rapidly renewable and typically requires little in the way of inputs of fertilizer, water, or other resources. Mycelium, the root structure of fungi, is subject to increasing applications in construction and consumer products as well. It has been used for wood replacement as well as a concrete substrate. This is only a brief list of the bio-based materials for construction materials.

174 Chapter 7 Materials use in buildings: Choosing the best options

FIG. 7.13

Bamboo floating floor materials include solid bamboo and engineered layered bamboo that has a higher hardness rating than hardwoods. It also makes a nice desktop surface when writing a book about green building.

7.9 Advanced manufacturing and construction methods

The construction of buildings, their engineering, and the processes and materials are detailed in many textbooks (Mehta et al., 2018). Magwood provides a guidebook for the home designer/builder to use in approaching a more sustainable building process (Magwood, 2017).

Recently, offsite manufacturing of assemblies or entire structures has gained a new following. Construction and form factors of complete kits range from arched cabins to domes to more traditional rectangular structures. An upside of manufactured housing units or manufactured components of housing is that material waste can be minimized vs the onsite "stick-build" construction process.

This world is diverse in terms of new players in manufactured housing. Some are fabricating buildings offsite with minimal assembly onsite. An example is Nexii Building Solutions, Inc., which manufactures integrated building panels with a proprietary low-carbon form of concrete (Nexiite) for onsite construction with nearly zero waste. Buildings are essentially built offsite and assembled rapidly onsite. Applications have included wildfire-resilient residential homes, commercial buildings including restaurants, retail, and hotels, and a retrofit of an existing commercial building (www.nexii.com). Another manufacturer is Zenni Homes, which makes individual manufactured units that can also be stacked up to five units high to make multifamily communities for urban areas.

Other new techniques include 3D printing a building onsite with a large-scale printer. Additive manufacturing using 3D printers has taken hold in the construction industry as well vs subtractive manufacturing which cuts away excess from a solid piece, additive manufacturing wastes far less material. 3D-printed houses certainly have advantages over stick-built structures in terms of labor, construction costs, scalability (e.g., wall thickness), and reduced waste generation. The material savings result from concrete placed only where and in the quantities needed for structural stability. 3D-printed buildings also offer design innovations not possible with traditional construction, such as curved walls.

7.9 Advanced manufacturing and construction methods

3D printing is becoming advanced enough that whole building panels and even entire structures can be 3D-printed onsite. Construction times are measured in days to a week or so, rather than months of a conventional stick-built house. Though initially expensive, projected construction costs can eventually scale down, similar to or cheaper than conventional builds. Barriers to manufactured buildings include zoning restrictions, HOA rules, cost structures, scalability, as well as a perception of manufactured homes as lacking durability.

3D-printed homes are being built and permitted by companies such as Icon Technology, Inc. and SQ4D. An example of an onsite 3D-printed residential building with 3 bedrooms and 2.5 bathrooms is shown in Fig. 7.14. "House Zero" is a 186 m^2 in floor space (2000 ft^2) structure built onsite in early 2022 in Austin, TX. It is the first production single-family house that was printed by Icon Technology with a 3D printer using a proprietary concrete mix called Lavacrete, which has more insulating properties than ordinary concrete. The curvy walls were 3-D printed onsite, and the wall printing process took about 10 days. The approach offers the potential for much shorter and less expensive costs to build. Currently, beyond the foundation and walls, all the windows, doors, roofing structure, insulation, electrical, and plumbing are done conventionally, meaning the costs are far more than the 3D printing. The 3D printing of homes is not limited to high-end homes; the same company, Icon Technology, printed six small houses and a community center in Community First built for formerly homeless people in Austin, TX (Fig. 7.15).

FIG. 7.14

Among the first 3D printed code-compliant residential houses built by Icon Technology, Inc. and located in Austin, Texas.

The University of Maine's Advanced Structures & Composites Center recently built the Bio-Home3D, from all bio-based materials, including mostly wood waste materials, that also offers the possibility of multiple recycling processes. Companies such as Azure are using recycled materials (e.g., plastic waste) as feedstock for their printing operations. Advantages include substantial time and cost savings as well as building-integrated high insulation values with little thermal bridging. Another company, Tecla, produced a 3D-printed house using onsite clay as the raw material. The use of 3D printed structures using wood-like material grown in the lab is another new direction (Beckwith et al., 2022). It offers the promise of sustainable construction material with negligible land use and no susceptibility to pests, climate change, or other ambient stressors. This area is likely to see many new innovators and more materials, energy, and cost-efficient construction.

FIG. 7.15
Affordable 3D printed code-compliant tiny house in the Community First! Village built by Icon Technology, Inc. and located in Austin, Texas.

Image by University of College/Shutterstock.

7.10 Tall buildings and green building

The discussion of tall buildings fits well with the materials chapter in that materials use becomes paramount with increasing building height (Denison and Beech, 2019). The green facets of tall commercial buildings parallel those of smaller buildings, though with a nonlinear amplification with size among other unique challenges (Al, 2022). The typical construction of tall buildings for residential or office space incorporates large masses of high-embodied carbon materials, particularly concrete and steel, plus aluminum and copper, among others. As buildings become very tall, this is an exponential relationship between structural materials needs with increasing height due to increasing foundation and wind loads.

Modern supertall buildings of the past couple of decades often pursue green building features to varying degrees, as was done with the construction of the Taipei 101 and One World Trade Center buildings (Denison and Beech, 2019). Other major sustainability challenges with supertall buildings include cooling increasingly tall glass-clad buildings exposed to amplified solar gain and delivering people and materials to such heights (Al, 2022). Construction, particularly of supertall buildings, is complicated and energy-consumptive, as it takes powerful cranes and pumps to deliver the materials to such heights. The operations have unique challenges, such as large loads on elevators with rapid deliveries to great heights. The following example estimates the elevator electricity consumption of a supertall building (Example 7.3).

EXAMPLE 7.3 Approximate elevator power requirements in a supertall building

Problem: The "Mile High Illinois" tower proposed by Frank Lloyd Wright in the 1950s was to be a mile high with 18.5 million square feet of floor space. Estimate that each of its 15,000 occupants (75 kg plus another 150 kg per passenger of associated elevator mass) rides the elevator halfway up the tower three times per 8-h workday. Assume a 75% efficient elevator system and that the going-down consumption is negligible. Compared to a typical office building's total electric consumption per floor area.

Given: Mile-high Illinois Tower (proposed), 15,000 workers; 75 kg per person plus 150 kg of elevator mass on average per rider; three roundtrip rides each day; $\eta = 0.75$.

Find: Power for elevator operations.

Assume: 10 h/d year-round.

Solution:

$$P = \frac{\left(\frac{dm}{dt}gh\right)}{\eta} = \frac{(15{,}000\,(225\,\text{kg}))\left(\frac{3}{\text{workday}}\right)\left(\frac{\text{workday}}{8(3600)\,s}\right)\left(9.8\,\frac{m}{s^2}\right)(2640\,\text{ft})\left(\frac{1\,m}{3.28\,\text{ft}}\right)}{0.75}$$

Elevator: $P = 3.7\,\text{MW}$ which assuming $10\,\frac{h}{d}$ operations year rounds equates to:

$$\text{Elevator}\,E = P(t) = 3.7\,\text{MW}\left(10\,\frac{h}{d}\right)\left(\frac{365\,d}{\text{year}}\right) = 13{,}500\,\frac{\text{MWh}}{\text{year}}$$

The average US office building benchmark is 22.5 kWh/ft² annually according to the US DOE.

$$\text{Buiding Total: Annual}\,E = 0.0225\,\frac{\text{MWh}}{\text{year}-\text{ft}^2}(18.5 \times 6\,\text{ft}^2) = 416{,}250\,\frac{\text{MWh}}{\text{year}}$$

- The elevator is not a small load but still only a few percent of the average office building electric consumption. Much like a hybrid electric vehicle, newer elevators can recapture electric energy while descending (rather than frictional losses as heat). Speaking of heat, the need for additional space cooling is likely in an elevator shaft due to the heat load of these losses as well as passengers.

Tall buildings do have distinct advantages, including the small land footprint and the decreased surface area to volume ratio, reducing heat transfer. They lend themselves to urban density and thus can reduce the transportation emissions associated with tenants. Compared to single-story buildings, fewer materials per level are needed (e.g., only the top floor needs a roof), with the tradeoff of greater structural steel and foundational materials as the tower height increases.

The place of the mid-rise building becomes clear (Fig. 7.16), as does the retrofit of existing towers. The greenest tall tower is the already existing tower, and renovation of existing office space into residential is increasingly of interest with the surplus of office space beginning during the pandemic (Al, 2022). It becomes challenging regarding interior lighting, ventilation, plumbing, and many other details that distinguish residential buildings from office buildings. Rather than retrofit though, unfortunately, it has not been rare that existing towers have been razed to build taller or more ornate buildings with incumbent materials use, waste, and reemitted embodied carbon!

FIG. 7.16

Mid-rise Bullitt Center building in Seattle, WA built in 2013 as the world's greenest office building at the time

Image by Joe Mabel licensed under the Creative Commons Attribution-Share Alike 3.0 Unported license via Wikimedia Commons). The building's website details its many sustainable attributes (https://bullittcenter.org/).

7.11 Case study: Green renovated tall building Empire State Building in New York

In this book, the Empire State Building (ESB) is often referenced in problems as a benchmark for comparison to gauge the scale of large volumes or masses (e.g., how many ESBs could we fill with the annual CO_2 emissions of coal-fired power plants?). Considering that the ESB is an iconic and historic art-deco-era skyscraper, it is only fitting to examine its renovation as a case study (Fig. 7.17). The 102-story building was originally constructed in 1930–31, amazingly in a little over a year. At the time it was the world's tallest skyscraper (380 m rooftop and 438 m total with antenna), remaining the tallest until 1970, when it was eclipsed by the World Trade Center twin towers. ASCE designated the ESB as one of the original "Seven Wonders of the Modern World." The ESB renovation project can be contrasted with the approach of a nearby skyscraper: the 707 ft (215 m) Union Carbide building at 270 Park Avenue in New York. The latter was certified LEED platinum in 2012 but then was demolished in 2019 to make way for a new taller tower for the JP Morgan Chase headquarters.

Renovations to improve ESB energy use and operations began in 2010 led by a consortium of public, private sector, and nonprofit groups. It shows that a century-old building can be effectively renovated, serving as a model for other historic tall buildings detailed in the state energy agency "playbook" (https://knowledge.nyserda.ny.gov/display/EBP/The+Empire+Building+Playbook). Despite little thought at construction to its environmental impacts, it is worth asking what the building has going for it. To start, it is already built, and the structural materials used, cost, and emissions are long-ago sunk costs. The ESB is not a modern-day glass curtain building, helping with the thermal envelope and unwanted solar gains at the cost of some reduced daylighting, of course.

The renovation focused on the building shell, including refurbishing all 6514 windows in the building, reducing the u-value of the windows and hence energy loss through them by 75% while reusing 96% of the demolition waste onsite. The windows were removed, glass panes reused, and an additional low-E coated pane was added to each unit. The renovation installed elevator systems that recapture energy normally lost from friction. The lighting systems were all upgraded to LED or other efficient technologies. This included the iconic tower-top exterior decorative lights that are color-changing LEDs for light shows, which replaced incandescent floodlights. Indoor lighting levels are automatically adjusted for varying daylight, so only the electricity needed is used. The building's chiller and climate control system were upgraded, maintaining much of the infrastructure through renovating for more efficient systems. Overall, according to the ESB website, the building has reduced its energy use by 38% compared to pre-renovation, and it achieved LEED gold certification for existing buildings while maintaining its National Historic Landmark designation. The reduction in operational CO_2e emissions is even larger at 54%. The ESB has used Renewable Energy Credits (RECs) for offsite wind power for 100% of its electric use since 2011. More details are given in an onsite hands-on display of sustainable aspects of the building as well as on its website (https://www.esbnyc.com/about/sustainability) (Example 7.4).

> **EXAMPLE 7.4 Materials mass in tall tower**
>
> **Problem:** Research the Empire State Building size data. The exterior is clad with 200,000 ft³ of Indiana limestone. Estimate the fraction of the overall building mass that the cladding represents. What is the total mass loading on the soils if it were all distributed uniformly on the horizontal footprint of the building (not realistic but just for information's sake)?
>   **Given:** ESB with 200,000 ft³ of Indiana limestone cladding
>   **Find:** Cladding mass fraction; soil loading
>   **Assume:** uniform horizontal footprint

Continued

> **EXAMPLE 7.4 Materials mass in tall tower—cont'd**
>
> **Solution:**
> From the ESB building fact sheet, we find the following (https://www.esbnyc.com/about/sustainability).
> Volume of building $= 37{,}000{,}000\,\text{ft}^3$
> Area footprint $A = 79{,}288\,\text{ft}^2$
> The volume of Indiana limestone $V = 200{,}000\,\text{ft}^3$
> Assume a limestone density $\rho = 144\,\text{lbs/ft}^3$
> The total mass of the ESB building $M = 365{,}000\,\text{tons}$
> Soil Loading of Building $= 365{,}000\,\text{tons}/79{,}288\,\text{ft}^2 = 4.6\,\text{tons/ft}^2$
> Mass of limestone cladding:
>
> $$M = \rho V = 144\,\frac{\text{lb}}{\text{ft}^3}(200{,}000\,\text{ft}^3)\left(\frac{\text{ton}}{2000\,\text{lb}}\right) = 14{,}400\,\text{tons or about 4\% of the total mass}$$
>
> If constructed in the modern era, the use of more local materials would help reduce the carbon footprint (not to mention that of all the Italian marble imported for interior use).

FIG. 7.17

The Empire State Building in New York City with the Chrysler Building in the background.

Image by Eduard Moldoveanu/Shutterstock.

Finally, in light of the new artificial intelligence writing tools available, the following paragraph on the ESB renovation was produced using a generative AI tool. These tools will change the landscape of written communications most certainly. Although it remains to be seen, perhaps the job of textbook writers may become obsolete or at least fundamentally changed to curator and editor of sections and chapters in a book. The paragraph was produced on 6 June 2023 with the prompt "Summarize the Sustainability Features of the Empire State Building Renovation" using Microsoft's Copilot add-on to Microsoft Word that uses Chat-GPT4. It produced a short form, a long form, and a bulleted list of key highlights of the ESB renovation. I will leave it to the reader as to which of the two summaries reads better!

"The Empire State Building, an iconic landmark, has undergone a significant transformation to become a leader in energy efficiency and sustainability. This historic skyscraper's green retrofit, completed in 2010, included a comprehensive upgrade of its chiller plant, utilizing much of the original materials for a modernized system. The building's 6514 windows were enhanced with an additional pane of coated film to minimize heat gain during summer and heat loss during winter, contributing to a substantial reduction in energy usage. Insulation behind radiators was improved to prevent heat loss, and the elevator system was revamped with regenerative braking technology, which recaptures electricity, further reducing the building's energy consumption. These innovations have resulted in approximately a 40% decrease in energy use and over $4 million in annual savings for the building owners. The retrofit has not only cut costs but also attracted high-value tenants, boosting the building's financial performance. The Empire State Building's commitment to sustainability extends to its indoor environmental quality, with measures like the use of no/low VOC materials, prohibition of Red List materials, and regular indoor environmental quality testing. Minimum Efficiency Reporting Value (MERV) 13 filters ensure clean air circulation and access to daylight and views are available on every tenant floor. The building also practices green cleaning and integrated pest management, emphasizing health and wellness for its occupants. This transformation has positioned the Empire State Building in the top 20% of all Class A commercial assets in terms of efficiency, setting a precedent for other buildings to follow in the pursuit of sustainability."

7.12 Case study: Newly constructed green tall tower in Austin, TX

The 70 Rainey Street Tower built in 2020 is a key modern example of a sustainably built mixed use tall building in Austin, Texas. The building is a 34-story tower designed for residential units with some commercial on lower floors (e.g., a restaurant, bar, and patio on the lowest levels). The building was featured on the Austin Cool House Tour hosted by Austin Energy Green Building in 2020. Page Architects in Austin designed and built the 164-unit condo building, of which 10 were allocated as affordable units. The building was planned using Autocase's triple bottom line software analysis of building impacts, and a few of the green highlights are discussed below.

The 70 Rainey Street building is in the Rainey Street district in downtown Austin near the Colorado River, a National Historic District dating to the late 1800s and largely residential with bungalows. The site is situated among public park areas to the west and south. It is surrounded by many of the original century-old bungalows, some of which have been converted to clubs or restaurants as part of a burgeoning nightlife scene here.

The area features a substantial urban tree canopy, and maintaining this was a project goal in creating a "park in the sky." The parking levels are hidden by screens, planters, and vegetation to integrate with the urban canopy, which in more conventional construction may have been an open parking garage (Fig. 7.18). The parking levels include EV charging options in garages, showers, and bicycle parking. The park-like green space amenities level is at the top of the metal-clad parking level and includes a dog park and other green roof vegetated areas. It forms a shared "green space" that integrates well with the surrounding tree line and also includes lounge and dining space, a theater, shared pools, fitness, and yoga rooms. Site views are paramount, and the residential apartment levels were rotated by 14 degrees to face the Colorado River. Projected and recessed balconies create shady outdoor areas for residential units. The proximity to Ladybird Lake as well as hike and bike trails made bike storage and showers for residents and staff a necessary amenity.

The building is part of a district cooling loop, requiring fewer onsite mechanicals, especially on the roof. In this hot location with significant urban heat island effects, solar ban R100 Low-E windows with a solar heat gain coefficient (SHGC) of 0.27 were used. The high-performance glazing was mixed with opaque panels with high R-values to maintain views and create a better thermal envelope. Balconies provide shading for those below to allow more window area. The building features an Energy Recovery Ventilation (ERV) system for low-energy ventilation and helps to reduce peak demand from air conditioning loads. Building automated systems, including smart thermostats, Energy Star LED lighting throughout public spaces, low flow plumbing fixtures, and plentiful native and adapted plants watered via drip irrigation.

Among materials, 65% by cost were sustainably sourced materials that were either locally sourced or recycled. During construction, 84% of C&D waste was diverted from landfills. Low-emitting materials indoors, no formaldehyde, and low-VOC coatings and glues were used. Also, maintenance closets were isolated to reduce indoor air quality impacts. Concrete with fly ash in the cement was specified for the structural supports. The building is a key example of a tall tower designed with sustainability in mind.

FIG. 7.18

Sustainably built tall tower at 70 Rainey Street in Austin, Texas, exterior (*left*) showing the parking levels sitting below the rotated residential levels and (*right*) the amenities level *green* roof sitting on the top level of the parking levels.

7.13 Chapter summary and conclusions

All construction materials involve substantial impacts including energy use, water use, and waste generation as related to their extraction of raw materials, processing, administration, and transport. The array of choices for alternative building materials is vast, and a high-level survey shows key categories with references for further information.

Much of the focus to date has been on operational energy use in buildings. As these systems continue to become more efficient, the importance of embodied energy and embodied carbon increases. Embodied energy and carbon (the energy and emissions associated with producing and delivering a material) are of growing importance.

Compared to the ~270 million tons per year in the US in 2015 for municipal solid waste, construction waste adds up to over twice that mass at ~550 million tons per year. Rather than straight demolition and landfilling of a building, deconstruction offers a way for hazards to be remediated and materials from an obsolete building can be repurposed.

As a key construction material, the production scale of concrete makes it the second most important material produced globally after treated water. The production of steel and iron is of similar magnitude. In terms of climate impact, concrete has a large carbon footprint. If it were a nation, cement would be the third largest emitter after China and the United States. In terms of greenhouse gases, the emissions associated with concrete production are up to 8% of total global emissions, depending upon the estimate.

The impact of the production of concrete, and in particular its cement component, is undeniable and considered a "difficult to reduce" category along with steel and iron (Davis et al., 2018). Concrete's carbon intensity is a combination of CO_2 emissions from fuel combustion plus those resulting from calcining the limestone. Portland cement is the key contributor to the carbon footprint of concrete. A rough metric dependent upon fuels and processes, each ton of Portland cement is responsible for a ton of CO_2, and roughly half is from the fossil fuel combustion in the high-temperature kiln and half from the curing process.

Improvements in reducing overall concrete mass and energy use include reductions in Portland cement and various uses of recycled materials such as fly ash, silica fume, or blast furnace slag. Much of the focus is on supplementary cementitious materials (SCMs) that can displace some of the Portland cement. Although Portland cement is the largest concern with concrete, the vast water, sand, and aggregate use raises concerns about sustainably producing and using these materials.

An application of carbon-sequestering materials (those whose carbon storage exceeds their embodied emissions) is the use of cross-laminated timber or mass timber. These are increasingly used in both residential and tall buildings. Other novel materials are under research or implementation, including bamboo, hemp, straw bale, and cob construction. Other builders use a concrete mix and 3D print the structures onsite. Manufactured structures and buildings are becoming more viable and sophisticated.

Transporting local materials involves far fewer emissions typically than shipping from halfway around the world. Using recycled materials helps divert what would otherwise be a waste stream that would end up in a landfill or rubble heap. Recent legislative efforts, including the Inflation Reduction Act, signed into law in summer 2022, have focused on the emissions reduction associated with construction materials, most notably concrete and steel.

7.14 End of chapter exercises

(1) **Concepts:** What are five potential reuses, recycling options, and uses of waste materials associated with concrete?

(2) **Concepts:** Describe five ways in a sentence or two each to reduce the environmental impacts of concrete in your building.

(3) **Concepts:** Compare the sustainability of flooring materials. Use strand woven bamboo, cork, reclaimed hardwoods, and ceramic tile. Consider materials, embodied energy, production, transport, regeneration, recycled content, recyclability, costs, and durability. This sounds like a terrific opportunity to make a comparison table!

(4) **Concepts:** What is fly ash, what is it used for, and why is it considered green?

(5) **Concepts:** In this chapter, we looked at 3D-printed houses. Explain the construction process. Describe or draw a diagram of a typical wall section. What are some of the pros and cons of 3D-printed concrete structures?

(6) **Concepts:** Discuss the pros and cons related to the sustainability of a tall building vs a shorter building of the same square footage. Consider land use, material use, embodied energy, and operational energy use.

(7) **Problem:** You are building a new house. Using the site boundaries as your control volume, perform a conceptual mass and energy balance. Show a diagram and write an equation.

(8) **Problem:** Calculate the cinder block mass of the $8'' \times 8'' \times 16''$ cinder block (30 lbs apiece) building shown in Fig. 7.8. Use an 11.5-ft exterior height from the foundation to the wall top.

(9) **Problem:** Using the data in Table 7.1, calculate the embodied energy of concrete from its components in both BTU/ton and MJ/kg. How closely does it match the values given in the table for overall concrete embodied energy?

(10) **Problem:** With the global rate of concrete production, how quickly would one fill Lake Michigan? How long would it take to fill the Empire State Building using reasonable assumptions? What is the mass density of the Empire State Building compared to bulk concrete?

(11) **Problem:** In 2017, EPA estimates 94 Mt of solid waste were recycled or composted resulting in a reduction in GHG emissions by 184 Mt/year CO_2e. Equate to the number of cars removed from the roadways assuming 25 mpg, 8.9 kg CO_2/gal gasoline, and 12,000 mi/year driven.

(12) **Problem:** Sand and gravel (aggregate) production has a best global estimate of 40 Gt/year production rate. This is twice the sediment delivered annually by the world's rivers. Assume an average spherical sand grain size of 100 µm (beach sand size). Assume a sand density of 1.5 times liquid water. How many sand particles does this represent and compare to the number of stars in the universe?

(13) **Problem:** Compare the average CO_2 storage in a wood constructed home to the annual CO_2 emitted by the average passenger vehicle. Do not forget to subtract the embodied emissions of building the home. Do some research, as both numbers have quite large ranges. Assume it is a gas automobile with fleet average fuel efficiency, burning pure octane, complete combustion, and fuel with a specific gravity of 0.7.

(14) **Problem:** The US produced $\sim 80 \times 10^6$ tons of cement in 2015. This required ~ 0.5 quads of energy and was worth US$9.8 billion. Compare the energy intensity of cement production (GJ/$) to the overall US economy's energy intensity (you might have to do some research here; cite your sources).

CHAPTER 8

Building thermal envelope: Constructing a tight barrier between the indoors and outdoors

Learning objectives

(1) Apply the laws of physics to understand the movement of air, water, and energy through a structure and form the basis of building science.
(2) Define and understand the importance of the thermal envelope in buildings.
(3) Identify and quantify the most common thermal leakage points in the thermal envelope.

8.1 Defining the thermal envelope of a building

A structure is intended to shelter its occupants from the elements. The building envelope (or thermal envelope) is the barrier between the conditioned space and the ambient environment. It consists of a thermal boundary (to reduce conduction and radiation) and an air movement barrier (to reduce convective losses via air infiltration and exfiltration). The thermal envelope controls the movement in and out of the building of energy, water, and air. Here we will use the principles discussed previously related to mass and heat transfer and explore some of the techniques and technologies that help provide a tight building.

The effective separation of the indoor and outdoor environments by a thermal boundary reduces the building's ongoing energy usage and cost. It will also enhance occupant comfort and health. Despite some pitfalls to avoid, an improved building envelope can enhance building longevity by preventing condensation and thus mold and mildew issues, among other problems.

The building envelope follows much of the physical boundary of the building but will deviate in specific areas. For example, unconditioned, vented attics or crawlspaces exist outside the thermal envelope. The ideal is to locate any climate control systems within the thermal envelope. Otherwise, losses to the environment are inevitable, even with well-sealed, insulated distribution systems. An example of a weak point (entry doors) in a thermal envelope is shown in Fig. 8.1.

186 Chapter 8 Building thermal envelope

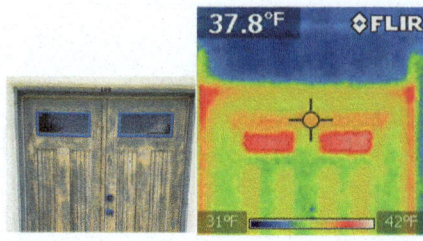

FIG. 8.1

Infrared image from the exterior during winter of double wooden entry doors with single pane glass windows. The doors and the frame around them are significant areas of thermal loss (*yellower colors*). Rigid foam board insulation is under the stucco exterior surface except for the area around the door, which also shows significant heat loss.

The thermal-fluid properties of the building envelope are a complex area of building science. Likewise, building science is a complex field that is beyond the scope of an introductory textbook to cover all the intricate details. Many resources exist for detailed installation and retrofit of buildings to enhance the thermal envelope by insulating, air sealing, and creating other thermal barriers (Building Science Corporation, Green Building Advisor www.greenbuildingadvisor.com, Fine Homebuilding, www.finehomebuilding.com/project-guides/insulation). These references give much more complete and detailed consideration of construction details and practical techniques.

8.2 Heat and mass transfer in buildings

The fundamental concepts of heat transfer as applied to buildings as discussed earlier in the book are fleshed out for building structures here. The reader can refer back to the chapter on energy science to review basic physics. A more extensive treatment as applied, particularly to heat transfer in buildings, is provided in books dedicated to building physics (Moss, 2007). Here, we examine key green construction materials, including major interior and exterior components of residential and commercial construction, with a focus on insulation properties and techniques to minimize heat transfer.

Whether in winter or summer, the efficiency goal in tightening the thermal envelope of a building is slowing the rate of heat transfer via all three pathways: radiative heat loss, conduction, and convection. Radiative heat loss is reduced via reflective barriers, including the low-emissivity coatings on windows that serve to reflect infrared (IR) wavelengths. Conductive heat loss relies upon using thicker materials that have a low thermal conductivity k (higher R-value).

The convective losses in a building structure are driven by the pressure gradient between the inside and outside of the building modified by the building leak tightness (Fig. 8.2). The driving force can be the effects of wind on the structure, the stack effect of warm air rising and exiting the top of the structure, and mechanically driven convection due to exhaust fans, HVAC systems, clothes dryers, or other systems. Reducing convective losses involves air sealing and preventing infiltration and exfiltration. Convection drives the exchange of air with the ambient and is a major pathway for energy loss. Exfiltration is the movement of conditioned indoor air outside, while infiltration is the opposite. The "leakiness" of a building enables both. The driving force is the pressure differences between the indoor and outdoor environments as a result of wind effects, the buoyancy of warmer air and the stack effect, and mechanical systems that can pressurize or exhaust air (Straube, 2017).

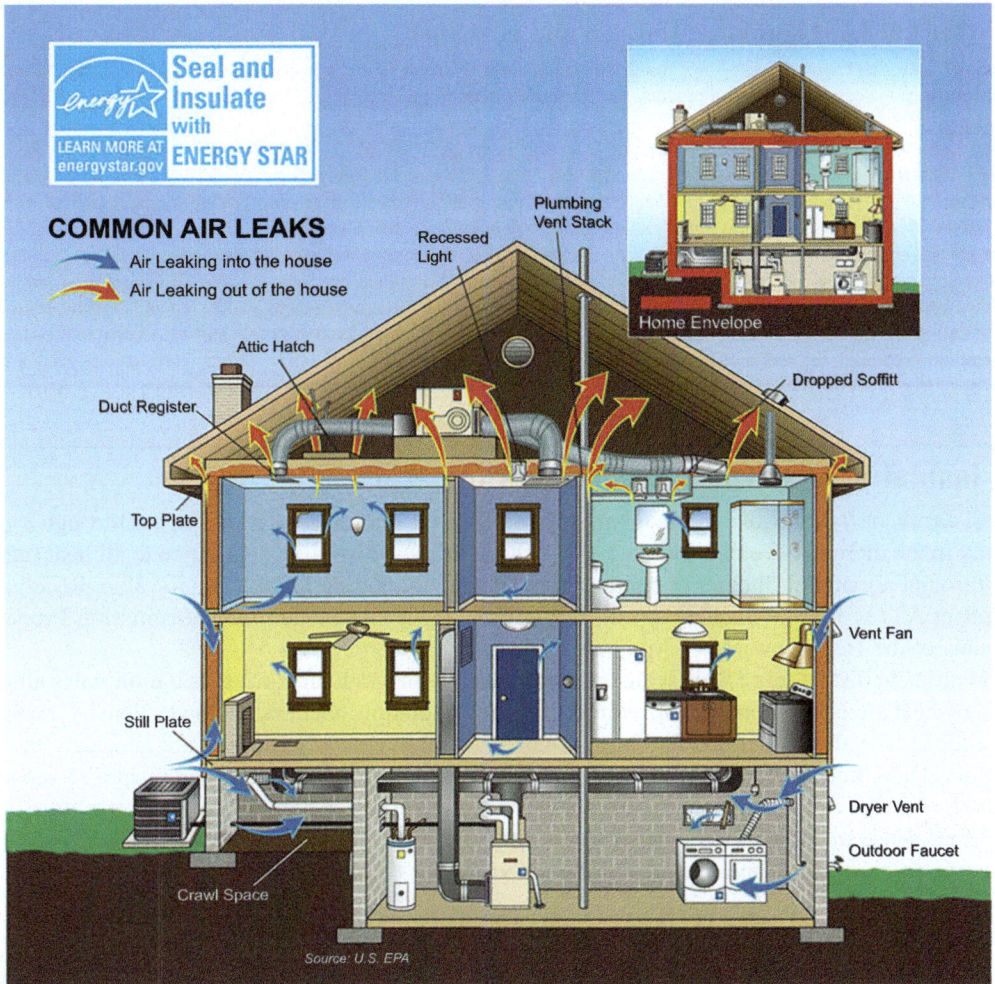

FIG. 8.2

Common thermal leakage pathways in the thermal envelope of a building structure cause a "stack effect" of upward motion of air leakage (public domain image US Energy Star/DOE/EPA).

Conventional wisdom says a building "needs to breathe," but this is largely a fallacy (Bailes, 2022). Indoor fresh air exchange is important for indoor air quality. Likewise, in most cases, it is beneficial for sufficient moisture permeability in wall structures so that building materials have a direction for drying. However, the real mantra in green building is to "build it tight and ventilate right." In other words, a tight thermal envelope is important with the planned introduction of ventilated fresh outdoor air (Example 8.1).

> **EXAMPLE 8.1 Tightening the thermal envelope**
>
> **Concepts:** Describe the processes of heat transfer as an attic and the room below it heats on a sunny summer day. What are three measures to reduce this heating process, and how does each work?
>
> **Solution:**
> - All forms of heat transfer occur in the summer as an attic heats up. The sun's radiant energy heats the roof shingles. This heat is transferred to the roof structure by conduction. Heat in the roof structure is then reradiated into the attic, warming the air in the attic. This heat is then transferred by conduction through the ceiling joists, insulation, and ceiling drywall into the conditioned space. Convection can cool attics by allowing cooler ambient air to enter through soffit and gable vents and exhaust through ridge vents. Within the attic space, convective currents may also form within the insulation.
> - The attic can be kept cooler by a higher reflectivity roof surface, a radiant barrier to prevent transmission of radiation from hotter to cooler surfaces, air sealing of the thermal boundary (ceiling) to prevent infiltration and exfiltration, more effective attic insulation to reduce conduction through the ceiling and attic ceiling rafters and increasing convection or ventilation through the roof assembly to reduce the temperature.

8.3 Insulating materials and *R*-value in series

The resistance, or *R*-value, of a thermal barrier quantifies the resistance to heat transfer through a given pathway in the thermal envelope (Fig. 8.3; Table 8.1). *R*-value quantifies resistance to all heat transfer types through a material, including conduction, convection, and radiation. The *R*-value procedure is detailed in ASTM C518, "Standard Test Method for Steady-State Thermal Transmission Properties by Means of the Heat Flow Meter Apparatus."

The ultimate insulator is a well-sealed vacuum panel with a radiant barrier, as it eliminates all three means of heat transfer—conduction, convection, and radiation. Generally, lower-density materials have a higher *R*-value. Materials that have small voids or stagnant air spaces are generally the best insulators. Think light, fluffy materials such as fiberglass batts, mineral wool, rigid or spray foam products, and loose-fill cellulose. Generally, higher *R*-value materials also minimize noise transmission as well. More dense materials such as stone, masonry, concrete, and in particular metals are relatively conductive (high *k*-value) and thus have a low *R*-value per depth of material.

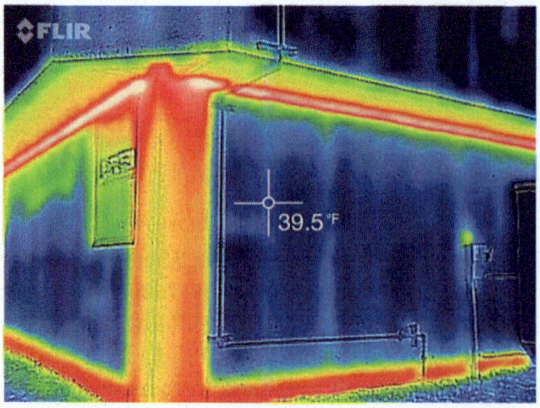

FIG. 8.3

Infrared detail showing typical areas of heat loss (low *R*-value areas show up as warmer *yellow* and *red areas*) including windows, framed-wall members, corner losses through the concrete structure, joints between vertical and horizontal surfaces, concrete slab foundation, and cantilevered floor in a mid-20th Century concrete laboratory building on the campus of New Mexico Tech.

Table 8.1 Average thermal resistance (*R*-values) in US customary units for common building materials (from ColoradoEnergy.org) (varies by manufacturer, installation, etc.).

Material	*R*-value (ft^2-°F-hour/BTU)
Outside air film on wall (winter)[a]	0.17
Inside air film on wall (winter)[a]	0.68
½" Plywood sheathing	0.63
¾" Plywood subfloor	0.93
¾" Hardwood flooring	0.68
Tile, Linoleum	0.05
Carpet (fibrous pad)	2.08
Carpet (rubber pad)	1.23
3 ½" Fiberglass batt	13
5 ¼" Fiberglass batt	21
12" Fiberglass batt	38
3 ½" Softwood stud (2×4)	4.38
5 ½" Softwood stud (2×6)	6.88
½" Drywall	0.45
Concrete block 4"	0.8
Concrete block 8"	1.11
Concrete block 12"	1.28
Brick 4" common	0.8
Brick 4" face	0.44
Single pane glass	0.91
Double pane ½" airgap	2.04
Double pane ½" airgap Low-E	3.13
1 ¾" Hollow core door	2.17
1 ¾" Solid core door	3.03
2 ¼" Solid core door	3.7
Metal door with 2" of urethane	15
Air spaces ½" to 4"	1.0
Material	***R*-value per inch**
Loose-fill fiberglass	*R*2.2–4.3/in.
Loose-fill cellulose	*R*3.6–3.9/in.
Fiberglass batt	*R*4/in.
Rock wool batt	*R*3.5/in.
Polyurethane foamed in place	*R*6.25/in.
Foil-faced polyisocyanurate board	*R*7.2/in.
Extruded polystyrene	*R*5/in.
Expanded polystyrene	*R*4/in.
Poured concrete	*R*0.08/in.
Autoclaved aerated concrete	*R*1.05/in.
Vermiculite	*R*2.13/in.

[a]*Air film R-values change with season and surface orientation.*

A cross-section of a wall showing key components of the thermal envelope is shown in Fig. 8.4 for typical construction. The figure shows an analogous electrical circuit diagram where thermal resistances for each component are represented by an electrical resistor. The wall assembly must perform several functions beyond the structural support of the roof. Besides insulating, the wall cross-section provides a rain barrier, an air barrier to prevent infiltration/exfiltration, a moisture barrier to prevent water vapor migration, a drainage plane to drain liquid water, a thermal boundary, and finish layers on both the interior and exterior as well.

Many options exist for improving thermal boundaries using advanced framing, and one common option is also shown in Fig. 8.4. The use of deeper framing members spaced more widely allows enhanced insulation in the wall cavity and reduced thermal bridging through structural members. The latter can also be reduced further by using a thermal break such as aerogel on the faces of structural members (particularly important if they are metal). Other advanced framing techniques include double-stud exterior walls with offset studs as well as structural insulating panels (SIPs), which will be discussed further below.

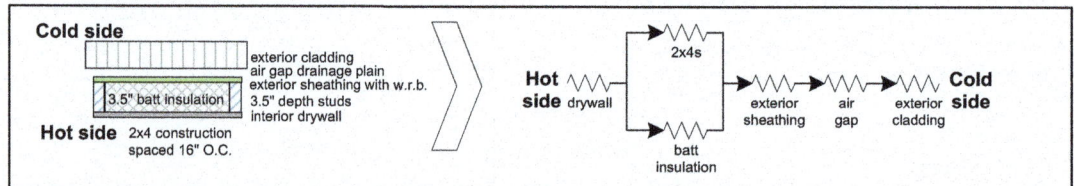

(a) standard 2x4 structural framing

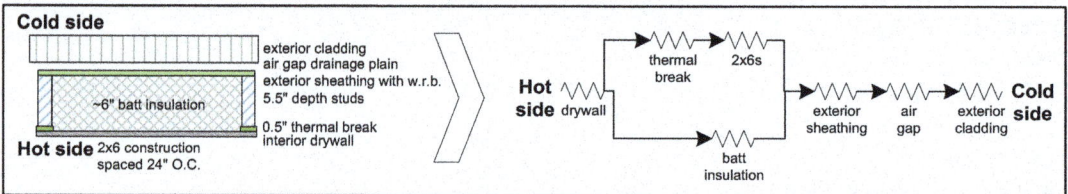

(b) advanced energy efficient 2x6 framing

FIG. 8.4

Wall cross sections from above for (A) standard 2×4 framing and (B) advanced, energy-efficient 2×6 framing showing the components needed for a thermal barrier, air barrier/moisture barrier/liquid water barrier provided by a water-resistant barrier (W.R.B.), and an air gap and drainage plane. Also shown in the analogous electrical circuit diagram for each case.

The thermal transmittance (U) and thermal resistance expressed as an R-value are reciprocals. The U-value is a function of the material's conductivity coefficient (k) and the depth of the conducting layer (l) in a conduction-only simplified case (Eq. 8.1):

$$U = \frac{k}{l} = \frac{1}{R} \qquad (8.1)$$

Two key distinct geometries of heat transfer—series or parallel resistors—are possible through the structure's thermal envelope. The series configuration is the simplest, where heat is only transferred through materials one following the next (and recall that the net heat transfer is always from the hotter to the cooler body). The composite R-value of a material composed of multiple layers that are perpendicular to the direction of heat transfer can be thought of as resistors in series (Fig. 8.4). The composite R-value is the sum of the R-values of the layers that form each layer i of the material and add linearly (Eq. 8.2). The example that follows shows the process (Example 8.2).

$$R_{total} = \sum_i R_i \qquad (8.2)$$

EXAMPLE 8.2 R-values of wall composite

Problem: What is the approximate R-value of a wall composed of a 4″ concrete block that has 2″ of rigid polyisocyanurate board in the middle and a ½″ sheet of drywall on the inside?
Given: 4″ concrete; 2″ polyiso; ½″ drywall.
Find: R-value of composite wall.
Assume: uniform profile.

Solution:
R_{total} is R_{series} of the components.
From R-value data:
$R_{polyiso}$ is ~7.2/in.,
R_{block} is 0.8R for 4″ block,
$R_{drywall}$ is 0.45R for ½″ drywall.
Calculate R_{total} as the sum of its components:
$R_{tot,series} = \Sigma R_i = 0.8R + \left(\frac{7.2R}{in} \times 2\,in\right) + 0.45R = 15.65R$

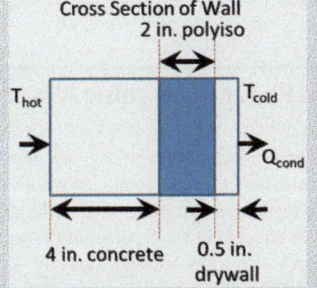

Cross Section of Wall
2 in. polyiso
T_{hot}, T_{cold}, Q_{cond}
4 in. concrete, 0.5 in. drywall

Notably, the R-values of materials are not entirely static in time and with respect to environmental conditions. Environmental conditions and materials aging can alter the insulating value of materials (usually downward). The impact of wetness is a prime example of the variability of R-values, where materials that become wet lose considerable thermal resistance (and can degrade due to rot). Some materials perform slightly better in insulating in cold vs warm climates, and some can have lower insulating values with extreme conditions (e.g., in extreme cold, R-values can change). Other examples of changes include the settling of loose-fill cellulose or other loose-fill materials into more compact layers with lower R-values and the off-gassing of blowing agents that contributed to thermal resistance in foam insulation.

Calculating R-values of components and composite wall structures in a building is important to quantify the need for space conditioning and Heating, Ventilating, and Air Conditioning (HVAC) system size. Manual J calculational models are available to calculate heat loss and gain for an entire building in a given location. The user inputs construction details, including dimensions and best knowledge of R-values for windows, walls, doors, attics, etc. The program uses the nearest meteorological data for finding the design temperatures for the hot and cold seasons. Typically, the (near) maximum heat gain in the summer and the (near) maximum heat loss in winter must be compensated for with HVAC systems, which will be discussed in the next chapter. The specification an energy-efficient thermal envelope can reduce the need and costs for installing larger HVAC systems.

8.4 Calculating parallel pathway *R*-values

The composite *R*-value of a material with multiple parallel pathways for heat transfer is analogous to parallel electrical resistors, modified with a face surface area weighting factor (Fig. 8.4). The most straight-forward approach is to find the *U*-value (reciprocal of *R*) of each material, weight each parallel pathway *U*-value by the face area fraction, and then add them to form a composite *U*-value (Examples 8.3 and 8.4). The composite *R*-value then is the reciprocal of the composite *U*-value (Eq. 8.3).

$$U_{\text{tot}} = \sum_i (\text{AreaFrac}_i \times U_i) = \frac{\sum_i (\text{Area}_i \times U_i)}{\text{TotalArea}} \tag{8.3}$$

EXAMPLE 8.3 *R*-value of vermiculite-filled CMU

Problem: What is the approximate *R*-value of an 8″ CMU (concrete masonry unit, use a nominal block size of 8″ × 8″ × 16″) that is filled with vermiculite? Use the nominal *R*-value of the 8″ block of *R*-1.1 for the block (entire depth). Assume 2″ webs on all sides. Also, assume the *R*1.1 is effectively the *R*-value of the web pathways themselves (i.e., a block of *R*1.1 in parallel with two 4″ cavities filled with vermiculite). This is a simplification, and a more detailed calculation is in the exercises.
Given: *R*1.1 block, parallel with two 4″ cavities filled with vermiculite
Find: Effective *R*-value of filled block
Assume: Uniformly filled cavities; ignore mortar joints

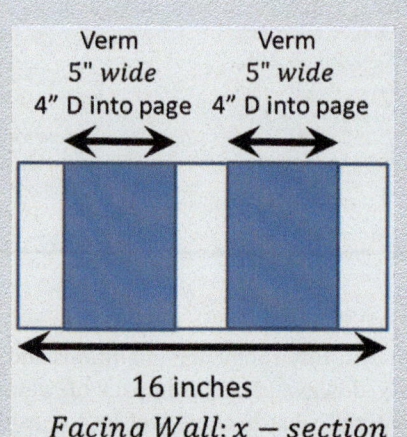

Solution:
R_{total} is R_{parallel} of the CMU block in parallel with vermiculite-packed cavities
From data on approximate *R*-values, R_{verm} is ~2.1/in.
Calculate *U* first with the face area fraction of each material:

$$U_{\text{total}} = \sum_i \text{AreaFrac}_i \times U_i = \left(\frac{6}{16} \times \frac{1}{R_{\text{cmu}}} + \frac{10}{16} \times \frac{1}{R_{\text{verm}}}\right) = \left(\frac{6}{16} \times \frac{1}{1.1R} + \frac{10}{16} \times \frac{1}{4\,\text{in.} \times 2.1R/\text{in.}}\right) = 0.415$$

$$R_{\text{total, parallel}} = \frac{1}{U} = \frac{1}{0.415} = 2.4R$$

This compares to an approximate *R*1.1 for the CMU unfilled. Obviously, filling with a loose-fill material helps the thermal resistance (doubling it or halving the heat transfer). This would still be a poorly insulated wall, as *R*2.4 is pretty minimal.

EXAMPLE 8.4 *R*-value of parallel configuration

Problem: Find the *R*-value of a wall composed of continuous 0.5" drywall, wall interior (w.i.) is 2" ×6" framing at 24" on center O.C. spacing plus rigid polyurethane board of 6" depth in cavities and a continuous exterior cladding of 4" brick? Assume 7.2R/in. for polyiso and 1.25R/in. for the studs. Draw a diagram facing the wall with height *z* and assume nominal framing member sizes given.

Given: 2"×6" with 24" spacing; Polyiso; ½" drywall and 4" brick
Find: Nominal sizes for studs 2"×6", 1.25R/in. pine, and 7.3R/in. polyiso
Assume: Uniformly filled cavities; nominal dimensions are actual

Solution:

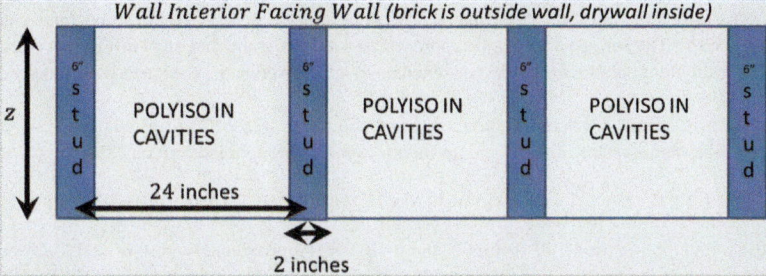

$R_{total} = \Sigma R_{series} = R_{drywall} + R_{wallinterior} + R_{brick}$
First, find the wall interior $R_{w.i.}$ from the reciprocal of $U_{w.i.}$.

$$U_{w.i.} = \frac{U_1 A_1}{A tot} + \frac{U_2 A_2}{A tot} = \left(\frac{1}{6''*1.25R/in.} \times \frac{2z}{24z} + \frac{1}{6''*7.2R/in.} \times \frac{22z}{24z} \right) = 0.0323$$

$$\text{or } R_{w.i.} = \frac{1}{U_{w.i.}} = \frac{1}{0.0323} = R\,30.9$$

$$R\,total = R\,0.45 + R\,30.9 + R\,0.44 = R\,31.8$$

This is a well-insulated wall section. The reduced thermal bridging with deeper studs and wider spacing plus the deeper insulation in the wall cavities all help this.

8.5 Estimating heat loss through the thermal envelope

To maintain a thermal steady-state, heat loss through the thermal envelope must be compensated for by mechanical means such as HVAC systems. The steady-state rate of heat loss through the thermal envelope can be estimated with the following equation, where *U* is the composite thermal transmittance of the entire structure in BTU/(hour-ft²-°F), A is the area in ft², and ΔT is the temperature difference across the surface in °F (Eq. 8.4):

$$\dot{Q} = UA\Delta T \qquad (8.4)$$

This can also be calculated similarly in SI units of W/(m²·K), m², and K, respectively. Integrating the rate of heat transfer over time gives the total heat transferred (Example 8.5).

> **EXAMPLE 8.5 Estimating global building conductive heat loss**
> **Problem:** Use a global total building area of 235,000 km² to estimate the space conditioning power consumption globally. Assume there are one billion buildings with a 4 m height and that each has a square profile. Assume an average R-value of 12 (in US customary units). Use a global average ambient temperature of 12°C and an indoor temperature of a comfortable 22°C. Estimate the global space conditioning power use and compare it to a top-down estimate.
> **Given:** 235,000 km² of building area, 4 m height,
> **Find:** Power for space conditioning
> **Assume:** $R12$, $T_{in} = 22°C$, $T_{out} = 12°C$
>
> **Solution:**
> $R12$ is approximately $R_{si} = 2$ such that $U = 1/R = 0.5 \, W/(m^2\text{-}K)$
> Floor area $= 235 \times 10^9 \, m^2$, which is equivalent to a square of dimension 484,767 m on a side.
> Assume 1 billion square structures of 235 m² apiece. The floor area of each structure can be approximated as a square with a dimension of 15.3 m. The surface area can be approximated as four walls that are 4 m × 15.2 m with a ceiling area of 235 m². We can ignore the floor area losses (some heat exchange) and just consider the surface area exposed to the ambient air temperature.
> $SA = 4 \, (4 \, m)(15.2 \, m) + 235 \, m^2 = 478 \, m^2$ of surface area per house, or $478 \times 10^9 \, m^2$ global total.
> Use an average global temperature of 12°C and an indoor comfortable temperature of 22°C for $\Delta T = 10°C$.
>
> $$\dot{Q} = UA\Delta T = 0.5 \frac{W}{m^2 \cdot K} \times 478 \times 10^9 \, m^2 \times 10 \, K = 238 \times 10^{10} \, W = 2.38 \times 10^{12} \, W = 2.38 \, TW$$
>
> The global steady-state power use in all sectors is about 18 TW. Approximately 40% of that is for buildings, and approximately one-third of that is for space conditioning. This gives a space heating power use of (18 TW) (0.4) (0.33) = 2.38 TW!
> This "back of the envelope" calculation was somewhat serendipitous in that it worked with this first set of numbers, although they are generally reasonable. The need to heat (or cool) would vary significantly depending on location, season, and other factors. This R-value is often the code requirement for walls. Older structures would be worse and newer structures better. No explicit consideration was made of convective or radiative losses although the R-values in theory include them. As a first-order global one, though, it is not a bad estimate!

The local climate information can be integrated into this calculation as well by incorporating the heating degree days (HDD) for the site. One source of climatological data for energy purposes is www.degreedays.net. Recall the degree days at the site are found from the deviation from a baseline temperature, often taken as 65°F. So, if the mean temperature for the day is 32°F, it will add 33 HDD for that day to the seasonal total. Analogously for cooling degree days, CDD, the deviation in the opposite direction is examined. Using the annual HDD, the following equation can be used to estimate annual space heating requirements (Q), where U is the composite conductivity of the structure through the surface area A (Eq. 8.5). This can be divided by system efficiency to give the input energy needed.

$$Q = UA \, (\text{HDD}) \left(\frac{24h}{day}\right) \tag{8.5}$$

8.6 New innovations in insulating materials

The choices in insulating materials continue to evolve broadly and rapidly in terms of the types, materials, installation options, and effectiveness of insulating materials. The current dominant insulation choices include fiberglass, foamboard, onsite blown materials, and loose-fill cellulose.

New materials include many different form factors and are derived from animal (wool), mineral (rock wool, mineral wool, fiberglass), plant-based (hemp wool, recycled denim, cellulose), petroleum-based (spray and rigid foam products), and even fungi (mycelium). These choices vary immensely in job-specific applicability, insulating value, cost, market availability, longevity, and health considerations. The market is still dominated by the ubiquitous fiberglass batt insulation due to its cost effectiveness. A few main newer options are discussed below, and it should be emphasized that some some of these will not persist in the market while new options will appear regularly.

Though it has a long history, mineral wool and other mineral-based materials have new and increasing applications for insulation. The material can be spun into fiberglass batt-like material or rigid panels. It has the advantages of being fireproof, having a reasonably high R-value per unit depth, and having low-embodied carbon. The manufacturing cost-effectiveness and health effects are not yet entirely characterized, however.

One of the most effective insulators developed over the last several decades is aerogel, which has the highest R-value per in. ($R10+$/in. in US units) of commercially available material due to its minimal thermal conductivity (~ 0.013 W/m K) (Cuce et al., 2014). Aerogel was developed decades ago for space applications, though more recently it has become of interest to building applications and consumer products. Aerogel is a silicate material that is extremely low density and thus is mostly air. An interesting application by Thermablok is to use an adhesive layer that is the width of a standard stud and ¼″ thick with $R10.3$/in. on the face of metal studs as a thermal break to reduce thermal bridging. Though still expensive, aerogel is making its way into more commercial products and applications in building-insulating materials.

Structural insulating panels (SIPs) are prefabricated wall sections that typically sandwich a rigid foam board insulating layer between outside layers of sheathing, such as Oriented Strand Board (OSB) (Fig. 8.5). Insulated concrete forms are a similar application where the foam board forms the outside shell with connecting structures in between them. The concrete for a foundation wall is then poured into the interior space and allowed to cure. Unlike wood forms, the foam board is left in place as foundation wall insulation. Exterior Insulating Finishing Systems, or EIFS, have evolved significantly over the last decades as well. It is an outside the existing wall system for insulating under an exterior cladding, typically a synthetic stucco-type material. EIFS can work for recladding existing buildings as well and are also a continuous insulating layer on the exterior of a wall structure.

Novel bio-based insulating materials have emerged in recent decades. For example, lignin foams have been developed to replace polyurethane foams that are difficult to recycle (Sternberg and Pilla, 2023). Another interesting product is the use of hemp fiber insulation by companies such as

FIG. 8.5
Insulated concrete forms (ICFs) comprise the basement walls, and structurally insulated panels (SIPs) comprise the first-floor walls of this residence under construction. Both assemblies minimize thermal bridging in a structure.

Image courtesy of Dr. Mark Cal.

Hempitecture. The fibers are very resilient, and the plant grows in a rapidly renewable manner. Mycelium, the rootsy structural component of fungi such as mushrooms useful for their insulating properties, is also in nascent building products. Bio-products also include a "low density" fiberboard that can replace traditional wood products. It features trapped air formed into semirigid wood fiber panels that can reach $R6$ per inch plus are waterproof and permeable to water vapor to allow drying. These are renewable more quickly than woody materials and have a lower carbon footprint. It remains to be seen which of these products will occupy a significant market role.

8.7 Windows and doors: Minimizing heat transfer while allowing access, light, ventilation, visibility, and views

Windows (including skylights) and doors, by their nature, are typically one of the weakest links in the thermal envelope. However, what would a structure be without them? The best thermal envelope would omit windows and use a single exterior door. This would be a dark and arguably depressing interior space lacking any views, ambient light, fresh air introduction, and needing copious supplemental artificial lighting. Windows and doors are the portal from the built environment to the natural world and critical to the transmission of light, ambient air, viewsheds, and visibility to an indoor environment devoid of these.

Windows have improved remarkably in the past 50 years in terms of their resistance to all means of heat transfer. Glass panes, due to their conductivity and thinness, are poor insulators (Table 8.1). This is overcome by using multiple panes with effective sealing and low-emissivity-coated glass. The typical single-pane window of the mid-20th Century and earlier has now been replaced by dual-pane or even triple-pane windows.

Improvements in modern windows include:

- Multipaned windows with two or three glass panes to reduce heat transfer through the window
- Optimized gaps between glass panes to minimize convection between the panes.
- Window frames are designed with effective air sealing, dead air spaces, and multiple barriers preventing heat transfer
- Low-emissivity coatings that act like an infrared mirror, minimizing radiant heat transfer through the window
- Low-conductivity materials, including fiberglass or composites, compose the frame or as thermal breaks in the window unit
- Lower-conductivity inert gases such as argon can fill the gaps, reducing conductive losses
- Advanced "smart" glass varieties that include autochromic glass that self-adjusts their transmission based on solar intensity
- Effective window sizing, placement, orientation, and shading will enhance or reduce solar gain depending on the design, as will be discussed later in Chapter 13 with passive solar

The key energy concerns with windows in a structure are (a) proper placement and orientation, and (b) using suitable units for the climate zone. This is quantified by a low U-value, high visible transmittance (VT) for maximum daylighting and views, low infrared transmittance for minimizing heat transfer, low ultraviolet (UV) transmittance for protection of humans and materials, and an appropriate solar heat gain coefficient for the situation (SHGC) (Table 8.2). The latter depends on window size, placement, orientation, site location, and climate, plus occupant needs and preferences. A large picture window facing west in a warm, sunny climate zone would benefit from low SHGC, while a small south-facing window in a northerly climate would benefit from a high SHGC.

Illustration of the improvements in window performance is shown by the window profile of a single pane of glass and a dual-pane, low-e coated window (Fig. 8.6). A summary of window transmission properties is given in Table 8.3 for common window types ranging from single-pane to triple-pane to low-e coated windows. Both the increasing number of panes and the coating types help reduce infrared (IR) heat transfer, minimizing UV transfer while giving some reduction in visible light transmission as well.

Table 8.2 The key parameters that quantify the performance of windows.

Parameter	Description	Range	Notes
SHGC	Solar heat gain coefficient	0–1	The fraction of the incident solar spectrum transmittance through a window. Goals depend on climate and window orientation.
VT	Visible transmittance	0–1	In most cases, high values of VT are desired except in glare or locations where some tinting is desired
IR	Infrared transmittance	0–1	Low values of IR transmittance are desired in almost all cases to reduce heat transfer in all seasons
UV	Ultraviolet transmittance	0–1	Low values of UV transmittance are desired to prevent UV damage to interior finishes and materials
U-value	Thermal transmittance	$U > 0$	Low U-values are desired for reducing energy loss, where modern dual-pane windows have U-values ~0.3 in US units of BTU/(h·ft^2·°F)
Leakage	Air infiltration	Leakage > 0	Air leakage rate of air through the window when properly installed (lower values are better)

FIG. 8.6
Window profiler showing the transmission properties of a single-pane of glass (*left*) and a low-E coated dual-pane window (*right*).

Table 8.3 For several common window types, the key parameters that quantify the performance of windows include transmission decimal fraction in the ultraviolet (UV), visible, and infrared (IR) ranges and solar heat gain coefficient.

Parameter	UV transmission	Visible transmission	IR transmission	Solar heat gain coefficient (SHGC)
Single-pane (clear glass)	0.72	0.87	0.80	0.87
Dual-pane (clear glass)	0.48	0.78	0.64	0.77
Triple-pane (clear glass)	0.37	0.70	0.51	0.60
Hard-coat low-E	0.42	0.73	0.55	0.71
Double-silver low-E	0.16	0.70	0.08	0.32
Triple-silver low-E	0.03	0.55	0	0.27
"Advanced comfort"	0.16	0.68	0.08	0.28

Fig. 8.7 shows thermal images from the interior both on a cold winter day of a (A) typical vinyl-clad double-pane, argon-filled low-e coated window from the (A) inside and (B) outside, and a raised panel wood door with a single-pane windows at the top from the (C) inside and (D) outside. Indoor and outdoor temperatures are approximately 65°F and 25°F. Note the colder areas around the window frame, particularly near the corners. Also, note the thermal reflection of the photographer due to the low-e coating reflecting long wavelength IR radiation! On the interior, the glass area is ~58°F and the frame is in the lower 50s °F, whereas wall temperatures are ~63°F. In a double-hung window that slides

vertically, small air leaks occur, particularly due to the movable window frames. The wooden double doors in Fig. 8.7 show both conductive losses through the wood panels (note the colder recessed panels, which are thinner) as well as some infiltration of cold air at the edges. Still, the largest losses per unit area—shown as the coldest, darkest blue areas—are through the single-pane small windows near the top of these doors.

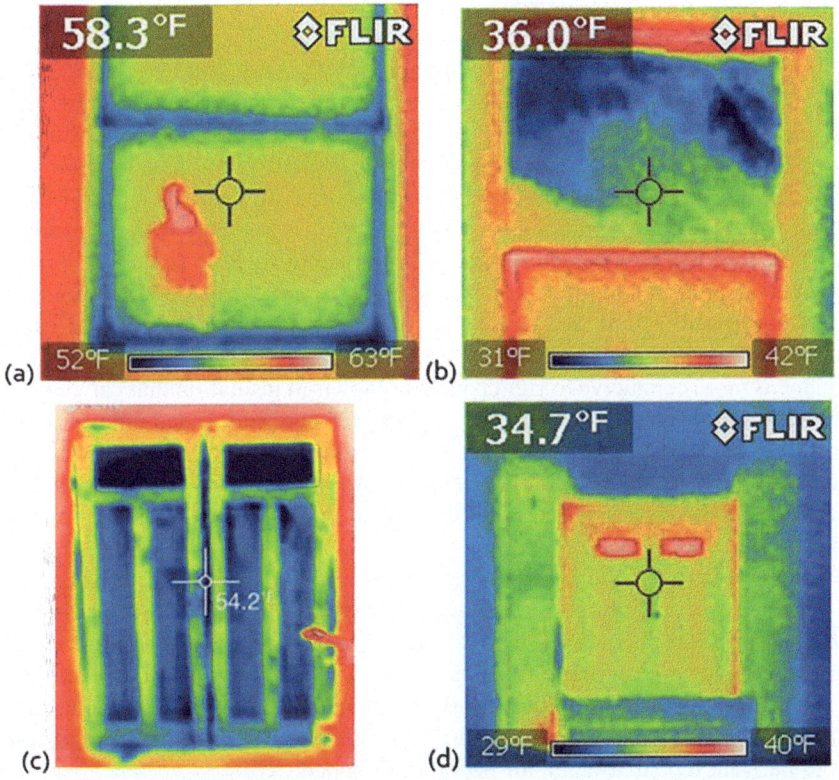

FIG. 8.7

Infrared interior view on a cold winter day (exterior temperature ~25°F, interior temperature ~65°F) of a typical double-hung vertically sliding double-pane window ($u=0.31$) from (A) the inside (not the IR reflection of the photographer) and (B) the outside, and a recessed panel wood double doors with a small single pane window at the top from (C) the inside and (D) the outside door.

8.8 Roofs and attics including "cool roof" technologies

Per square foot, attics are arguably the most accessible and cost-effective area for insulating and effective both in summer and winter. Due to a stack effect, the loss of exfiltrating buoyant warm air through ceiling leaks as well as conduction through the ceiling makes it important in the winter

(Fig. 8.2). Conduction plus radiation downward from a typically hot attic in the summer makes it an equally important pathway for heat gain.

This raises the question of where to insulate and vent in the attic, and one should think about its use and heat transfer. Is the attic now or in the future desired as a conditioned space? Are there currently HVAC or other mechanicals in the attic? If the HVAC system or significant ductwork is in the attic, or if there is unavoidable air exchange with the main living area, it may make more sense to bring it into the conditioned space by insulating the roof plane.

If the attic is a buffer space that need not be conditioned, insulating at the attic floor level (after air sealing) is generally preferred to insulating in the roof truss cavities. The quality and quantity of insulation are often improved above the ceiling versus below the roof. Any insulating material will tend to fall toward the floor with gravity anyway, so any loose-fill material requires attic floor installation. It is easier to reinsulate or add insulation to an attic floor with loose-fill materials as compared to the roof plane. It can also be easier to achieve high R-values with attic floor insulation, where even approaching ~1 m of insulation is possible in some cases as compared to insulating in the roof bay cavities. A challenge with insulating attic floor bays (and roof bays in some cases too) is the areas where the walls intersect the roof if these are pinch points. In a retrofit situation, there can be little that one can do in these narrow areas, which are often a thermal weak spot.

The roof plane is more conducive to rigid panels or blown-in insulation that can remain suspended. Roof SIP panels are quite effective as continuously insulated and well-sealed units with a considerable depth of insulating material (6–10 in.) allowing insulation at the roof plane. These are a particularly viable choice for new construction where the attic is within the thermal envelope, which extends to the roofline rather than the ceiling joists on the attic floor that separate the attic from the upper story of the building.

Venting of the roof structure is important and differs whether it is part of the thermal envelope or outside of it. Ventilation of attic spaces is to reduce attic overheating, particularly during summer when the hot attic can conduct, convect, and radiate additional heat into the interior conditioned space. Passive is preferred as active attic ventilation (e.g., attic exhaust fans) can cause problems as it depressurizes the attic, enhancing air exfiltration through openings in the ceiling. When insulation is in the roof cavities, ventilation is particularly important but can be difficult. When insufficient, it can lead to roof ridge rot due to buoyant moisture-laden air becoming trapped.

In all but the most northerly heating-demand climates, light-colored or preferably white roofs facilitate moderate attic temperatures. The higher reflectivity drops roof and attic temperatures by up to 20°F (10°C), according to Florida Solar Energy Center (FSEC) research. Super white coatings and paints are currently under intensive research that could lead to the reflection of incoming solar radiation and high emissivity that enhances IR emission.

8.9 Basements and crawlspaces

Basements and crawlspaces are other often necessary "thermal evils" in a building. They are widely considered unpleasant spaces unless you desire poorly accessible, poorly conditioned, humid, lumpy, poor air quality spaces with near darkness and often strewn with construction debris. They can, however, be a handy location for the mechanicals of the building and storage, particularly basements. Crawlspaces do offer a way to elevate the living space above grade and provide a semiaccessible space for plumbing, electrical, one that offers far easier modification than a slab foundation. However, there is a reason their square footage is not included as "livable space," though sometimes basements are very much lived in. Appropriately, they are well-known for various creepy-crawler insects and the low point to which any liquid leaks will drain (curious to know if your dog had any unreported accidents? Check the crawlspace).

Like the attic, the crawlspace offers two major options for insulating: (a) bringing the crawlspace into the conditioned space by insulating the exterior perimeter, or (b) insulating the cavities between the floor joists separating the crawlspace from the occupied level above. The first choice is almost always preferred and is best if the foundation walls themselves are brought into the conditioned space for added thermal mass. Adding insulation is typically easiest and most effective with the highest continuity on the exterior around the foundation, second on the interior of the crawlspace wall, and lastly the floor joist cavity option. In most cases, the crawlspace can be sealed as tightly as possible, particularly if a vapor barrier is present on the ground surface, and includes air sealing all penetrations such as hose bibs.

This raises the question of venting crawlspaces with its energy penalty. Radon, water vapor, and other gases from the soil can migrate from the soil, while mold, mildew, and rot issues can develop in the often-moist environment. The venting of crawlspaces was intended to reduce humidity and radon issues, which can largely be controlled more effectively and with less of an energy penalty with an air-sealed, appropriate membrane barrier on the ground. The best approach is a completely encapsulated crawlspace that isolates the conditioned space from the soil with a sealed membrane (Bailes, 2022). The space should be well-insulated and air-sealed in the thermal envelope as well. In areas with acute radon, a passive or active exhaust system drawing from underneath this membrane can help reduce radon levels. For example, after the author upgraded the insulation and air sealing in a home along the Front Range of Colorado, the radon levels reached a very elevated 10–12 PCi/L. A radon mitigation system including membrane, perforated pipes in the crawlspace, and an exhaust stack with a 20 W fan reduced this by an order of magnitude and kept it well below the EPA threshold of 4 PCi/L. These systems do work, even in a high-radon area.

8.10 Thermal mass and thermal bridging

Thermal bridging is a common unwanted occurrence in wood and metal-framed building construction (Fig. 8.8). The structural members often have low thermal resistance and serve as "bridges‘ for energy loss. Solutions include installing a thermal break on either side of the structural member, adding a continuous layer of insulation outside of the structure, using dual-framed walls with offset studs, and other possible mitigation measures. The thermal bridging through the webs in cinder blocks makes the drill and fill method of insulating the cavities of cavity or block walls only marginally successful (Example 8.6).

FIG. 8.8

Thermal image of the exterior of a university building wall on a chilly winter morning (~30°F). Warmer colors indicate areas of heat loss, and cooler temperatures are indicative of less heat transfer from inside the heated building. The visible wall structure shows the thermal bridging across the metal studs in the wall. Also note the warm metal window frames, likely poorly insulated areas above the windows, and heat transfer through the concrete slab at the base.

EXAMPLE 8.6 Thermal bridging and comparing loss pathways

Problem: Let's compare the thermal losses from the components in the wall shown in Fig. 8.8. Take a 10 m length of the example wall structure and use a height of 4 m. The 2 m × 2 m windows are spaced 10 m apart and are characterized by a 5 cm × 5 cm metal frame holding a 2.5 cm thick glass pane with $U = 5$ W/m·K. Assume the concrete slab depth through which it conducts is 0.25 m tall, and the conducting layer can be assumed to be 0.25 m deep along the entire bottom of the wall.
(a) Calculate the conductive heat loss on a winter day through the entire wall, including the slab, and each component of the wall (windows, studs, slab, and wall cavities).
(b) Find the effective R-value of the composite wall.
(c) If a ½" layer of aerogel had been installed on the stud faces, how much reduction in total heat loss would have occurred?

Given: 10 m × 4 m wall section with three windows each 2 m × 2 m glass plus a 5 cm × 5 cm Al frame.
Find: Thermal power transmitted by component
Assume: $R18$ in 6-in. wall cavities, $T_{in} = 22C$, $T_{out} = -3C$; 16-in. (40 cm) spacing between studs; R-value for metal studs is ~0.1R/in.; edge loss only through slab

Solution:
Conductivity Coefficients:
The conversion from R in US customary units to R_{si} is a divisor of 5.678
Glass as a single-pane window is approximately $U = 2.5$ W/(m²·K) for a commercial ½" pane

Continued

> **EXAMPLE 8.6 Thermal bridging and comparing loss pathways—cont'd**
>
> Metal window frame: $(0.1R/in.)(2\ in.) = R\ 0.2$ such that $R_{si} = 0.035$ or $U = 1/0.035 = 28.4\ W/(m^2\text{-}K)$
> Metal studs: $(0.1R/in.)(6\ in.) = R\ 0.6$ such that $R_{si} = 0.106$ or $U = 1/0.106 = 9.46\ W/(m^2\text{-}K)$
> Wall Area: $R18$ in cavities is approximately $R_{si} = 3.17$ such that $U = 1/R = 1/3.17 = 0.315\ W/(m^2\text{-}K)$
> Concrete slab: $0.25\ m$ of concrete has $R = 1$ such that $R_{si} = 0.176$ or $U = 1/0.176 = 5.678\ W/(m^2\text{-}K)$
>
> Surface Areas:
> Assume single-member framing on top and bottom of windows
> Window glass: $A_{glass} = 3\ (2\ m)\ (2\ m) = 12\ m^2$
> Window frames: $A_{frames} = 0.05\ m\ (2\ m)\ (4\ sides)\ (3\ windows) = 1.2\ m^2$
> Metal studs: $A_{studs}\ (0.05\ m)\ (4\ m)\ (1000\ cm/40\ cm) + 6\ (2\ m)\ (0.05\ m) = 5.6\ m^2$
> Wall area: $A_{wall} = 10\ m\ (4\ m) - 12\ m^2 - 1.2\ m^2 - 5.6\ m^2 = 21.2\ m^2$
> Concrete Slab: $A_{slab} = (0.25\ m)\ (10\ m) = 2.5\ m^2$
> Total Area: $A_{total} = 42.5\ m^2$
>
> Conductive Losses:
> Ambient temperature of $-3°C$ and an indoor comfortable temperature of $22°C$ for $\Delta T = 25\ K$.
> Window glass: $dQ/dt = UA\Delta T = 2.5\ W/(m^2\text{-}K)\ (12\ m^2)\ (25\ K) = 750\ W$
> Window frames: $dQ/dt = UA\Delta T = 28.4\ W/(m^2\text{-}K)\ (1.2\ m^2)\ (25\ K) = 852\ W$
> Metal studs: $dQ/dt = UA\Delta T = 9.46\ W/(m^2\text{-}K)\ (5.6\ m^2)\ (25\ K) = 1324\ W$
> Wall area: $dQ/dt = UA\Delta T = 0.315\ W/(m^2\text{-}K)\ (21.2\ m^2)\ (25\ K) = 167\ W$
> Concrete slab: $dQ/dt = UA\Delta T = 5.678\ W/(m^2\text{-}K)\ (2.5\ m^2)\ (25\ K) = 358\ W$
> Total heat loss $= 3.45\ kW$
>
> Effective R-value:
>
> $$U = \frac{\left(\sum_i U_i A_i\right)}{A_{total}} = \frac{((2.5)(12) + 28.4(1.2) + (9.46)(5.6) + (0.315)(21.2) + (5.678)(2.5))}{42.5} = 3.25\ W/(m^2 - K)$$
>
> $$R_{si} = \frac{1}{3.25} = 0.308\ \frac{m^2-K}{W}$$
>
> $$R = 1.75 (ft^2 - °F)/(BTU/h)$$
>
> Entire wall structure: $dQ/dt = UA\Delta T = 3.25\ W/(m^2\text{-}K)\ (42.5\ m^2)\ (25\ K) = 3453\ W$
> The heat loss using the effective R-value of the entire wall matches the sum of the components. In this case, the thermal bridging dropped the effective R-value of the wall insulation by 90%. Though an extreme case (a 75% reduction is not uncommon), this demonstrates the importance of thermal bridging through conductive pathways with relatively small surface area components.

Conversely, thermal mass can be a beneficial feature in buildings for stable thermal control (Fig. 8.9). It uses a high-heat capacity material and a large mass to store thermal energy. It will not cool or heat a space but can dampen the temperature fluctuations indoors, improving comfort. It is a useful part of a passive solar approach as well. If the occupants have a tolerable temperature range of comfort, it can reduce or eliminate the need for active temperature control in the shoulder season, particularly in climates with large day-to-night temperature swings, such as desert regions of the western Unites States.

FIG. 8.9

The San Miguel cathedral in Socorro, NM, built between 1615 and 1626 and expanded and renovated several times since, demonstrates a structure built for longevity, the use of local materials including hand-carved timber vigas and massive adobe walls approximately 5 ft thick that provide thermal mass and greatly reduce the need for space conditioning (http://www.sdc.org/~smiguel/about.htm).

8.11 Other perils, problems, and possibilities in insulating the building envelope

The mantra worth repeating regarding the building thermal envelope is to make it tight and ventilate right. Ventilation techniques will be discussed in more detail in the next chapter. Changing the thermal properties of your interior space, walls, and structure is likely to affect the physics of the building in many positive ways but also can have downsides. The building is a system, and changes in one area often have consequences in other, sometimes in unintended ways. A few of these are worth discussing.

One detail is that many solutions in one region or climate are not effective elsewhere. Using summer night-time ambient cool air flushing works well in a dry climate prone to large temperature swings from day to night. Contrarily, relying on this approach in the Mississippi River delta is not a good plan, as the night-time temperatures rarely drop to cool conditions, and the introduction of humid ambient air makes the HVAC system have to work harder with latent heat loads.

8.11 Other perils, problems, and possibilities in insulating the building envelope

Warm air exfiltration in winter up through the ceiling can lead to the widespread problem in northern snowy climates of ice damming on roofs. Conditioned air from below warms the roof enough to melt snow, which then re-freezes further down on the overhang. Multiple freeze–thaw cycles form an ice dam, which prevents runoff and can degrade shingles and cause roof leaks.

Attic hatch doors are notorious thermal leak spots that are often poorly or uninsulated (Fig. 8.10). The loss mechanisms are often all three heat transfer mechanisms: convective, conductive, and radiative leakage. This allows cold winter air infiltration, conduction through a thin drywall layer, and radiative transfer due to the large ΔT. The improved insulation involved a few layers of rigid foam board, sealing with spray foam, and using a gasket around the frame.

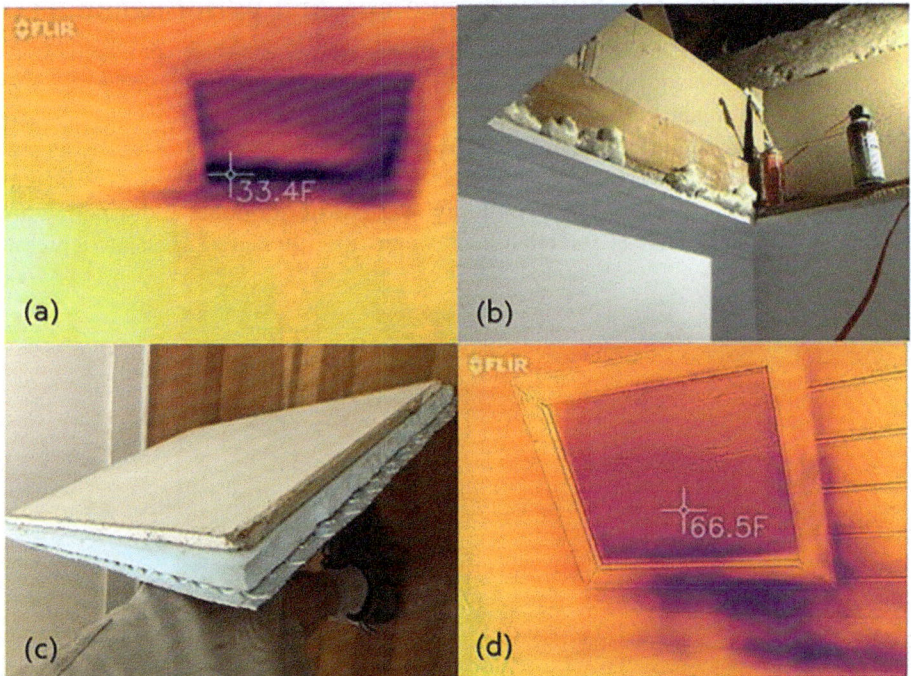

FIG. 8.10

Infrared detail from indoors on a cold winter morning (approximately 20°F) shows a poorly insulated attic hatch (A). After air sealing (B) insulating, a similar picture on the right shows this has been "warmed up" (D). The hatch was air-sealed with spray foam, a gasket was installed around the perimeter, and the hatch was insulated with rigid foam board and a reflective layer plus a fiberglass batt laid over the top (C).

Running ductwork or plumbing in wall cavities, attics, or other spaces that are outside or at the periphery of the thermal envelope can lead to energy losses at best and frozen pipes and plumbing disasters at worst. Isolated rooms are often ones that are difficult to condition and keep comfortable. The room over the garage is one such example that has only one side connecting to the home and is usually at the end of HVAC ducts. Even with insulating well on the five exposed sides, these rooms are prone to being cold in the winter and hot in the summer.

Another problem mentioned earlier is that improving our thermal insulation has increased problems in some cases with water vapor condensation in wall assemblies. Water shedding is one of the key aspects of a building. The thermal envelope has to integrate well with the goals for shedding liquid water and allow for drying of wall structures. Interface areas are the focus, e.g., parapets, chimney penetrations, roofline intersections, any low-slope areas, among many others.

Over time, gravity takes a toll, and the settling of insulation layers reduces its effectiveness. This can be a particular peril with loose-fill blown-in insulation in wall cavities. A full-wall cavity at the outset may have an uninsulated space after some years of settling.

8.12 Tools for testing the building envelope and performance metrics

Building airtightness is one of the key measures of building energy use (Chan et al., 2013; Sfakianaki et al., 2008). It also affects indoor air quality including indoor $PM_{2.5}$ concentrations and radon levels (Carlton et al., 2019; Collignan et al., 2012).

One of the most common building diagnostics is the blower door test kit which tests building air infiltration (Fig. 8.11). The blower door system is a sealed frame and fan assembly to be temporarily installed in an open exterior door. After the building's other exterior doors and windows are closed, the system fan is turned on and is sufficiently powered to either pressurize or depressurize the

FIG. 8.11

Blower-door test device for testing the air leakage in the thermal envelope. The fan is blowing out, depressurizing the indoor environment.

building enough to measure its leakiness. The typical test is to depressurize the building to a pressure difference of 50 Pa with ambient by blowing the fan mounted in the doorway outwards. The flow rate through the fan is balanced and equivalent to the flow rate of air into the building from all the leak points. This is called the ACH50 test, and the air changes per hour for the building is a metric that characterizes the leakiness of the building. A similar test is applied to building ductwork using a "duct-blaster" test system.

Blower door testing two ACH50 or less can be considered a tight house. Where does that unconditioned air leak in? Many times, the attic, foundation, and crawlspace are places with leaks that do not enhance indoor air quality (Example 8.7).

> **EXAMPLE 8.7 Blower door testing**
>
> **Problem:** Using a blower door with a pressure drop of 50 Pa, students measured a flow rate of 3000 cfm in testing an old and drafty rental house. The house is approximately 1400 ft² in size. Estimate the air changes per hour (ACH50) for this structure. Is this a leaky building?
>
> **Given:** $Q = 3000 \, ft^3/min$; Floor $A = 1400 \, ft^2$
> **Find:** Air changes per hour (ACH50)
> **Assume:** $z = 10 \, ft$
>
> **Solution:**
> $$V = \text{volume} = 1400 \, ft^2 \, (10 \, ft) = 14{,}000 \, ft^3$$
>
> $$ACH50 = \frac{Q}{V} \times \frac{60 \, min}{h} = \frac{\left(3000 \, \frac{ft^3}{min}\right)}{14{,}000 \, ft^3} \times \frac{60 \, min}{h} = 12.86/h$$
>
> This is a very leaky structure, as an average house is in the range of ACH = 4. This building could undoubtedly use some improved air sealing and weatherization.

The blower door test is a useful diagnostic for whole building performance. However, it is expensive, requires technical training, and is cumbersome in the field. It requires trained technicians and perhaps a day's effort to deploy, setup, conduct, verify, remove the blower door unit, and analyze the data. The blower door kit purchase costs several thousand US dollars or more for commercial-scale units. It is also less specific as to where individual leak points exist. Highly qualitative smoke generation and tracing can be used to find some of the leaks, but only where easily accessible by technicians. More advanced imaging techniques are needed to, throughout a structure, visualize, record, and ideally quantify and thus prioritize discrete leak points (Barreira et al., 2017).

Blower door testing for a very well-sealed house strives for ACH50 of 2 or less. As a rule of thumb, one can take the blower door measured CFM50 and divide by 20 to get an approximate equivalent "UA" for the air leakage. Put another way, each CFM50 of measured air leakage creates about the same heat loss as $1 \, ft^2$ in an *R*-20 envelope.

Thermal imaging of a building, as has been shown in numerous images in this book, is also vital for identifying leak points in a building. The camera images the building in the infrared region of the spectrum, showing the areas of heat loss/gain. A perfectly insulated structure would show a thermal envelope with an exterior temperature equal to the ambient temperature and an interior surface temperature equal to the room temperature. Real-world buildings will show interior surface cold spots and exterior surface warm spots during winter, showing building heat loss (and vice versa during summer, showing heat gain).

One of the most useful modeling tools is the Manual J calculation, of which numerous software implementations exist, the basic versions of which are free and web based. The Manual J calculation takes as inputs the building thermal envelope components (wall sections, windows, doors, and their R-values) and a location to use seasonal weather data to calculate an estimate of the required sizing of a space heating and cooling needs for the given building.

8.13 Chapter summary and conclusions

This chapter focused on the building thermal boundary which also serves as the air and moisture barriers of the structure. The building thermal envelope is the boundary that separates the inside conditioned space from the outside world. Although it usually follows the building's exterior surface, it does deviate in cases such as unconditioned attics and crawlspaces that are outside the thermal envelope. All three heat transfer mechanisms operate across the thermal envelope. Improvement of the thermal envelope is the best way to reduce the building energy use in both the cooling and heating seasons.

Roofs, attics, crawlspaces, and basements all have particular issues related to the thermal envelope to not create new problems by addressing an existing one. Windows and doors are often thermal weak points, though they occupy only a small fraction of the shell area, mitigating the overall effects.

Windows are a requisite architectural element for both the exterior and interior views and functionality. Window technology has improved dramatically over the past decades. The use of low-conductivity materials, low-emissivity coatings, dead air spaces, and/or thermal breaks in the structures, air sealing of the unit, and weather-stripping all contribute to reducing the U-value (conduction coefficient) of the window unit. Despite improvements, however, windows remain less thermally resistive than an effectively insulated section of wall and are often a weak point.

The walls and ceiling of the structure resist heat flow based on their conductivity, air sealing, and radiant properties. The rate of heat transfer through a general wall section is a function of the temperature difference across the wall, the thermal transmissivity of the wall, and the interface area of the wall with the environment. Any of these parameters can be reduced to minimize the heat transfer.

R-values (reciprocal of the U-value) quantify the resistance to heat flow of a building component. Increasing the R-value decreases the rate of heat transfer through the thermal envelope. Calculating an estimated R-value of a wall assembly includes first determining if components are in series or parallel. For the series situation, the composite R-value is the sum of the R-values of the layers that form the material and add linearly. The parallel situation is an analog of parallel electrical resistors. Area-weighted U-values are summed for a composite U-value for the parallel situation.

New materials and techniques are under development for more cost-effective insulation with lower embodied emissions. These include bio-based materials such as hemp and mycelium. One interesting application is with aerogel insulation with an R-value of $R10$/in.; it can thus be applied to stud faces to greatly reduce thermal bridging. Here we covered only the most important facets of the thermal envelope, but suffice it to say that many resources exist detailing the thermal envelope, including Green Building Advisor and Building Science Corporation's websites.

8.14 End of chapter exercises

(1) **Concepts:** Insulation Choices Questions (mark one the best choice for each).
    (a) The best choice for insulation that has the highest R-value per inch:
        fiberglass batt loose-fill cellulose polyiso mineral wool SPF XPS EPS aerogel SIPs

(b) The best choice for insulation that has structural strength:
fiberglass batt loose-fill cellulose polyiso mineral wool SPF XPS EPS aerogel SIPs
(c) The best choice for insulation that has recycled content:
fiberglass batt loose-fill cellulose polyiso mineral wool SPF XPS EPS aerogel SIPs
(d) The best choice for insulation that has a continuous integrated air barrier, including around penetrations
fiberglass batt loose-fill cellulose polyiso mineral wool SPF XPS EPS aerogel SIPs
(e) The best choice for insulation that has the highest R-value per inch for a rigid board:
fiberglass batt loose-fill cellulose polyiso mineral wool SPF XPS EPS aerogel SIPs
(f) The best choice that will both insulate and completely fill and thus stop air movements through plumbing penetrations:
fiberglass batt loose-fill cellulose polyiso board mineral wool SPF XPS EPS aerogel SIPs
(g) The best choice that is fireproof:
fiberglass batt loose-fill cellulose polyiso board mineral wool SPF XPS EPS aerogel SIPs

(2) Concepts: Which is not a potential benefit of adding windows, doors, and skylights to a structure?
   (a) Daylighting
   (b) Ventilation
   (c) Improved thermal insulation
   (d) Passive cooling
   (e) Passive heating
   (f) Views

(3) Concepts: Which of the following are **potential** problems associated with well-insulated and air-sealed homes (select all that apply)?
   (a) Draftiness
   (b) Back drafting of combustion appliances
   (c) Increased radon concentrations
   (d) Mold and mildew issues
   (e) High energy bills
   (f) Uncomfortable spaces

(4) Concepts: What is thermal bridging? Describe how one can reduce thermal bridging; describe briefly. Describe in a sentence three ways to reduce thermal bridging in a wall structure.

(5) Concepts: Which of the following wall construction methods does not reduce thermal bridging over a conventional framed stud wall?
   (a) 2×6 @ 24″ O.C. spacing with OSB sheathing
   (b) 2×4 @ 16″ O.C. spacing with rigid foam board sheathing
   (c) 2×4 @ 16″ O.C. spacing with aerogel-faced studs
   (d) Steel stud framing
   (e) Double 2×4 wall framing

(6) Concepts: What is the purpose of a Low-E coating on a window?
   (a) Keep the heat out in cold weather
   (b) Keep heat in during hot weather
   (c) Reduce visible light passing through the glass
   (d) Reduce UV radiation passing through the glass
   (e) Reduce infrared radiation passing through the glass
   (f) Reduce convective losses

210 Chapter 8 Building thermal envelope

(7) **Concepts:** Examine the following structure shown in Fig. 8.12 from the inside during the cold season. Diagnose the energy problems you can see. The exterior of the house is also shown and appears in good condition. Discuss three insulation solutions. Which would be the best from an energy perspective? Which would be the most practical and cost-effective?

FIG. 8.12

Block construction from the 1950s era is shown on a winter day with an ambient temperature of ~30°F.

(8) Concepts: Discuss the latent vs sensible cooling loads for a building in southern California vs southern Louisiana. Which will dominate in LA (Los Angeles) vs LA (Louisiana)?

(9) Concepts: What are three advanced wall framing techniques and why can they be considered green?

(10) Concepts: What are three differences between SIPs and ICFs, including the materials used in these items?

(11) Concepts: Define the building envelope and describe four common holes in it.

(12) Problem: Draw a diagram of problematic areas of heat loss through a building-attic hatch. Consider all three heat transfer mechanisms.

(13) Problem: Advanc-R VIP Vacuum panels from Panasonic, Inc., offer a thermal conductivity of 0.002 W/m-K with a thickness of 1 in. Find its R-value in US customary units for the 1-in.-thick panel.

(14) Problem: Calculate an R-value for the two wall sections shown in Fig. 8.4 assuming typical pine studs, fiberglass batts, 4″ brick exterior, ½″ air gap, ½″ standard drywall, and ½″ plywood sheathing. Assume Thermablok aerogel ¼″ insulation on the stud faces in the second case. Use lumber's actual dimension versus the nominal (e.g., a 2×4 is actually $1.5'' \times 3.5''$).

(15) Problem: We have a block/brick wall that was retrofit insulated on the inside of the wall. The insulation was added with 2″ thick rigid polyiso boards ($R6$ per inch) fit between 2″ wide x 2″ deep softwood studs spaced every 24″ on center along the cinder block wall. The exterior of the wall is 4″ thick common brick ($R0.2$/in.) outside a 4″ thick cinder block (assume half the R-value for a full 8″ cinder block). The continuous brick and block layers are separated by a ½″ air gap ($R0.9$). The interior of the wall was finished with ½″ drywall over the studs/polyiso.
  (a) Draw a diagram looking at this wall from the top.
  (b) Calculate the R-value of this assembly.

(16) Problem: According to the National Hemp Association, hempcrete's U-value can range from 0.16 to 0.52 W/m²·K and depends on thickness, temperature characteristics, and other variables. Find the R-value in US customary units of 4″ of hemp insulation.

(17) Problem: Calculate the thickness of softwood lumber, poured concrete, carbon steel, and copper that would be required to equal the R-value of 6″ thickness of blown cellulose in a wall. Hint: for copper, it is almost a mile!! Calculate this for granite ($k = 3.1$ W/m·K).

(18) Problem: An elevated chicken coop is $3 \, \text{ft} \times 3 \, \text{ft} \times 3 \, \text{ft}$ and contains four chickens each generating 10 W of waste heat. The coop has been insulated all around perfectly with 2″ of EPS foam board that has an effective R-value of $R8$ in British units. What is the steady-state temperature inside the chicken coop structure? Outside temp is 0°C. Assume conduction losses only.
  (a) An elevated chicken coop is $3 \, \text{ft} \times 3 \, \text{ft} \times 3 \, \text{ft}$ and contains four chickens, each generating 10 W of waste heat. The coop has been insulated all around perfectly with 2″ of EPS foam board that has an effective R-value of $R8$ in US customary units. What is the steady-state temperature inside the chicken coop structure? The outside temperature is 0°C. Assume conduction losses only.
  (b) You add a 100 W light bulb to help keep the birds warm. What is the new equilibrium temperature?
  (c) An incandescent light bulb is being used to keep a chicken coop warm at night. Does it make sense for the owner to replace the incandescent bulb with a more efficient LED bulb to accomplish this task? Why or why not?

(19) Problem: You have a building with an interior volume of 29,250 ft^3. During an ACH50 test, you measure an infiltration flow rate of 1900 CFM. Find the ACH50 for this structure.

(20) Problem: The data below were measured for the decay of indoor PM$_{2.5}$ in the structure after a cooking event that produced substantial particulate matter. Presume that the only loss is through exfiltration (no deposition or chemistry). In this case, the governing equation becomes a first-order loss where $C = C_o \exp(-ACH^*t)$. Calculate the actual ACH (natural conditions, no pressurization or evacuation of the structure). Compare to the above answer using the rule of thumb that ACH$_{actual}$ = ACH50/20 (Sherman, 1987).

(21) Problem: An insulated cube-shaped cooler with R50 vacuum panels has dimensions of 368 mm × 388 mm × 538 mm. How long will it keep your frosty beverages below 40°F? The total volume is 70 L. At the beginning, assume 35 L or half the volume frozen is 32°F cooler packs, and half is liquid beverages at 32°F. Assume water properties for both the beverages and ice packs.

(22) Problem: An owner upgrades his insulation from R-2 to R-6. Another owner upgrades from R10 to R-20 while a third owner goes from R20 to R60. Which will see a larger decrease in energy costs with all else being equal (square footage, surface area, temperatures)?

(23) Problem: A building is constructed with 8"-thick cinder block walls, which are determined to be quite chilly in the winter.
  (a) Calculate the conductive heat transfer dQ_{cond}/dt through the walls for the following conditions: $T_{outside} = 10°F$, $T_{inside} = 70°F$, $U = 3.5$ W/(m^2-K), and total surface area can be represented by a rectangle 12 ft × 228 ft.
  (b) Using your engineering judgment. From a heat transfer and indoor temperature stability perspective, is it better to retrofit insulation on the exterior face, inside the cavities, or on the interior surface of this wall, and why? Assume all else is the same and any method of insulation is similar effort and capital costs.

(24) Problem: You have a 2" × 6" steel stud with a U-value of 50 W/(m^2·K). Find its R-value in the typical imperial units used.

(25) Problem: You have a wall structure with the following: 2" × 6" steel studs as above with 24" O.C. spacing, wall cavities filled with 6" of R6 per inch polyiso foam board, a 4" brick layer on the outside, a 1" stagnant air layer behind the brick (R0.8), and a 0.5" layer of drywall on the inside.
  (a) Calculate the effective R-value of the assembly. Comment on the effects of the steel structural members on the overall effectiveness of the insulated wall.
  (b) Instead, you use 6" wood studs rather than the steel studs. Recalculate.
  (c) All else being equal, how much have you reduced heat transfer by using wood rather than metal studs?

(26) Problem: What is the R-value of a wall composed of 0.5" drywall, 2 × 6 framing at 24" on center spacing, rigid polyurethane board of 6" in the cavities, and an exterior cladding of 4" brick?

(27) Problem: What is the R-value of a wall composed of 0.5" drywall, 2 × 6 framing at 24" on center spacing, cavities filled with cellulose of 6" thickness (presume full coverage and no settling), and an exterior cladding of 6" brick? You can use the nominal dimensions of the given components. Draw a diagram of the wall section and show your parallel and series resistances. Assume full-depth studs.

(28) Problem: A wall has 4" concrete brick cladding with a 1/2" air gap behind it. Rather than a stud wall, the interior wall is a SIP composed of 6" rigid foamboard with ½" plywood sheeting on the inside and outside surfaces. The interior wall is ½" sheetrock. Draw a diagram from outside to inside looking down from above to show the layers of the wall. Calculate an R-value for the wall assembly. Would you characterize this as a well-insulated wall? Is it above typical code requirements for an external wall?

(29) Problem: From a heat transfer perspective, which walls would you rather have and why? A wall with 2×4 pine studs with 16" O.C. spacing, fiberglass filled cavities, and 2" of EPS outer sheathing or a 2×6 pine studs with 24" O.C. spacing, fiberglass filled cavities, and ¼" aerogel ($R12$/in.) on stud faces. Neglect any other wall components or edge losses, with all else being equal. Assume full-depth studs.

(30) Problem: You have two types of insulation, with one that is rated in metric units. All else being equal, which would you rather have the walls of your building for maintaining a comfortable temperature and minimizing loss? R-value $= 12\,\text{h·ft}^2\cdot°\text{F/BTU}$ or U-value $= 0.4\,\text{W/m}^2\cdot\text{K}$ and why?

(31) Problem: R-value and heating calculations.
  (a) Calculate the R-value of a $2'' \times 4''$ spaced 16" on center wall with fiberglass batts vs a $2'' \times 6''$ spaced 24" on center wall. Hint: just consider one repeating unit of each. You can assume 4" and 6" in. wall depth.
  (b) The house can be considered a box that is $40' \times 40'$ with an 8-ft wall height. What is the seasonal difference in heating needs in BTU? Assume the site is characterized by 3220 HDD in a year. You can neglect any corner framing non-idealities and windows. Assume all else is equal.

(32) Problem: The outdoor temperature is 32°F in winter. You turn down the thermostat in your room from 77°F to 66°F. How much would you estimate your heat bill to decrease? Consider conduction losses only and presume no other changes.

(33) Problem: It is a cold winter day with average temperatures indoors and outdoors of 20°C and $-15°$C. A window measures 36 in. $\times$ 50 in., and there are 20 of these windows in a structure. The U-factor for this single-pane metal window is $1.3\,\text{Btu/h-ft}^2\text{-°F}$. Assume conduction only.
  (a) Estimate the heat transfer (in joules per day) through these 20 windows when the inside temperature is 20°C and the outdoor temperature stays at $-15°$C.
  (b) The U-factor of replacement windows (double-pane, argon-filled, and low-E vinyl frames) is $0.31\,\text{BTU/h-ft}^2\text{-°F}$. How much heat is transferred in a day in this case?

(34) Problem: You have a double-pane window with a U-value of $0.31\,\text{BTU/h-ft}^2\text{-°F}$. You add a cellular shade to the window with $R4$. What is the new overall R-value? How much have you reduced heat loss through 20 of these windows?

(35) Problem: A Solatube skylight is given a spec of $U = 1.3\,\text{W}/(\text{m}^2\,\text{K})$. Convert this to an R-value in typical US customary units.

(36) Problem: It is a cold winter day. A building with the profile shown earlier in Fig. 7.8 is insulated with an R-value of 10 in the attic (the roof profile shown has 2-ft overhangs all around). If the owners upgrade the insulation levels to R-60 (hour-ft^2-°F/BTU), what are the before and after heat losses through the ceiling (conduction only)? Assume the area above the exterior walls cannot be insulated.
  (a) Through the existing ceiling, estimate the heat transfer (in joules per day) when the inside temperature is 20°C and the outdoor temperature stays at $-15°$C.

(b) Through the reinsulated ceiling, estimate the heat transfer (in joules per day) when the inside temperature is 20°C and the outdoor temperature stays at −15°C.

(c) If this is the situation for 2 months per year, how many therms of natural gas per year would be saved? Assume a 70% efficient gas heater.

(37) Problem: In the previous problem, you looked at the change in heat transfer with upgrading attic insulation in a ~2350 ft^2 attic. Estimate the seasonal heat load reduction in therms (assume 70% efficiency) by upgrading the attic insulation from $R10$ to $R60$. The site has 3420 HDD annually.

(38) Problem: Using the annual heating degree days from your location, estimate the annual heating load for the building with the following characteristics: 12-ft-tall walls with a perimeter of 250 ft (walls have a ½ layer of stucco, 2″ of EPS foam, and a 6″ cinder block). The ceiling is insulated to $R60$ and has an area of 2350 ft^2. 20 windows as described in the problem above. Compare to the actual usage of 245 therms (445 therms before upgrades).

(39) Problem: Minimizing the surface area to volume ratio is one way to reduce heat loss/gain through a structure. You are building a 2000 ft^2 floor area structure. Calculate the SA/V ratio of making the home a half dome, a square with 8′ ceilings, a 4:1 rectangle with 8′ ceilings, a square with 12′ ceilings, a 4:1 rectangle with 12′ ceilings, a 2-story box with 8 ft ceilings, and 2 story box with 12-ft ceilings. Ignore the floor area.

(40) Problem: There is a small modification to the CMU example problem with the vermiculite-filled blocks. We can calculate the R-value by building up the block from the concrete (assume $R0.1$/in.) and an air gap ($R1$). Compare to the vermiculite-filled block calculated similarly. Does it make a significant difference?

CHAPTER 9

Electro-mechanical systems and appliances in buildings: Evaluating alternatives

Learning objectives

(1) Evaluate the many vital roles energy plays in building operations.
(2) Interpret the physical processes related to heating, ventilation, and cooling (HVAC) systems.
(3) Understand the major mechanical, electrical, and plumbing (MEP) systems and their characteristics.

9.1 Overview of energy systems in buildings

Among the most important, complex, and costly systems in building management are the mechanical, electrical, and plumbing (MEP) systems in a building. Key among these are the heating, ventilation, and air conditioning (HVAC) systems, and detailed technical information exists on these systems (Reddy et al., 2017).

According to the United States Energy Information Administration (EIA) in 2001, the average home annually uses 81 MMBTU (81E6) of total energy from all energy sources, including electric, natural gas, fuel oil, and wood combustion. Energy Use Intensity can provide performance metrics to gauge building energy use. Metrics for a typical 2000-era building for total energy use are 300 kWh/m^2/year (or 100,000 BTU/ft^2/year), while a current high-performance building uses ~100 kWh/m^2/year (Kibert, 2016).

The largest share of energy use for homes and commercial buildings is for space heating on average, at about one-third. Cooling, lighting, and ventilation add up to another third or so (Fig. 9.1). In the remaining one-third, the other large categories include cooking, refrigeration, water heating, computing, and all other uses.

The alternative systems that are feasible are a function of the energy services that are available. Does the building have gas service or a propane tank, electric service, or is it off-grid? Is there a readily available source of biofuels onsite or nearby? Electrification of all the systems in a building is a current emphasis and is a shift that is occurring. The nonprofit Electrify Now is one of the proponents of this approach. This can work with off-grid buildings but is far simpler to accomplish with grid power available. Grid power is becoming cleaner with lower GHG emissions, while simultaneously, buildings are more often featuring onsite renewable sources. Systems for space heating, water heating, and clothes drying are all possible using electric sources, most often in the form of heat pumps, which will be discussed below in detail.

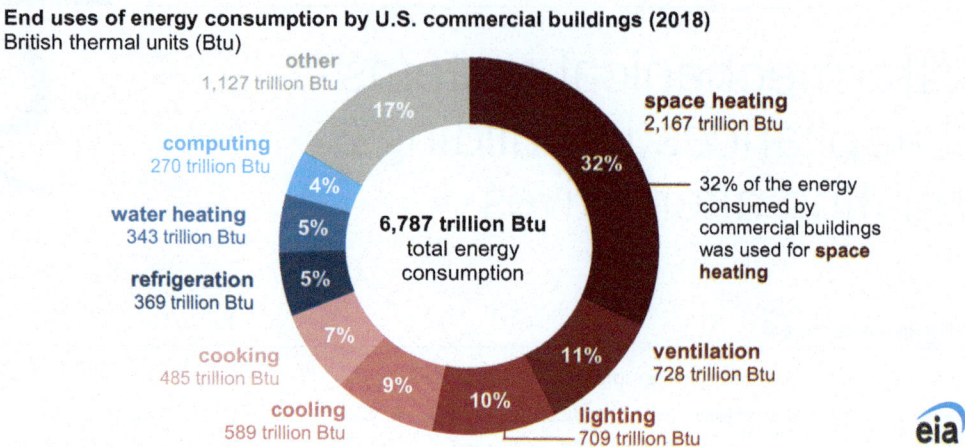

FIG. 9.1

Energy end-use categories for commercial buildings (US EIA).

9.2 Climate control systems for conditioned spaces

The advancement of indoor climate control systems, generally known as heating, ventilating, and air conditioning (HVAC) systems, is one of the most consequential building technology developments of the 20th century using a variety of technologies (Fig. 9.2). HVAC advancements have decreased the deaths and suffering due to excessively hot weather (as well as cold).

Reliable heating systems have allowed the further settlement of cold climates and have made open indoor fires (and their health and safety implications) mostly obsolete for cooking and heating. Heating before the industrial era relied upon open combustion of wood or coal, which had acute air quality impacts indoors and in urban areas. In more recent times, space heating has relied upon closed system combustion of any number of fuels—solid, liquid, or gaseous—and can also be done simply, though expensively, via electric resistance heating.

Refrigerated air systems have allowed the greater settlement of many warm climate areas of the world, including the American South. Cooling spaces (and water) is typically more difficult and costly than heating them, which historically can simply rely on fire from one source or another. The method of most common and growing use for cooling is refrigeration-based systems, also called air conditioners.

Air conditioning has not come without a cost in terms of energy use, refrigerant management issues, climate, and air quality impacts, as well as indoor air quality concerns. In fact, improving refrigerant use and leakage is ranked in the top three individual solutions, capable of ~90 Gt in CO_2 equivalent emissions reduction (Hawken, 2017). The study by Hawken is a comprehensive assessment of ~100 realistic, scalable measures for potential greenhouse gas reductions throughout society. Reducing emissions of hydrofluorocarbons (HFCs) now in common use in refrigerators and air conditioning systems, has become vital in terms of greenhouse gas emissions as well as the energy use from compressor systems. Indoor climate control systems are worth further exploration for their role in building emissions. The reader is forewarned that this is only an incomplete introduction to a topic that has much engineering depth and is covered more thoroughly in references focused solely on HVAC.

9.2 Climate control systems for conditioned spaces

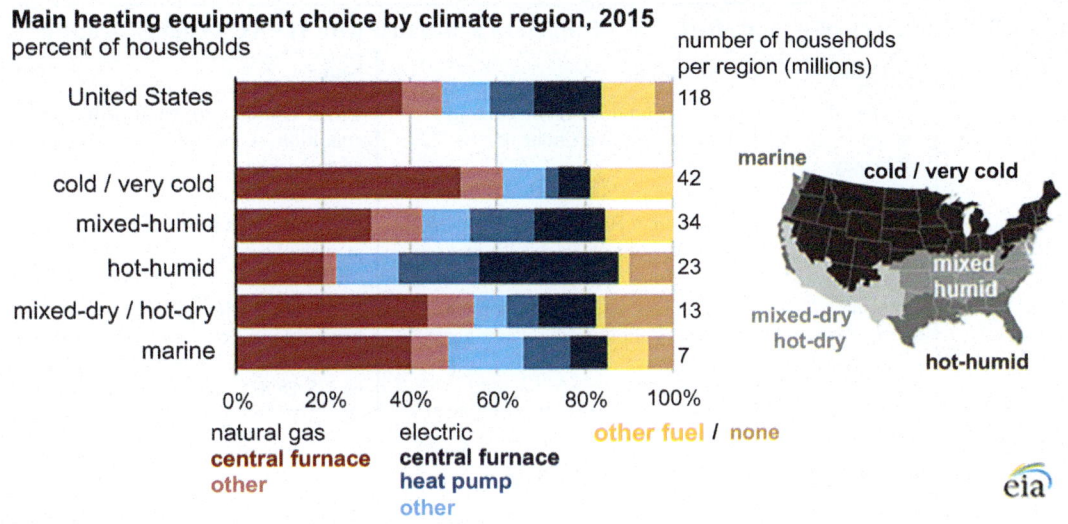

FIG. 9.2

Distribution by region of home heating systems in 2015 (public domain US EIA image).

Earlier in Chapter 2 on thermodynamics, we defined the unitless COP for heat pump systems as delivered cooling energy (Q_{cold}) divided by the electrical consumption (W) (Eq. 2.24). The seasonal energy efficiency ratio (SEER) is the most common efficiency metric reported for air conditioning systems. Other metrics for HVAC systems include CEER, EER, HPSF, plus a number of others (Table 9.1). This is an area of constant evolution as well as continuing improvements in performance and efficiency.

Table 9.1 Summary of major efficiency parameters associated with HVAC systems.

Parameter	Definition	Description	Notes
COP	Coefficient of performance	Unitless measure of cooling energy delivered divided by electrical energy consumed	This is the thermodynamic parameter most often used for heat pumps of all varieties
SEER	Seasonal energy efficiency ratio	Seasonal total BTUs of cooling produced per watt-hour of electricity	SEER is most commonly used with central air conditioning (refrigerated air) systems
SEER2	Seasonal energy efficiency ratio v2	Seasonal total BTUs of cooling produced per watt-hour of electricity	SEER2 protocols were developed to better represent real-world operating conditions vs the idealized conditions of SEER1
EER	Energy efficiency ratio	BTU/h of cooling output per watt of electrical consumption	EER is an instantaneous value specified at an ambient temperature of 95°F

Continued

Table 9.1 Summary of major efficiency parameters associated with HVAC systems—cont'd

Parameter	Definition	Description	Notes
CEER	Combined energy efficiency ratio	Seasonal total BTUs of cooling produced per watt-hour of electricity	CEER is used with window unit air conditioners, and CEER >9.8 qualifies as an energy star
HSPF	Heat seasonal performance factor	Season total BTUs of heating output per watt-hour of electricity in	Measure of the heat pump's efficiency at space heating
U.E.F.	Uniform energy factor	The BTUs of water heating per BTU of energy input to the water heater	Rating used with water heaters beginning in 2017. Determined based on a 24-h uniform test cycle specified by US DOE

9.2.1 Advent of refrigerated air heat pumps

Because of their importance to civilization and the environment, it is worth discussing the advent of refrigeration systems. The use of refrigerated air conditioning systems was an early 20th-century development that made urban life in warm climates much more feasible and comfortable. Willis Carrier is widely credited with inventing the refrigerated air system in the early 20th century, though many engineers and scientists contributed. It has allowed the growing popularity of the sunbelt states in the United States and other warm climate regions that were previously considered excessively hot.

A typical refrigeration system for space conditioning is composed of indoor and outdoor heat exchangers (or "coils"), a throttling or expansion valve, and a compressor (Fig. 9.3). One of the coils is operated to condense vapor into liquid, releasing heat to the surroundings, while the other evaporates the liquid phase refrigerant into vapor, thus extracting heat from the surroundings. The system works on the large energy exchange related to the latent heat of vaporization for the working fluid in the system.

During the summer, a low temperature refrigerant in the liquid state is reduced in pressure through an expansion valve or throttle. It then passes through the indoor coil, which serves as an evaporator as it vaporizes the refrigerant, absorbing heat from the surroundings and thus cooling the room. As an analogy, think of releasing the pressure in a bike tire and the cooling it causes on the valve, the exiting air, and surroundings. The vapor is now at low pressure and high temperature as it exits the evaporator and passes outdoors into the compressor. Here it is pressurized prior to passing through the condenser, where it releases heat into the surroundings. If designed for such with a reversing valve, a heat pump system is also capable of space heating by reversing the flow of refrigerant through the system. In the end, these heat pump-type systems are using work input to move heat either into or out of the conditioned space (Fig. 9.3).

Though focused commercial efforts at improving air conditioning efficiency continue, the nature of compressor-based systems requires significant energy use. Thermoelectric systems are another possibility, though these are mainly used in small BTU/h applications (e.g., drink coolers) and not of significant enough efficiency or output for building scale use presently. Desiccant-assisted systems can help to reduce the indoor RH, advantageous in situations where the latent cooling load is significant (i.e., high humidity areas). Evaporative coolers, viable in very dry climates, will be discussed below as well.

9.2 Climate control systems for conditioned spaces

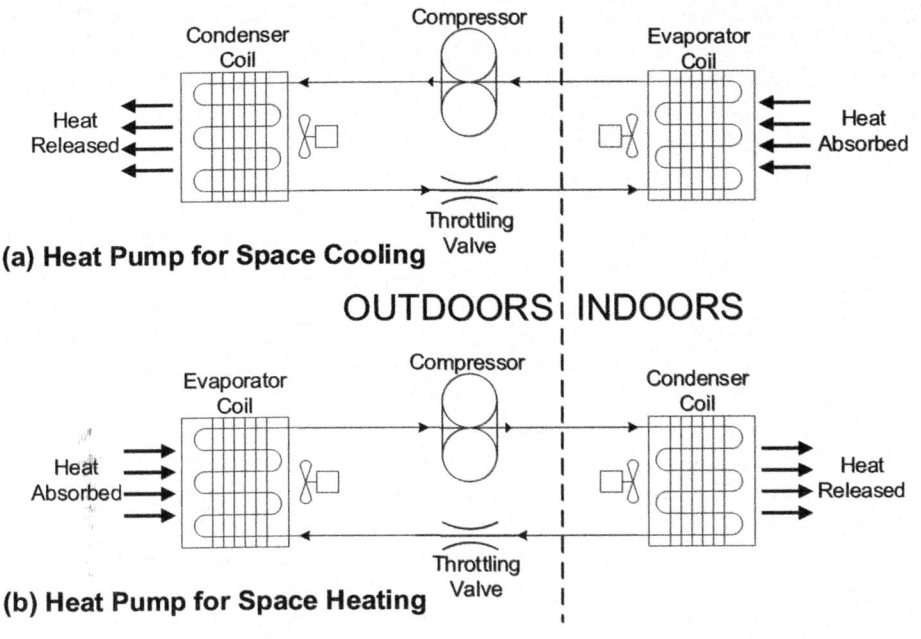

FIG. 9.3

A typical heat pump system for (A) space cooling and (B) space heating.

9.2.2 Refrigerants, stratospheric ozone depletion, and global warming potential

An innovation that made refrigerated air conditioning more practical and economical was the synthesis of chlorofluorocarbon (CFC) refrigerants that replaced more toxic and flammable compounds like ammonia and propane used previously. Initially, CFCs were considered ideal as they were inert, nonflammable, nontoxic, and had very favorable thermal properties for use in air conditioning systems. Unfortunately, this stability led to their harmful effects on the atmosphere. The long lifetime allowed their eventual migration to the stratosphere, where the intense UV light exposure would cause their photochemical degradation (Seinfeld and Pandis, 2016). Halogen radicals produced by their breakdown then catalyzed the destruction of ozone in the stratosphere. Stratospheric ozone is essential to protect life on the planet from short-wavelength ultraviolet radiation. Recall there is also a problem with too much ozone in the troposphere, as discussed previously in the section on air quality. The mantra with ozone is "it's good up high, it's bad nearby."

The primary ozone depletion offender was the ubiquitous use as a refrigerant of CFC-12 known by the trade name Freon (Table 9.2). Despite the phaseout of Freon, the IPCC, in its 2005 Special Report on Ozone-Depleting Gases and Climate, estimates approximately half of the world's refrigeration systems still rely on CFC-12 as the refrigerant, which has slowed recovery of the stratospheric ozone layer. Refrigerant R22 (which replaced R12 Freon) is still plentiful but ceased production in 2020 in favor of the less ozone-depleting R410A. These changeouts to lower Global Warming Potential (GWP) refrigerants take time and effort, which is an important consideration for climate change.

The common hydrofluorocarbons (HFCs) now in use (e.g., R-134a) have zero to far lower ozone-depleting potentials (ODP) but are still potent greenhouse gases with global warming potentials (GWP) three orders of magnitude greater than CFC-12. So, despite the relatively small mass produced, they have outsized effects on climate if released into the atmosphere (Table 9.2).

We have transitioned from refrigerants with high ozone-depleting potential, such as chlorofluorocarbons (CFCs), to much lower ones with zero ODP (Table 9.2). Now we are moving from the replacement HCFCs to HFCs and other lower GWP refrigerant gases such as ammonia, propane, and even carbon dioxide itself. Lower global warming potential (GWP) refrigerants are emerging from the R&D labs to product implementation. For example, a gradient window heat pump system from a startup uses R32 refrigerant in a window unit heat pump with a high CEER of 15 BTU/H/Watt (https://www.gradientcomfort.com/). Many of the heat pump systems have moved to R32, which has a GWP of approximately 675, which is about two-thirds less than the more conventionally used R410A, R22, and R134A.

The Montreal Protocol is the international agreement that has phased out ozone-depleting compounds, mainly CFCs, as refrigerants in compressor systems. It has been described by former Secretary of the UN Kofi Annan as the most successful international agreement to date. It has been a model of success for international agreements in solving global environmental problems. Despite this success, replacement compounds are still potent greenhouse gases (Table 9.2). Recently, the Kigali Amendment, signed in 2016 and implemented in 2019, has put into action a plan to reduce the greenhouse gas footprint with some of the current replacement refrigerant compounds.

Table 9.2 Common refrigerant compounds, their ozone-depleting potential (ODP, where CFC12 = 1.0), and their global warming potential (GWP, where CO_2 = 1.0).

Refrigerant	Ozone-depleting potential (ODP)	Global warming potential (GWP)	Notes
R12 (CFC12 or Freon)	1.0	10,900	Freon has effectively been phased out as a refrigerant
R22 (HCFC22)	0.05	1810	
R23 (HFC23)	0	14,800	Although HFCs eliminate the ozone depletion potential lacking chlorine, many are potent greenhouse gases
R410A	0	2088	
R134A	0	1430	Currently in phaseout
R32	0	675	Increasingly used by manufacturers
R744 (CO_2)	0	1	Some commercially available systems
R290 (propane)	0	3.3	Flammability issues
R717 (ammonia NH3)	0	0	Toxicity issues

Refrigerant management, containment, and recovery are considered low-hanging fruit for reducing greenhouse gas emissions (Hawken, 2017). It will continue to become more important as we further electrify our energy end-use, including heat pumps for space heating/cooling, water heating, and clothes drying rather than combustion appliances. An estimated 10% of systems are classified as leaky, and over 90% of leaked refrigerants are due to "catastrophic leaks" that involve a loss of 50% or more of a refrigerant charge (Harrod and Shapiro, 2021). Leaking systems also have shortened lifespans and run less efficiently. It requires skill and attention to assemble a refrigeration system like a heat pump that is devoid of leaks and stays that way. The keys are properly sealed fittings and a robust testing protocol—including a pressure test, bubble leak test, vacuum decay test, and final bubble test—during commissioning to identify the leaks before charging (Harrod and Shapiro, 2021). Because of the importance of refrigerants, there are significant LEED program credits for proper management of refrigerants, as discussed later.

9.2.3 HVAC system sizing and ducting

Proper HVAC system sizing is critical to both energy efficiency as well as system performance. HVAC sizing is an approximate science; it is a function of not only the thermal envelope and its peculiarities, but also the very local climate, site characteristics (e.g., shading, air movement, and site topography), and other parameters. An HVAC system that is undersized will not adequately heat or cool a structure. An oversized system is one that will run on short, inefficient cycles, which wears the system more rapidly.

Among contractors, the rough rule of thumb of $\sim 500\,\text{ft}^2/\text{t}$ of cooling for an average building almost always oversizes HVAC equipment. The prevailing thought is that larger equipment means larger profit margins and minimizes the risk of call-backs over complaints of insufficient cooling or heating. The penalty is overly expensive equipment, lower efficiency, and more wear. An oversized unit will cool more quickly and thus "short cycle," resulting in suboptimum efficiency, often reduced dehumidification, and more wear on equipment.

The ASHRAE Manual J and Manual S calculations are key steps in sizing systems. Manual J takes as input the local climate variables (typically accessing the closest meteorological monitoring station data) as well as building thermal envelope characteristics input by the user. Manual J calculation then estimates the size of the heating and cooling systems to meet the needs of maintaining a comfortable indoor environment. The Manual S calculation is the next step that helps select the right equipment using the manufacturer's OEM data in comparison to the Manual J estimate.

Ductwork is also a key element of a centralized HVAC system. Ductwork delivers the conditioned air throughout a building and also returns air to the HVAC system (typically through an air filtration system) for reconditioning. Ductwork design requires significant expertise to make a balanced system that effectively delivers and returns air throughout the building to avoid hot and cold spots. There are tools similar to Manual J and S calculations that assist with ductwork design.

9.2.4 High-efficiency combustion systems

The heating systems in both commercial scale and residential buildings are predominantly still combustion-based furnaces, typically using natural gas or propane in more rural areas. The systems have evolved over the decades from "gravity" furnaces of a century ago to much higher-tech and more efficient systems. The higher-performance systems have come at the cost of higher initial costs and increased complexity.

Older designs used indoor air for combustion and exhausted combustion products through an atmospheric vent. These had the potential for back-drafting with indoor depressurization. Closed combustion appliances have a much-reduced probability of carbon monoxide poisoning as ambient air is brought in and exhausted through another vent, all in a closed combustion cell. The key risk for closed systems from carbon monoxide poisoning is from leaks in the heat exchangers.

The most efficient combustion systems are what are termed modulating-condensing furnaces (mod-cons) (Fig. 9.4). Combustion air is introduced to the closed combustion system with an induced draft blower, typically. The thermal energy released from combustion is nearly all recovered by adding a downstream secondary heat exchanger to maximize extraction of the thermal energy of the combustion products. The systems are efficient enough that the exhaust gases cool to the condensation point, and thus condensed water must be removed with a condensate pump. The flue gas is cool enough that it can be exhausted outside through a horizontal PVC pipe rather than a chimney with a metal flue.

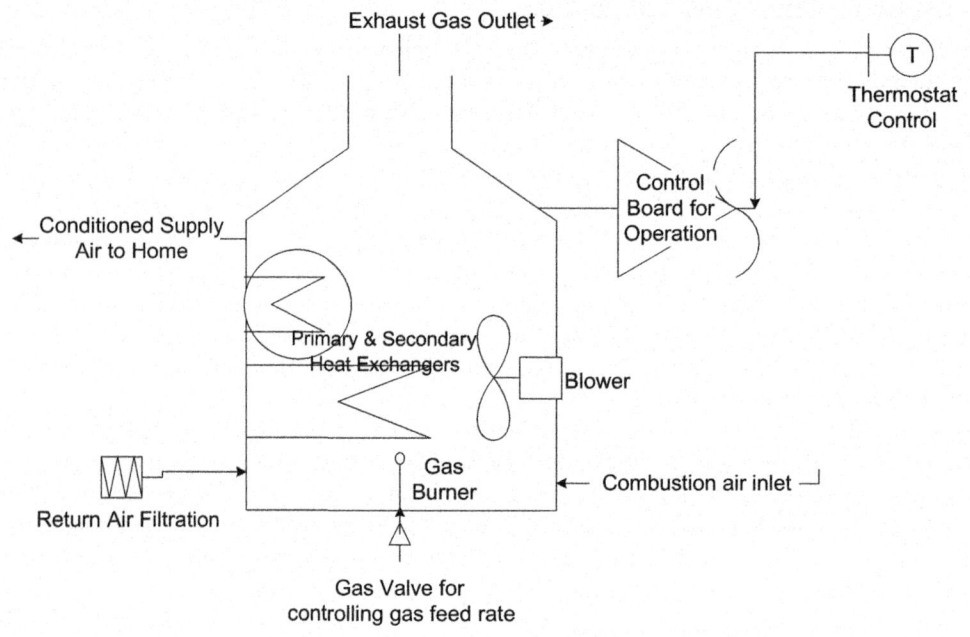

FIG. 9.4

Typical closed modulating-condensing "mod-con" combustion gas furnace. The modulating gas valve and secondary heat exchanger result in A.F.U.E. efficiency approaching 100%.

Control systems have evolved to be an important determinant of indoor comfort, energy efficiency, and indoor air quality (Dounis and Caraiscos, 2009). Traditional thermostats were all-on when heat was called for and all-off when not (with an appropriate dead band range). Pulse-width modulating controllers are one such advanced control that can help modulate the burner output as well as the blower speed over a much wider range. This can make the system both more efficient and more comfortable, with a "gentler" and longer heat cycle replacing the short blasts of hot air.

The efficiency gains of the mod-con gas furnaces are measurable (Fig. 9.5). The author replaced a 20-year-old typical 78% efficient natural gas furnace with a 97% A.F.U.E. mod-con in 2009. The reduction in energy use can be seen from the slope of the graph showing therm usage vs monthly heating degree days as measured in Colorado. The monthly heating use in therms is clustered near zero for HDD < 200. However, as the monthly heating degree days increase during the winter-time heating season, differing usage profiles emerge. The slope of the mod-con furnace gas usage is about 25% less than the conventional furnace. The linear model fits the data in both cases reasonably well with R^2 values >0.9.

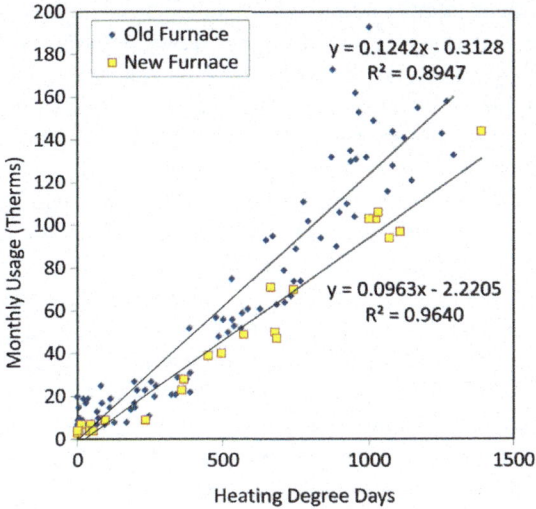

FIG. 9.5

Comparison of monthly natural gas use vs heating degree days (HDD) for a standard 78% efficient furnace that was replaced by a 97% efficient mod-con unit. Note the ~25% difference in slope values corresponding to the reduced energy use.

The point should be made that an ~100% efficient gas furnace or electric resistance heater will still use three times the purchased energy of a heat pump with a COP of three. One must consider the upstream generation, though. The emissions are reduced proportionally if the upstream electricity production is via a low-carbon energy source.

9.2.5 Air source heat pumps and mini-split systems

Air source heat pumps use ambient air as the heat source or sink. They can be air-to-air or air-to-water systems, the latter of which often use hydronic systems for distribution throughout the building. The air-to-air systems can use building ductwork for distribution and may or may not have a conventional backup system such as a gas furnace or electrical resistance heat strips for winter weather extremes.

Mini-split systems are a category of air source heat pumps becoming increasingly popular and particularly well-suited to low- to net-zero energy use, all-electric buildings. The mini-split systems are often ductless with a heat exchanger and blower mounted through the thermal envelope high on an exterior wall. Mini-splits are well-suited to retrofit applications where ductwork is not available. They are often used for zoned applications servicing a room or area of a building. Construction additions to a building where they are more cost-effective as an add-on system rather than trying to expand an existing central traditional ducted HVAC system.

As with the typical heat pump, the mini-split system includes two heat exchangers, one in the conditioned space and the other in the outdoor environment (Fig. 9.3). Whether the system is heating or cooling the indoor environment is determined by the direction of refrigerant flow through the system. As a room air cooler, the evaporator is inside and the condenser is outside, and they swap places for heating mode.

Increased efficiency of heat pump systems has resulted from inverter-based systems that have variable-speed compressors that can be ramped up and down. Previously, heat pumps were limited to operations down to approximately the freezing point of water, at which point a backup system (e.g., resistance heating, gas furnace) would be required. However, cold-climate heat pumps have given options for reasonably efficient operations even in the coldest climates, such as Maine, Minnesota, and Alaska. In addition to being used for both space heating and cooling, mini-splits are versatile in that they do not require but can be used with ducting systems and without. Their applications have expanded to multizone systems as well. They will likely play a dominant role in the future HVAC market with continued evolution, efficiency, and cost-effectiveness.

The performance of mini-split systems is affected by improper refrigerant charge as well as air flow outside of the proper range. Although continually improving, current highly efficient heat pumps have COPs of ~3–4.5 with some variation for heating and cooling and as a function of ambient temperature. Heat pump systems have even been developed for domestic hot water purposes as well as clothes dryers (Fig. 9.6). These units are available commercially and are making market in-roads. An alternate heat pump approach for space conditioning discussed in more detail in the later chapter on renewable energy systems is ground source heat pumps, also known as geothermal systems. These use the earth (or sometimes a body of water) as the source and sink for heat.

FIG. 9.6

Combination clothes washer/dryer that uses a heat pump to dry clothes. The unit recirculates dry air and avoids taking indoor-conditioned air to dry the clothes and exhaust out of the building. For a full wash/dry cycle, its consumption was 0.9 kWh and 13 gal of water.

9.2.6 Evaporative coolers

Evaporative coolers (or sometimes called swamp coolers) have long been used in dry, warm regions such as the desert areas of the Middle East and southwestern United States. They have fallen out of favor, replaced by heat pumps or refrigerated air conditioning systems. Nonetheless, evaporative coolers are still quite effective in the right environment. The concept is simple: introducing very dry ambient outdoor air into the home after first passing through a wet medium. Evaporation of the liquid water into the air passing over the wetted media occurs in the device. As heat is extracted from the dry air via the evaporating water, it provides a cooling and humidifying effect to the air pumped through the cooler and into the house.

Modern units are quite effective in arid environments and use a small fraction of a traditional compressor-based HVAC system (Fig. 9.7). Without a compressor, the electrical consumption is 10%–20% of a compressor-based refrigerated air system. For example, the author uses a Bonaire 4500 in the main living areas. It is improved over the "old rusty box" type evaporative coolers with a thinner profile model that has a noncorrosive composite case and uses a denser cellulose media for improved contact between the air and water. It is plenty capable of lowering the indoor temperature by 5–10°F to make a hot indoor temperature of 80°F a tolerable low to mid-70s. The seasonal consumption was a mere 258 kWh to cool a 2700 ft^2 structure in the summer heat of the southwest United States. They are highly efficient and greatly reduce cooling-related electricity consumption, as shown in Example 9.1.

FIG. 9.7

A newer evaporative cooler design from Bonaire, Inc. is constructed of a composite nonrusting cabinet, has a more effective evaporation honeycomb media, and with its low profile can be wall or window mounted. The drain bleed-off line going into the rain barrel helps to minimize the effects of hard water mineral deposits.

> **EXAMPLE 9.1 Evaporative cooler output and power consumption**
>
> **Problem:** Evaporative cooling is one of the most energy-efficient means of space cooling in the right environment. Estimate the cooling capacity (BTU/h and kW) of an evaporative cooler with an airflow rate of 4500 CFM at Standard Temperature and Pressure (STP) that evaporates 100 gal/day of water. Ignore the small, sensible heating contribution. Estimate the ΔT of the airstream. The unit consumes 250 W of electricity. Assume no losses and use the constant volume-specific heat for air. What is the approximate seasonal cost and consumption if the unit is used for 3 months in a year?
>   **Given:** Evap; 4500 CFM; 100 gal/day; 250 W
>   **Find:** BTU/h and kW; ΔT; water and electric consumption
>   **Assume:** 3 months continuous; STP298K; C_v, air = 0.7 kJ/kg·K
>
> **Solution:**
>
> $$P_{cool} = \frac{dm}{dt} \text{LatentHeatVap} = \left[100 \frac{gal}{day}\left(\frac{day}{24h}\right)\left(\frac{3.79 kg}{gal}\right)\right]\left[\left(\frac{2260 kJ}{kg}\right)\left(\frac{kBTU}{1055 kJ}\right)\right] = 34\frac{kBTU}{h} = 10 kW$$
>
> $$10 kW = \frac{dm}{dt} C_{air}(\Delta T) = \left[4500\frac{ft^3}{min}\left(\frac{min}{60s}\right)\left(\frac{m}{3.28 ft}\right)^3\left(\frac{1.2 kg}{m^3}\right)\right]\left(\frac{0.7 kJ}{kgK}\right)\Delta T = 10\frac{kJ}{s}$$
>
> $$\Delta T = \frac{10\frac{kJ}{s}}{\left(\frac{2.55 kg}{s}\right)\left(\frac{0.7 kJ}{kgK}\right)} = 5.6 K$$
>
> $$P_{elect} = 0.25 kW (3 months)\left(\frac{24h}{day}\right)\left(\frac{30 days}{month}\right) = 540 kWh \text{ at } \frac{\$0.15}{kWh} \text{ is } \$81 \text{ for the season}$$
>
> Tap water consumption is 9000 gal over the season, which adds another $20–25.
>   A cooling output of 10 kW for an electrical consumption of 0.25 kW is a rather good exchange rate (COP=40!) and why the evaporative coolers are such low power draw and efficient coolers in a dry environment. A large enough temperature drop of the air may become problematic at very high outside temperatures, in which case there will be a tradeoff between air throughput and air ΔT.

On the plus side, evaporative coolers are relatively simple devices with a water pump, a blower, a minor amount of plumbing and controls, and some high surface area media. They can be user-serviced, though this can be more problematic for rooftop units. They do have downsides and are generally not effective in humid conditions. They have water issues including consumption, mineral deposits, potential biological growth and associated odor, potential leaks, and the addition of humidity to the conditioned space (usually undesirable in the summer). They do require some maintenance, such as periodic replacement of the media, pumps, and possible water path clogging issues, especially with hard water. Introducing ambient air into the building is usually desirable unless outdoor ambient air quality is poor, such as during periods of wildland fire impacts.

9.2.7 Ventilation systems for space cooling and heating

Ventilation of buildings is important for indoor air quality and avoiding "sick" building syndrome, minimizing contagion spread, though all with a nonnegligible energy cost. Ventilation has taken on new importance following the pandemic era to help mitigate indoor disease transmission as well as indoor air quality issues. When outdoor air quality is poor, for example during a wildfire event, ventilated air requires filtration.

9.2 Climate control systems for conditioned spaces

Via nighttime cool air flushing, whole house fans can be an adequate partial or complete approach to space cooling in the right environment, such as mountain, arid, and northern regions. In a humid climate, the outdoor temperature (a) does not cool down sufficiently to make them effective, and (b) latent cooling loads are added by bringing in moist outdoor air. This makes these systems less useful in the moister eastern half of the United States and useless in the high temperature, high humidity environment of the southeast United States.

Buildings with modest cooling loads due to their climate, cool nighttime temperatures, and/or a tight thermal envelope are good candidates. These high-flow, though efficient, fans are mounted in the ceiling or in the attic and draw the warm air out of the building through a dampered, insulated vent in the ceiling. The fans introduce cool ambient air, reducing the indoor temperature via many air exchanges. Typically, they push exhaust air into the attic and out the attic vents, also dropping attic temperature. Tamarack, QuietCool, and Airscape are some of the major manufacturers of whole house fans. The fans can move a few thousand cubic feet per minute of air for a few hundred watts as they use efficient fans developed for server farms.

The author's home, a converted small schoolhouse/church, is an example. The building, as discussed elsewhere in the book, is a high-thermal-mass and moderately well-insulated building. The arid, high-altitude location cools typically into the 60–70°F during summer nights (Fig. 9.8). The use of air flush cooling using zoned whole house fans is sufficient for all but the hottest periods, as shown in the early June period of 2020 (Fig. 9.8A). The indoor temperature can be dropped by 10°F during the night while during the day the house is closed up and stays below ~80°F even with ambient temperature approaching 100°F. On the hottest days of 5–6 June 2020, the evaporative cooler in the living room was run to moderate the indoor temperature as indicated. In Fig. 9.8B, the ventilation approach can be inverted on sunny winter days if the outdoor temperatures reach ~70°F. The house is opened in the afternoon, particularly the southern windows, and fans are run to bring in warmer outdoor air. During the fall and spring, the high thermal mass of the building moderates diel indoor temperature swings enough that the building can be passively allowed to float with the ambient with no indoor climate control (Fig. 9.8C). In that case, thermal inertia was even sufficient through a couple-day intense cold front on 9–10 September 2020, when ambient temperature dropped to ~40°F.

The key parameter for a ventilation-based cooling approach is moving air, which is much less expensive than heating or cooling air. For cold-air flushing, frequent air exchange with the outside is key. One exchange of cool outside air is far from sufficient to cool the interior mass of the building. The parameter of interest is the air changes per hour (or ACH), which is defined as the volumetric flow rate (Q) divided by the interior volume (V) (Eq. 9.1).

$$ACH = Q/V \qquad (9.1)$$

In a planned ventilation scenario, larger ACH is desirable, contrasting with air leakage or unplanned ventilation, where the ACH of this process should be minimized. Ventilation can be a push, pull, or, most appropriately, a balanced system. An important consideration in ventilation systems such as whole-house fans is mass balance and the need for makeup air. If these are run without sufficient fresh air entry into the building, bad things can happen, such as back-drafting combustion appliances or introducing sewer gas inside. The air must come from somewhere, and a tight thermal envelope means that location may be nonideal.

Other ventilation means seek to cool humans convectively by moving indoor air. Indoor air-conditioned temperature setpoints can be higher in the summertime because of convective and

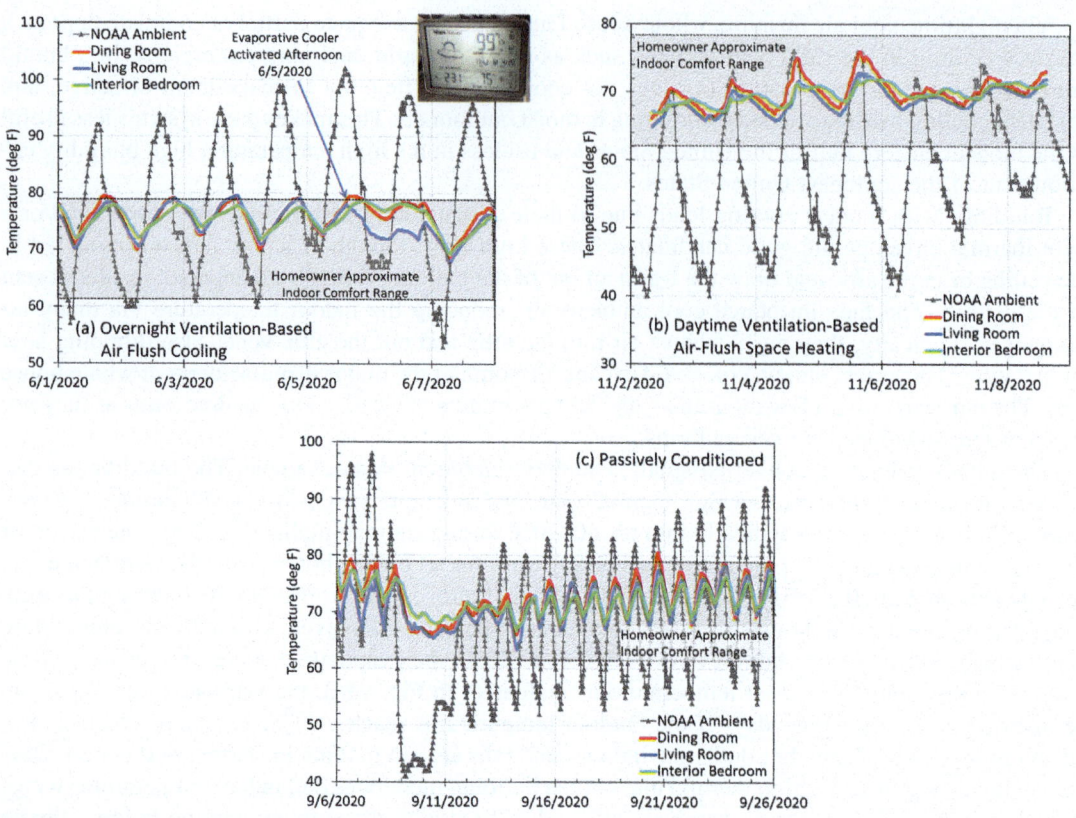

FIG. 9.8

Temperatures of a high thermal mass reasonably well-insulated building that allows (A) ventilation-based nighttime cooling during the warm season as confirmed by indoor-outdoor temperature sensors showing 75°F and 99°F, (B) space heating during the cold season, and (C) "thermal floating" with passive indoor thermal management during the shoulder season and even during cold-outbreaks of a few days.

evaporative cooling on the skin's surface. Ceiling fans have evolved to use more efficient DC motors driving larger swept area sizes, slower rotation speeds, and to provide more airflow per unit of power consumption (CFM/Watt), known as efficacy. An illustration of the improvement in efficiency is shown by the thermal image in which the more efficient fan is running at a lower surface temperature (i.e., less waste heat) (Fig. 9.9). The SMC AC motor 52″ fan shown on the left uses 1.2A at 120 VAC. Power is approximately 144 W on high, while the 60″ DC motor fan from Fanimation, Inc. shown to the right in Fig. 9.9 uses 32 W on high.

The air movement is larger for the DC fan as well. Currently, DC motor fan's efficacy often exceeds 200 CFM/Watt, as this unit, for example, moves 5833 CFM on a high setting while consuming 32 W, yielding ~184 CFM/W (Table 9.3). Both the flow rate and power use are a function of the fan speed, and the lower fan speeds have a higher efficacy. They are also much quieter and lack the humming

9.2 Climate control systems for conditioned spaces

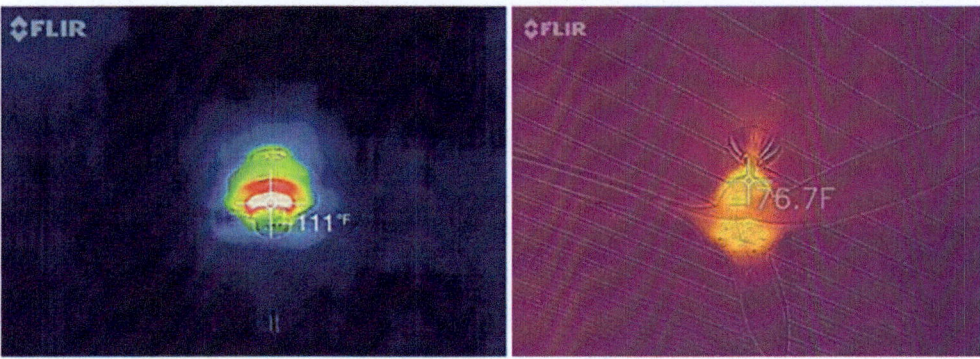

FIG. 9.9
Infrared comparison of a circa 1990s AC motor ceiling fan vs a ~2015 DC motor fan after running on a high setting. The thermal signature and lower temperature of the newer fan show the improved efficiency of the DC unit.

Table 9.3 DC motor ceiling fan specifications.

Speed	Max RPM	Q (CFM)	Amps (120 V)	Watts	Efficacy (CFM/W)
High	132	5833	0.43	32	184
Medium	94	3966	0.15	10	412
Low	51	1977	0.05	3	743

Many older AC motor ceiling fans were on the order of 50 CFM/W.

operation of AC motors. Other newer design high volume low speed ceiling fans, such as those from Big Ass Fans, use large fan diameters to move air more uniformly in a room rather than higher speed (and noisier) more localized fans.

Ceiling fans are effective ways to convectively mix air in a space and reduce the vertical thermal gradient. The use of a ceiling fan in the winter can have minor benefits too from breaking up the thermal stratification of a room (Fig. 9.10). The hotter air that collects near the ceiling can be brought back down to the surface, particularly with very high ceilings. This can reduce space conditioning needs modestly in the winter by gently convecting warmer air down from the ceiling level. A very low fan speed and air flow directed upward so that the ceiling blunts any wind chill effect both help. This can reduce the temperature gradient driving conduction and convection losses through the ceiling.

Powered attic ventilation fans that strictly exhaust hot attic air through the roof are generally thought to add little benefit and are potentially a detriment. They can drop attic temperatures, but they also depressurize the attic, leading to potential loss of conditioned air through the ceiling. The better approach to attic thermal management is using a white or highly reflective roof exterior surface combined with upgraded attic insulation and air sealing to mitigate the effects of a hot attic on the conditioned space below.

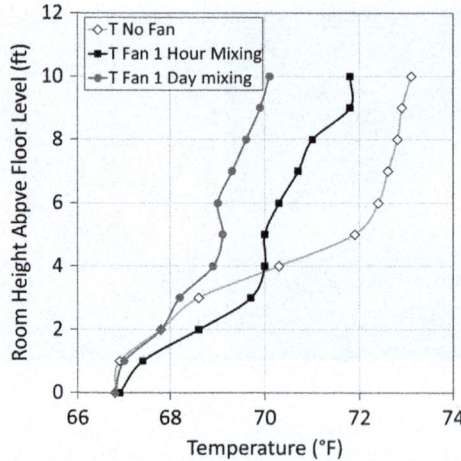

FIG. 9.10

Vertical temperature uniformity during winter in a room that is unmixed, mixed for 1 h with a ceiling fan on low, and mixed with the same fan settings for 24 h.

9.2.8 Heat and energy recovery ventilators (HRVs and ERVs)

As discussed in the chapter on the building thermal envelope, the mantra for the building envelope is to "make it tight and ventilate right." One drawback of tight thermal envelopes is a heightened need to introduce fresh air into the indoor environment.

Solutions include heat recovery ventilators (HRV) and energy recovery ventilators (ERV). HRVs and ERVs are essentially heat exchangers. In the case of the ERV, they also exchange moisture, which makes them somewhat more effective. The systems bring in fresh outdoor air while exhausting stale indoor air. The flows pass each other through a heat exchanger to minimize the energy penalty of doing so. The ERV thus covers both sensible and latent heat loads. The HRV may be sufficient in a dry location where latent heat effects are minimal. An HRV can also act as a sink for moisture if the outdoor air is exceptionally dry and indoor humidification is desirable.

9.2.9 Combined systems and cogeneration

Innovative HVAC systems are constantly entering (and in some cases exiting) the marketplace. Combination or "Combi" boilers provide both water heating needs and space heating through hydronic heaters (water to air heat exchangers) or radiant systems such as radiant floors. A newer example is a combined space and water heating system using heat pump technology. Such systems provide domestic hot water as well as indoor climate control in both summer and winter. One startup company pursuing this is Harvest Thermal, which combines a heat pump system, hot water thermal storage, and a heat exchanger to transfer heat to the building air handler.

Combined heat and power systems (often termed cogeneration when it is on a larger scale) are often implemented on commercial, industrial, or utility scales. The fuel for the main unit, or "prime mover," can be fossil (often natural gas), hydrogen, biogas, or other renewable fuels. The advantage is that the waste heat from a typical electrical power generator can be used for purposes such as water and space heating (Fig. 9.11). These systems often work well on the scale of college or industrial campuses where distribution distances are reasonably short. Such district systems are amenable to microgrid applications. Whereas a typical Rankine cycle electrical power generator is 30%–40% efficient, the CHP systems often will achieve 70% plus efficiency by reusing the waste heat normally dumped to the atmosphere or a water body.

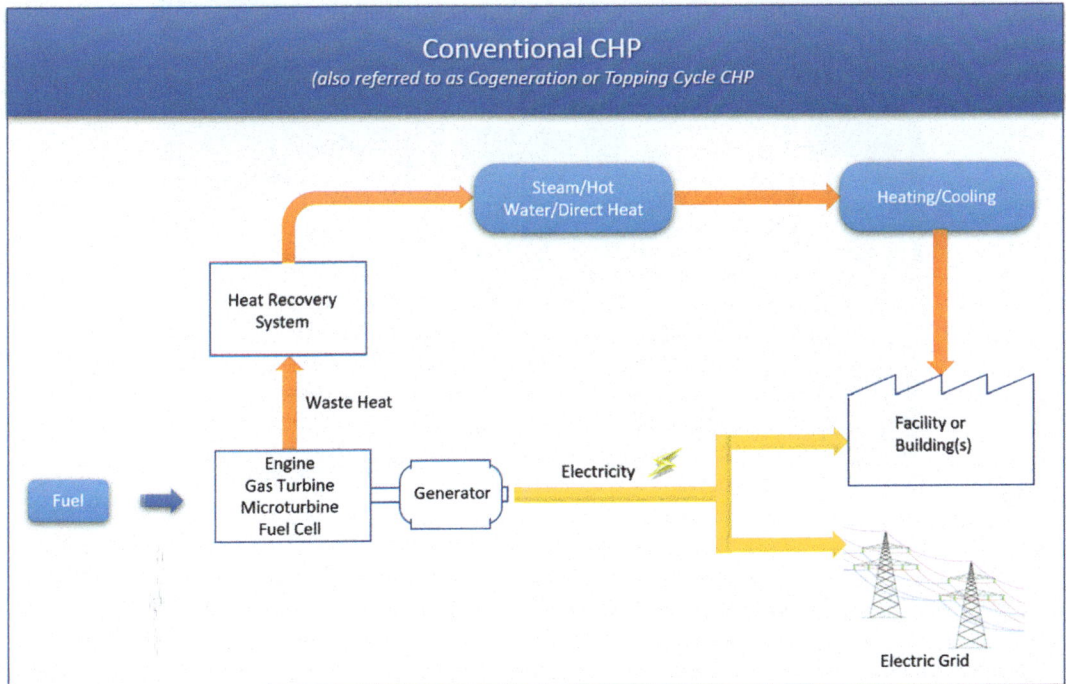

FIG. 9.11

Example application of combined heat and power (CHP) systems that cogenerate electrical power and provide space heating or cooling (public domain figure from the United States Department of Energy).

An example of cogeneration application of combined heat and power systems on a community scale is the Calgary District Heating facility that has a maximum thermal output of ~60 MW output (Fig. 9.12). It generates electricity with natural gas turbines and uses the waste heat to distribute hot water to neighboring municipal buildings in the district, including the city municipal plaza and building, the library, and Bow Valley College. The utility operator estimates that the greenhouse gas emissions are one-quarter of standard coal-generated electricity, according to the facility website (https://fvbenergy.com/projects/enmax-district-energy-system/). It saves the connected buildings the mechanical and maintenance footprint of running stand-alone systems. It also saved the city from replacing seven obsolete boilers in the municipal building.

232　Chapter 9 Electro-mechanical systems and appliances in buildings

FIG. 9.12
Calgary District Heating facility is located in downtown Calgary, Alberta. The stacks are from four natural gas boilers. The facility uses combined heat and power with roof-mount solar PV integrated as well. It recirculates heated water to nearby buildings through insulated underground pipes.

Waste heat recovery is a related application that can take the heat from processes, cooling water, or even wastewater flows and recover the energy from this for process use, preheating of air and water flows, or other reuses. These flows may be at temperatures of 100°C or less and thus not viable for use for electricity generation. The application to data centers and cryptocurrency mining operations is more common with the surge in electric use with these facilities. This will likely be an area of continued development.

9.3 Case Study: Laboratory building district chiller systems control

Commercial buildings use much larger-scale HVAC systems for indoor climate control. Often buildings of this scale use district chiller systems that can service a number of buildings or a campus. Such systems use sophisticated control systems, requiring highly trained installers and operators. One innovation and efficiency measure with the commercial scale systems is the use of variable-speed drives (VSD) in the air handler part of the HVAC system.

The monitoring of real-time data in building management allows effective and early identification of building performance issues. The New Mexico Institute of Mining and Technology used a common source of chilled water for the space conditioning needs in a subset of buildings on campus. The data in Fig. 9.13 show the temperatures of the Central Plant supply and return temperatures. The chilled water flow is used to reject heat from building exteriors to keep spaces cool as needed. A new building, the New Mexico Bureau of Geology (NMBG), was keeping temperatures comfortable on this system, but the runtime of the chilled water system and its energy use were higher than expected. Monitoring of the supply temperature of water to the NMBG showed that it was significantly higher than the campus supply temperature and followed the diurnal trends of the central plant's return temperature. The observation led to the identification of the problem: the building supply and return flows had been mistakenly swapped at installation during the construction phase. The building was pulling its chilled water from the return side (at a temperature 10–15°F above the supply side) and dumping the warmer water back to the supply side. The problem was subsequently fixed, and the operational times of space cooling in the building as well as energy use decreased significantly upon the replumbing of the system correctly. The online real-time monitoring of this data facilitated the identification and remedy of the problem.

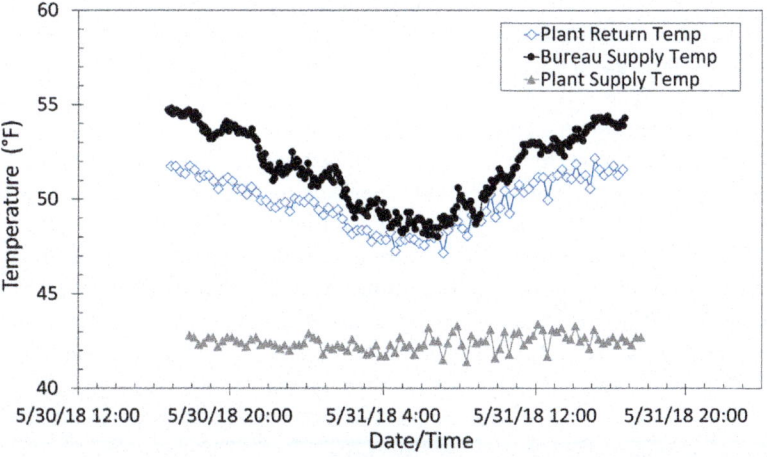

FIG. 9.13

Temperature profile of New Mexico Tech chiller plant supply and return temperatures where chiller water is used for space cooling in numerous campus buildings. Also shown is the supply temperature to a new building on campus, the New Mexico Bureau of Geology. Monitoring this temperature led to the discovery of a construction mistake in swapping the inlet and outlet lines for this building.

9.4 Water heating systems-storage tank and tankless on-demand systems

Water heating technology has also evolved over the past decades to become much more efficient. The old-school tank water heaters with an atmospherically vented, fixed-output natural gas burner and standing pilot light, though simple, are becoming obsolete. Likewise, electrical resistance water heaters are expensive to operate on the grid since roughly two-thirds of the thermal energy is lost at the power plant in typical Rankine cycle operations (more efficient with Brayton or combined cycle power plants). These have been replaced by on-demand, no-standby-loss water heaters (both fuel-burning and electric) and more recently by heat pump water heating options (air source and ground source).

Generally, the on-demand "tankless" water heaters are more efficient than storage tank water heaters as they avoid standby losses. This advantage, however, is reduced by the fact that the on-demand units have been observed to emit more uncombusted methane than tank water heaters (Lebel et al., 2020). These units have a high thermal output in BTU/h since they must heat the water quickly as it passes through the water heater. Thus, they require upsized gas lines, upsized exhaust flues, and other concerns such as the potential for sediment formation and precipitation at the heat exchanger, especially in areas with hard water.

Many of the concurrent developments in indoor climate control systems using heat pumps have been applied to water heating systems. New applications include heat pump water heaters, in which instead of burning a fuel or using electrical resistance heaters, a heat pump cycle is used to heat the water. These electric heat pump systems have been developed over the last two decades to offer a more efficient all-electric option. The COPs of such a system are in the range of three to four, meaning three to four units of thermal energy are moved into the hot water tank for every one unit of electrical energy consumed. The maximum Uniform Energy Factor U.E.F. for a conventional gas or electric water heater is 1.0, and real-world units are 0.95 or less, whereas some of the heat pump water heaters currently available exceed 4.

A number of the large commercial and residential water heater manufacturers, such as A.O. Smith and Rheem, now offer heat pump water heaters. There are new to the United States market, higher COP heat pump systems such as the split systems offered by Sanden, the $SanCO_2$ heat pump, which uses carbon dioxide as the working fluid. Such devices have COPs of five in typical operations, though with a higher capital and installation cost. There have been more extensive applications in Japan for such systems where mini-split air-to-air heat pumps were first introduced. There are many details of installing these that bear consideration (e.g., a source of "warm" air into which to extract heat, where to dump the cooled air, and others), but are the likely future of water heating in the all-electric home (Example 9.2).

EXAMPLE 9.2 Water heating costs

Problem: According to the US Department of Energy, "At an average of 64 gallons of hot water per day, US households spend between $400 and $600 a year for water heating needs." From first principles, calculate if this seems reasonable for gas and electric resistance water heating. Compare this to the cost of a high-efficiency air source heat pump water heater with a U.E.F. of 3.5. Find the CO_2 emissions associated with each option with grid-supplied energy.

Given: 64 gal/day hot H_2O & $600/year
Find: Verify and find CO_2 emissions of CH_4 vs resistance vs heat pump
Assume: $0.14/kWh; $1.40/therm; U.E.F. = 3.5

Solution:

We will use the "MCAT" equation for sensible heating of water $E = MC\Delta T$ and express it on a differential basis and use an efficiency of 85% for electricity and 60% for gas. These are dependent on the variable rates for electric and gas usage. Here we will use the national averages of $0.14/kWh and $1.40 per therm. For emission factors, we use 0.37127 kg/kWh for US grid emissions and 5.3 kg/therm for natural gas combustion.

Continued

> **EXAMPLE 9.2 Water heating costs—cont'd**
>
> $$P = \frac{dM}{dt} \times C \times \Delta T = 64 \frac{\text{gal}}{\text{day}} \left(\frac{8.34 \text{ lb}}{\text{gal}}\right) \left(\frac{\text{kg}}{2.2 \text{ lb}}\right) \left(\frac{\text{day}}{24(3600)\text{s}}\right) \times \left(\frac{4.186 \text{ kJ}}{\text{kgK}}\right) \times (120F - 55F) \left(\frac{5K}{9F}\right) = 0.424 \frac{\text{kJ}}{\text{s}} = 0.424 \text{kW}$$
>
> $$\text{Electric Cost} = P \times t \times \frac{1}{\eta} \times \text{Electric Rates} = 0.424 \text{kW} \left(\frac{365 \text{ days}}{\text{year}}\right) \left(\frac{24 \text{ h}}{\text{day}}\right) \times \frac{1}{0.85} \times \frac{\$0.14}{\text{kWh}}$$
>
> $$= 3718 \text{ kWh} \times \frac{1}{0.85} \times \frac{\$0.14}{\text{kWh}} = \frac{\$612}{\text{year}}$$
>
> $$\text{Electric } CO_2 \text{ Emissions} = 3718 \text{ kWh} \times 0.37127 \frac{\text{kg}}{\text{kWh}} = 1380 \text{ kg } CO_2$$
>
> $$\text{Gas Cost} = P \times t \times \frac{1}{\eta} \times \text{Gas Rates} = 3718 \text{ kWh} \times \frac{1}{0.6} \left(\frac{3.6 \text{ MJ}}{\text{kWh}}\right) \left(\frac{\text{MBTU}}{1055 \text{ MJ}}\right) \left(\frac{\text{therm}}{0.1 \text{ MBTU}}\right) \times \frac{\$1.40}{\text{therm}}$$
>
> $$= \frac{211 \text{ therms}}{\text{year}} \left(\frac{\$1.40}{\text{therm}}\right) = \frac{\$296}{\text{year}}$$
>
> $$\text{Gas } CO_2 \text{ Emissions} = 211 \text{ therms} \times 5.3 \frac{\text{kg}}{\text{therm}} = 1122 \text{ kg } CO_2$$
>
> $$\text{Heat Pump Cost} = P \times t \times \frac{1}{\eta} \times \text{Electric Rates} = 0.424 \text{kW} \left(\frac{365 \text{ days}}{\text{year}}\right) \left(\frac{24 \text{ h}}{\text{day}}\right) \times \frac{1}{3.5} \times \frac{\$0.14}{\text{kWh}}$$
>
> $$= 3718 \text{ kWh} \times \frac{1}{3.5} \times \frac{\$0.14}{\text{kWh}} = \frac{\$149}{\text{year}}$$
>
> $$\text{Heat Pump } CO_2 \text{ Emissions} = 1062 \text{ kWh} \times 0.37127 \frac{\text{kg}}{\text{kWh}} = 394 \text{ kg } CO_2$$
>
> The heat pump water heater is the clear winner in both costs to operate and lower emissions. This would be an even stronger case if the electricity was generated via renewable sources. The natural gas water heater is a much less costly to run option and has lower emissions than the electric resistance water heater on grid power.

9.5 Lighting

Lighting accounts for a significant fraction of building electrical use, ranging from ~10% in the residential sector and sometimes exceeding a third of the energy in the commercial sector office building. A succession of technological "revolutions" has occurred since the 1980s with lighting technology. The traditional incandescent lamp, simple, cheap to purchase, and effective (mostly as a heater), has been replaced first by fluorescent lamps, particularly the compact varieties, and more recently by light-emitting diode lamps (LEDs). LEDs are currently the mass-market technology of choice and have evolved in terms of their efficiency, output, lifetime, and quality of light. Quantum dot LEDs such as those made by Osram, Inc. are a newer technology that can maximize efficacy with high color rendering index (CRI) and low "blue light" generation while maintaining a 3000 K effective temperature.

The primary advantage of the newer technologies has been efficiency and longer life, particularly with LED bulbs. The innovations that have allowed LEDs are ways to produce white light (LEDs are single wavelength) and scaling up of power output in a compact form factor. Increased light output has benefited from improved thermal management for LEDs. Although LEDs are more efficient than incandescent (on the order of 25% vs 5%), they still produce significant heat. LEDs have become the dominant technology as costs have become more competitive while light quality is on par with the best incandescents in terms of Color Rendering Index (CRI).

There are other concerns with lighting besides energy use. It is desirable to minimize light spillage into the ambient and avoid light pollution. Thus, directing the lighting flux downward for outdoor lighting is important. Light pollution is undesirable for astronomy and also has consequences for the sleep cycles of sensitive species, including humans.

Control systems are the other key components extending beyond the lamps themselves (Fig. 9.14). Controls from the residential to commercial scale allow automation and adjustment based on ambient light levels. Even simple switching systems and timers in homes can turn on, turn off, and modulate light levels. Systems are optimized to allow diffuse natural light while providing efficient, high-CRI artificial light at an appropriate intensity when needed (Example 9.3).

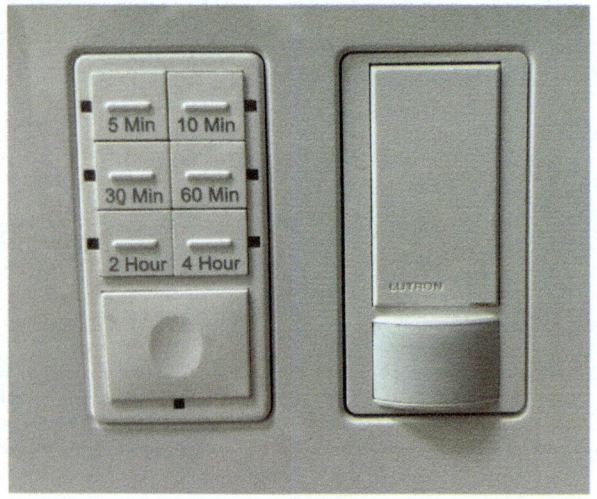

FIG. 9.14

Switches that help moderate energy use include adjustable dimmer switches, timer switches (*left*), and occupancy/vacancy sensor switches (*right*). These can be used to modulate output, switch off exhaust fans or other loads, and turn on and/or off lights in rooms based on occupancy.

The efficiency of lighting ranges from ~5–10% for incandescent bulbs to ~25–50% for LED bulbs. Lighting is often quantified by its efficacy, the light output the bulb generates divided by its electrical power consumption. The common metric for lighting efficiency is in lumens/watt. This is the lumens of light intensity output per watt of electrical consumption. Typical ranges for some common lighting technologies are given below in Fig. 9.15. The more recent LED lamps are becoming yet more efficient.

EXAMPLE 9.3 Lighting

Problem: An incandescent bulb can be considered a maximum of 10% efficient (the other 90% is, guess what?). It is operating in a room temperature room.
(a) What happens to the other 90% of the energy input? Make a diagram of the energy inflows and outflows to the bulb shown.
(b) Find the effective surface temperature of a 100-W light bulb. For this calculation, you can assume radiation only. Assume the emissivity of the surface of the bulb is 0.5. Make any other necessary and reasonable assumptions.
(c) Define the efficiency of this process and suggest a way that one could measure the efficiency of this light bulb.

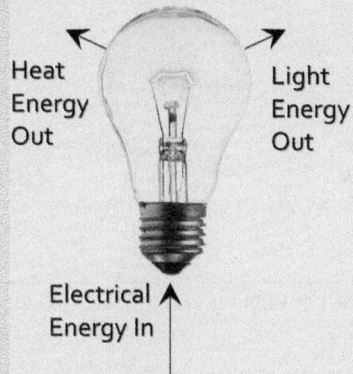

Given: 100 W
Find: $T_{surface}$
Assume: $\epsilon = 0.5$; sphere D = 10 cm; $T_{room} = 293$ K
Solution:

Most of the energy in the light bulb is directly turned into radiant energy in the infrared (i.e., heat). Assume the bulb can be represented by an effective 10-cm sphere.

$Q = \sigma \epsilon A (T_{surface}^4 - T_{room}^4) = 100$ W (1-Eff) = 90 W
$Q = 5.67 \times 10^{-8}$ W/(m²K) (0.5) (4)π(0.05 m)²($T_{surface}^4$ - 293K⁴)
$T_{surface} = 573$ K

Efficiency (%) = Radiant energy as light/electrical energy input × 100

One can measure the electrical consumption of the bulb and then measure the light intensity by integrating over all directions and visible wavelengths with an intensity meter. Alternatively, one could measure the heat generated directly while allowing the visible light to escape (perfect absorber in the nonvisible and perfect transmitter in the visible).

Lighting options often interact with the thermal envelope of the building as well. An example is shown of recessed can lights in Fig. 9.16. Insulating around these fixtures is difficult, particularly when they are near the periphery of the building. Furthermore, they often are not particularly well air-sealed. More advanced, better sealed options exist for recessed can lights.

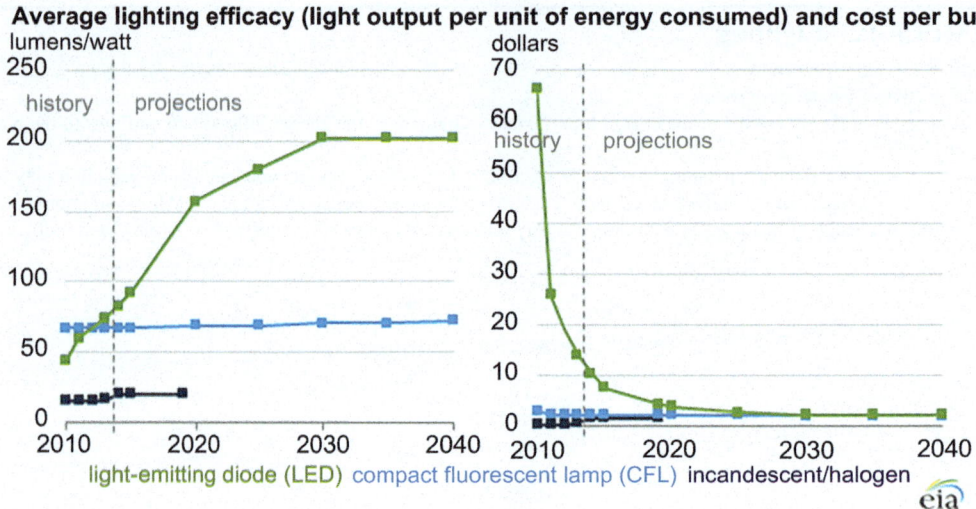

FIG. 9.15

Lighting efficacy (lumens/watt) and cost per bulb for LEDs, CFLs, and incandescent bulbs over time (public domain from US EIA).

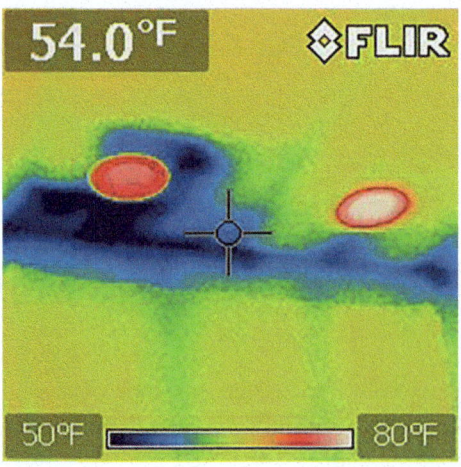

FIG. 9.16

Thermal image of recessed can lights showing the nonideal insulation around these fixtures.

9.6 Electrical appliances and phantom loads

A range of innovations are occurring with other categories of electrical loads. At the same time, we are adding more numerous electronic devices to our buildings and lives. One example that is facilitating the transition from combustion to all-electric operation is induction stoves. Rather than burning gas or using electrical heating coils, the induction units heat the contents of a properly designed magnetic skillet or pot by inducing a magnetic field. Small single-burner units are available for under $100 as well that can be used to displace much of the combustion needs without the cost of full replacement (Fig. 9.17). The gas control knobs for the burners underneath were removed as a user error resulted in the burners being ignited, which started the bottom of the induction burner to melt. This could be a case study in user interface and human factors engineering!

FIG. 9.17

Single-burner induction cooktops used to displace much of the gas stovetop use. Note the removal of the control knobs from the gas burners on the left to prevent accidental ignition of the induction plate from turning on the controls of the gas burners.

Many electronics are not completely off when powered off (Fig. 9.18). Phantom loads are an area that electronic manufacturers are working to reduce, and improvements have happened in the last decade. The consumer can reduce phantom loads by use of power strips to turn off the connected load at the socket, smart power strips that turn off peripherals when a main load is powered down, using timer switches for automatically turning on and off loads (e.g., night light), and simply by unplugging chargers when not in use.

FIG. 9.18
Thermal and visible images of man's best friend, "Lobo" (R.I.P.). Note the phantom load of charger bricks showing up warmer than the room temperature even while on standby. One of these warm power bricks is a timer for turning off a load, though the timer itself consumes electricity. Also note the warm temperature of the body, cooler at the extremities such as ears and paws.

9.7 Demand management and load shifting

Historically, the electrical grid is designed with grid load centers to be responsive and match the power supply to the power demand across the system in real time. For instance, in sunny, summer late afternoons when building air conditioners are running at high capacity and the workforce begins arriving home and upping residential demands, the grid can be strained to provide that short-term peak power demand. This historically relied upon bringing online "peaker" plants, typically natural gas conventional turbines that are often smaller units and less efficient than baseload generators.

Another approach is what is called demand side management, using an entire range of manual and automatic, voluntary and mandatory, and behavioral and engineered approaches to reshape the demand curve. The goal is to flatten the demand curve by moving nonessential uses to off-peak hours. The concept of managing demand, particularly as it applies to electricity, has become an industrial sector all its own. The major goals of this include:

- Using many of the efficiency techniques already described throughout the book, reduce the overall peak kW demand (peak shaving) and kWh energy consumption of a facility.
- Shaping the demand curve throughout the day and season to better match the supply curve.
- "Load shifting" of high-power loads to times when the overall grid demand is lower and electricity is less costly (e.g., overnight time periods).

Load shifting becomes particularly important as more nondispatchable renewable energy sources are added to the grid. The "duck curve" is a problem that has emerged with systems that are increasingly dependent upon solar PV production. Solar PV production peaks near solar noon and drops off in the later afternoon. As PV production fades in the later afternoon and demand continues or increases, the net production needed from other sources follows a profile that oddly resembles the profile of a duck from the side (Fig. 9.19). Shifting loads away from the normal late afternoon peak to the middle of the day when solar is available or overnight when demand is historically low in most areas can both help shave the peak demand.

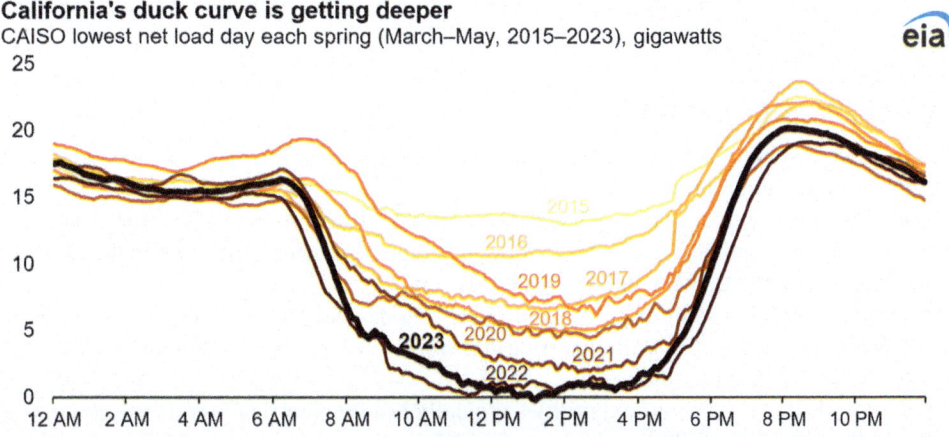

FIG. 9.19

California's net electric demand curve (subtracting out the renewable energy production) on the lowest demand day each year from 2015 to 2023) (public domain image from US EIA with data from California Independent Systems Operators, CAISO).

9.8 Chapter summary and conclusions

The mechanical, electrical, and plumbing (MEP) systems in a building have high capital and ongoing energy costs. In particular, the heating, ventilation, and air conditioning systems (HVAC) are particularly intensive. The second most intensive is often the water heating system. More efficient systems for conventional gas-fired space and water heating are standard. These include modulating-condensing gas furnaces, similar advanced gas boilers, and on-demand gas and electric water heating systems that avoid the standby losses of tank water heaters.

Ventilation can be an important technique to cool, heat, and introduce fresh air into a building. Air flushing is the strategy for cooling in locations with large day-to-night temperature fluctuations, like the desert southwest and mountainous areas. Whole house fans are effective in cool nighttime environments, including the dry climates of the western United States.

The imperative is moving HVAC systems, as much as practical, to an all-electric building obviating indoor combustion sources. The advantages of this are both the health and safety of the occupants and potential expanded renewable electrical sources, whether onsite or grid-supplied.

In recent decades, all-electric heat pumps have evolved to be more efficient, able to operate effectively with more extreme hot and cold temperatures, and with more flexible and effective operations. The classic refrigerated air conditioning system is now capable of operating for space heating in winter by reversing the cycle. Mini-split heat pump systems allow more flexible retrofits with or without ducting and zoning of indoor climate control. Improved performance allows newer heat pumps to operate down to very low ambient temperatures, albeit with lower efficiency as quantified by a lower COP. The other major heating appliances in a home, water heaters and clothes dryers, also have high-efficiency heat pump options now commercially available. This is an active area of ongoing development and promises to help electrify everything in a modern building and avoid indoor combustion sources.

9.9 End of chapter exercises

(1) **Concepts:** Discuss three ways that air conditioning (aka refrigerated air) adds to global warming. In a paragraph list and describe them.

(2) **Concepts:** Consider a condensing gas furnace. What concept related to thermodynamics does it rely on to achieve high efficiency? How does this type of furnace contribute to the efficiency of a heating system? What is the approximate efficiency?

(3) **Concepts:** Your air conditioning system fails. You decide that you can open the door of your refrigerator to cool your apartment. Will this work to keep your apartment at least somewhat cooler, and why or why not?

(4) **Problem:** kWh is what electrical utilities typically bill for electrical use. Is it a unit of energy or power? Determine the BTU equivalent of 1 kWh.

(5) **Problem:** You accidentally leave on a light fixture that uses 200 W of power while you are gone on vacation for 2 weeks.
  (a) How much energy did you use in kWh?
  (b) If electricity costs $0.10 per kW h, how much did this add to your electricity bill?

(6) **Problem:** How effective would a dog be as a heating unit for humans? Take a typical beagle at 25 lbs. Approximate this beagle as a cylinder that is 2 ft in length and 1 ft in diameter, of which half is exposed to the human. Take the surface temperature of the beagle to be 95°F and the effective surface temperature of the human as 88°F. Assume an emissivity of 1.0. What is the BTU/h rating of this heater?

(7) **Problem:** Consider conduction now for the canine heater. A beagle has an exterior (skin + fur) thermal conductivity of 0.05 W(m K), and the estimated length of the dog's fur is 7 mm. The core

temperature of the dog is 38°C, and the room temperature is 20°C. The dog can be approximated as a closed cylinder of 50 cm long by 20 cm.
   (a) Estimate the dog's conduction from the interior to the exterior.
   (b) Estimate the thermal output to the surrounding air (20°C) due to radiation. Assume a black body radiator and assume the same temperatures.
(8) **Problem:** You put 12 of your favorite beverages on room temperature into a refrigerator. How long will it take to cool this to 40°F? Assume water properties, and your refrigerator can extract 250 W of thermal power from its interior. Assume no other heat loss and neglect any mass outside of the liquid. Does this seem reasonable, and why or why not?
(9) **Problem:** Confirm or deny. A "ton" of air conditioning capacity equals 12,000 British thermal units, or BTU per hour (this curious term is derived from the amount of heat that is required to melt a ton of ice in 24 h).
(10) **Problem:** It is a winter day and has just dipped to the freezing point of water (0°C). You decide it is a good day for a soak in the hot tub. (a) How much heat (MJ) is required to raise a hot tub (400 gal) from the temperature of 0°C to a piping hot 50°C? (b) How much additional energy is required to raise the temperature from 50°C to 100°C, the boiling point of water? (c) And how much additional energy is required to boil away all of this water remaining at 100°C? (d) If the tub is heated by electricity, how many kWh of electricity would it require to raise the temperature from 0°C to 100°C and then boil the entire tub of water? The heat capacity and latent heat of water may be of use.
(11) **Problem:** You leave your house for a trip for 3 weeks knowing that your standby steady-state power use is 1200 W.
   (a) How much energy did you use in kWh?
   (b) If electricity costs $0.13 per kW h, how much did this add to your electricity bill?
(12) **Problem:** A 75-gal (US) electric water heater provides 9000 W of power to heat water. If the heater is filled with water at an initial temperature of 50°F, how long will it take for the water to reach 140°F? Neglect the effect of heat loss.
(13) **Problem:** It is noticed that a neighbor leaves on continuously a fluorescent fixture with three 40 W bulbs and an incandescent fixture with two 60 W bulbs. Bulbs are rated by their electrical consumption.
   (a) How much electricity is used in a year (kWh), and how much does it cost at $0.15/kWh?
   (b) If coal at 25 MJ/kg is burned at 33% efficiency, how much did we burn in a year?
   (c) How long is a simple payback time if you could replace these with three 16 W LED tubes and two 10 W LEDs, which have a capital cost total of $32?
(14) **Problem:** You have a (a) heat pump for space heating, (b) a refrigerator, and (c) a heat engine. These are operating between a hot reservoir of 500°F and a cold reservoir of 60°F. Find the Carnot Efficiency, or COPs, of each device. Would a refrigerator with $Q_{cold} = 2$ kW and a compressor performing work = 1 kW be possible?
(15) **Problem:** Do a bit of research and find comparable laptop and desktop computers. Find the power consumption of each (steady state). Do they give a standby power use, and what is it if so? Calculate the energy use of a laptop vs a desktop computer. How many kWh in a year does each use, making any reasonable assumptions?

(16) Problem: You are using a gas stove to heat a 5-gal crockpot of water. You start from ice at −20°C. The effective efficiency of the gas stovetop can be taken as 45%.
  a) Calculate the thermal energy required in megajoules (MJ) that go into heating the ice by boiling away all the water.
  b) How much natural gas did you burn in cubic meters (at STP)?
  c) If the waste heat goes into heating the air in the room, how much did you increase the temperature of the room, if the room is 15 ft × 20 ft × 10 ft (assuming STP)? It is clearly going to be less than what this calculation shows; why?

(17) Problem: Internet claim, confirm or deny: "For every 6 h that a 20 W energy-saving LED bulb is used in preference to a 100W incandescent bulb, it saves its own weight in oil." Make any appropriate assumptions.

(18) Problem: A ceiling fan operates 6.4 h per day on average. Its electric consumption is 20 W. Find the annual cost to run this fan, assuming electricity costs $0.12/kWh.

(19) Problem: Find the average annual gas and electric use for a house in the United States. Convert into total GJ per year. Compared to the 600 GJ of embodied energy to build a home, how many years before the operational energy use surpasses the embodied energy of construction?

(20) Problem: A natural gas home heating system has an output of 65,000 BTU/h. Assume the natural gas retail rate is $0.75/therm, while the current cost for electricity is $0.13/kWh. You can assume the minimum gas furnace efficiency of 78% and an effective resistance heater efficiency of 100%.
  (a) What are the annual costs of heating with the gas furnace assuming the units run for six cold-weather months at a 50% capacity factor?
  (b) What are the annual costs of heating using electrical resistance heating?
  (c) Why is such a cost difference with the electrical efficiency so high, and when a large part of electricity generation is now via natural gas? Is there a better electrical option that may be useful and more efficient?

(21) Problem: Find the carbon footprint. The average US home uses 10,632 kW-h of electricity, 68 million BTU of natural gas, and 620 gal of gasoline annually. Use the national average CO_2 emissions per unit of use.

(22) Problem: A home in the southwest United States uses approximately 300 kWh for cooling a home using evaporative cooling. The same home uses approximately 300 therms of natural gas for heating it in the winter. Which is a greater energy expenditure (compared in MJ)? Compare again, considering the electricity is provided by a typical thermal efficiency power plant. Compare costs with energy prices of $1/therm and $0.15/kWh. Comment on the priorities for energy use reduction.

(23) Problem: A small portable cooler freezer is the most efficient spare freezer capacity and uses an average steady-state power draw of 8 W. How much energy does it use (MJ and kWh) and how much does it cost to run over a year where electricity costs are $0.14/kWh?

(24) Problem: You are running an electric heater off of 120 V AC power. The heater resistance is rated at 10 ohms. Find the current and power draw of the heater. How long would it take to heat a 10 × 10 × 10 ft room from 61°F to 70°F, assuming 20% heat loss? You can take a constant volume heat capacity of air of 0.7 kJ/(kg-°C) and an air density of 1 kg/m³. Assume the walls are mass-less and no heat or mass loss from the room.

(25) Problem: According to Statista, globally the average number of electronic devices per person is currently 3.6 in 2023. If the standby power use of each device is on average 1 W, how many kWh and large power plant outputs is this equivalent to? How many additional US households of electrical consumption does this add (assume 800 kWh/month for an average household)? Assume they run constantly.

(26) Problem: About 2.2 million bulbs illuminate Belgium's roads, and with 186 bulbs per square mile, the country is the unrivaled leader in Western Europe! Assuming the streetlights are 230 VAC and 2.5 A. What is the power draw of this lighting system, and how many large power plants does it represent? How many kWh are consumed per year?

(27) Problem: Compare the carbon footprint of various stovetop options. You are boiling completely 2 L of water on your stovetop starting from room temperature.
  (a) Find the energy needed to do this in MJ and kWh.
  (b) Find the CO_2 footprint using the natural gas combustion equation and assuming the midrange of 25%–40% for natural gas direct heating stovetops.
  (c) For electric stovetops, induction, in general, is 80%–85% efficient, and 65%–75% for electric resistance cooktops. Using mid-range efficiencies, compare the carbon footprints of induction vs electrical stovetops using (a) coal, (b) natural gas, and (c) solar PV electrical generation. You can assume CO_2 emission factors for coal electric as 955 g/kWh, as gas-electric 430 g/kWh, and solar PV 133 g/kWh (from embodied emissions).
  (d) Comment on the results. Which option provides the lowest carbon footprint?

(28) Problem: You decide to run your house off the grid using a standalone diesel generator. How many gallons per year would it require to generate the average US household electricity consumption? What are the resulting CO_2 emissions in tons per year? Make reasonable assumptions for the chemical formula for diesel and the energy content and efficiency of a diesel generator.

(29) Problem: The currently accepted estimate for the social cost of carbon is $50/t. What are the costs associated with the lifetime emissions of a natural gas furnace (assume 1000 therms/year, 80% efficiency, and a 25-year lifetime)?

(30) Problem: It is proposed to heat a house using a heat pump.
  (a) Draw the heat pump box diagram of heat and work flows from hot and cold reservoirs.
  (b) Write an equation for the COP for the heat pump system for heating a home. Think about what you are trying to produce and what you have to supply.
  (c) The heat transfer from the house is 15 kW. The house is to be maintained at 24°C while the outside air is at a temperature of −7°C. What is the minimum power required to drive the heat pump?

(31) Problem: I have come up with a brilliant idea to go off-grid. I am going to power my electricity needs with C-sized batteries as I can get them at Mega-Lo-Mart, and they only cost $1 apiece! If the voltage is 1.2 V DC, the C-sized battery can provide about 1 A of current for 8 h. What is my cost per kWh? Compared to what the typical utility electricity cost, did I make a good choice?

(32) Problem: A home is heated with (assume 100% efficient) electric heat at a total energy expenditure of 2.5×10^5 kWh per year at a cost of $0.10 per kWh. If the heating system is replaced with a heat pump having a coefficient of performance of 8, what are the cost savings per year? Draw a diagram of a heat pump showing the two heat flows and one workflow.

(33) Problem: A gas furnace can be up to 97% efficient. Based on Carnot's relationship, if the low temperature is the ambient temperature of 10°C, what is the upper temperature to achieve this? Is this reasonable? Why is it possible for a gas furnace to achieve such high efficiency?

Chapter 9 Electro-mechanical systems and appliances in buildings

(34) Problem: You install a security streetlight that uses 500 W of power. It turns on from dusk till dawn, which over the course of a year can be considered 12 h per day. How much electrical energy in kWh does it use in a year? How many kilojoules is this? How many BTUs? If electricity is $0.12/kWh, how much did this cost?

(35) Problem: You are considering a gas fireplace with a 30,000 BTU/h thermal output to the room. Assume typical electrical resistance efficiency.
  (a) What amperage of an electric heater (at 240 V) would provide the same thermal output?
  (b) With electricity costing $0.14/kWh, how much would it cost to run the electric heater continuously for 24 h?

(36) Problem: You use an electric space heater in your garage for 8 h. Its power consumption is 1500 W. How much electrical energy did you use in kWh?
  (a) How many joules is this?
  (b) How many BTUs?
  (c) If electricity is $0.12/kWh, how much did this cost?

(37) Problem: Estimate the electrical energy consumed by brewing a pot of coffee and keeping it warm for 2 h. Make any reasonable assumptions necessary.

(38) Problem: You are presently heating with electric baseboard heat costing $2000 annually at $0.14/kWh. You replace it with a standard efficiency gas furnace. Natural gas presently costs $0.50/therm, and a therm contains 100,000 BTU.
  (a) Make a reasonable assumption for the equipment efficiency of the baseboard heaters and a standard efficiency gas furnace.
  (b) How many therm/year of gas will you use, and what are your annual cost savings?
  (c) Why is the electric option so costly?

(39) Problem: A natural gas furnace used as a home heating system is rated to provide 80,000 BTU/h of useful heat to a home. The furnace is 85% efficient at converting the input fuel's energy content into heat delivered to the home (the rest is waste heat up the chimney).
  (a) Is this an endothermic, exothermic, or nonthermic chemical reaction happening in the furnace?
  (b) What is the energy content of the fuel that is provided (Q_{in}) to the furnace in BTU/h?
  (c) How much heat Q_{waste} is being wasted up the chimney in BTU/h?
  (d) If the furnace runs 50% of the time for 1 month, how much natural gas in cubic feet. Will it consume? Remember, 1 cuft of natural gas contains 1027 BTU.
  (e) How many therms of natural gas is this?

(40) Problem: GHG emissions from refrigerant leaks vs operations for mini-split heat pumps are a concern and can eliminate their efficiency advantage. One of the most common refrigerants used in refrigerated air conditioning systems, including mini-split heat pumps, is R410A. According to ASHRAE, the GWP of R410A is 2088. Calculate the equivalent CO_2 emissions from this as compared to the emissions associated with the annual operations of the heat pump for space heating with $COP_{heating} = 3$ and 12,000 BTU/h of cooling output. Assume a charge of 3 lbs of the refrigerant and it all leaks. Assume it operates continuously for 3 months/year and that grid electricity with CO_2 emissions of 1 lb/kWh runs the unit.

(41) Problem: You have a string of incandescent lights on your patio. Assume they are lit nightly from 5 pm till midnight. The light string consists of 24 sockets, each with an 11 W incandescent bulb. You decide to upgrade your energy efficiency with a 24-socket light string with LED bulbs that

9.9 End of chapter exercises

are 1 W apiece. The LED light string costs $40. How long until this replacement cost is paid back through electricity savings? Use mid-range electricity costs of $0.15/kWh.

(42) Problem: A gas home clothes dryer has a flow rate of 150 cfm, taking conditioned room air, heating it, and exhausting it. There are two main aspects to its energy use. The first is taking this room air and heating it to the dryer's operational temperature. Second, the home is now devoid of some volume of conditioned air that it must replace with outdoor air (leaking in through penetrations in the thermal envelope) and then reheat or cool it depending on the season. Assume it operates for 1 h for a load of laundry. Assume STP298. Assume a gas appliance with 70% efficiency. Take 130°F as the typical operating temperature of a clothes dryer, 70°F as the house interior temperature, and a winter day with a temperature of 30°F. Neglect any other power use besides heating the air used in the dryer (and reheating the compensating ambient air). Investigate how many therms a gas dryer consumes per load.

(43) Problem: You have a building that has a 2400 ft^2 floor area and an outside wall height of 11'. You have two whole house fans ventilating the space (on high setting). How many air changes per hour does this represent (you can estimate from the volumetric flow rate divided by the volume)? The fans are turned on at 10 pm and off at 8 am. How much energy was consumed (kWh and in joules)? What does it cost per day if the electricity use rate is $0.14/kWh? Fans are each in a 248 W steady state, and each drives a flow rate of 2280 actual CFM. Where would you locate these fans for the most effectiveness?

(44) Problem: The average energy use of a modern-day refrigerator has decreased by a factor of 3 in the last 30 or so years. Calculate the energy payback time of a typical Energy Star refrigerator compared to the beasts of circa 1975. Do some research to find the consumption of vintage and modern refrigerators and the embodied energy in manufacturing one.

CHAPTER 10

Water, infrastructure, and buildings

Learning objectives

(1) Understand the vital role water—in all its phases—plays in the natural world, civilization, and buildings.
(2) Investigate water treatment objectives and processes on the utility-to-building scale.
(3) Identify areas for water conservation and on-site treatment options, including rainwater, blackwater, and graywater systems.

Water is central to the purposes and functioning of buildings, from shedding precipitation and keeping its occupants and contents dry to its use in landscaping and plumbing systems. Chapter 4 on global environmental problems gave a picture of hydrology, water quality, and potable water use on the global scale. A fundamental understanding of water treatment objectives, challenges, and processes is vital for understanding building-scale water management and treatment. For the nonspecialists in water, the utility-scale water treatment provides a window to the complexity, scale, and energy requirements of water treatment. This chapter drills down to the individual building-scale water use. Those in the water field will of course take a much deeper dive on water treatment systems, and this chapter provides only a wide-ranging view of this sector for the nonspecialist engineering student. A later chapter provides insight into water use on the site and the landscape associated with buildings.

10.1 Water quality parameters and pollution

It is important to understand the general water quality problems and the treatment systems that exist to understand the challenges and see how these can be potentially used on a smaller scale. Threats to water resources are both to the quantity and quality of both freshwater resources. The availability of freshwater resources is under continuing threat due to several inter-linked issues:

(1) Growing population and prosperity increasing water resource demands for industrial, agricultural, and potable uses,
(2) Areas, particularly in urban-industrial regions where contamination has constrained water uses,
(3) Increasing warming and severity of drought conditions are causing added stressors to water supply security in arid regions, and
(4) Amplified precipitation quantities are associated with increased severe weather conditions.

The improvements in water quality in many places in the last century are quite striking. Unlike the middle of the 20th century, we no longer have water bodies so polluted that the surface catches fire, as was not uncommon in the industrial regions of the United States and elsewhere. Nonetheless, water quality issues are myriad and vary regionally, affecting surface water in lakes (both fresh and saline), rivers, and groundwater resources. A key differentiator is that of a point source or an area source (also applicable in the air quality world). Classic examples include the discharge point of a wastewater treatment plant or that of an industrial wastewater source. The other end of the spectrum is large-area sources like agricultural fields and fertilizer runoff. Depending on the scale, a building site will act more like an area source of water pollution rather than a point source.

One key link between water quality and the construction of buildings is the construction runoff. This is due to soil erosion and the generation of debris flows related to construction and demolition debris that can contaminate stormwater systems and waterways. The removal of vegetation on a building site exacerbates these problems.

In measuring water quality, the chemical, physical, and biological characteristics of the water are the key indicators of potability. Micropollutants in water, in general, even at typically low concentrations, have caused widespread issues of degraded water quality impacting marine life and biodiversity, as well as posing problems for human exposures and drinking water quality concerns (Schwarzenbach et al., 2006). Thousands of chemical species have been identified, ranging from heavy metals, petrochemicals, persistent organic compounds, chlorine-containing compounds, pesticides and herbicides, an overabundance of fertilizer species, and numerous others. Primary drinking water standards issued by the USEPA set limits on several hundred individual chemical compounds. Their form is typically either a maximum contaminant level (MCL) or a specified treatment technique (TT) as regulated by the US EPA.

A wide range of parameters, a majority of which are contaminant concentrations, determine water quality and its potability (npwdr_complete_table.pdf (19january2021snapshot.epa.gov/sites/static/files/2016-06/documents/npwdr_complete_table.pdf)). The Clean Water Act and Safe Drinking Water Acts provide much of the context and detail on water quality standards of which there are 92+ presently (Cooper, 2015). The general trend has been adding new compounds (e.g., in 2024 *per-* and polyfluoroalkyl substances, or PFAS) and tightening standards. It is beyond the scope here to examine these all, but it bears discussion of some of the key parameters and groups of compounds, bearing in mind that often that the dose makes the poison. Most all of these (with the exclusion of toxic compounds and radioisotopes from industrial uses) can be in wastewater from a building site with major groupings, including (with some overlaps):

(1) **Total suspended solids/turbidity:** Solids suspended in water and may contain toxins
(2) **Total dissolved solids/salinity/strong acids & bases:** Water-soluble compounds, including many salts as well as strong acids and bases, affect water quality
(3) **Biochemical oxygen demand/organic compounds:** The organic material in human waste serves as the food for microorganisms that consume this waste and reduce dissolved oxygen
(4) **Biological contaminants:** Various microorganisms are known disease transmission agents
(5) **Excess nutrients:** An excess of nitrogen and phosphorus in natural waters leads to eutrophication or harmful algae blooms
(6) **Toxic organic compounds:** These include many aromatic compounds having benzene rings, including dioxin, PCBs, and select pesticides and herbicides
(7) **Toxic inorganic compounds/metals**: Lead, arsenic, cadmium, and many other metal species are toxic in sufficient concentrations, as are compounds that are beneficial at low levels, such as fluoride and chlorine

(8) Chlorinated compounds/disinfectant by-products: These include substituted organic compounds such as various pesticides and herbicides, as well as disinfectants and disinfectant by-products in sufficient quantities

(9) Radioisotopes: These include both naturally occurring isotopes such as uranium, radon, and its daughter products as well as isotopes produced by humans

(10) Pharmaceuticals: Various synthetic pharmaceuticals, including endocrine-disrupting compounds, have effects on human and environmental health

Several emerging contaminants are not currently regulated but are subject to concerns, including microplastics now found globally, endocrine disrupting compounds, and novel pharmaceuticals, among others. Though treatment is challenging in that these compounds are often part per billion or trillion concentrations, science is advancing on knowing the consequences of these compounds on the ecosystem and human health. Research is also progressing on treatment and degradation methods as well.

Fertilizer runoff has contributions from the landscaping around buildings. The inputs of fertilizers to water bodies, most notably nitrogen and phosphorus, and other synthetics have resulted in downstream damage at the outlet of major rivers like the Mississippi River as well as contributed to the increase in harmful algal blooms (HABs) (Vitousek et al., 1997). The human "fixing" of nitrogen (into ammonia-based fertilizers and NOx emissions from fossil fuel combustion) now exceeds the natural N-fixing. The Häber-Bosch process, while enabling industrial-scale agriculture in many regions, has also been a perturbation of the nitrogen cycle comparable to humans' perturbation of the carbon cycle via fossil fuel use. The excessive inputs of these fertilizers cause eutrophication of water bodies, leading to HABs, depleted dissolved oxygen, potential fish kills, and other harm to the flora and fauna of lakes, streams, and coastal estuaries. The Great Lakes and the outlet of the Mississippi River are areas of particular concern. The problem of eutrophication from the excessive inputs of N and P is difficult to assess and limit the contributors as they are dispersed and highly variable area sources (Carpenter et al., 1998).

Saltwater resources, although plentiful in our oceans, are threatened by biodiversity issues and unsustainable fishing, ocean plastics, and other contaminants, the warming of the oceans due to global warming, the related acidification due to increased CO_2 concentration, and resulting impacts such as coral bleaching. Excess plastics accumulating in ocean waters is an issue that has emerged in the last few decades. Many of the pieces are shredded bits on the order of millimeters in size. The preponderance and persistence have led to problems related to ingestion as well as other issues. Though generally unreactive, its persistence in the environment as well as the steep increase in its use and improper disposal have led to the issue. Convergence regions in the ocean and other water bodies are areas where plastics accumulate.

A key water quality parameter, of course, is the acidity as measured on the pH scale (Eqs. 10.1 and 10.2) as a function of negative log base ten of the H+ concentration, where neutral water is pH 7.

$$\text{pH} = -\log_{10}[\text{H}^+] \tag{10.1}$$

$$\text{pH} + \text{pOH} = 14 \text{ or } [\text{H}^+][\text{OH}^-] = 10^{-14} \tag{10.2}$$

An underappreciated consequence of anthropogenic CO_2 emissions is the slow but steady ocean acidification due to the dissolution of CO_2 into water as the basis of the carbonate system. The acidity of the world's oceans and major water bodies is an area of concern related to the increasing concentration of atmospheric CO_2, which is shifting the equilibrium of the carbonate system in the acidic direction, as discussed in Chapter 4. The carbonate system is driven by the dissolution of CO_2 and dissociation of carbonate species with the following reactions, the endpoint of which is the formation of limestone (Eq. 10.3 through Eq. 10.6):

$$CO_2 + H_2O \rightarrow H_2CO_3 \qquad (10.3)$$

$$H_2CO_3 \rightarrow H^+ + HCO_3^- \qquad (10.4)$$

$$HCO_3^- \rightarrow H^+ + CO_3^{2-} \qquad (10.5)$$

$$Ca^{2+} + CO_3^{2-} \rightarrow CaCO_3 \qquad (10.6)$$

10.2 Potable water treatment

Although admirable ancient engineered systems were constructed, such as the Roman aqueducts, the modern water treatment field emerged with wastewater-related disease outbreaks in London and other industrializing cities in the 19th century. Episodes in industrializing nations in the middle 20th century, such as the Cuyahoga River catching fire several times, led to further developments in water quality standards, treatment, and regulations. The focus is on providing potable water and treating wastewater both to minimize disease transmission risks and to remove contaminants of a physical, chemical, or biological nature causing harm to human populations or the surrounding environment.

Potable water and wastewater treatment systems have been among the most effective improvements in human health and longevity. Global treated water production totals more than 4 trillion m^3 annually. The field of water and wastewater treatment is broad, and only a few salient points can be highlighted here. The reader is directed to dedicated textbooks to further explore this topic (Davis, 2019; Hammer and Hammer, 2011). Many environmental engineers work in ensuring the proper treatment of municipal drinking water as well as treating wastewater, and the larger-scale systems are a vital part of our national infrastructure. They also serve as a model for smaller, building-scale treatment.

The treatment of drinking water and wastewater became an acute concern with cholera outbreaks in 19th-century London. The work of Dr. John Snow, often considered the "founder" of modern environmental engineering and epidemiology, linked a cholera outbreak in mid-19th-century London to a water supply well that was located too close to wastewater discharge (Snow, 1936). Dr. Snow mapped and traced the outbreak to the Broad Street water supply pump, contaminated by a wastewater outflow (Johnson, 2006). Before this time, there was less credence given to water-borne transmission of diseases of the digestive tract. Here we will look at the anatomy of centralized water treatment facilities attached to a distribution network.

A time history of United States water withdrawals (i.e., treatment and use of potable water) is shown in Fig. 10.1. The trend up through 1975 was increasing water usage in all categories. The growth was particularly pronounced with thermoelectric power plants and agricultural irrigation, the two of which dominate the fractional composition. We may think we are creating water for human consumption, but in the end, much of it is for power plants and agriculture! Beginning around 1970, concerted efforts to conserve water resources began. Total water use flattened for several decades despite increased population, industrial and agricultural production, and power generation. Beginning in 2005, the increasing

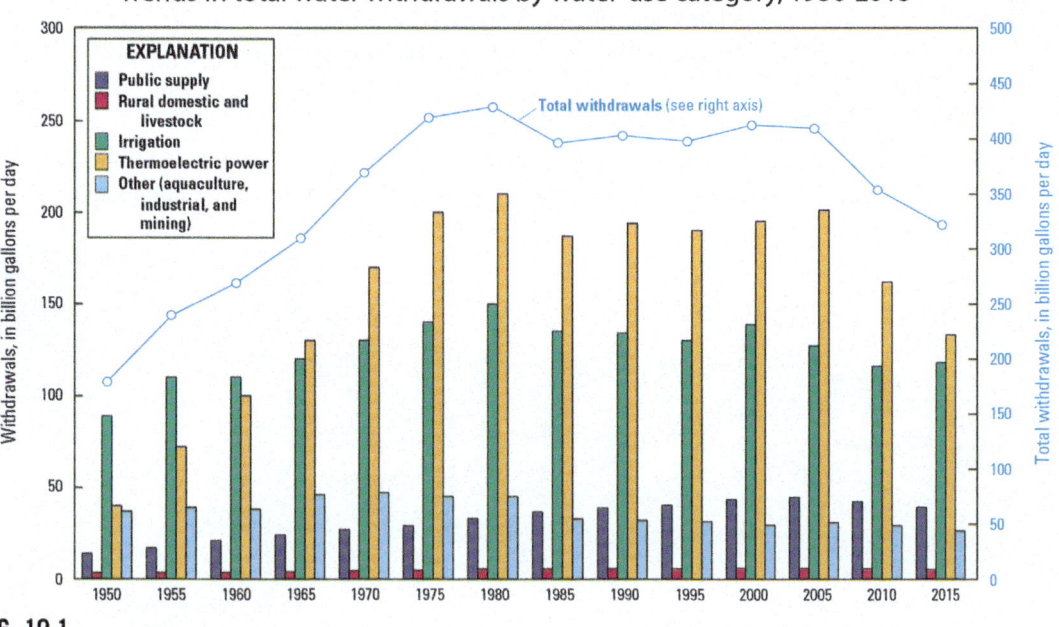

FIG. 10.1

United States municipally treated water withdrawals 1950–2015 (Department of the Interior/USGS, public domain).

retirement of steam-cycle coal-fired power plants is abundantly clear, and overall withdrawals have declined for 20 years as a result. Natural gas generators and renewable sources of electricity such as wind and solar require minimal treated water for their operations.

Municipal and any onsite water treatment systems are required to comply with primary and secondary standards for nearly 100 contaminants as specified in the Safe Drinking Water Act (Fig. 10.2). Primary standards are set to ensure that risk levels from water-borne contaminants are minimal. Maximum contaminant levels (MCLs) are set to ensure adequate protection of public health. Some of these are set to zero, but many are at a contaminant concentration that is considered de minimis risk based on the best available science. Contaminants that are regulated range from heavy metals to biological contaminants to radioactive elements to both naturally occurring and synthetic organic compounds, including herbicides and pesticides.

Technology has advanced dramatically over the last century to routinely produce domestic potable water for billions of consumers globally (Hammer and Hammer, 2011). Ideal potable water has the following characteristics (Chapra, 2018):

- in equilibrium with atmospheric O_2 and CO_2 with minimal other dissolved gases such as H_2S,
- low in suspended solids (i.e., low turbidity) and dissolved solids, including hardness species
- low in fertilizer species such as phosphorus and nitrogen,
- has pH near the neutral range,

- low in pathogen microbial content and biochemical oxygen demand,
- devoid of plastics, nanoparticles, and pharmaceutical species,
- devoid of toxic contaminants, including heavy metals, radionuclides, and organic carbon species, and
- palatable temperature and lacking taste and odor effects.

The source of the water is important to the water treatment process, with surface water and groundwater as the key distinction. Groundwater vs surface water sources demand different treatment approaches.

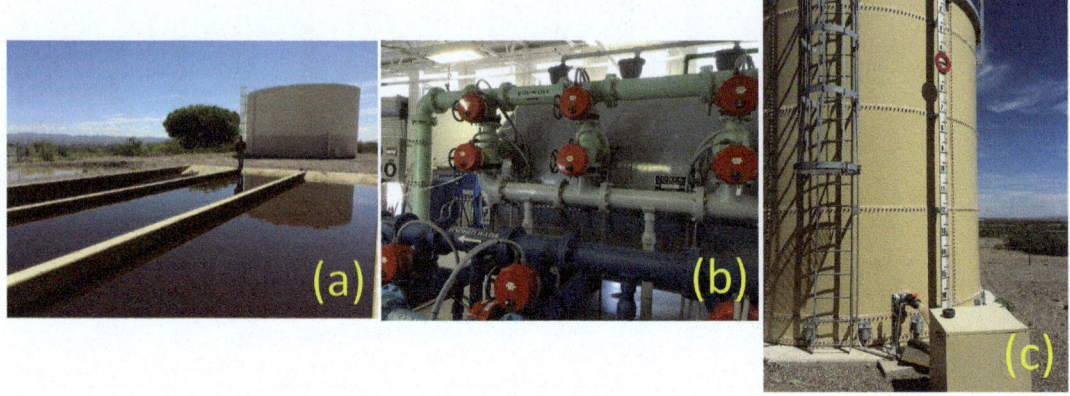

FIG. 10.2

Components of a small potable water treatment plant include (A) evaporation ponds for the retentate, (B) multimedia, and multichamber water filtration system with backwash, (C) water storage, and tanks for providing the system pressure and consistent supply (Socorro, NM municipal water treatment facility).

Water treatment systems involve several stages or unit processes to treat the water to suitable potable characteristics based on its physical, chemical, and biological properties (Hammer and Hammer, 2011). A critical parameter for a given water treatment process is the hydraulic residence time (HRT), given as tank volume (V, m^3) over volumetric flow rate (Q, m^3/h). Typical HRTs for unit operations in water treatment systems are minutes (filtration) to days (biological treatment), dependent on the process.

$$\text{HRT} = \frac{V}{Q} \quad (10.7)$$

The removal of suspended and dissolved solids is a high priority, typically using settling tanks or clarifiers to allow the settling of particles and aided by using coagulants to form "flocs" that will settle out. The key comparison is the overflow velocity (OFR) of the settling tank vs the settling velocity of the particles. The overflow velocity is a ratio of the volumetric flow rate to the surface area (SA) of the clarifier (Eq. 10.8).

$$OFR = \frac{Q_{\text{water}}}{SA_{\text{tank}}} \quad (10.8)$$

Using Camp's Analysis with many simplifying assumptions, the settling velocity of the particles is a function of the gravitational constant (g), the difference in particle and fluid densities (ρ), particle diameter (D), and fluid viscosity (μ) (Eq. 10.9) (Example 10.1).

$$V_s = g(\rho_p - \rho_f) \times \frac{D^2}{18\mu} \tag{10.9}$$

EQUATION 10.1 Terminal settling velocity (applicable to water and wastewater treatment)

Problem: $D = 10\,\mu m$ particles ($\rho = 1.5\,g/cm^3$) are settling out in water. T is 20°C and P is 1 atmosphere. You can take the viscosity of water as $0.001\,kg/(m\text{-}s)$.
(a) Calculate the terminal settling velocity (cm/s) of a particle with the above characteristics.
(b) A sedimentation tank treats a wastewater flow of 0.5 million gal/day (MGD). The tank is a circular tank with a diameter of 100 ft and a depth of 10 ft. The second rectangular tank is 50 ft × 200 ft × 10 ft deep. Which tank will be more effective at settling the above particles (and show why)?

Given: $D = 10\,\mu m$ particles; $\rho = 1.5\,g/cm^3$; $Q = 0.5$ MGD; tank1(cylindrical, $D = 100$ ft, $d = 10$ ft); tank2(rectangular, $50 \times 200 \times 10$ ft)
Find: V_t, OFR of both tanks
Assume: viscosity $= 0.001\,kg/(m\text{-}s)$

Solution:
(a) Calculate terminal velocity using Camp's relation:

$$V_t = g(\rho_p - \rho_f) \times D^2/(18\mu)$$
$$= 9.8\,m/s^2 \times (1500 - 1000)\,kg/m^3 \times (10 \times 10^{-6}\,m)^2/(18 \times 0.001\,kg/(m\text{-}s)) = 2.72 \times 10^{-5}\,m/s = 0.0027\,cm/s$$

(b) Find the overflow velocity for each and compare:

$$OFR1 = Q/SA = \frac{5 \times 10^5\,gal/day \times (3760\,cm^3/gal) \times (day/(24 \times 3600\,s))}{\pi(50\,ft \times 30.5\,cm/ft)^2} = 0.003\,cm/s$$

$$OFR2 = Q/SA = \frac{5 \times 10^5\,gal/day \times (3760\,cm^3/gal) \times (day/(24 \times 3600\,s))}{50\,ft \times 200\,ft \times (30.5\,cm/ft)^2} = 0.00234\,cm/s$$

The first tank has an OFR that is too high to settle out these particles, while the second tank will settle them out effectively (though slowly).

Chlorine is often used for disinfection purposes in many municipal water systems before distribution. Chlorine has residual disinfection properties as compared to other disinfectants such as ultraviolet light or ozone. Chlorine, along with some of its derivative compounds called disinfection by-products, are compounds that are regulated.

Despite the tremendous progress in reducing the water-borne transmission of diseases, outbreaks still occur in the developed and developing world. The outbreak of cryptosporidium in Milwaukee, Wisconsin, in the 1990s is a case study of a water-borne illness transmitted via public drinking water (Mackenzie et al., 1994). Future similar episodes are likely, particularly in an accelerating climate change scenario, but diligence toward water treatment will minimize the threats of water-borne disease outbreaks.

10.3 Wastewater treatment

The cholera outbreaks of the 19th century and the resulting sewage system to address problems in London's Thames River, often called the "Big Stink," is one of the first applications of engineering to wastewater treatment (Chapra, 2018). According to the US EPA, 16,000+ municipal-scale wastewater treatment plants (WWTP) operate in the United States, serving about 75% of the population (the remainder is served by onsite septic systems or smaller-scale lagoons or constructed wetlands). The combined flow of wastewater treated by municipal plants is approximately 32 billion gallons per day (US EPA). The vast majority (80%) of facilities treat less than 5 million gallons per day (MGD) apiece. The flow of wastewater from a building is treatable onsite, particularly when the blackwater and graywater are treated separately as discussed later.

A typical process flow diagram for a wastewater treatment plant shows a complex process (Fig. 10.3). Notably, it incorporates many similar treatment processes to drinking water treatment with the key addition of activated sludge digestion of organic wastes in wastewater (Cooper, 2015). The plant begins filtering out large detritus with a bar rack and a grit chamber for larger suspended solids. A primary clarifier is used to settle out mid-sized suspended solids. Secondary treatment consists of the activated sludge digestion process. Natural decomposition of the organic material in the water is accelerated by selected microbes optimized for decomposition activity and the health of the microbial community. This is followed by another clarifier to settle out the suspended material, where a fraction of the viable organisms is recycled back into the reactor. The tertiary treatment typically includes a disinfection step using chlorine or UV light before returning the water to the receiving water body. It may also include advanced treatment to remove contaminants of high concentration or particular concern, such as N and P fertilizer species, persistent organic pollutants, heavy metals, and other biohazards. Advanced treatment techniques include membranes, filtration, and activated carbon adsorption, among others. Sludge from the various clarifier steps is dewatered and must be treated before its disposal or potential reuse.

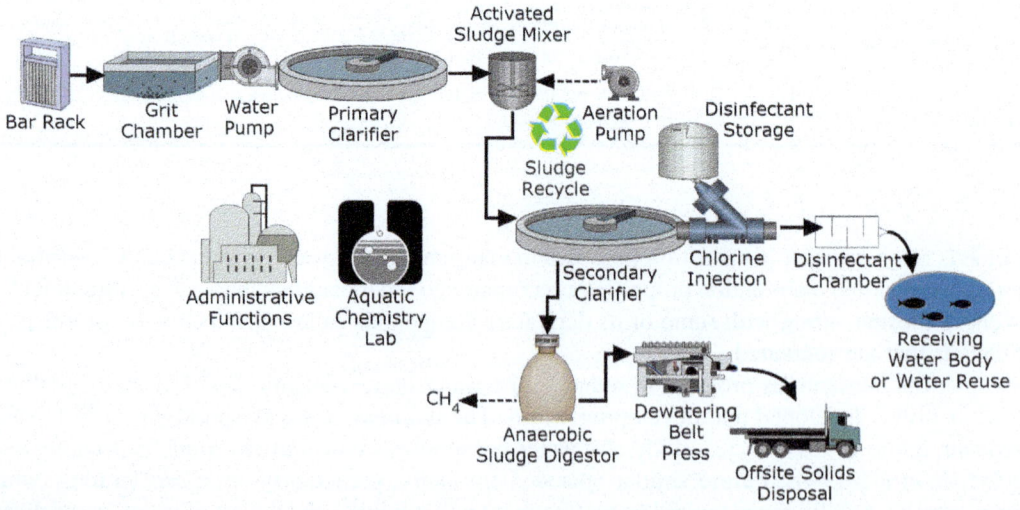

FIG. 10.3

Flow diagram showing the components of a typical wastewater treatment plant.

Alternately to discharging to a water body, wastewater reuse has increased for multiple reuses and often requires further treatment to meet potability standards. The treatment at this point will be a function of the reuse application, including agricultural use, groundwater recharge, industrial water use, graywater reuse, or potable water production as discussed later. "Earthship" buildings beginning in the 1970s were designed for multiple water reuses, where collected rainwater could be used for graywater uses such as bathroom sinks, followed by onsite greenhouse or aquaculture food production, then finally used for toilet flushing where onsite septic systems would treat it.

The selection of the individual unit processes used in a given wastewater or municipal water system is beyond the scope here and is covered in many textbooks and is the subject of many semester-long courses (Cooper, 2015; Davis, 2019; Hammer and Hammer, 2011). In the end, it is a complex function of capital and operational costs, inlet and outlet water quality and flow rate, on-site resources available, and many other variables.

Wastewater treatment plants (or industrial sources discharging to receiving water bodies) emit numerous species that are limited by a National Pollution Discharge Elimination System permit. These are many of the same pollutants discussed above, including total suspended solids (TSS), total and fecal coliform, fertilizer species of nitrogen and phosphorus (Fig. 10.4), and a particular focus of wastewater treatment systems, the biochemical oxygen demand (BOD) of the discharged water.

FIG. 10.4

Algae bloom on a pond in a natural area in Missouri in December 2023 that is fertilized from excessive inputs of nitrogen and phosphorus. This is an issue with both wastewater and stormwater runoff. Advanced wastewater treatment systems can reduce this problem, while runoff reduction of fertilizers can also help.

BOD, a surrogate for biological hazard, results from the microbial degradation of organic matter exerting a demand on dissolved oxygen (DO) as these organisms proliferate while degrading this organic matter. The BOD remaining to be exerted in a water sample L is described by a first-order reaction (Fig. 10.5) or exponential decay function according to Eq. (10.10), where k is the first-order decay constant, t is time, and L_0 is the ultimate BOD of the water sample.

$$L = L_0 \exp(-kt) \tag{10.10}$$

The exerted BOD (y) of the sample is the difference between L and L_o, and BOD is often categorized on a 5-day basis by a BOD_5 test (Fig. 10.5).

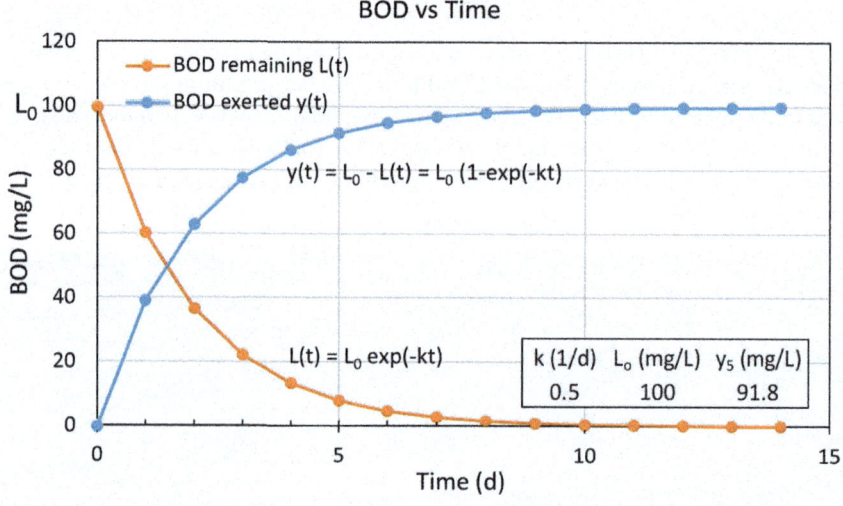

FIG. 10.5

BOD exerted and remaining over time with a decay constant of 0.5/day and ultimate BOD $L_0 = 100$ mg/L.

10.4 Stormwater management

Tying directly to buildings and their sites, stormwater runoff is a major infrastructure and management challenge, particularly with the enhanced precipitation associated with storm systems as temperatures increase (Fig. 10.6). Management of stormwater runoff is best accomplished in decentralized, distributed systems close to the source of the water rather than downstream engineered treatment facilities (Sorvig and Thompson, 2018).

The progression of human development has resulted in a decreased ability of lands to absorb liquid water. Stormwater management has taken on a new urgency in the effects of increasingly extreme precipitation events combined with increasing low permeability, paved-over areas of urban centers. In major cities around the world, this has increased the risk of flooding events due to runoff constraints and subsequent flooding. One of the most prominent United States cities where this has occurred is

10.4 Stormwater management

FIG. 10.6

Even in dry climates, the transport of runoff away from buildings and foundations is important, as shown here on the campus of New Mexico Tech. Note the intrepid agave growing up from the gap in the concrete—a plant that is well-adapted to its environment!

Houston, Texas, a city built amidst the low-lying bayou of southeast Texas. The major storms of the last decade, including Hurricane Harvey and Tropical Storm Imelda, in combination with the increasing impermeable surfaces, contributed to severe, recurrent flooding events.

The rational equation describes the flow rate of stormwater (Q, ft³/s) given a precipitation rate (i, in./h), a surface area (A, acres), and a surface runoff coefficient (C, unitless and between 0 and 1 dependent on surface permeability) (Eq. 10.11).

$$Q = 1.008\, CiA \qquad (10.11)$$

The constant that converts the units is essentially 1 and is often ignored. The runoff coefficient quantifies the site's surface permeability, and some typical values for selected surfaces are given in Table 10.1.

Table 10.1 Typical runoff coefficients (the fraction that runs off from 0 to 1) for various surfaces (American Association of Civil Engineers, ASCE) (Cooper, 2015; Mihelcic and Zimmerman, 2014).

Surface type	Runoff coefficient
Open undeveloped lands	0.02
Agricultural fields	0.05
Low-density residential 2/acre	0.14
Mid density residential 5/acre	0.25
High density residential 15-acre	0.47
Industrial	0.9
Commercial	0.95

As one alternative to reduce impermeable surfaces, rain gardens feature a depressed zone where stormwater runoff from impermeable surfaces can collect and percolate into the soils more readily. This helps mitigate local stormwater peak flows and provides some level of filtration of stormwater runoff. The depressed area contains plantings that are more tolerant of the wetter soils that come from this. They often feature perennials and grasses and are left "naturalized" rather than mowed like a lawn and thus provide species habitat. Although these are most useful in wetter regions, drylands that receive infrequent but heavy downpours can also find these useful. Raingardens can be thought of as a miniature "constructed wetland." On a larger scale, this practice is called bio-retention as it allows stormwater retention via biological features. Working with site characteristics and topography is a key factor for water management, and there is a tradeoff between urban density efforts and maintaining substantial permeable surfaces in an urban area.

There are improved approaches to reduce the impermeability of urban areas, including:

- Minimization of pavement and other impervious surfaces
- Permeable concrete with grass infill in low-load applications
- Wetlands preservation, including using constructed wetlands where appropriate
- Collection of rainwater for landscape watering needs

Reducing the runoff of stormwater and providing potable drinking water begs the question of whether onsite rainwater harvesting can address both. The collection of rainwater for landscape use will be discussed in more detail in Chapter 11 on outdoor spaces. Elevating this to drinking water quality is a considerably taller order, more often in many off-grid buildings.

10.5 Systems promoting more sustainable water use

Increasing demands and freshwater constraints combined with more severe drought impacts have led to greater interest in more sustainable systems. Exacerbation of climate extremes has amplified interest in such alternate sources as well. Blue-green infrastructure refers to the parts of the built environment focused on sustainable water use and management. The opportunities span from the individual to regional-scale potable water sources. The urgency of water availability issues has spurred international agencies such as the United Nations to advocate for the exploration of unconventional freshwater resources, including deep groundwater, offshore resources, water reuse from wastewater, desalination of ocean water or other saline sources, harvesting meltwater from ice mass, air harvesting of atmospheric water vapor, and potential transport pipelines overland to arid regions (Qadir et al., 2022). Growing interest in desalinating saline aquifers or ocean water has accelerated in arid regions. Desalinated seawater is prevalent in regions in the Middle East, though it is very energy intensive and costly.

Recently, reclaimed waters from wastewater treatment, industrial water treatment, oil and gas production, or other mining operations and other sources have become other sources for domestic treated water. The additional treatment is dictated by the end use of the water, where potable water requires the highest treatment while landscape reuse or groundwater aquifer recharge requires less.

In addition to the role of federal agencies such as the USEPA, state and local municipalities have taken up management for wise water use. Many municipalities have been successful in reducing their per capita water use over the last several decades (Fig. 10.7), which has helped ground and surface water resources recover (Fig. 10.8).

The term low impact development (LID) is an approach that focuses on stormwater management as part of the site development. To protect aquatic ecosystems, LID tries to mimic natural hydrological processes to reuse stormwater where practical and enhance infiltration and evapotranspiration elsewhere. Successful efforts have begun in cities around the world, particularly those in arid regions. One effort has been wastewater reuse, including for landscape use or even, with further treatment, reuse as municipal drinking water. This is easier with graywater (e.g., from sinks and laundry) vs blackwater (having high organic matter, e.g., from toilets and kitchen sinks).

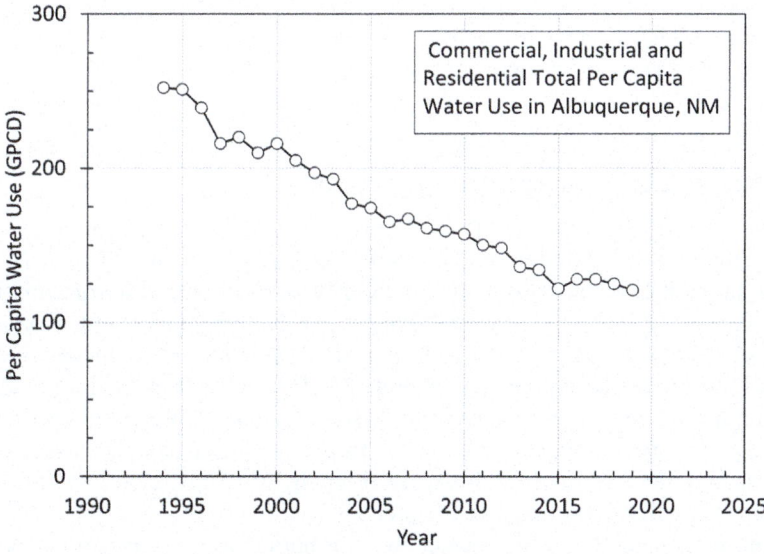

FIG. 10.7

Data on daily per capita water use (GPCD) from the city of Albuquerque, New Mexico, including residential, commercial, and industrial users. Beginning in the mid-1990s, the City of Albuquerque/Bernalillo County ramped up water conservation efforts through incentive programs (data from the city of Albuquerque, NM).

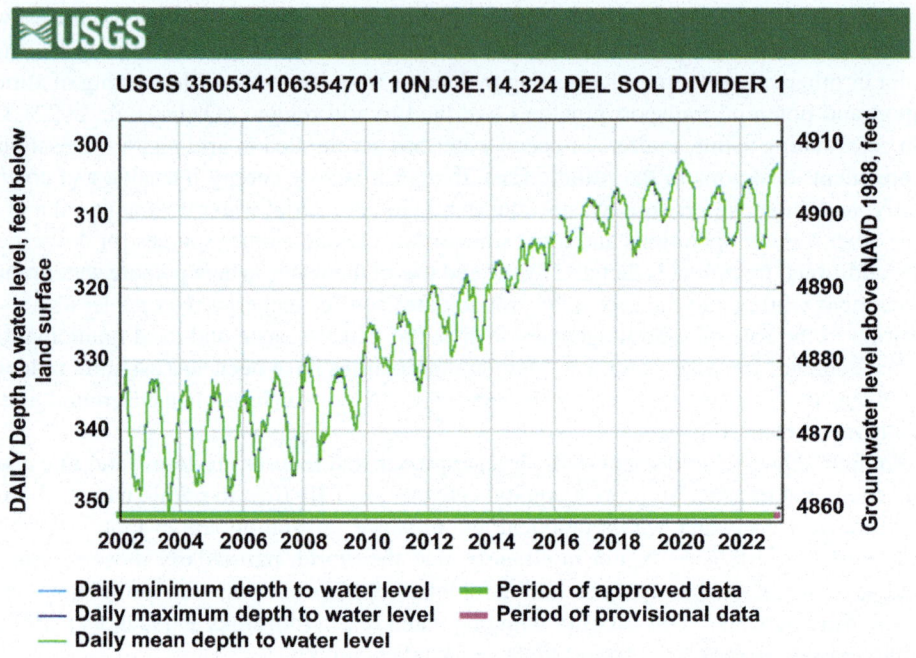

FIG. 10.8

An example USGS water table height measurement in Albuquerque, New Mexico, shows a sustained recovery of groundwater resources beginning in the early 2000s (Department of the Interior/USGS, public domain). This is reflected in multiple groundwater monitoring sites.

10.6 Onsite black and graywater systems

Wastewater treatment systems minimize the risks, particularly biological, to humans and the surrounding environment due to contaminated water. Graywater and blackwater are differentiated by the quality of the waste and, in particular, the organic matter and potential for pathogen transmission. Graywater systems allow use of wastewater from residential or commercial sites that is low in organic matter and biological risks. Graywater originates from bathroom sinks, laundry, and other low-biochemical oxygen demand (BOD) sources and requires less treatment before it can be reused for landscape needs. It can be used onsite for nonpotable uses. Graywater systems can be whole buildings or can service one bathroom, for example, where graywater is collected from a sink drain and used for toilet flushing with only a small pump and filter required for this reuse (Ludwig, 2015). A related system is the rainwater harvesting system that will be discussed in Chapter 11.

The portion of wastewater that is blackwater contains human waste or is from the kitchen sink and thus is prone to high concentrations of organic matter. Many rural buildings are served by an onsite system for treating blackwater: septic systems. These flow-through tanks detain the wastewater solids, and it somewhat mimics the functioning of a wastewater treatment plant by providing a biological treatment of the wastewater. The septic tank is typically anaerobic and serves to store and provide breakdown of some part of the solids. The rest must be cleaned out and removed for offsite treatment periodically. An aerobic septic system that injects air will have microbes that more completely degrade the organic wastes detained in the septic tank. The liquid overflow at the tank outlet is distributed over a broad underground area called the leach field, where any remaining organic material will be filtered and decomposed by soil microbes as it percolates toward the aquifer. These systems have been used for centuries for treating wastewater onsite, particularly for off-grid buildings where no viable municipal system is available. As an alternative to a leach field, a constructed wetland that mimics nature's treatment process can be used for the effluent from the septic tank. The US EPA and other state and local regulatory bodies provide robust permitting and design criteria for any blackwater treatment system.

As discussed earlier in the section on cogeneration, it is also worth reiterating that wastewater systems are increasingly the focus of energy recovery systems to extract the heat from flows of wastewater, cooling water, or other elevated temperature water or gas flows. Section 10.9 discusses the multiple ways that water and energy flow are connected.

10.7 Water management on the building scale

To this point, we have looked at larger-scale water treatment and conservation. Many opportunities exist for the sustainable use of water on the building scale, both in the interior and exterior environment. The Earthship architecture mentioned earlier and common around Taos, New Mexico, is a case study of efficient water use. Many use some variety of a system with captured rainwater that is filtered and used for domestic, nonpotable applications first, before trickling through the greenhouses in the sunny southern exposure of the house, then feeding toilets, and finally passing through an onsite septic system whose leach field supports landscape plants.

Buildings and their sites can contribute to select water quality problems, many related to stormwater and runoff. The construction site is particularly susceptible to issues of runoff and the associated soil erosion from such sites due to soil disturbances and typical truck traffic. Management approaches, simple barriers, berms, and control measures can reduce these runoff issues and are usually required for new construction and renovation sites. The use of potable water and the generation of wastewater, both blackwater and graywater, are key facets of any building.

Water systems can be scaled such that individual buildings from the residential scale and up can implement from "final polishing" filtration systems to collecting rainwater to managing stormwater to greywater reuse. Though systems exist for treating drinking water standards as well as wastewater treatment (e.g., septic tanks), most buildings that are not remote will likely have "on-grid" access to potable water and wastewater treatment via drain lines for sewage delivery to a municipal wastewater treatment plant.

Plumbing fixtures have been far more water-efficient over the past several decades. Toilet water use has decreased from 5 or more gallons per flush (GPF) to typical values of 1–1.5 GPF with current standards in place. Showerhead flow rates have similarly dropped with the current typical use of 2 GPM using aeration and improved designs. Some designs of toilets or urinals are waterless. The former can be stand-alone composting toilets, and the latter have dense liquid layers that allow the flow of urine through them while preventing sewer gas from back-flowing into the building.

The USEPA Watersense program gives a good overview of the approaches and systems available for water conservation (https://www.epa.gov/watersense). The program estimates savings of over 6 trillion gallons of treated municipal water over the last ~16 years. Like the Energy Star program from DOE, the program features labeling devices that meet standards for water use efficiency. Toilets, sinks, showers, clothes washing are all given water sense metrics.

Water has additional considerations for individual buildings, some of which are discussed here. Moisture management in a structure is an area of building science subject to considerable complexity. Building materials will function best if they are allowed to dry, particularly wall assemblies and insulation layers (Kolbert et al., 2022). Wall systems are prone to mold, rot, and degradation if they are not able to dry reasonably fast. Also, insulation loses much of its effectiveness with high moisture content. Rainscreens, water resistance barriers, drainage planes near wall exteriors, and proper sealing of penetrations all provide a barrier between water intrusion.

Landscape water management tools that will be discussed in the next chapter include constructed wetlands, drip irrigation, xeriscaping efforts, and rainwater harvesting that can all be used to enhance water management (Lancaster, 2019). A diagram of the key components of the system for the capture and treatment of rainwater is given in Fig. 10.9. Reuse for landscape is much more common, though use for potable water production is feasible and useful for off-grid applications. The treatment to drinking water standards will require permitting, a licensed operator, and a treatment system. The latter is typically a scaled-down version of the water treatment shown in and using filtration in a media filter, use of an activated charcoal trap, and disinfection using chlorination, ozone addition, or UV light exposure (Novak et al., 2014).

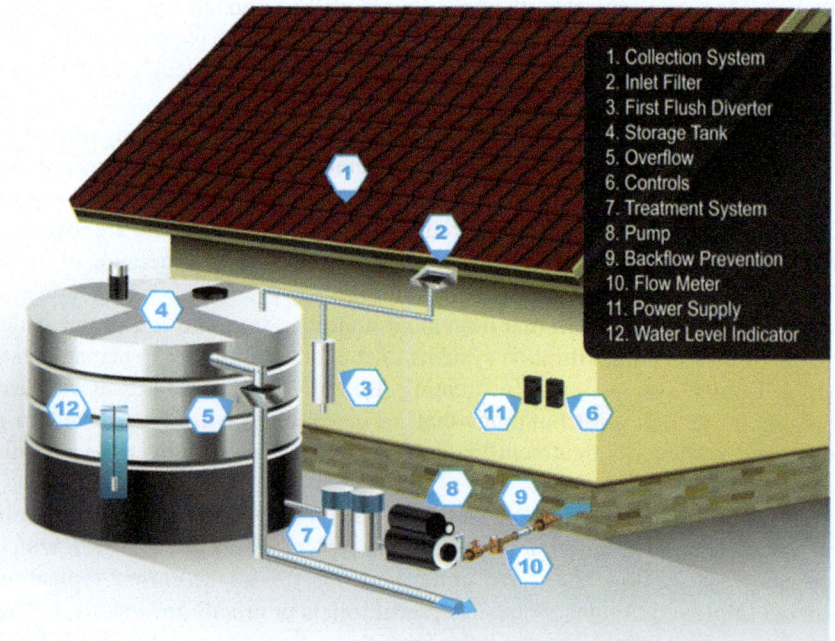

FIG. 10.9

Components of a rainwater capture system for indoor potable water applications. The treatment system will likely include a media filter and activated carbon trap for removing contaminants from the water (public domain image from the US Department of Energy).

10.8 Other water problems in buildings

Hazards with water (e.g., from water leaks) in the indoor environment relate to materials staying wet. A constantly wet environment is subject to the formation of mold or mildew. Some varieties are hazardous to humans, such as black mold. Human health hazards relate to the variety, quantity, and exposure to the biohazard.

Rain exclusion is a primary defense of the cladding of a building. A water-resistant barrier and a drainage plane are important components of the building's wall structure. Likewise, roofing material, underlayment, and flashing around openings help exclude water. The main goal is the exclusion of liquid water from the structure and a way for it to drain and dry if liquid water does reach it. The hazards can also occur from water vapor hitting a cold surface and condensing. This occurs when the dry bulb temperature of moisture-laden air drops to the level of the dew point temperature. It is facilitated by a high RH (i.e., dew point temperature), and the larger the drop in dry bulb temperature, the greater the potential for condensation.

10.9 Water-energy nexus: Water treatment requires energy and vice versa

Freshwater is a small fraction of global water resources, and the part that is not locked up in the (declining) frozen water mass is limited, as discussed previously (Cooper, 2015). Water is "free" from the sky and flowing and stationary water bodies. However, its delivery, treatment, distribution, possible heating, and wastewater treatment all require energy, mostly electricity. This connection is called the water-energy nexus. An obvious residential connection is the production of hot water for domestic use (Example 10.2).

The treatment of water requires significant electrical generation and/or fuel combustion. Municipal water and wastewater pumping and treatment consume ~2% of total United States electricity, and this use has increased by >50% in two decades (EPRI, 2013). For individual states such as California, the electricity cost is far larger due to large pumping and transmittance costs to arid regions.

Electricity use accounts for around 80% of municipal water processing and distribution costs (Copeland and Carter, 2017). Roughly 3300–3600 kWh is required to treat and pump 1 million gallons of water. Pumping costs are one of the greatest contributors in the form of electricity to run pump motors. The pumping costs vary based on water availability, source, and quality. Groundwater sources require typically more pumping but less treatment than surface waters. Saltwater treatment is the most energy-intensive source (Example 10.2).

> ### EXAMPLE 10.2 Water-energy nexus
> **Problem:** Estimate the energy use and CO_2 released from the typical individual's shower. Assume a tank system and three alternatives: direct natural gas water heating, electrical resistance, and electric heat pump water heaters. Electricity is generated by a 50% efficient combined cycle natural gas power plant.
> **Given:** $T = 10$ min; $c = 4.18$ J/(g-K); $T_{cold} = 55°F$; $T_{hot} = 115°F$, 2 gpm, $\rho = 1.0$ g/cm^3
> **Find:** E(J), CO_2(kg)
> **Assume:** 10 min shower, gas water heater $\eta = 0.6$, electric resistance heater $\eta = 0.9$, HP COP = 3.
> **Solution:**
> $$M = Q \times t \times \rho = \left(\frac{2 \text{ gal}}{\text{min}}\right) \times \left(\frac{3.76 \text{ L}}{\text{gal}}\right) \times 10 \text{ min} \times \left(\frac{1 \text{ kg}}{1}\right) = 75.2 \text{ kg}$$

Continued

> **EXAMPLE 10.2 Water-energy nexus—cont'd**
>
> $$Q_{out} = MC\Delta T = 75.2\,kg \left(\frac{4186\,J}{kg\,K}\right)(60°F)\left(\frac{5°C}{9°F}\right) = 10.5\,MJ$$
>
> Now we need to find the input energy for each option.
>
> Gas water heater: $Q_{in} = \dfrac{Q_{out}}{\eta} = \dfrac{10.5\,MJ}{0.6} = 17.5\,MJ$
>
> Electric water heater: $Q_{in} = \dfrac{Q_{out}}{\eta} = \dfrac{10.5\,MJ}{0.9} = 11.7\,MJ\left(\dfrac{kWh}{3.6\,MJ}\right) = 3.25\,kWh$
>
> Heat pump water heater: $W_{in} = \dfrac{Q_{hot}}{COP} = \dfrac{10.5\,MJ}{3} = 3.5\,MJ\left(\dfrac{kWh}{3.6\,MJ}\right) = 0.97\,kWh$
>
> This is end use energy the customer would be billed for. Recall that the upstream fuel needed for electric is 2 times larger than the electrical consumption due to power plant thermal losses. The heat pump would still have the lowest use whether considering end use or primary energy.
>
> Using typical EPA emission factors:
>
> About 117 pounds of CO_2 are produced per million British thermal units (MMBtu) equivalent burning natural gas. Grid-based electricity produces 2.23 lb CO_2 per kWh billed of coal-generated electricity and 0.91 lb/kWh for natural gas turbines.
>
> Gas water heater: $CO_2 = 17.5 \times 10^6\,J\left(\dfrac{BTU}{1055\,J}\right)\left(\dfrac{117\,lb}{1 \times 10^6\,BTU}\right)\left(\dfrac{kg}{2.2\,lb}\right) = 0.88\,kg\,CO_2$
>
> Electric resistance water heater using gas turbine electrical generation: $CO_2 = 3.25\,kWh\left(\dfrac{0.91\,lb}{kWh}\right)\left(\dfrac{kg}{2.2\,lb}\right) = 1.34\,kg\,CO_2$
>
> Heat pump water heater using gas turbine electrical generation: $CO_2 = 0.97\,kWh\left(\dfrac{0.91\,lb}{kWh}\right)\left(\dfrac{kg}{2.2\,lb}\right) = 0.441\,kg\,CO_2$
>
> The heat pump also generates the lowest CO_2 emissions for gas turbine-based electricity.

The combined use of electricity for water and wastewater treatment constitutes ~2%–4% of United States electricity consumption plus a smaller amount of fuel consumption (Mihelcic and Zimmerman, 2014). According to the US EPA, 39.2 billion and 30.2 billion kWh of energy are used annually to treat drinking water and wastewater, respectively. Total municipal plant energy requirements have a broad range from ~40 $(10)^4$ to 200 $(10)^4$ kWh/MGD of plant capacity. The smaller-scale lagoon, septic, and wetlands treatment systems are about a factor of 5 less energy-intensive though have a higher land area footprint and less controlled water treatment outcomes (Mihelcic and Zimmerman, 2014). With a typical WWTP, the activated sludge process is roughly half the energy used for the wastewater treatment plant between water pumping and aeration pumps.

The production of treated potable water and treated wastewater both involve carbon emissions from the energy used in these processes. Though these vary considerably with the energy efficiency and grid mix at a location, approximate values for these are 0.46 and 0.38 kg of CO_2,e per m^3 of treated municipal water and wastewater, respectively (Zib et al., 2021).

An example conceptual problem below illustrates the primary treatment processes and energy use. Notably, the energy intensity of the operations of the WWTP dominates its life-cycle energy use (95%) compared to the construction and demolition-associated energy (Mihelcic and Zimmerman, 2014) (Example 10.3).

EXAMPLE 10.3 Wastewater treatment

Concepts: Document the significant energy inputs (and outputs) from a typical wastewater treatment plant.

Solution: As shown below, any introductory environmental engineering or wastewater treatment textbook can give the reader an overview of the associated components and processes for wastewater treatment.

Energy use includes major pumping operations indicated (the water pumping would exceed the aeration costs, as you can calculate later in the problems (Fig. 10.10).

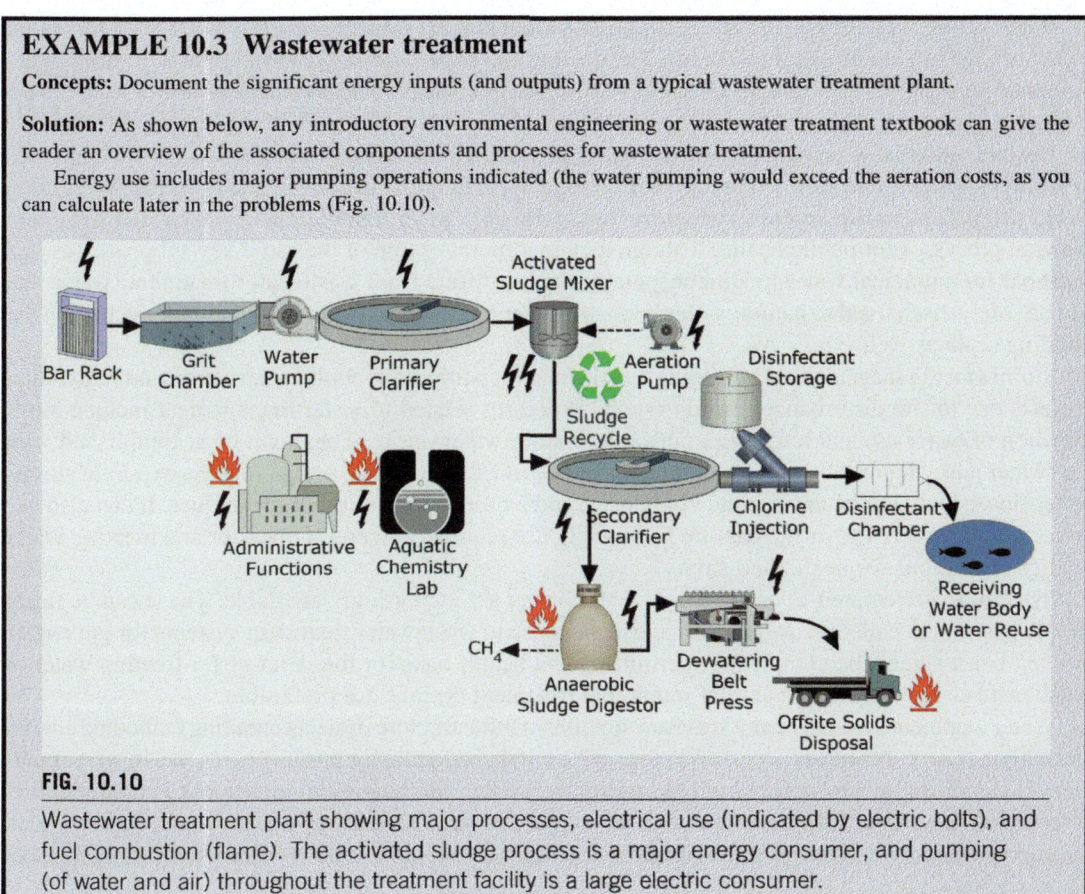

FIG. 10.10

Wastewater treatment plant showing major processes, electrical use (indicated by electric bolts), and fuel combustion (flame). The activated sludge process is a major energy consumer, and pumping (of water and air) throughout the treatment facility is a large electric consumer.

10.10 Chapter summary and conclusions

Water's vital role in buildings involves potable water use, wastewater generation and treatment, stormwater management, as well as a role in building performance. Each municipal system meets stringent standards codified by the USEPA in the Safe Drinking Water Act, among other statutes. Buildings are key portals for both potable water and wastewater generation; they also need to consider stormwater management. Water of sufficient quality and quantity is required for life, and the treatment of municipal potable water and wastewater is arguably one of the top contributors to improving public health in the 20th century. Water is also a key concern with water vapor transport and thus the comfort and efficiency of buildings. Water pollution originates from both point sources (e.g., an industrial effluent pipe draining into a waterbody) as well as area sources (e.g., runoff from agricultural lands).

Much of the western United States and other mountainous arid regions rely on winter mountain snowpack for its potable water resources. The decline and shifted timing (early and fast) of runoff from mountain snowpacks (i.e., rain vs snow) have resulted in critical water availability issues in the West and other global regions. Water management is becoming increasingly challenging going forward.

Despite remarkable progress in water quality in the last ~50 years, water contamination continues to be an ongoing threat from chemical and biological pollutants. Moreover, many in the developing world still lack adequate access to water treatment. Potable water treatment and wastewater treatment are vital processes for public health. Though they incorporate some of the same key subprocesses, the end goal for municipal water treatment (potable water) differs from wastewater treatment (water that can be safely discharged to natural water bodies). Water treatment systems may be implemented on the building scale as well.

Stormwater management is a building-level concern, particularly with construction and renovation projects due to site disturbance. Building-level processes related to water management include minimization of water use and retaining graywater or stormwater onsite. These can be accomplished with graywater reuse opportunities, with diversions and permeable surfaces, preventing the export of stormwater flows, low-flow plumbing and appliances, and a broad range of other techniques. It can also use collected rainwater. The onsite capture for potable use requires a licensed operator and meeting water quality regulations for health and safety.

Systems to collect, reuse, or even treat water onsite are increasingly available. These can be fairly simple systems to collect rainwater for landscape reuse to wastewater treatment systems for graywater or even blackwater. The complexity, permitting and health risks for the latter or for treating water to potable standards are often such that municipal treatment options are preferable.

Water treatment systems of any scale are significant infrastructure projects entailing embodied energy and carbon (e.g., concrete and steel). Whether at a centralized facility or onsite, energy use in water treatment is significant in the form of both electricity use (water pumping, mixing, aeration) and direct fuel combustion. Likewise, the production of energy requires a large quantity of treated water, depending on the energy source. This energy-water nexus is important for buildings and civilization on a larger scale.

10.11 End of chapter exercises

(1) **Problem:** The USEPA provides the following data on the water wasted per dripping faucet: At 70 drips/min, the water wasted is 3679 gal/year. Using appropriate assumptions, what droplet size did they assume?

(2) **Problem:** You have an indoor temperature of 22°C and an RH of 60%. If interior air migrates through a wall structure and hits a surface that is 15°C, will it condense? The psychrometric chart in Fig. 12.1 may be helpful here.

(3) **Problem:** Use the Antoine equation (Chapter 2) to calculate and plot the saturation water pressure as a function of dry bulb temperature from T from 273 to 350 K.
  (a) Give the value for the saturation water vapor pressure (mBar) at $T = 298$ K. You can take the global average temperature as 14°C.
  (b) Make an estimate of how much "juicier" a rainstorm could be based on the change in saturation water vapor pressure for a world that warms by 2°C and 5°C, assuming the actual water vapor concentration scales with the saturation water vapor pressure.
  (c) Compare the Antoine vs Clausius-Clapeyron Equation for P_{sat} for water. How well do they agree?

(4) Problem: A municipality needs a pump to lift 5 million gallons per day (MGD) of water by ΔZ of 350 ft to reach a water tank. The pump is 70% efficient. What is the power requirement for the pump in HP? What does it cost to run this pump for a year if electricity is \$0.14/kWh?

(5) Problem: One can consider the human population a stock or reserve of water in the earth system. Estimate the mass of water (gigatonnes) and volume (km^3) contained in the global human population and compare it to that in the liquid phase freshwater resources on the planet. Use a population of 7.35 billion and a mass fraction of water of 60%. Compare this to the volume of water in lakes and rivers (9.3×10^4 km^3).

(6) Problem: The output of a power plant is 500 MW_e, and the efficiency of the plant is 33.3%, it uses a river to reject waste heat. The temperature change of the river cooling water is limited to 20°C. Find the flow rate of cooling water required to remove all of the waste heat.

(7) Problem: Running water for even 5 min uses as much energy as running a 60-W light bulb for 14 h, according to the US Environmental Protection Agency. Assume a 2-gpm showerhead and a 60% efficient water heater that heats incoming water from 50°F water to 113°F. Confirm or deny.

(8) Problem: According to the Electric Power Research Institute (EPRI), public water supply utilities consumed 39 billion kWh of electricity in 2011. Calculate the number of large 1000 MWe power plants needed to supply this, assuming a capacity factor of 0.75. What fraction of total annual total US electrical consumption does this represent?

(9) Problem: Water is often bought and sold in an odd unit, an acre-foot AF. Show by calculation the approximate gallons and kg of water represented by an AF. How much precipitation on a 2000-sqft roof area would be required to produce an AF?

(10) Problem: You have a gravity-driven waterfall of interest for energy generation. Find the potential energy of 2000 m^3 of water at a height of 20 m. If this is converted entirely to KE at the bottom of the fall, what is the water velocity at the bottom?
If the efficiency of the hydro-generator unit is 80%, how long could you light a 100 W light bulb?

(11) Problem: A building is t-shaped with the following dimensions (looking down on the building from the top: Top of T: 54 feet; Top left and right sides of T: 27 feet; Bottom sides of T: 40 feet; Bottom of T: 34 feet). See the diagram from earlier in Fig. 7.8. Draw a diagram of the building roof profile. Find the monthly average precipitation data from Socorro, NM or a town of interest from the Western Regional Climate Center or another source like NOAA or NWS. How much rainwater annually can be collected? Make a monthly time series plot of water (gallons) that could be collected off this rooftop surface. For rainfall collection, discuss when it would or would not matter if the roof is pitched.

(12) Problem: Approximately 193.6 million pounds of toxins are released annually into US waterways. Assume the volume of water on earth is 1.365×10^9 km^3. Of this, 2.5% is freshwater. And of this 0.3% is surface water that is not in the frozen state. The surface occupied by the US land area is 2%, and assume this scales to the fraction of freshwater within the United States. What is the added concentration to surface freshwater in μg/L and parts per trillion by mass? Assume no degradation and all the pollutants end up in the surface freshwater of the United States.

(13) **Problem:** An underground passively chilled device was invented to harvest water vapor from the air, even in arid environments. It uses cooler soil temperatures to drop the air flowing through it below the dew point temperature. It was said to produce 14 gal/day of clean drinking water.
  (a) Estimate the air flow rate needed through the device to produce this much water, assuming $T = 75F$, $RH = 30\%$. The saturation water vapor pressure at 75F is 30 mBar.
  (b) Assume a 10 cm-diameter neck through which the air flows. Estimate the Reynolds number for this airflow where the air viscosity is 0.0000182 kg/(m-s).

(14) **Problem:** Find a conversion factor to convert inches of rainfall on a 1 ft^2 surface to find collected water in gallons. Find the conversion factor for 1 cm of rainfall on a 1 m^2 surface into L of collected water. Think about the roof area for rainwater collection—do you use the total surface area or cross-sectional area of the roof?

(15) **Problem:** Given the volume of Greenland and Antarctic ice sheets are 2.85×10^6 and 25.7×10^6 km^3, respectively.
  (a) Estimate the rise in sea level if both melted completely (a long-term process).
  (b) The Arctic Sea ice volume is tiny (20,000 km^3 on average, varies from about 11,000 km^3 at its minimum in September to 28,000 km^3 at its peak in April) compared to the Antarctic and Greenland ice sheets. Since 1980, its average decline has been about 10,000 km^3. The Arctic will likely have an ice-free summer by mid-century. Given this projected loss of 20,000 km^3, how much does this contribute to sea level rise (pretend it is land-ice and thus will contribute from displacement)?

(16) **Problem:** Temperatures at the Greenland summit in mid-August 2021 rose above freezing for the third time in less than a decade. The warm air fueled an extreme rain event that dumped 7 billion tons of water on the ice sheet. It was the heaviest rainfall on the ice sheet since record-keeping began in 1950, according to the National Snow and Ice Data Center, and the amount of ice mass lost on Sunday was seven times higher than normal for this time of year. How many times would the precipitation (a) fill Navajo Lake (2×10^{12} L) and (b) the reflecting pool (25.6×10^6 L) on the National Mall in Washington, DC? (c) How long would this supply the water needs of Albuquerque metro (1×10^6 residents) using reasonable assumptions?

(17) **Problem:** Compute the power requirements for a desalination plant (using distillation) for a city of 5 million people. You can assume the process is 80% efficient. Do not just look it up; use first principles thinking about what the process entails from a physical standpoint and making reasonable engineering assumptions.

(18) **Problem:** You decide you want a lush green lawn on your 0.3-acre lot. You can take the average weekly rainfall needed as 2″ for bluegrass. Don't forget to subtract the rainfall amounts during those months.
  (a) If the irrigation season is 7 months (April to October) in Socorro, New Mexico, how much water will you use to sustain it?
  (b) Compare to the average household water use (assume Albuquerque standards of 127 gal/person/day and 2.5 people per household).
  (c) Now find the energy footprint of this new lawn. You can assume that it takes 1.1 kWh/100 gal of water to treat it. Compare to the actual annual average electric use for a New Mexico household.

(19) **Problem:** You are curious about this water-energy nexus and how it applies to your home. You take an average 10-min shower with the current standard showerhead at 2.5 GPM and a temperature of 115°F. Your gas water heater takes incoming water at 50F. Assume overall delivery of 65% efficiency. How many therms of gas does this take? You decide to detain the water in the tub and let it warm your home's interior air (its winter). You have a 2000 sqft house with 8 ft ceilings and assume STP298. How much will the hot water heat the room air if it exhausts all of this heat into the room air? Does this seem reasonable and why or why not?

(20) **Problem:** You measure wastewater to have a decay constant of 0.25/day. It starts with an ultimate BOD of 500 mg/L. Find the Y5 for this wastewater. Make a plot of exerted and unexerted BOD as a function of time.

(21) **Problem:** Texas and Louisiana were inundated in 2017 with approximately 33 trillion gallons of water from Hurricane Harvey (measured by the compression of Earth).
   (a) Find the mass and the number of Olympic swimming pools (50 m × 25 m sized and 2-m deep swimming pool) that one can fill with this volume of water.
   (b) Navajo Lake is the second-largest waterbody in New Mexico. It is in Northern New Mexico and is 24.38 mi^2 and 109 ft as an average depth. How many times could it be filled?

(22) **Problem:** I've come up with a super rad idea, no pun intended! Let's nuke the next approaching hurricane and destroy it before it makes landfall! This will obliterate it or at least knock it far off course. Take a big hurricane like Harvey, which dropped about 33 trillion gallons of water. Take an average-sized nuclear weapon like the W88 warhead that releases 2000 TJ. Does this seem like a viable plan, and why?

Green outdoor spaces: Harmonizing the built and natural environments

CHAPTER 11

Learning objectives

(1) Find value in developing an aesthetic connection to nature through regenerative landscaping.
(2) Appreciate the outdoor environment as an extension of the indoor environment.
(3) Recognize techniques for minimizing landscape energy, water, and materials needs.

As discussed previously, site considerations, and most notably the landscape, are often not top of mind for engineers when designing buildings. Rather, it is often an afterthought or a tidying up exercise that is often left to the landscape contractor after the construction concludes.

However, the site and landscape are an integral part of the building and, in the best cases, even blend seamlessly with the indoors (Sorvig and Thompson, 2018). Site planning, construction, and landscaping are key determinants of energy, materials, and water flows associated with a building. Moreover, many scientific studies indicate that the sensory experience in an outdoor environment improves human mood and productivity and affects the physiological metrics related to stress (Coventry et al., 2021). The COVID-19 pandemic and recovery from it have only underscored the importance of outdoor spaces to human well-being. New paradigms include, much like with agriculture, not just reducing impacts but pursuing "regenerative" landscape management that restores degraded lands to improved ecological functioning. Thus, the engineer must understand and appreciate the outdoor environment of the building as a partner to the function of the indoor environment.

11.1 History and importance of landscape design

Frank Lloyd Wright (among others) was an early influential green architect in the early 20th century, especially with respect to integrating the building and site. Wright, through what he termed "organic architecture," focused on integrating the building and site, used local materials, connected indoor and outdoor spaces, used passive air conditioning, and brought in copious natural daylighting to his designs (Fig. 11.1). Wright's most famous building, the Fallingwater (shown earlier in Fig. 1.4) house near Pittsburgh, Pennsylvania, was so well integrated into the waterfall landscape that it became the centerpiece of the house.

FIG. 11.1

Frank Lloyd Wright's "organic architecture" approach was seminal in integrating the site and structure. Shown in the Dana-Thomas House in Springfield, Illinois, built in the Prairie School Style (low and wide with large overhands). Some of the organic features include integration with the existing landscape (in this case downtown urban), natural daylighting, wide overhangs for limiting summer direct sun, reuse and renovation of an existing building, and preservation of the site and structure for a century and a half.

Recognition and adjustment of urban landscaping practices have taken hold over the last few decades to adopt a more sustainable landscaping approach (Hurd et al., 2006). This has been spurred by incentives from municipalities to reduce landscape water use, particularly in water-constrained ecosystems. Sustainable landscaping efforts can be summarized with the following key attributes (Loehrlein, 2021):

(1) **Plant Selection and Location**: Choosing and placing plants that are native or well-adapted to the climate increases species diversity, enhances the site's wildlife habitat, and decreases wildfire potential.
(2) **Resource Efficiency**: Sustainable landscaping can enhance building energy and water use efficiency and help to minimize and treat stormwater.
(3) **Emissions Reduction**: Sustainable landscaping reduces water, air, and climate-related emissions due to landscape maintenance.
(4) **Plants as Remediators**: Plants through phytoremediation and use in constructed wetlands can reduce soil contamination issues and enhance soil fertility, reducing the need for synthetic fertilizers or other lawn chemicals.

For residential lots in particular, landscape use of water is often the largest, if not the dominant, water use category (Hilaire et al., 2008). Water use for a two-bedroom, two-bathroom 1500 ft^2 house on a

9000 ft^2 lot on the arid Front Range of Colorado is shown in Fig. 11.2. The wintertime minimum is <2000 gal per month, while in summertime the water usage often exceeds 10,000 gal per month. This is despite efforts xeriscaping approximately half of the lot and using rainwater collection. Landscape water use is of vital importance, and even more so in the arid regions of the United States.

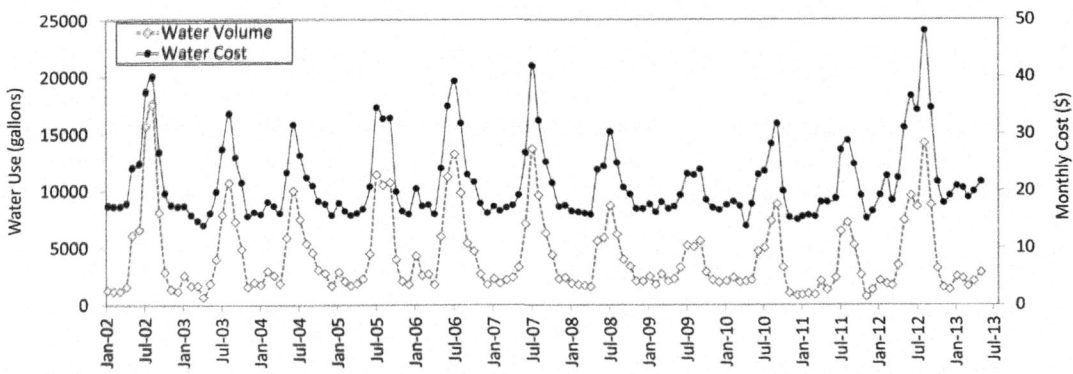

FIG. 11.2
Water usage for a residence in Fort Collins, Colorado, with two occupants. Despite efforts to xeriscape the lot, the summer landscape usage still dominated the annual usage.

11.2 Demographic trends and green sites

Beginning in the industrial era, the shift from an agricultural to an urban-industrial society featured a shift to both smaller home sizes and much smaller plots of land. The farmhouse was replaced by the small urban dwelling often in a "company town" with much smaller lots needed for an urban family.

The urban population fraction and its density have continued to increase in the United States and globally. Resultingly, the average lot size has generally declined while building size has increased in the United States since 1990 (Fig. 6.6). The average residential home size has a floor space of 1761 ft^2 (164 m^2), though this varies only by ±20% by state (Fig. 6.6). With increasing building square footage, the number of occupants has decreased, translating into more interior and less exterior square footage per occupant. Simultaneously, the market entry cost has created a problem of affordability of small starter homes, especially in major urban areas. Very recently, the push toward higher urban density paused due to the COVID-19 pandemic and lockdowns. It remains to be seen if this trend continues or is a short-term anomaly.

A larger building footprint and a smaller lot size both lead to smaller yard sizes as shown in Fig. 6.6. Part of this is driven by the ~$85,000 average cost of a residential lot circa 2020, according to the National Association of Home Builders (NAHB). Lot cost ranges from a factor of two smaller to a factor of six larger depending on local economic conditions and population density.

In some urban areas, the dwelling for a single-family residence can occupy more than half the lot. Most lots, especially in suburban or rural areas, though, are still a majority yard versus building. Thus, the green building imperative must also carefully consider the yard and landscaping, and particularly how they integrate with the building.

11.3 Site planning, engineering, and minimizing construction impacts

There are many facets to site planning, setup, and management that are relevant to green building renovations, particularly with new construction. Many of these involve mitigation of disturbances to the site. As discussed elsewhere in the book, the "Sustainable Sites" program addresses the ecological aspects of the site selection and design process by establishing prerequisites and providing credits for specific efforts. They include (Loehrlein, 2021):

- Protecting existing site functions, including avoiding farmland conversion
- Preventing disruption of aquatic ecosystems and floodplain functionality
- Maintaining viability of existing soils and vegetation, particularly native and adapted species
- Providing habitat and ecosystem functionality, particularly for threatened and endangered species
- Optimizing new vegetation using native and adapted species and avoiding invasives
- Appropriate plantings can enhance building energy efficiency and wildfire safety

In planning site development, another key concern is using site-based or locally obtained materials to construct the site topography and hardscape. Preserving topsoil as well as balancing the cut-and-fill materials from the site as much as possible helps to minimize importing or exporting earth from the site. According to the USEPA, solid waste generation features a significant contribution of landscape materials. This can be amplified at a construction or renovation site. Rather than landfilling or even composting offsite, this material can be shredded for site mulching needs. Likewise, planning from project onset for the minimization of construction waste and its recycling is beneficial waste minimization.

Much like an energy audit, a landscape audit can provide a snapshot of the sustainability and roadmap of future improvements of an existing site (Loehrlein, 2021). Developing a landscape plan for the site, akin to Fig. 11.3, is helpful to the planning process, especially for a newly built site. Depending on the size of the lot, a zoned site plan can help to minimize irrigation needs.

Grouping plants with similar water, sun, and soil needs is most effective. A "rough" or naturalized outer zone can be left to generally take care of itself without the need for water, fertilizer, or heavy maintenance. Zones nearer the dwelling can be established and grouped based on the water needs of the plantings. The area between the sidewalk and street is often called the "Hellstrip" by landscapers. This area is a challenging environment for plantings due to watering difficulty, the surrounding large impermeable, heat-retaining surfaces, and surface traffic. Areas for edible plants and fruit trees represent a higher water zone that typically needs more light as well for fruiting plants. Low-water ornamental plantings are good choices for intermediate needs areas surrounding the indoor and outdoor living spaces and connecting to the rougher periphery.

11.3 Site planning, engineering, and minimizing construction impacts

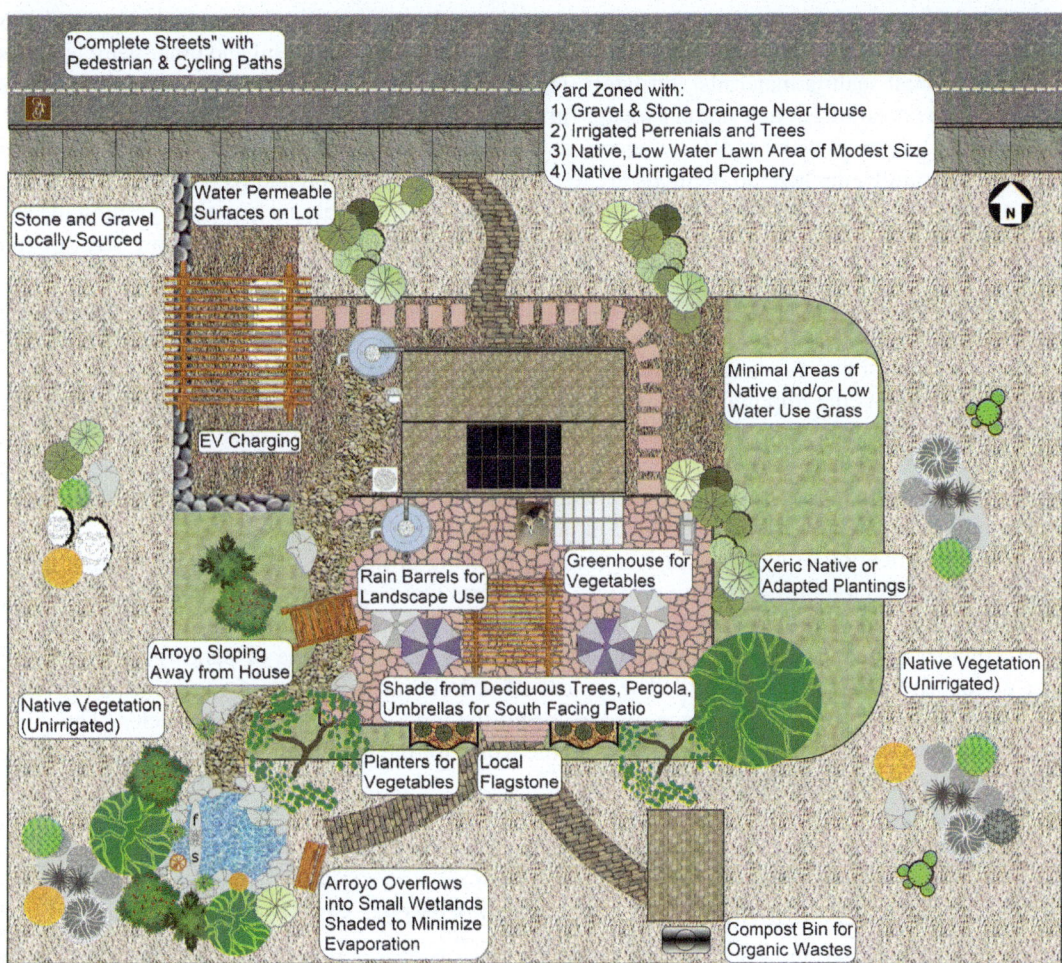

FIG. 11.3

Some of the key features of a sustainably designed site for an arid US location.

Working within the constraints of the existing site grading is desirable in most cases to minimize disturbing the hydrology, topography, and ecosystem functioning. A top priority for site planning is reducing stormwater runoff, preventing soil erosion, and decreasing contaminants from leaving the site. Also important is to minimize the impervious surfaces and paved surfaces, such as dark asphalt, which contributes to both an urban heat island effect and incrementally to regional warming effects as previously discussed in Chapter 7 (Fig. 7.11).

Stormwater runoff is strongly dependent on on-site characteristics, including soil properties, vegetative cover, and surface permeability. Materials such as asphalt and concrete are generally of very low to zero permeability, while undeveloped lands are higher depending on soil properties (Cooper,

2015). Furthermore, whatever contaminants we dump onto the soils or street surfaces end up as part of the stormwater flow or percolate into the groundwater. This includes motor oil and other vehicle fluids, fuel spills, asphalt sealing materials, fertilizers, pesticides and herbicides, other household chemicals, improperly disposed of pet and human waste, fluid leaks, and others.

Managing too much water is also a concern at times in some areas. To prevent this from adding to surface water runoff, the general methods are to promote water retention and its permeation in appropriate areas. On-site methods to handle this include increasing permeable surface area, grading the surface away from structures, water storage in gravel areas such as arroyos or dry stream beds, or rain barrels, and, in a low area, swales or rain gardens planted with hydrophilic plants.

Prudent planning and mitigation efforts can reduce the urban heat island effect by up to 4.5°C (Haddad et al., 2024). Reducing urban heat island effects depends strongly on the surface material reflectance and emissivity. More reflective and high-emissivity surfaces will shed rather than store incoming solar energy, the importance of which grows in hotter climates. This affects energy use, local air quality, climate, and personal comfort. One of the worst materials is asphalt, as it has a high absorption (low reflectance) of 0.1 or lower and considerable thermal mass. Many more options are emerging for asphalt, concrete, and roofing materials and coatings that reflect incoming solar radiation helping reduce urban heat island effect and regional warming.

The site engineering of a landscape is a civil engineering project on a small scale but often much more intricate than large infrastructure projects. It involves locating roadways and pathways, building and outbuildings, and working with the site soils, grading, topography, and hydrology (Strom et al., 2013). Another concern with site planning and land use is the prevalence of parking areas. Parking space might be termed a necessary evil. The typical office building, however, often devotes more space to a worker's vehicle than office space, underscoring the land use footprint for vehicles. The code-mandated parking areas are being relaxed around the world for the use of more flexible and multimodal approaches to urban transportation.

11.4 Community, commercial, and other shared scale green spaces

Many communities, particularly urban areas where space is at a premium, are finding new ways to share city and school parks, community gardens, shared vegetable plots, dog parks, and other common areas. Apartment dwellers, who are less likely to have outdoor spaces available, are prime beneficiaries of these resources. Such amenities are also where pedestrian and cycling facilities can be incorporated into the infrastructure of a development. This reduces transportation-related emissions and costs while improving humans' physical and mental health.

Outdoor areas are also important for commercial and office landscapes. As an example, Phoenix Plaza, also known as the "Boat Building" in Hartford, Connecticut, is a retrofit of a degraded hardscape with a green roof area above the building parking to provide an urban "park-like" environment (Fig. 11.4). The benefits include helping reduce the urban heat island effect previously discussed, moderating building utility costs, providing greenspace for humans and habitat for wildlife, and reducing stormwater runoff with the green roof above the lower parking levels. Visually, it breaks the monotony of the urban "concrete jungle." This is something difficult to accomplish by simply putting a row of alike trees along a streetside.

FIG. 11.4

Greenspace and green roof associated with a National Historic Register office tower the Nassau Financial Group headquarters in Hartford, Connecticut, USA, constructed in 1963 and renovated in 2010 achieve LEED silver status with a green roof and open green space added for employees in 2013 (from the company website at https://nfg.com/iconic-boat-building-in-hartford.html).

Community gardens are another option for shared space that can be a boon for urban residents lacking yard or garden space. As an interesting reuse project, the architectural mock-ups that are often used with larger housing developments (and often landfilled after the sales phase ends) are being reused or up-cycled for use as meeting spaces, semienclosed classrooms, and the like in urban community gardens by such organizations as Testbeds (https://www.testbeds.org/).

11.5 Case study High Line rail line repurposed into a linear park in New York city

Multiple studies have shown the value of urban park areas to the economic and social well-being of urban residents. In urban areas, green spaces are often at a premium beyond the legacy parks established decades or centuries ago, like Central Park in New York.

High Line Park in New York, NY, is an adaptive reuse project establishing a raised linear park redeveloped in 2009 from a derelict elevated former freight rail line in the Chelsea district of Manhattan's west side. The current park serves to connect several neighborhoods and destinations, such as the Whitney Museum and Chelsea Market in New York City (Fig. 11.5).

The rail line, originally constructed in the 1930s, was designed to reduce traffic congestion and enhance safety by reducing conflicts and accidents with increasing vehicular and pedestrian traffic. It connected numerous commercial and industrial buildings with trains that allowed transport of raw materials and finished goods without disruptions from daily commuters and truck traffic below. It started to become obsolete in the postwar era as freight transport by truck on the interstate highway system dominated (Lindner et al., 2017; Sternfeld, 2023).

Much of the rail line was demolished, and even the section that became the High Line Park was scheduled for demolition. Preservation efforts saved a part of the elevated overpass, which became the redeveloped park. It is a city-owned public facility that includes walking path connections and naturalized areas vertically separated from vehicular traffic below. In addition, it features landscaped gardens, city skyline views, food vendors, public art, and a sculpture park along the pathway. The High Line redevelopment served as a catalyst for this area's private renovation projects as well.

FIG. 11.5

The wintertime pedestrian scene on the linear, elevated High Line Park which is an adaptive reuse project in the Chelsea neighborhood of New York City featuring walking paths, landscape and naturalized areas, and public art spaces.

The development and changes in the neighborhood prompted discussions of gentrification and the double-edged sword of the renovation process and what it can do to less affluent communities. Since such projects often raise adjacent real estate values, equity efforts are now afoot to require a certain fraction of local business involvement, maintain an ample supply of affordable housing, provide jobs training and youth centers, and genuinely involve and respect existing residents (e.g., honor the history with public art) with such green infrastructure projects. The project website also details the history and amenities of the park (https://www.thehighline.org/). Nevertheless, it provides a prime outdoor example of adaptive reuse of an industrial structure for community connectivity. It has been used as a model in other urban areas.

11.6 Appropriate plants and their benefits to site and soil health

The benefits of plants are numerous; they are an integral part of the site plan shown in Fig. 11.3. Plants are a vital component of the landscape for soil conservation. Plantings will all have associated maintenance (as does bare ground), but proper selections can help minimize the inputs of energy, water, nutrients, herbicides, and pesticides. Landscapers like to talk about the right plant in the right place. The considerations are mature plant size, shape/structure, water, nutrients, and light needs, all with respect to the space occupied.

The concept of permaculture is to landscape as whole building as a system is to green building. Permaculture can be summarized as working with the site, landscape, topography, climate, flora, and fauna of your site rather than against it—imitating nature rather than battling it. Plants chosen appropriately can even have a restorative effect on a site. The concept of bioremediation employs biological organisms to treat contaminated sites. Phytoremediation is a subset of this approach which uses selected plant species for uptake and bioaccumulation of particular contaminants, such as heavy metals, in soils.

The benefits of a rain garden were discussed in the last chapter related to water management and stormwater runoff minimization. The riverine plants are those that thrive in standing water, while the riparian plants are those on the shores. These gardens, in wet environments, are best populated by water-tolerant plants such as some grasses, sedges, willows, rushes, ferns, and cattails that can tolerate wet soils or standing water.

The traditional turfgrass lawn, often dominated by Kentucky bluegrass, covers approximately 50 million acres of land and accounts annually for 3 trillion gallons of water and 70 million pounds of pesticide and herbicide use, while consuming 200 million gallons of gasoline for mowers annually, according to the NRDC (https://www.nrdc.org/stories/more-sustainable-and-beautiful-alternatives-grass-lawn). Additionally, the ecosystem habitat benefits are slim to none from a monoculture of turfgrass. Plant selection is a key factor in making a range of alternatives successful, from xeriscaping in arid regions to no-mow prairie-like lawns to edible landscapes.

Plants suited to the site and climate have numerous benefits, as summarized below:

(1) Habitats for Native Fauna: They provide species habitat protection, preserve native species, and provide pollinators opportunities to thrive. Promoting the benefits of native wildlife and pollinators

including food sources for the fauna even in urban areas is a benefit. Using native and climate-adapted species of plants provides habitat for native wildlife and pollinators.
(2) **Carbon Sequestration:** Carbon sequestration is quasi-permanent removal of CO_2 from the atmosphere and conversion to solid or liquid phase. Plants, particularly trees, have environmental attributes, including sequestering carbon in their tissues. Beyond reducing fossil fuel use, forest and agricultural management are tools that are most relevant to climate mitigation. Urban trees in the lower 48 states store an estimated 700 million tons of carbon (Loehrlein, 2021).
(3) **Minimize Urban Heat Islands:** Local cooling effect can be enhanced with appropriate tree species via transpiration, channeling of breezes and/or providing wind blocks, and direct shading. These all help moderate the local ambient and indoor structure temperatures on-site. Deciduous trees provide a shade canopy in summer while allowing more sun in the winter when they lose their leaves.
(4) **Foster Healthy Soils and Urban Ecosystems**: Plants foster healthy soils and even serve for the phytoremediation of disturbed soils to prevent contaminants from reaching the groundwater. Maintenance of a healthy landscape will also reduce needs for additional fertilizers, pesticides, herbicides, and little additional inputs.
(5) **Provide On-site Views:** Views inside to out and vice versa are impacted by the site plantings. Thus, landscape plantings can enhance the economic value of a site as well as contribute to the well-being of occupants.
(6) **Frame or Demarcate Outdoor Spaces:** Outdoor 'rooms' and their various applications are framed and defined by the plantings on-site, which serve as room dividers.
(7) **Local On-site Food Production**: Plantings not only provide beauty, habitat, and shade but also food in some cases enhance water and soil management at the site.

Appropriate and preferably native trees in the landscape provide many benefits, including shade, privacy, removal of air pollutants, carbon sequestration, and, in some cases, can benefit water conservation by reducing evaporative losses through their cooling shade. However, the urban tree canopy is under greater stress due to increasing extremes of temperature and precipitation that are occurring globally (Esperon-Rodriguez et al., 2022). This certainly sets up a "Catch-22" situation where one of the tools that can help mitigate climate change speed and impacts are under threat themselves from a changing climate.

As another feedback system, soil characteristics and health are intimately tied to the plant ecology of the site. First, plants physically preserve the topsoil on a given site where, without their presence, the soil would erode away through the action of wind and water. Soil characteristics in turn dictate what plants will survive or thrive in a given environment. Soil pH in the west is generally alkaline, while in the east, particularly bogs, swamps, and pine forests are ecosystems that feature more acidic soils. This plays a key role in the plant choices for a given site.

Key soil properties include the proportions of sand to clay to silt to gravel to organic matter. Critical soil nutrients include nitrogen, phosphorus, and potassium (NPK) as three primary nutrients listed on fertilizers as percentages. Unless soil tests reveal a severe deficiency, lower numbers and more balance in a complete fertilizer that has slow release is preferable to avoid shock from overfertilization. The number of products available made from waste agricultural materials (e.g., corn gluten) and/or organic processes has increased markedly in the last 20 years. Organic fertilizers have advantages: less water soluble and slower release, preventing fertilizer burn, lasting longer, and preventing leaching into groundwater. Watch out, your dog or other critters may find these organic amendments quite tasty!

Soil amendments are critical for many plants in the western soils, which are very lean and typically alkaline. Soil amending is often necessary for new planting areas, but the goal should be to minimize its ongoing usage. Closed loop system where grass clippings, leaf litter, plant trimmings, and other materials can be returned to the soil after mulching and/or composting to enhance soil fertility. This saves landfill space, materials, collection energy and expenses, and provides a valuable raw material such as organic compost or mulch.

Covering the bare ground with mulch, which often can be produced partially on-site from trimmings, is important to water conservation, soil preservation and health, weed minimization, soil surface temperature reduction for light mulches, and provides food, habitat, and site beautification (Loehrlein, 2021). Mulches can be organic matter (like bark or wood mulch) or inorganic (like gravel in arid regions). Composting kitchen/yard waste is helpful in this process and in closing the loop on waste production.

Invasive species as mentioned earlier (Fig. 4.26) in the West like salt cedar (tamarisk), cheatgrass, Siberian elm, Russian olive, and others should be avoided, particularly near riparian areas where they can spread quickly and choke out the native willows, cottonwoods, and other species. Salt cedar as a halophyte promotes saline soils and can cause the lowering of the water table in a water-stressed environment, according to the National Invasive Species Information Center (https://www.invasivespeciesinfo.gov/terrestrial/plants/saltcedar). Characteristics that make nonnative plants undesirable: noxious weeds pushing out natives, a weak structure that drops branches, penetrates sewer lines, grows on top of existing vegetation, strangling it, spreads rampantly, and impacts soil health. The invasive species vary regionally as well; the classic example is the kudzu vine in the southeastern US. A more complete listing of invasive species is provided in Loehrlein (2021) and other landscape and horticulture-oriented books.

11.7 Xeriscape: Low-water-use landscaping

As illustrated earlier in Fig. 4.13, precipitation across the United States ranges from less than 10 in. per year in the desert southwest to over 50 in. per year in some locations in the Southeast United States and Pacific Northwest (Fig. 4.13). Mountain locations in wet regions can exceed 100 in. per year, as topography plays an important role in precipitation regionally. In general, the west side of mountain ranges is wetter due to the orographic lifting of air masses that generally travel west to east in the continental United States. Outside of the Pacific Northwest and some mountain locations, the United States can be roughly separated into the drier, sunnier west with lean, alkaline soils and the wetter, cloudier east with richer, more acidic soils. This is a key driver of plants that can thrive in these areas.

The increasing severity, effects, and duration of droughts make plant choice and viability ever more important in arid regions that on average receive less precipitation. Decreased winter snowpack and more winter rain in areas dependent upon snowfall runoff is a similar concern linked to a warming planet.

Xeriscaping centers on minimizing the need for water use by appropriate design and plantings at a site (Fig. 11.6). Xeriscape often focuses on low-water use plants appropriate to the site. This is a vital practice in the water-constrained western United States and other arid regions worldwide. Despite its importance, the research literature on the practice is scant (e.g., a web of science topic search turned up only 30 journal articles). Research has shown that not only does xeriscaping in arid western US landscapes save water and utility costs, but it also reduces the maintenance costs and time spent on yard work and is generally preferable to a conventional turfgrass lawn (Hurd et al., 2006).

Native and adapted plants are the most successful and least resource-intensive choices. The mantra in the landscaping world is that xeriscape is not "zero-scaping" of a site. Xeriscaping includes the following key principles, some of which have already been highlighted (Weinstein, 1999):

(1) Developing a landscape plan for the site, similar to Fig. 11.3.
(2) Reducing large nonessential areas of turfgrass and other high-water-use plantings.
(3) Improving soil as appropriate by aeration and soil amendments.
(4) Selecting native or adapted plantings appropriate to the location.
(5) Mulching to reduce evaporative losses and control weeds.
(6) Creating an efficient irrigation and operations and maintenance plan using drip and below-ground watering versus above-ground broadcast water application.

FIG. 11.6

Native and adapted species (*left*) such as the Maximilian sunflowers in the foreground, the ice plant (*Delosperma*, which of note some varieties are invasive in coastal California) in the lower left corner, and the Texas sage (*Leucophyllum*) in the background are well-adapted to dryland conditions and fit well in these landscapes. The use of gravel and/or shredded woody mulch is an effective substitute for vast areas of thirsty turfgrass. Also shown is solar-charged, low-intensity landscape lighting that can be used for lighting pathways and landscape features. A small PV cell in the lid of the canning jar charges a battery beneath it which illuminates an LED light in the glass jar at night. On the *right* are sage and ornamental grass varieties that have much lower water use than turfgrass.

One of the most important questions to consider is: is it a wet or dry climate, and does it freeze or not? It makes as much sense to plant vast tracts of Kentucky bluegrass in Arizona as it does to try to grow cactuses in Seattle. Kentucky bluegrass requires significant inputs, and other more adapted grasses such as buffalo grass, grama, and fescues are less needy (Loehrlein, 2021).

Some of the most water-stressed cities in the desert southwest, such as Las Vegas, Nevada, have banned the use of turfgrass lawns as well as spray irrigation on lots of new developments (parks and schools excluded). Substituting ground covers other than water-thirsty Kentucky bluegrass lawns has increased. Creeping thyme, specific clover varieties, and selected succulents such as low-height iceplants (*Delosperma*) and other "stepables" can form a green mat that tolerates some foot traffic.

Plants mainly regulate the flow of water via the stomata openings in their leaves. As the plant dries, these stomata naturally close reducing the transpiration of water. Furthermore, low-water-use plants reduce water needs by reflective silvery-gray coloring, low surface area to volume ratios in succulent leaves (the extreme version of this being the barrel cactus), waxy hydrophobic coatings that reduce water loss, and osmotic regulation to control water loss or dormancy in drought conditions, among others. Some plants even create microclimates near their surface with hairy surfaces or needles, which create more humid conditions at the surface, limiting water loss. Dryland plants will also have a deep taproot or a surface root system that extends wide. These are all key characteristics of plants in arid regions.

All trees require water but can serve to reduce water needs of the understory as well as provide shade and passive cooling in hot climates. Windbreaks, to shelter the structure from winter winds, can be constructed at an appropriate distance from the structure with denser evergreen species. Drought-tolerant tree varieties, often smaller in size, of many of the hardwood trees of the Eastern half of the nation (e.g., Tartarian maple, scrub oak). Ash, cottonwood (cottonless varieties to minimize allergies), western catalpa, mulberry (fruiting versus the pollen generating), desert willow, and New Mexico privet, among others, are trees that can tolerate alkaline western soils.

Perennials include native and adapted species that tolerate low precipitation, drought conditions, high direct solar exposure, and cold, dry winters. Many people are unaware of the winter watering needs of plants in arid regions, but the further to the southwest of the country one resides, the more it becomes important for plants to over-winter. High Country Gardens and Denver Botanic Gardens "Plant Select" program provides great guidance for gardening in this challenging region. Local nurseries are also an invaluable resource for discovering what works in a given location.

Much like the adaptation of our physical infrastructure, the adaptation of both human and natural landscapes is required in a changing climate. The urban canopy will have to adapt with a new focus on climate extremes and adaptable cultivars of trees and other ornamental plantings (McPherson et al., 2018). The multipurpose tree that can survive in climate extremes while providing shade, fruit, and aesthetic values without much supplemental input of water or fertilizer, such as the mulberry tree, will become of higher value (Sharma and Zote, 2010). It should be noted that the pollen of male (nonfruiting) mulberry trees can cause allergies in susceptible individuals and is thus not sold in nurseries.

11.8 Case study: Landscaping of the "Hellstrip" of an arid region

The right-of-way area on the periphery of a lot between a sidewalk and a roadway is often termed the "Hellstrip" by landscapers. It is often sunbaked and dry, trampled, and subject to various inputs of solids, liquids, and gases from passing traffic. As such, it is often hard to create any significant landscaping.

An example Hellstrip project creates a "linear park" on a residential scale and provides a case study of local materials, materials reuse, low-embodied carbon concrete, and waste reuse (Fig. 11.7). This was a DIY project by the author, so labor was all "in-kind," not to mention it obviated visiting the gym for weight training!

FIG. 11.7

Linear park created on the "Hellstrip" of a southwestern US residential lot using a mortared stone landscape wall. The wall used primarily on-site and waste materials, and a low-carbon custom mortar mix was used.

The first goal was to use only on-site or waste materials though this proved unrealistic with the need for cementitious materials for the concrete (Table 11.1). A dry-stack approach was considered, but the smooth, round river-type rocks that were onsite made this a challenge, and thus mortared construction was used.

Table 11.1 Ways to minimize the environmental impacts of concrete.

Goal	Example
Minimize the concrete specified and use an appropriate mix	High-strength concrete does not need to be used for nonstructural applications
Minimize the use of Portland cement in concrete	Substitute hydrated lime and other supplementary cementitious materials
Minimize concrete waste	Reuse of concrete waste for aggregate in the new mix
Reuse of materials	Use of used rebar and other repurposed metal reinforcements
Locally derived materials	Use of sand and rocks washed onto the streets from monsoon rains
Minimize water use	Use wash water to keep the concrete wet during curing

The project also featured the following components:

- Rubble concrete foundation: including 47 total wheelbarrows of concrete for the foundation and well over 100 in total. Roughly half was waste materials, including chunks of reused concrete, gravel-packed containers, bottles, and cans, and various other dense, rigid materials.
- Low-carbon custom cement containing approximately one-quarter Portland cement, one-quarter hydrated lime, one-quarter pulverized limestone, and one-quarter supplementary cementitious materials (SCMs) (silica, diatomaceous earth, zeolites, and others destined for disposal).
- Scrap rebar and other long metal pieces for tying together the courses both vertically and horizontally.
- Stones were mostly onsite from a former rock wall and miscellaneous excavation projects.
- The sand was collected from monsoon rain runoff that collects in the gutters and streets.
- Aggregate was gravel that was leftovers from landscaping projects and crushed concrete rubble.
- Excavated soil was relocated to raised bed planters to balance the cut-and-fill activity.
- Waste materials included approximately 30 glass jars and 20 plastic jugs of various sizes filled with onsite sand and other rubble, waste and one large shock absorber dropped in the gutter, and waste materials from a student concrete mixing project.

All this was completed with only the purchase of 6 bags each of Portland cement (~500 lbs total), 8 bags of hydrated lime (~500 lbs), 8 bags of barn lime (pulverized calcium carbonate, 100 lbs), and 4 bags of diatomaceous earth (50 lbs) with a total cost <$500.

Existing plantings along this "Hellstrip" park were worked around during construction (another reason for its serpentine pathway). New additions to the area included a Purple Robe Locust tree (drip-irrigated) and unirrigated agave, prickly pear, ocotillo, and cholla cacti.

11.9 Outdoor rooms: A low energy extension of the indoors

Connecting the indoors with outdoor environments is a central theme of a well-designed building linked to its site. The organic architecture approach of Frank Lloyd Wright is a predecessor to the modern approach of bringing the outdoors inside. The focus was on integrating a building with its surrounding landscape. Such connectivity to nature is well-known to enhance occupant and user satisfaction with using the spaces.

The outdoor "rooms" can be the least expensive and impactful spaces, even if not available year-round. The materials cost to build these and the operational costs to use them are often quite low to negligible once built. By their nature, outdoor spaces allow a closer connection to nature—its flora and fauna plus its sights, sounds, fragrances, airiness, and light.

Most every climate zone can benefit from outdoor spaces at least during a large part of the year. Transitions from indoors to outdoors can be continuous and nearly seamless when well-designed. Outdoor rooms are important to extend the space, particularly for smaller residential buildings and the "tiny homes" movement with indoor constraints. The site plan shown in Fig. 11.3 has several outdoor rooms including the back patio area and the area around the water feature.

11.10 Permaculture, regenerative landscaping, and producing food onsite

Permaculture from permanent agriculture approaches the landscape as a productive ecosystem. A classic book on the permaculture approach is Gaia's Garden by Toby Hemenway (Hemenway, 2009). On the other end of the water spectrum are many of the fruit and vegetable species that are water-thirsty but provide vital local produce. Plants like humans are ~75% or more water by mass. Typical Kentucky bluegrass turf is a monoculture that requires up to 40 in. of water per year. If the majority of this is from irrigation as in arid regions (plus weekly mowing, fertilizing, and weed prevention), why not plant fruits and vegetables where there is produce (Fig. 11.8)? In fact, there is a classic book and an organization titled as such (Flores, 2006) (https://www.foodnotlawns.com/). With a well-developed landscape plan, a largely xeric yard with selected areas on drip irrigation for vegetables and fruits, also described as foodscaping, is entirely possible. This can also be applied to commercial lots, particularly for a restaurant or other food-oriented business onsite.

The permaculture movement has intersected with the regenerative farming movement to create a regenerative landscaping movement. Regenerative landscaping borrows from the parallel effort at regenerative agriculture, which focuses on microbiology and soil improvement as a way to solve multiple ecological and other ills (Ohlsen, 2023). The goals are better landscape management, more nutritious food production, healthier soils with better water retention, reduced pesticide, herbicide, and fertilizer use, and a diversity of plants and the microbiological species in soils. The potential for greater soil sequestration of atmospheric CO_2 is also estimated in the tens to 100+ gigatonnes and thus can have climate stabilization benefits as well.

It should be noted that the climate benefits of "urban agriculture" (UA) are far from assured. A life-cycle study from the University of Michigan found on average the greenhouse gas footprint of UA vs conventional food production (Hawes et al., 2024). The study found greenhouse gas benefits are only for certain crops, such as tomatoes, versus their greenhouse-produced conventionally produced counterparts. There are many upsides to urban gardening and food production including community involvement, outdoor and physical activity, and control of the inputs of water and chemicals. However, depending on the details, greenhouse gas emissions may not be less for UA-produced food compared to conventional agriculture.

FIG. 11.8
Growing food rather than lawns is a better application of water than vast expanses of lawn, particularly in arid environments. Fruits and vegetables are by nature water-intensive, but limited drip irrigation and mulching help conserve water.

11.11 Site water management systems: Rainwater, graywater, constructed wetlands

Water management was discussed in detail previously in terms of site development and indoor use, and here we review a few additional aspects related to the landscape. Too much or too little water are both potential problems, even at the same site. In exceptionally wet areas, constructed wetlands or rain gardens for promoting water infiltration are useful in some applications. Constructed wetlands are a larger-scale feature for stormwater and runoff management, applicable to neighborhood scale or larger lots.

In dry regions, xeriscape efforts can be combined with on-site rainwater collection or greywater reuse. Several companies produce rainwater collection and storage devices, such as Bushman and Tijeras Rain Barrels. The benefits of rainwater harvesting extend from decreased potable water use for landscaping to reduced stormwater management, providing a water source, and enhancing habitat for local flora and fauna.

Rainwater collection is a useful technique for providing irrigation water for landscape and garden use. Typical 50-gal rain barrels can quickly overflow with even modest rainfall events when collecting from a roof-sized surface area. Larger storage tanks or underground cisterns are used in larger, commercial-scale facilities. Rather than an above-ground tank, vendors such as Aquascape offer an underground water storage system paired with a pondless water feature such as a small waterfall or fountain system (https://www.aquascapeinc.com/rainwater-harvesting-system). Other issues of note to avoid include the formation of mildew or algae in tanks or serving as breeding grounds for mosquitoes. These can be addressed by frequent use of the stored water and mosquito dunks which kill the larvae. Another ancient technique derived from the arid regions of the world is the use of ollas for landscape watering of edible and ornamental vegetation. These permeable clay water containers slowly weep water from the permeable clay vessels directly to the plant root zone. The whole field of gardening and agriculture is devoted to methods to reduce impacts.

Even with conservation efforts, many sites will need additional irrigation with potable water. Drip emitters, soaker hoses, timers, and soil moisture sensors are all available as tools to assist in these regions. The replacement of overhead broadcast spray irrigation with targeted, dripper-based, and/or underground watering systems can drastically reduce water consumption. Many of these water-saving techniques are illustrated as part of the site plan in Fig. 11.3 (Example 11.1).

EXAMPLE 11.1 Rainwater collection

Problem: You have a 1500 ft^2 (single level, so floor and roof area) house. You live in a region with annual precipitation of 15 in. Assume a uniform vertical flux of rain droplets.
a. Determine the annual collection of rainwater, assuming a collection efficiency of 90%.
b. Where would the losses leading to less than 100% efficiency come into play (describe three) loss?
c. From the big picture first-order factors, would it help your collection effectiveness to change from a flat roof to a steeply pitched roof?

Given: 1500 ft^2 $z = 15$ in.
Find: Mass collected
Assume: Uniform rate and 90% collection
Solution:
$$\text{Vol} = \text{XSA} \times \text{Precip Depth} = (1500\,\text{ft}^2) \times \left(\frac{15}{12}\text{ft}\right) \times \left(\frac{7.48\,\text{gal}}{\text{ft}^3}\right) = 14{,}025\,\text{gal intercepted}$$

Collection volume is $0.9 \times (14{,}025\,\text{gal}) = 12{,}623\,\text{gal}$ collected

Splashing off of the roof, horizontal wind perturbations, evaporation from the roof surface, evaporation from the rain barrel, possible gutter clogging, and/or overshoot of liquid flows from the gutters reduce collection efficiency.

The roof cross-sectional area, which is the key parameter, would not change, and so it would intercept the same precipitation flux. It may even reduce your collection efficiency minorly, as you may have more splash off the roof, overshot from rapidly flowing water into the gutters during heavy events, as well as more evaporative losses from the greater surface area.

11.12 Case study xeriscaped residential landscapes

In Fig. 11.9A, the front lawn of this house in Fort Collins, CO, which is on the Front Range of the Rocky Mountains, has been replaced by hardscape and low-water plantings such as Russian sage, sumac, and various evergreen species such as mugho pine and ornamental grasses that are watered by drip systems. A much lower water-needs ground cover, creeping thyme lawn, fills in the on-grade area surrounding the walkway. This replaced a typical Kentucky bluegrass lawn.

In Fig. 11.9B, the backyard of a residence in mixed-temperature arid central New Mexico rather than turf grass uses river rock gravel as the primary ground cover in pathways and mulch of shredded wood and bark in planting areas. Ground cover plants are primarily succulents and other perennials that are well-adapted to low-water conditions. The canopy is provided by a mulberry and a cottonwood tree, and smaller fruit trees, shrubs, and yuccas make up the mid-level plantings. The established deciduous trees provide shaded areas that become "bonus rooms" in the hot season.

Such xeriscape projects do not have to be all or nothing. A subsection of the lot, a front or backyard, or a contiguous area, may be selected for water conservation efforts. The Colorado case still has a manageable lawn area in the back for recreation, kids, dogs, and the like. Even within a yard area, select plantings may require more irrigated water (e.g., fruit trees in an otherwise low-water landscape). The latter can be water with drip and/or subsurface irrigation rather than broad-swath sprinkler systems. The upper canopy can help mitigate surface evaporation and provide a cooling effect around the building and its yard.

Xeriscaping does not have to entirely exclude turf areas, deciduous, shade, and fruit trees. The key is avoiding large tracts of water-thirsty turfgrass that add little except maintenance, water use, and the visuals of a lush green lawn. The fully established shade trees shown in the New Mexico yard do not require water beyond the incidental watering of the perennial beds.

FIG. 11.9

Two arid western US xeriscaped residential lots located in (A) the northern Front Range of Colorado and (B) central New Mexico, both about 1 mile above sea level. Outdoor design and landscaping can be used to create "outdoor rooms," spaces that can be used much of the year with little overhead cost.

11.13 Landscape maintenance and equipment use

A low-maintenance landscape is important too, as this will affect both energy use and ongoing maintenance costs. Wet periods with ample sunshine and less than scorching summer temperatures can often lead to more than a weekly mowing of a vast expanse of lawn with cool-season grasses such as bluegrass.

Emissions for gas-powered landscape tools are nonnegligible as they often have little in the way of emissions controls, and smaller machines are often powered by two-stroke engines. Most landscape engines, though small, use far fewer emission controls than passenger automobiles and thus have high emission factors (quantity of pollutant per horsepower per usage time). Small engine landscape equipment is often considered in the air quality management plans of urban areas, particularly those not in compliance with particulate matter, nitrogen dioxide, carbon monoxide, and ozone standards. States and municipalities are increasingly banning the sale of new gasoline-powered lawn mowers and other equipment with these issues in mind.

Increasingly, many tools now offer electric and rechargeable versions that are becoming more cost-effective than their gasoline counterparts. The advancement of battery technology has increased the run-time and decreased the costs of these tools. The current range of tools includes almost all major outdoor and power tools, including lawn mowers, leaf collectors/mulchers, tillers, chainsaws, and chipper-shredders. When combined with renewable energy production, they offer negligible emission options.

11.14 Green roofs

Green roofs are rooftop living features including vegetated roof coverings and rooftop garden spaces and were discussed briefly in relation to Fig. 11.4. They can help reduce runoff from impervious roof surfaces as well as reduce thermal losses and unwanted solar heat gain at the top of the building. Green roofs serve to reduce urban heat island effects as well. Though they can be done on residential dwellings, they are more common on commercial-scale buildings, as shown by the whimsical example of the California Academy of Sciences in Golden Gate Park in San Francisco, California (Fig. 11.10). The designs must make careful consideration of roof loads including water and snow weight and appropriate drainage and underlying membrane designs to prevent water infiltration into buildings.

The concept of a green roof has various levels of intensity of the biospheric component of the green roof ranging from a depth of a few centimeters of soil media plus succulent plants to large-scale plantings and multilayered media resulting in flora not unlike that in on-grade natural soils (Braham and Casillas, 2021). These in turn have varying runoff coefficients and associated building structural requirements for viability. Such details are critical for the effective, leak-free functioning of green roofs.

FIG. 11.10

The green roof at the California Academy of Sciences in San Francisco, California, comprised of many native species arranged in a whimsical though functional green roof.

11.15 Chapter summary and conclusions

Site characteristics and landscape design are commensurate with building design in sustainable construction. Though not often the focus of engineers, landscapes are key to materials, water, and energy inputs and outputs to the holistic site. The outdoors can provide "new rooms" that once built require little to no operational energy and can require minimal maintenance as well. Whereas the recent decades have shown increasing building size, the lot size, particularly in urban areas, has been shrinking on average since ~1990 owing to increasing urbanization as well as increasing land prices. Though the structure has been increasing its share of a given site, the landscaped area still typically dominates all but the most urban residential lots.

Integrating the structure and site is a classic technique that was central to the architecture of Frank Lloyd Wright in the early 20th century. Site planning and maintenance of existing topography, floodplains, aquatic features, and plant and habitat preservation are all important aspects. Balancing the movement of soil removed with those needed for fill at the site is beneficial.

Water use in buildings is clearly tied to landscape watering needs (and often dominates use for a residential lot). Water availability plays a critical role in sustainable landscape choices. Plantings are able to, at some level, exclude or admit sun, wind, and rain with placement depending on the needs of the particular site. Plant selection appropriate to the location and climate is critical. Thus, the first rule of site plantings is to work with your existing site and climate as much as possible. Second, designing a landscape around lower water use is central to the concept of xeriscaping. Xeriscape does not equate to "zero-scape" as landscape experts point out.

Although the climate can be challenging, as can the poor soils, the western US dryland plant selection is as diverse, attractive, and better adapted in terms of drought and fire risk, and is more in context than the traditionally popular ornamentals of wetter regions. Landscape planning can help delineate zones of plantings arranged by water needs combined with an efficient irrigation scheme. Rather than a monoculture of turfgrass, native and adapted species are more viable with fewer outside inputs of energy and resources. Other more adapted and lower water-need grass varieties are also available including grama and buffalo grasses. An appropriate landscape should be designed for minimization of maintenance as well.

Food production can also be appropriately integrated into the site landscape for added benefits. Mulches that help conserve soil moisture and reduce weed growth include wood shreds and chips, shredded leaves, shredded rubber, seeds and fruit pits, and inorganic covers such as gravel. Approaching a closed loop cycling of yard waste organic materials is particularly important in western landscapes, where soils are much leaner in organic matter than eastern soils.

11.16 End of chapter exercises

(1) **Concepts:** List six specific methods to reduce landscape water usage. Which of these do you think would be most effective and why?

(2) **Concepts:** Research the most viable plant choices for your local climate and soils. Come up with a list, pictures, and one-sentence description of two examples of (a) trees, (b) shrubs, (c) perennials, and (d) fruit-producing plants well-adapted to your local climate and soils.

(3) **Concepts:** Research the most prevalent invasive species in your region. What are they, their origin, their negative effects, and the control methods?

(4) **Problem:** Estimate from the data on the carbon footprint of the annual water consumption of the property illustrated in Fig. 11.2. Approximate the annual water consumption from the data. The CO_2 emissions factor for municipal treated water is $0.45\,kg/m^3$.

(5) **Problem:** Texas & Louisiana were inundated by extreme, prolonged rainfall from Hurricane Harvey.

(a) To estimate the rainwater volume dropped, consider a semicircle area that is 250 miles in radius that receives on average 20" of total rainfall from the storm. How many gallons of water did Harvey drop on this region? A square mile is 640 acres.

(b) Part of the flooding problem is due to the significant increase in impervious surfaces in metro Houston. Assume Houston's greater metro area occupies this entire semicircle. Its runoff coefficient changed from an average undeveloped site to an average residential single-family home region. How much more stormwater had to be managed during the storm due to this change?

(6) **Problem:** Estimate the energy released from Hurricane Harvey above in joules. Think about latent heat. If this energy release occurred over 4 days, what is the power in Watts for this storm?

(7) **Problem:** Walk-behind or push lawnmowers usually have a fuel tank $\sim 2 \pm 1$ qt. Generally, a push mower does about half an acre per tank according to gardentoolexpert.com. This will take

about 45 min. Likewise, a popular 56VDC battery-powered electric mower has about a 45-min runtime on a 5 A-h lithium battery. The main cost to operate a push mower is fuel (ignore other maintenance). Take an average of $2.80 per gallon for gas or 13 cents per kilowatt hour for electricity, to charge the battery for the electric mower. Assume the charging is 80% efficient. Estimate the mass of fuel, the cost of the primary energy use, and the CO_2 emissions to use these two options to mow 0.5 acres. Assume pure carbon for coal and 30 MJ/kg to generate electricity at 35% efficiency. Comment on the results.

(8) **Problem**: Find the heating degree days (HDD) and CDD (in °F) for your own location or use the following New Mexico locations: Socorro (KONM), Albuquerque (KABQ), Taos (KSKX), and Las Cruces (KLRU). Assume a baseline of 65°F. Plot the Socorro HDD and CDD as a function of day during 2017. Discuss the HDD and CDD from these sites in a paragraph. Any surprises?

(9) **Problem**: In the United States, turfgrass covers approximately 50 million acres of land and accounts annually for 3 trillion gallons of water, 70 million pounds of pesticide and herbicide use, and 200 million gallons of gasoline for mowers, according to the NRDC.
   (a) Estimate the water use in million gallons per day. Assume a 12-week growing season with an average of 1" water/week of irrigation water.
   (b) Municipal water treatment also uses 520 million MWh per year. Estimate 0.818 lb CO_2 per kWh in the United States. Find the carbon footprint of water treatment (total CO_2 emissions and per gallon of water produced considering the electric use and gas mower use).

(10) **Problem**: Use the following data: A 100-ft tree, 18 in. in diameter at its base, produces a mature leafy tree that produces 6000 pounds of oxygen. How many human respiration needs would that supply, making reasonable respiration assumptions?

(11) **Problem**: Take the average US home size of 2500 ft² and assume a box 50 ft on a side. Estimate overhang depth of 18". Calculate the volume of water collected in an average environment that receives 30 in. of precipitation annually. What is the mass of this (in kg)?

(12) **Problem**: Kentucky bluegrass requires roughly 40 in. of water per year to thrive. How much is needed to irrigate a 0.3-acre lot that has a 2500 sqft building but is otherwise covered with bluegrass? The area is in an arid region that receives 9 in. per year of precipitation.

(13) **Problem**: Besides providing shade, which has an obvious cooling effect, trees cause a cooling effect via evapotranspiration. A single tree can lead to 40 gal of liquid water being vaporized into the atmosphere. Take your 400,000 lb average house and calculate how much this will lower its temperature.
   (a) Assume only the house is cooled and it has a heat capacity of ½ that of liquid water.
   (b) Assume it cools only the air in a 2000 ft² house with 10 ft ceilings using the C_v for air of 0.7 J/kg/°C.
   (c) Which seems more reasonable?

(14) **Problem**: Revisit the last problem and calculate the effective cooling effect of the tree. It has been estimated that the average tree provides a cooling effect comparable to a several kW air conditioning unit. Confirm or deny (in rough numbers).

CHAPTER 12

Healthy indoor spaces: Maintaining indoor environmental quality

Learning objectives

(1) Recognize that sustainable buildings prioritize healthy indoor environments for occupants.
(2) Define a healthy and productive indoor environment including human comfort, thermal properties, humidity, air quality, light, and sound levels.
(3) Analyze approaches to minimizing indoor air quality and other environmental problems.

Early emphasis on green building programs focused on building energy use. A sustainable building does not make its occupants uncomfortable, unhealthy, and unproductive. Indoor environmental quality has received renewed attention by green building efforts including rating programs such as LEED and WELL-AP.

12.1 Human comfort and hygro-thermal properties

Maintaining human comfort is an obvious goal for a building structure. After all, we have constructed buildings to comfortably shield humans from the elements and the extremes of climate. Moreover, in the workplace, research has shown that discomfort leads to decreased productivity and lower job satisfaction and thus has economic value as well (Rupp et al., 2015).

Although humans have a great capacity for adaptation, they will be most productive, satisfied, and ultimately healthy when these parameters are within an acceptable though subjective range of comfort. Ergonomics and interior layout in the workplace or home is another vital contributor and beyond the scope of coverage here.

Human comfort is a science that examines what factors lead to the healthiest, most satisfied, and most productive humans (Rupp et al., 2015). Comfort expectations are somewhat malleable in that they depend upon factors that are both physiological and psychological. Behavioral changes play a role as well, and arguably we are accustomed to a much narrower range of comfort than our predecessors. This is a function of spending much more time indoors and growing expectations of a tightly controlled indoor environment. For example, a naturally ventilated building will have a wider range of comfort acceptability than a closed, air-conditioned building where control is expected (Brager and de Dear, 1998).

Even the definition of comfort is deceptively nebulous, and through long discussions and preference research, it has been defined in ASHRAE Standard 55 Thermal Environmental Conditions for Human

Occupancy. ASHRAE 55 incorporates the following parameters in describing hygro-thermal comfort: dry bulb and dew point temperatures, mean radiant temperature as discussed later, vertical temperature gradient, and wind speed. Two other important human factors are activity level and clothing properties. Several additional comfort considerations are temperature oscillations, temperature swings, and spatial gradients in temperature (ASHRAE, 2023).

In the grand scheme, the energy balance of the human body is the key concern. If the human is in equilibrium with the heat lost to the environment, balancing the heat generated from metabolic activities, they are more likely to be comfortable (Moss, 2007). An imbalance indicates that the person is losing more thermal energy and likely feeling cold or gaining more thermal energy and feeling hot.

Arguably, the most important parameter is room temperature (or dry bulb temperature), but other factors including air velocity, radiation intensity, sound and light levels, and indoor air quality all weigh heavily (Kruger and Seville, 2012). Also vital is the wet bulb or dew point temperature, which, combined with the dry bulb temperature, determines relative humidity. Light intensity and spectrum, noise and vibration, and heat transfer also contribute to comfort levels. The airiness, openness, and appropriately scaled spaces also lead to comfort and well-being. A human working while cramped into a closet or alone in an empty warehouse will not likely experience high job satisfaction.

Data show that the hygro-thermal conditions are typically top contributors to humans' overall comfort (Frontczak and Wargocki, 2011). Room air temperature vs the comfort range of the individual is typically the key consideration. A reasonable range of comfort is given as 18–25°C (65–77°F), though it is dependent on the individual. Human activity level will impact whether the room feels comfortable; the former is more comfortable for someone reading a book, and the latter is more comfortable for someone under physical exertion. All the factors combine as an air temperature as high as 25°C (77°F) is reasonably comfortable if the relative humidity (RH) is low, the air movement is substantial (e.g., fan ventilation), the radiant intensity is moderate or less, and the mean radiant temperature of the surroundings is lower than this. Likewise, an air temperature of 18°C (65°F) is comfortable with bright sunlight, little air movement reducing convective heat loss, and low RH.

Also vital to human comfort is the dew point temperature. The phrase "it feels close" denotes a dank, humid feeling and is when the dew point and dry bulb temperatures are close. Dry conditions with low relative humidity (RH) are when the wet bulb temperature is significantly below the dry bulb temperature, and saturated conditions are when they are equal. The mid-range relative humidity is considered the most comfortable, approximately 40% ± 10%. Too dry (dry and staticky) or too wet (damp and oppressive) conditions are both uncomfortable. At increasing temperatures, a lower RH mitigates the heat index and enhances comfort. The RH is determined by the dry bulb and dew point temperatures, as illustrated by the psychrometric chart in Fig. 12.1.

Humans' thermal comfort transcends the setting on the thermostat and is also a strong function of the insulation and air sealing of the building's thermal envelope. A surrounding thermal mass within the building envelope can be comfort-enhancing. It can moderate temperature swings in a structure which is particularly useful in regions that have large day-to-night temperature swings. The mean radiant temperature is the composite effective temperature of the solid surfaces in a building (walls, ceilings, floors, furnishings, windows, and doors) (Guo et al., 2020). This is the effective body to which the occupants are radiating energy to or from, seeking thermal equilibrium. Anyone who has sat next to an uninsulated wall or cold window in winter in an otherwise comfortable air-temperature room can attest to the surrounding mean radiant temperature. One can be in a 22°C room but will feel very chilly sitting close to a 10°C wall due to radiational losses. The proximity and area of windows are important to comfort as well due to radiative heating and cooling. Sitting near an unobstructed window area with sunny conditions enhances winter comfort. At night this will lower body temperature which may decrease comfort in winter. The diurnal comfort regime is the inverse in summer.

At a minimum, modern HVAC systems are designed to provide habitable conditions for the changing ambient environment. Modern controls such as proportional integral derivative control loops can control conditions well within the range of comfort, often within a couple of degrees and 5% or 10% in relative humidity of the set points (Example 12.1).

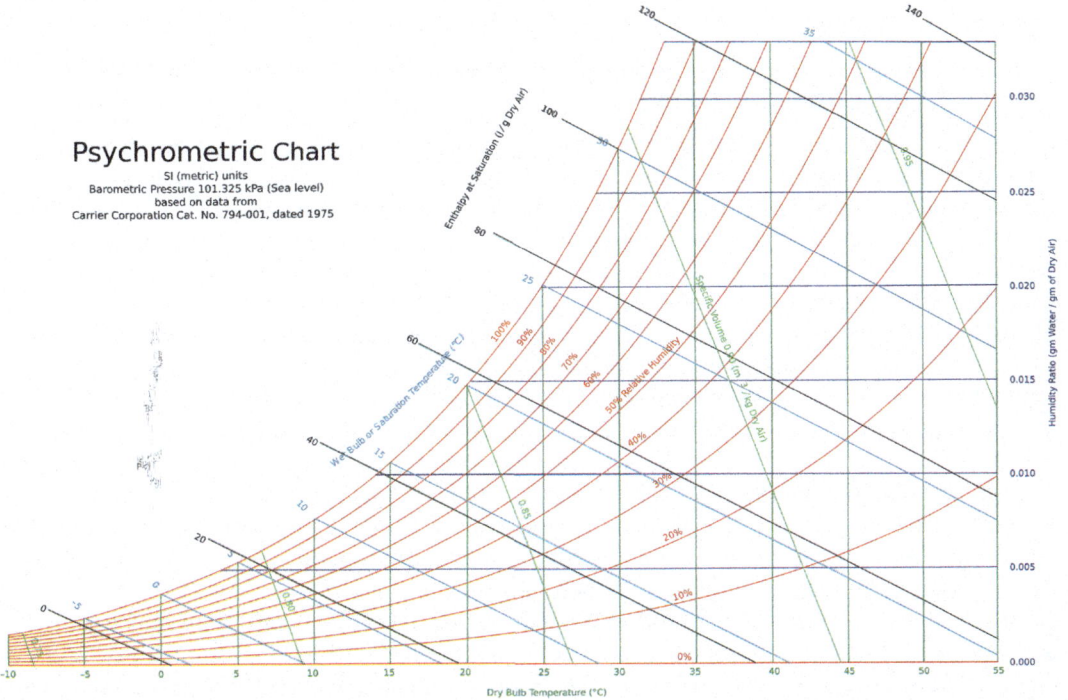

FIG. 12.1

Psychometric chart originally devised by the carrier corporation.

Image credit: ArthurOgawa, public domain via Wikimedia Commons.

EXAMPLE 12.1 Condensation

Concepts: Why does condensation form on the inside of a window in winter? What are possible solutions?

Solution: Condensation forms on the interior of a single-pane window in winter because the surface temperature is below the dew point for the conditioned air. Possible solutions are as follows:
- Raise the temperature of the interior surface of the window
- Replace the window sash with a double-pane unit
- Add a storm window and direct heated air to the window
- Increase the temperature inside the conditioned space
- Or reduce the relative humidity of the interior air.

12.2 Indoor environmental quality (IEQ)

Related, indoor environmental quality is a function of the building composition, systems, operations and maintenance, the activities in the building, and the outdoor environment (Kibert, 2016). One could have thermal comfort in a building with poor indoor environmental quality (e.g., high concentrations of $PM_{2.5}$, for example). Likewise, an open, unconditioned building in Duluth, MN may have great indoor air quality but would hardly be expected to be thermally comfortable in January. Including the importance of indoor air quality, indoor environmental quality has a wide range of factors discussed in the next sections.

12.3 Lighting characteristics

Indoor lighting levels are a key factor in ergonomics and occupant satisfaction. Depending on the activities, light levels that are too low or high can be detrimental. Appropriate light levels are dependent on the task and room characteristics. This does go well beyond just the intensity of light. Other factors related to the visual environment include window views, indoor visual clutter, glare, uniformity of illumination, and flicker of the light source. The quality of light in a building is important to occupant satisfaction and productivity. The color rendering index (CRI) is one measure of the light quality from artificial light sources. As it approaches 100, it is more like the solar spectrum that human eyesight has evolved to capture. The spectrum of solar radiation is also discussed with respect to climate in Chapter 4, window transmission characteristics in Chapter 8, and solar energy in Chapter 13.

The priority is to rely on natural daylighting as much as practical to moderate electric use and improve the spectrum of indoor lighting. Many approaches can be used including ample fenestration area, skylights, sun tunnels, light trays, a bright, reflective indoor color scheme, use of well-illuminated central atriums connecting interior spaces, and clerestory windows sometimes combined with "saw tooth" roof designs. The key goal is to promote diffuse natural light penetration deep into the building while mitigating glare from direct illumination (Kibert, 2016).

One negative regarding artificial lighting is the impact on the natural world from stray light largely from outdoor lighting of buildings, streets, and other sources. The Dark Skies initiative from the International Dark Skies Association is intended to minimize this impact which not only affects animal species expecting dark conditions at night but also astronomy. The effort is to minimize the adverse effects of light pollution, defined as "Any adverse effect or impact attributable to artificial light at night (https://www.darksky.org/about/)."

12.4 Indoor sound characteristics

Sound levels and characteristics are also important in maintaining productive workplaces. Noise pollution goes beyond just the decibel level and includes the frequency, intermittency, vibrations, and other characteristics of the noise source that determine its impacts. Noise pollution can be distracting and harmful in both the ambient and indoor environments.

Sound level is measured in pressure units (Pa) or more commonly as the sound pressure level (SPL) in the decibel scale due to the wide range of sound levels that humans experience, where SPL is the sound pressure level in decibels, P is the sound pressure (Pa), and P_o is the pressure at the threshold of human hearing (0.00002 Pa) (Eq. 12.1; Example 12.2).

$$\text{SPL(dB)} = 10\log\frac{P^2}{P_o^2} \tag{12.1}$$

> **EXAMPLE 12.2 Noise pollution**
> **Problem:** A mining operation is being monitored for noise levels and is limited to a boundary limit level of SPL = 55 dB. The current reading is 50 dB, after which a doubling of the pressure occurs. Do they violate the noise standard?
> **Given:** 50 dB currently, doubling of sound pressure
> **Find:** Standard violation (>55 dB)?
> **Assume:** No other noise contributions
> **Solution:**
>
> $$\text{SPL (in dB)} = 10\log\frac{p^2}{p_o^2}$$
>
> $$50\,\text{db} = 10\log\frac{p^2}{\left(2(10)^{-5}\text{Pa}\right)^2}$$
>
> $$p = \sqrt{\left(2(10)^{-5}\text{Pa}\right)^2 \times 10^{\frac{50\,\text{db}}{10}}} = 6.32(10)^{-3}\,\text{Pa}$$
>
> $$2p = 12.65(10)^{-3}\,\text{Pa}$$
>
> $$\text{SPL} = 10\log\frac{\left(1.265(10)^{-2}\text{Pa}\right)^2}{\left(2(10)^{-5}\text{Pa}\right)^2} = 56\,\text{dB}$$
>
> This violates the standard.

There is emerging evidence that chronic exposure to loud noise can also lead to human health impacts including cardiovascular effects. The mechanism is proposed to be the stress response of the body and its release of adrenaline, cortisol, and other stress-related chemicals. Generally, the higher the decibel level, the longer the exposure time, and the more intermittent the noise exposure (e.g., a train passing vs a constant highway rumble), the more it produces a human stress response.

Design choices can eliminate some of the problems associated with unwanted sound. Locating quiet office space next to the mechanical room often leads to problems. Also, this is one area where interior insulation can be beneficial in reducing sound transmission between rooms. Indoor sound levels can be reduced by using high STC-rated interior walls to reduce the sound levels transmitted. The STC rating is the approximate dB reduction through the material.

12.5 Indoor air quality

Ambient air quality and especially particulate matter concentrations are well known to impact human mortality and morbidity as related to cardio-pulmonary disease (Dockery et al., 1993; Liu et al., 2019; Pope et al., 2009; Samet et al., 2000). Humans, however, spend the vast majority of their time (>90%) indoors on average. Indoor air becomes contaminated with ambient air pollutants plus a myriad of indoor sources, most notably combustion sources, cooking, and cleaning processes (Farmer et al., 2019).

Air pollutants in the indoor environment have long been identified as health risks to occupants and a contributor in extreme cases to what has been termed "sick building syndrome" (Jones, 1999; Sundell, 2004). Indoor sources of air pollution include poorly vented/sealed combustion devices, cooking emissions, household chemicals both in their use and storage, biohazards such as mold and mildews, other allergens, environmental tobacco smoke (ETS), and soil release of pollutants including radon and its daughter products. Indoor air pollution also occurs from off-gassing from materials and cleaning products including volatile organic compounds (VOCs) and other semivolatiles (Nazaroff and Weschler, 2004).

Indoor air quality (IAQ) takes the starting point of the ambient (outdoor) air quality, time-lagged by a half hour or so, and then it is subject to additional emission sources inside of a building. Not surprisingly, many of the pollutants of concern are in common. Data are shown in Fig. 12.2, illustrating the effects of the indoor environment affected by poor ambient air quality due to biomass smoke from a nearby wildfire (Carrico et al., 2016). The indicator species for smoke impacts is ambient carbon monoxide which reaches several thousand parts per billion (ppb) vs typical background conditions ~200 ppb or less, as shown at the end of June 2012 (Fig. 12.2). The indoor $PM_{2.5}$ concentration as measured with a beta attenuation monitor (BAM) is also shown to frequently exceed $100 \mu g/m^3$. The daily profile (overnight peaks into the morning) was used to shift worker schedules later into the day at this facility. This is one of the emerging problems associated with increasing wildland fires linked to climate change and forest management discussed earlier, and that will also be discussed in Chapter 15, the book's final chapter on resiliency (UNEP, 2022).

Indoor pollutants include particulate matter (PM) and gas-phase pollutants, including the EPA criteria for ambient air pollutants: PM_{10} (particles <10 μm), $PM_{2.5}$ (particles <2.5 μm), nitrogen dioxide

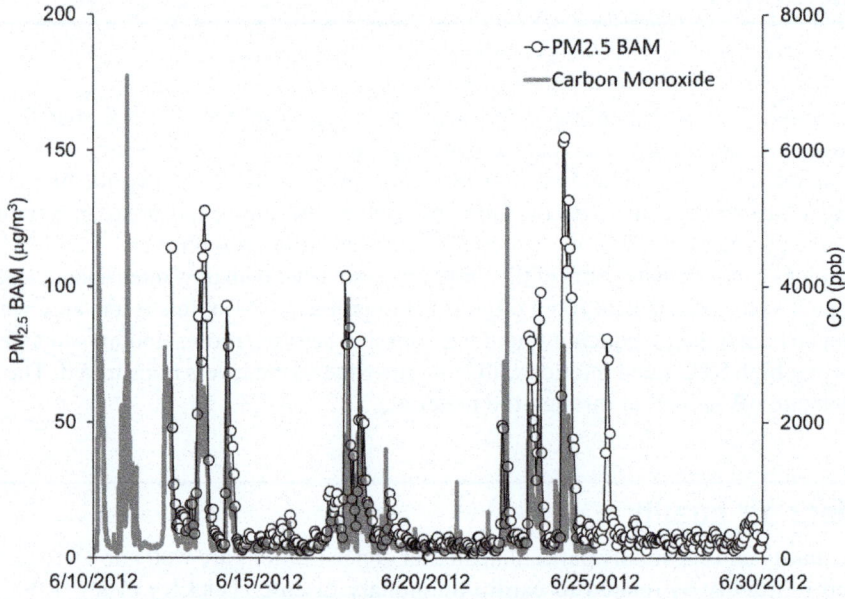

FIG. 12.2

Indoor concentration of PM2.5 and carbon monoxide in a commercial building in Fort Collins, Colorado, during a several-week period affected by the nearby High Park Fire in 2012 (Carrico et al., 2016).

(NO_2), sulfur dioxide (SO_2), ozone (O_3), and carbon monoxide (CO). Combustion sources are key sources of these pollutants. CO_2 only becomes a problem at ~1000 ppm or greater, and CO_2 can also be used as an "indicator" species for exposure to contagions and poor indoor air quality in general.

IAQ also includes semi- and volatile organic carbon compounds (VOCs) that off-gas from building materials or other sources such as solvents and cleaners. Due to their high vapor pressures, many are often associated with nuisance odors as well. They include several classes of organic compounds, including aldehydes (e.g., formaldehyde), aromatic ring-structured compounds (e.g., benzene), ketones (e.g., acetone), halogenated hydrocarbons (e.g., methylene chloride), aliphatic chain-structured compounds (e.g., ethane, butane, and propane), and alcohols (e.g., methanol and ethanol) (Kibert, 2016). Odor-producing compounds also include many sulfurous species (e.g., hydrogen sulfide, sulfur dioxide, mercaptans).

Cooking emissions include $PM_{2.5}$, nitrogen oxides (NO_x), sulfurous compounds, VOCs, and other emissions. The emissions of some of these pollutants are much greater with a gas or propane stove than with an electric unit. $PM_{2.5}$ emissions increase with higher temperature frying, particularly with food or oils that "smoke" when fried or grilled (Fig. 12.3). Most of this PM is submicrometer aerosols which are particularly respirable and able to penetrate deep into the lungs.

Traditionally, monitoring air quality has typically been limited to expensive ambient air quality measurement sites. Although these are still vital to diagnosing air quality events, smaller, less costly silicon-based sensor technology has enabled monitoring in more locations, including indoor environments. These sensors are improving in terms of sensitivity, accuracy, and specificity. The data shown in Fig. 12.3 used a PurpleAir indoor PM sensor (www.purpleair.com).

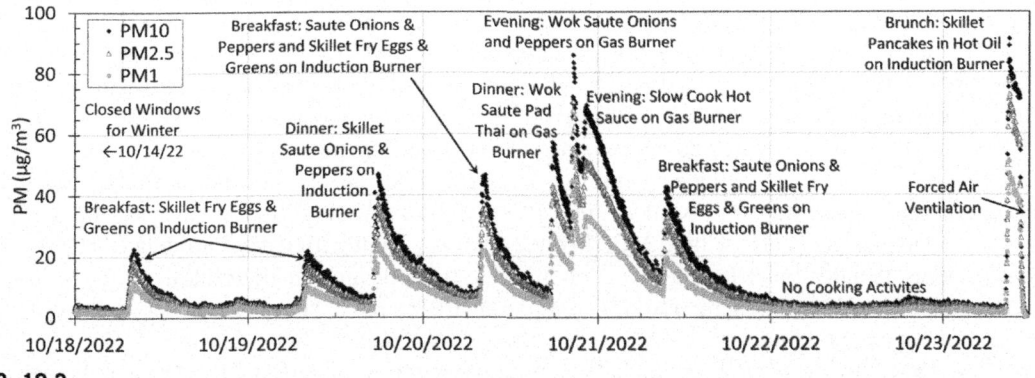

FIG. 12.3

Indoor particulate matter concentrations (PM10, PM2.5, and PM1) during a several-day period with a closed house showing the effects of cooking on indoor air quality (unpublished data from the author).

Extremely poor indoor air quality can lead to "sick building syndrome (SBS)," where multiple and sometimes the majority of occupants have multiple symptoms or a general malaise from spending significant time in such an environment. The exact IAQ cause(s) sometimes elude identification. A more limited occurrence is Building Related Illness (BRI) which affects a limited number of individuals with more consistent and traceable symptoms and causes (Kibert, 2016).

Based on mass balance, one can calculate a steady-state indoor concentration (C_{in}) of a pollutant given the volume of the building (V), the outdoor concentration (C_{out}), the air changes per hour (ACH), the source emission rate (S), and a first-order reaction rate constant (k) (Eq. 12.2).

$$\text{steady state } C_{in} = \frac{\text{ACH} \times C_{out} + S/V}{\text{ACH} + k} \tag{12.2}$$

The reaction rate constant k is a first-order reaction rate constant if there are other removal mechanisms (e.g., chemical reaction) besides outdoor air exchange.

As with ambient air quality, continued climate change will also affect indoor air quality; wildfire smoke, mold/mildew/allergens, urban heat islands, ozone formation, and off-gassing of semivolatile materials will all increase with higher temperatures. The interplay between a changing climate and air quality is an active area of research.

12.6 Mitigation approaches for indoor air pollution

The options for improving IAQ follow the general approaches to mitigating other environmental problems: reduce or eliminate the source, improve ventilation and exchange of outdoor air, alter exposure by changing schedules, improve air tightness (to reduce ambient air pollution infiltration and for moisture control), and install engineering controls such as filtration systems.

Indoor air quality concerns and stale air often raise the question of whether it is better to "let a house breathe" and keep ambient air exchange. Where would that unconditioned air leak in? Many times, the attic, foundation, crawlspace, and other unsavory places leak in ambient air that certainly does not enhance indoor air quality. The mantra "make it tight and ventilate right" is the better answer (Bailes, 2022).

The first line of defense is often maintaining an appropriate building air exchange rate in air changes per hour (ACH), where this flow of indoor air or fresh air is effectively cleaned via filtration or another method. However, if the outdoor air has elevated pollutant levels such as in an urban area or during a wildfire event, this can be detrimental, particularly if the HVAC system does not have adequate filtration of MERV12+. Minimum efficiency reporting value (MERV) ratings indicate filter efficiency. IAQ problems can also be mitigated to some degree with effective HVAC filtration systems; the best example is hospital systems that use high MERV-rated filters or High Efficiency Particulate (HEPA) filters. An energy penalty for bringing in unconditioned ambient air can be minimized using a heat recovery ventilator (HRV) or energy recovery ventilator (ERV) that exchanges heat between fresh outdoor air and indoor air exhaust.

Ways to remediate these problems include ventilation, engineered barriers, filtration, and other approaches. Filter systems can be in-room portable units or building-wide filtration associated with the HVAC system. The mitigation of indoor air quality problems is a multipronged effort focused on:

- Source replacement or elimination (e.g., swapping electric heat pumps for furnaces)
- Source control (i.e., reducing emissions by controlling and sealing sources)
- Introducing outdoor air, preferably heat exchanged and filtered through a high MERV filter
- Building-level media filter (e.g., high MERV filter to remove particulate matter and activated carbon to remove VOCs)
- Electronic air cleaning, such as UV-based disinfection systems for commercial/health care facilities

Several common particulate removal techniques are shared by ambient aerosol sampling and air pollution control. For example, inertial cyclones of different dimensional scales are used to remove large particles for both purposes. Particle removal mechanisms can be separated into dry and wet processes and are a distinct function of particle size (Cooper and Alley, 2011). The size range of the ambient aerosol may be considered from $D_P \sim 10$ nm to approximately 100 μm. Particles smaller than this are removed rapidly due to diffusion to surfaces, and particles larger are removed due to gravitational settling. The removal processes found in nature are similar to ones that have been employed in systems engineered to remove particulate matter from a gas stream. All removal processes rely upon settling, inertial impaction or interception, wet deposition, or diffusional losses. The processes may be enhanced by using electrical collection means such as in an electrostatic precipitator or by collecting particles in humid environments by collisions with droplets.

The accumulation mode contains the particles with diameters in the range $0.1 \, \mu m < D_p < 1 \, \mu m$. This population of particles is the longest-lived in the atmosphere and the indoor environment and is usually the most difficult to remove via filtration or other removal processes. Ultrafine particles are removed by surfaces due to their high diffusivity, while coarse particles are removed by gravitational settling. Accumulation mode particles penetrate deep into the lungs and are among the sizes that can carry contagious microbes in infectious quantities and thus are a hazard for respirable disease transmission (Seinfeld and Pandis, 1998).

The efficiency of a particle collection device (η) is often defined as the mass fraction of the particles removed from the gas stream (sometimes in total or sometimes as a function of size). The penetration is the fraction that passes through the device ($Pt = 1-\eta$). Less expensive, lower maintenance devices such as settling chambers and cyclones remove coarse particles with high efficiency but generally have low removal efficiency for fine-mode particles. More expensive pollution control devices such as electrostatic precipitators and filtration systems have efficiencies that can exceed 99% down to the submicrometer range. As a result, many times a coarse control device such as a cyclone will precede a more efficient device such as an ESP or filtration in an industrial setting. This also occurs with indoor devices that may have a coarse fiber filter upstream of a high-efficiency particulate (HEPA) filter. This approach conserves the expensive HEPA filter by removing the bulk of the mass contained in large particles first. A series of particle control devices have a combined Pt defined by the product given in Eq. (12.3). Optimizing particulate control involves assuring desired removal efficiency while minimizing pressure drops (pumping costs), capital costs, and operational logistical problems.

$$Pt_{\text{overall}} = \prod_{i=1}^{n} Pt_i \qquad (12.3)$$

The best systems for removing particles are most often filtration devices. These should be high MERV systems (12+) or HEPA systems. Mitigation of IAQ problems has been helped by new indoor air purification systems at the level of small room air cleaners to large building systems. The key is to find devices that rely on particle capture using HEPA filters and avoid ionization techniques that often produce significant ozone or other oxidants that negate the air cleaning efforts. The addition of an activated carbon fiber filter can remove the gas-phase pollutants that may be of concern such as VOCs and ozone. The room air devices are often specified as to their clean air delivery rates (CADR, typically in cubic feet per minute or CFM), which gives a metric of the flow rate that is effectively cleaned through the device. The rating multiplies the device flow rate by the particle mass removal efficiency for a given aerosol particle size distribution (subtracting out any removal that occurs through natural removal processes).

12.7 Radon and its mitigation

A specific concern with IAQ and particularly tight buildings is the infiltration and buildup of naturally occurring radon and its daughter products. Radon-222 is a gas, and its decay products are radioisotopes in the Uranium decay chain that includes alpha, beta, and gamma emitters. Radon migrates into a building through any passageways from the ground into the building. Radon tends to be higher in areas with high granite contents in the soils, such as the Rocky Mountains, as such areas generally have higher uranium in the soils. The EPA set a radon concentration remedial action level (4 pCi/L) above which they recommend a building owner mitigate the radon problem. Many municipalities require radon testing and remediation if levels are above the EPA action level.

The best remediation step is to eliminate the source by sealing off the connections between the soil and the building. A radon membrane sealed throughout the crawlspace with a radon evacuation system underneath is an effective solution. The removal system is typically a network of perforated pipes attached to a stack extending above the roofline. A small fan can add enough vacuum to remove the radon (and other soil gases) that are trapped under the membrane. As an example, with a house owned by the author on the Front Range of Colorado, radon measured before the mitigation system was 17 ± 4 pCi/L and after the mitigation system 1.4 ± 0.1 pCi/L. The mitigation system in this home reduced the concentration by over an order of magnitude. Sub-slab depressurization is another method that can help reduce indoor radon concentrations, especially when combined with effective sealing of the concrete slab and foundation. It creates a small vacuum under the concrete slab and exhausts this air through a stack before it can get into the house.

12.8 Biohazard concerns: Ventilation, air cleaning and disease transmission

Biohazards in the indoor environment include those on solid surfaces, waterborne, and airborne threats. Mold and mildew issues straddle all three of these phases as they form on solid surfaces, where sufficient water is available, and that can be transmitted through airborne pathways. Bioaerosols related to public health have recently been elevated as an indoor environmental quality concern. The intersection of public health, disease transmission, and particularly the transmission of COVID-19 concerns accelerated during the global pandemic beginning in 2020.

A key approach to minimizing many biohazards is avoiding dampness in building materials. Excluding water and allowing materials to dry when they become wet is the best defense against mold, mildew, and fungus growth inside buildings.

Beyond keeping things dry, biohazard mitigation is similar to minimizing the hazards of other indoor air quality problems or exposures, such as radiation exposure. The user can minimize the time spent exposed (spend more time outdoors instead of indoors), minimize the dose rate (test and isolate contagious individuals), enhance engineering controls such as HVAC ventilation rates and fresh air introductions, use engineered barriers as shielding such as filter masks, and other administrative controls such as quarantine rules where merited. Erring on the side of increased ventilation rates and greater ACH with the outdoor world will reduce indoor disease transmission risks. Research has shown

12.8 Biohazard concerns: Ventilation, air cleaning and disease transmission

the energy costs of filtration can be lower than ventilation depending on climate. The recommendation for mid-range indoor RH ~50% not only helps with comfort but also reduces viability of viruses in concentrated liquid droplets.

The indoor air quality devices that work well to reduce $PM_{2.5}$ also generally mitigate the risk of aerosol-transmitted biological hazards such as COVID-19. An indoor air filtration device is measured by its clean air delivery rate (CADR). A high CADR requires a high efficiency, MERV 12+, and a high throughput of air through the filter device. In terms of HVAC particulate filtration, beyond MERV 13, the costs go up and improved filtration falls off as per ASHRAE guidelines (ashrae.org/covid19). Greater MERV values show a flattening of the risk reduction and only increase costs due to increased pressure drops (Azimi and Stephens, 2013; Azimi et al., 2014).

Relying upon the gravitational settling of bioaerosols is not a great approach, except for avoiding the largest droplets produced by human coughing and sneezing. A plot showing the gravitational settling velocity in quiescent air of spherical particles as a function of particle diameter is shown in Fig. 12.4. The residence time varies over many orders of magnitude as a function of particle diameter. Using Stokes Law, the time to settle 2 m via gravitation is hours for submicrometer particles decreasing to minutes for 10 µm indicating the lingering nature of aerosols. Physical distance from the aerosol source can help reduce concentrations by gravitational settling by larger droplets and dilution for smaller aerosol particles. However, distancing in an indoor environment is not a bright line protection, as turbulent eddies can carry both droplets and aerosols >2 m, even up to 10 m or more depending on room air circulation (Bourouiba, 2020).

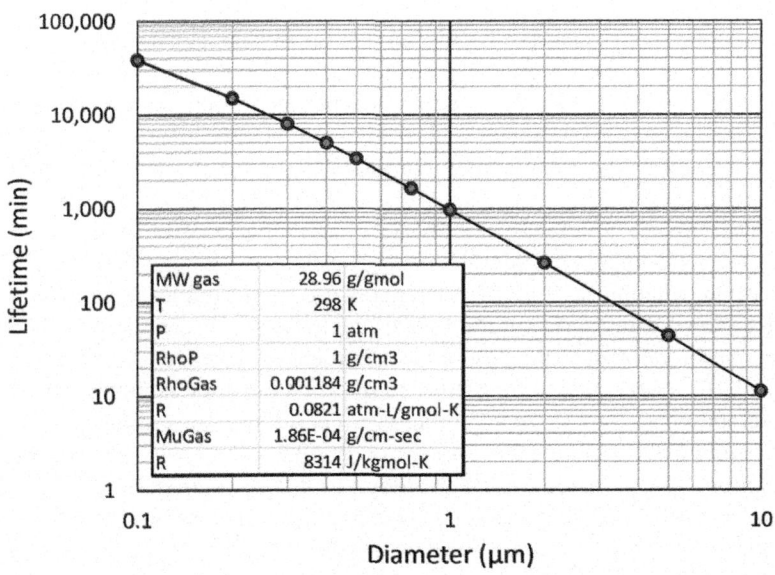

FIG. 12.4

Water droplet lifetime (2 m drop) via gravitational settling in quiescent air vs spherical particle diameter.

As a result of the global COVID-19 pandemic, the US federal government has launched a "Clean Air in Buildings Challenge." The focus is to assess the current status of buildings and upgrade the ventilation and air filtration systems to both reduce disease transmission potential as well as improve indoor air quality. The challenge is in concert with an interagency "best practices" guide published by EPA, CDC, and DOE. The effort made available COVID relief funding to help upgrade the ventilation and filtration systems in public buildings as well. The four concepts recommended are:

(1) "Create a clean indoor air action plan that assesses indoor air quality, plans for upgrades and improvements, and includes HVAC inspections and maintenance.
(2) Optimize fresh air ventilation by bringing in and circulating clean outdoor air indoors.
(3) Enhance air filtration and cleaning using the central HVAC system and in-room air cleaning devices.
(4) Engage the building community by communicating with building occupants to increase awareness, commitment, and participation."

The most effective approach for reducing the indoor exposure risk to COVID-19 and other respiratory diseases is an area of intensive research. The evidence is still emerging, but some lessons from the study of influenza transmission are applicable. Much of the viral load resides in aerosol particles with $D_p < 5\,\mu m$, thus indicating infection potential for minutes to hours after an infected individual is in a room (Milton et al., 2013). To date, the best approach appears to be multipronged, relying upon reduced occupancy, increased ventilation filtering and fresh air introduction, appropriate distancing and effective face masks, and screening and isolating of suspected infected individuals (Morawska et al., 2020).

The first and best step is to reduce or eliminate the source term by isolating and removing infectious carriers. ASHRAE gives the most effective approaches for reducing transmission of infectious diseases to include: dilution, airflow patterns, pressurization or depressurization, control of temperature and humidity, filtration, and disinfection strategies such as ultraviolet light exposure (ASHRAE, 2020). Other help comes from cleaning and disinfecting high-touch surfaces in buildings to prevent fomite transmission.

The recommendations for building operations for minimizing COVID-19 and other infectious disease transmission are somewhat in conflict with energy efficiency, though they dovetail with improving general promotion of indoor air quality. In general, the transmission of such diseases via HVAC systems through a known pathway is not the primary pathway for disease spread, as dilution, dispersion, and loss mechanisms all reduce the virus aerosol burden substantially as it passes through building HVAC (Schoen, 2020). A general goal for MERV13 filtration is a good target for maximizing removal while keeping pressure drop, system wear, and blower energy use constrained.

Fortunately, the optimal mid-range RH for comfort of 40%–60% is also the range that reduces microorganism viability as well. Increasing ACH from 2 to 6 is recommended in patient rooms where aerosol transmission of infectious diseases is considered a risk. The mechanisms are increased toxicity of salt droplets above the crystallization RH (vs some microbes being more stable in dry conditions), shorter settling lifetimes of droplets vs smaller dry particles, and improved mucus membrane response and virus clearing at intermediate RH (ASHRAE, 2020).

12.8 Biohazard concerns: Ventilation, air cleaning and disease transmission

As mentioned, ambient air quality becomes indoor air quality, with a minor time lag (<1 h). Penetration efficiencies for ambient aerosols in indoor environments are quite high (approaching 1) (Morawska et al., 2001), even with a tight building envelope. Proximity to a line source such as a freeway only increases the indoor particulate concentrations significantly with very close (15 m) proximity to these sources (Morawska et al., 1999). Indoor human activities, most prominently combustion activities, modify this background outdoor ambient air quality (Morawska et al., 2003). This is modified by any indoor sources of air pollutants or by removal mechanisms both natural and engineered (Hanninen et al., 2004). The influence of indoor relative humidity cannot be neglected as well in terms of occupant eye, nose, and throat irritations and thus comfort levels (Wolkoff and Kjaergaard, 2007).

Ventilation is critical to energy use, occupant comfort, and indoor air quality (Chenari et al., 2016; Dimitroulopoulou, 2012), and whereas stopping airflow and minimizing infiltration/exfiltration were a past focus, increased ventilation has been a more recent imperative (de Dear et al., 2013; Sundell et al., 2011). It has also been identified as the most important building factor though not unlimited in terms of biological transmission of disease (Li et al., 2007; Nardell et al., 1991). The latter is becoming an increasing focus in a postpandemic landscape.

The cessation of indoor public smoking in most jurisdictions has vastly improved indoor air quality around the world (Edwards et al., 2008). The working hypothesis is that healthy indoor environments provided by green buildings lead to more satisfied and productive workers (Rupp et al., 2015). Certainly, this has been found in some studies (Singh et al., 2010), although it is not a clearly demonstrable case universally (Altomonte and Schiavon, 2013; Paul and Taylor, 2008). Most likely this relates to attempting to link a highly subjective metric to one that has many separate metrics. Clearly, this is an area of further investigation and refinement (Al Horr et al., 2016).

Combustion associated with cooking activities is a continuing source of indoor contaminants, particularly high-temperature deep frying (See and Balasubramanian, 2008). The indoor air quality discussion would not be complete without discussing the use and effects of traditional cookstoves. Roughly half of the global population cooks with locally gathered biomass fuels on a fairly primitive cookstove typically located indoors, exacting terrible exposures to $PM_{2.5}$ and other air pollutants to the occupants (Petrokofsky et al., 2021). One area of intensive research is the effect of indoor cooking appliances and, in particular, that of biomass stoves used in many developing countries. These lead to exceptionally high indoor particulate matter concentrations, especially with solid fuel bio stoves (Chafe et al., 2014). Replacement and more efficient stoves can reduce these exposures significantly, particularly if switching to liquid or gaseous fuels (Grieshop et al., 2011); however, market penetration and adoption have been challenging (Lewis and Pattanayak, 2012).

Beyond air pollution, the lack of modern energy cooking systems also impacts the time use and forest sustainability in those regions where they are practiced. This has led to efforts for improved cookstoves that can both improve indoor air quality and, through improved efficiency, reduce fuel usage (Fig. 12.5). Though they are becoming more common, the adoption of these improved cookstoves has faced numerous barriers, from economic to logistical to social. Another viable alternative in many cases is to use a well-designed solar cookstove, which can serve many cooking needs with zero emissions and no fuel costs, as detailed on the Solar Cookers International website (https://www.solarcookers.org/). This fits with the impetus to eliminate indoor combustion sources with the multitude of liabilities associated with them.

FIG. 12.5

Improved design of a solid fuel cookstove under emissions testing. Indoor combustion products are a frequent source of indoor air contaminants. Cookstoves are relied upon by billions of people, particularly in the developing world.

12.9 Interior furnishings

The furnishings inside of a building are an entire other area related to green buildings that will only be touched upon here. Key concerns include the ergonomics of the workplace and the environmental impacts including off-gassing and the carbon footprint that accompanies the furnishings inside of a building. For example, new furnishings in particular will have VOC emissions in the indoor environment. Increasingly, certification programs such as Green Guard address the emissions associated with furniture and other materials used in the interior of the building.

Off-gassing of semivolatile species from materials, furnishings, coatings, adhesives, and other sources is an indoor air quality concern, as touched upon earlier. Many of the species of concern for off-gassing are semivolatile organic carbon compounds (SVOCs). These are hydrocarbon species that have high vapor pressures near ambient conditions and thus vaporize readily into gaseous form. Some of these compounds are known carcinogens or pose other health hazards (e.g., benzene and other aromatic ring-structured compounds, formaldehyde, and other hydrocarbons). Material choices are the key issue in minimizing off-gassing and exposure. Over the last 30 years, the VOC content of products, including furnishing, has been gradually and substantially reduced though hazards still remain.

Reusing office furniture is a robust industry that takes usable but used office furnishings for redeployment. This industry, though not new, has blossomed during and after the COVID pandemic and the adjustment of society to a greater work from home approach. Consulting firms such as Green Standards focus on the reuse and recycling of used office furniture. Numerous vendors are in this space, including Davies Office Furniture, Seatingmind, Reseat, Tenzin Norbu, Madison Seating, OHR Home Office Solutions, and numerous others. Many of these are in the New York area and offer used, refurbished office furniture that offers lower cost, less waste generated, and lower volatilized emissions in indoor environments. This business has picked up with the constant churn of startups in tech-dense areas such as Silicon Valley as well as the reshuffling that has happened with COVID and the greater prevalence of work from home.

12.10 Chapter summary and conclusions

A sustainable building is one that also prioritizes occupant health and well-being. A healthy environment is characterized by comfortable temperature and mid-range humidity, lack of indoor air contaminants, and appropriate noise and lighting characteristics.

Comfort is somewhat subjective to humans, but ultimately it is a function of air temperature, radiative intensity, air circulation, and the relative humidity of the surrounding air. An approximate comfortable temperature range is from the mid-60s to the high 70s°F with ~72°F as a midpoint comfortable temperature often used as the thermostat setting indoors. The relative humidity in the indoor space also plays a key comfort role. The optimal comfort range is in the middle ~40%, as too dry causes irritation and too wet causes clammy feeling.

The mean radiant temperature of the interior of the shell is affected by the thermal envelope, and, in turn, this affects human comfort. The mean radiant temperature characterizes the thermal envelope temperature surrounding the user and thus determines the radiation transfer between the occupant and the shell. Comfort also depends on the activity level of the occupant, as an active construction worker's comfort zone will be shifted to a lower indoor temperature range as compared with a sedentary office worker. Appropriate sound and light levels are also important, as is room ergonomics.

Comfort science has a psychological component, as humans are comfortable, to some degree, when the conditions are what is expected. Spending most of our time in a tightly controlled thermal environment raises that expectation. Comparatively, existing in an outdoor environment or relying on passive climate control engenders greater adaptability, opening things up to a wider "acceptable" range of conditions (Brager and de Dear, 1998).

Environmental hazards indoors include indoor air quality (IAQ) and relate to the (a) outdoor environment, (b) materials, (c) processes (e.g., combustion appliances), and the air exchange and cleaning devices inside the building. Generally, ambient air quality becomes indoor air quality, with a time lag of 0.5h, depending on how tight the building is. Important indoor pollutants include O_3, PM, NO_x, CO, and CO_2, VOCs, bioaerosols, disease vectors, allergens, radon, and other airborne contaminants. The VOC content of paints, coatings, and other materials has been greatly reduced over the past several decades, hence the lower odor associated with them. Evolving measurement techniques with IAQ sensor-based technology have enabled a much wider network of measurements both indoors and out, and much work is going into improving and validating these small, low-cost sensors.

General approaches include minimizing IAQ including COVID transmission indoors, which we all have been doing: (1) maximizing distance, (2) fresh air ventilation, (3) masking, preferably N95 fitting well (4) minimizing exposure time indoors, and (5) good hand hygiene and vaccination when available. These apply to other indoor air quality problems. Ventilation and filtration are often the best methods for reducing the risks of COVID-19 and indoor air quality problems in general. Air cleaning systems, particularly those associated with whole-building HVAC systems with MERV ratings up to ~13, are cost-effective for the reduction in disease transmission and removal of $PM_{2.5}$. The prioritization of occupant health and well-being will drive the industry toward continual improvements in the techniques for promoting IAQ and environmental health in general.

12.11 End of chapter exercises

(1) Problem: An air purifier claims to treat the air in a room up to 360 ft² floor space. Its maximum flow rate is listed as 80 m³/h. How many ACH are they using in their specification?

(2) Problem: Revisit the problem in Chapter 10 on condensation (Problem 10.2). Use the psychrometric chart figure in Fig. 12.1. Do you get the same result? Discuss.

(3) Problem: You are comparing the noise levels of 1-gal tank compressors to be used in your workplace. The pressure level is the key metric where doubling the pressure level is equivalent to a doubling of the intensity. Your choices are compressors that generate 56 dB, 69 dB, and a typical compressor that generates 90 dB. What are the differences in pressure (in Pa) between these?

(4) Problem: A new piece of equipment generates an SPL of $=100$ dB at 5-m from the source. A resident would be 500 m from the piece of equipment. Will this violate a noise standard of 55 dB? Sound intensity in watts/m² decreases with 1/distance². Take this as the reduction in sound pressure level as well.

(5) Problem: The indoor mass concentration of $PM_{2.5}$ is at 35 µg per actual cubic meter at $T=40°C$ and $P=0.83$ atm. You are collecting PM on a filter sample with a standard volumetric flow rate (STP is 1 atm, 25°C) of 16.7 standard liters per minute. You run a 24-h filter sample. How much mass do you collect on the filter?

(6) Problem: Estimate the total number of particles inhaled by a human over their lifetime at the ambient $PM_{2.5}$ annual standard. You can use a 300 nm diameter spherical particle as the characteristic particle size with a density of water.

(7) Problem: You are in a classroom with dimensions of 3 m × 5 m × 10 m. The outdoor $PM_{2.5}$ concentration is 20 µg/m³. There are smokers in the room with emissions of 12 mg PM/cigarette, with 10 smokers at 2 cigarettes/h. In this case, the removal rate (k) is 20% per hour due to the air filter in the room. Find an indoor steady-state concentration. The air changes per hour in the room are poor at 2/h.

(8) Problem: You are in an indoor environment. The building is rectangular and is 20 m W × 50 m L × 15 m T. You measure the flow rate of air out of the building as 10,000 CFM. There is a source of carbon monoxide that is emitting 1 g/min. There are no sinks for CO in the building. Find the steady-state concentration in micrograms per cubic meter in the building. You can assume outdoor concentration is negligible.

CHAPTER 13

Renewable energy systems: Advancing distributed energy production

Learning objectives

(1) Investigate the variety of renewable energy systems applications in buildings.
(2) Evaluate the advantages and constraints of renewable and energy storage systems.
(3) Recognize that the vast scale-up and materials needs for these systems are significant challenges.

13.1 Overview of renewable energy systems

First, one must define renewable energy: those energy sources with an inexhaustible source of energy or one that is rapidly replaced or renewed. It does not mean necessarily green or without environmental impacts although they are generally in the category of low impact. It also does not mean that they have zero carbon emissions, as most have at least a minimum embodied carbon emission.

One may wonder why we wait to put renewables into the mix until later in the book. Are not renewables poised to dominate the energy industry? Undoubtedly, renewable energy sources have and will make major inroads into our energy infrastructure with the urgency of our environmental problems. The vastly improving economics of what was once designated "alternative energy" sources all but guarantees accelerating deployment. More specifically, the steep and sustained reduction in costs for solar photovoltaic systems merits their particular emphasis. The US Energy Information Administration (EIA) projects that solar PV, which is now dominating new capacity additions in the United States, will become the largest renewable contributor before mid-century (Fig. 13.1). Moreover, solar PV is one of the distributed generation technologies most suitable for urban building-scale applications and quite amenable to retrofit applications.

As discussed in depth up to this point, our first imperative, however, is to reduce energy use and make our systems more efficient. This was described as a "negawatt" rather than a megawatt, as termed by efficiency pioneer Dr. Amory Lovins. A reduction in energy use is often the lowest-impact energy source. It is often been said in energy circles that "one must have their efficiency vegetables before their renewable dessert" or "reduce before you produce" (both unattributed quotes). Nonetheless, both sides of the equation must be approached: production and demand reduction. Though the falling prices of

314 Chapter 13 Renewable energy systems

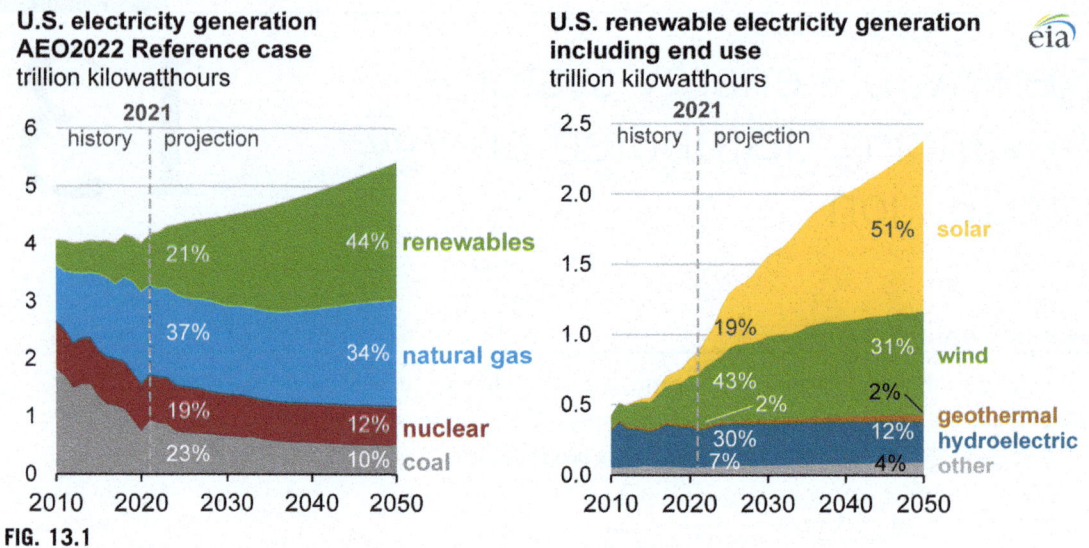

FIG. 13.1

US Energy Information Administration data showing current and predicted contributions to electricity production through 2050.

Public domain plot from US EIA.

renewable energy systems are challenging this notion, it still remains true that the lowest-impact kWh is the one that does not need to be produced. The other key point is that for renewables to continue scaling, building-integrated renewables is important despite utility-scale developments that are likely to dominate the market.

However, reduction of energy demand is only one side of the coin but not sufficient to address the liabilities of our current energy system. Societal energy needs, and in particular net-zero energy buildings, require not only outstanding efforts at efficiency but also renewable energy production systems. Such renewable energy systems include solar thermal and photovoltaics (PV), wind power, geothermal, microgrid applications, and microscale hydro.

Here we will examine the other most viable options, focusing on those that integrate within building systems. There are diversifying applications ranging from fuel cells, hydrogen, osmotic power generation, and many other small-scale energy systems that are potentially viable at an appropriate site. Although breakthroughs are possible, many of these will fulfill niche roles in the energy mix.

A single book chapter cannot adequately assess all the renewable energy systems under development. There are many textbooks, reference books, and other technical books that cover the renewable energy landscape (Andrews and Jelley, 2022; Dunlap, 2019; Gerring, 2023; Hinrichs et al., 2023; Kutscher et al., 2019). A big-picture summary of some of our major sources of electricity production and the pros and cons of those is given in Table 13.1.

Table 13.1 Comparison of attributes of major currently viable electricity sources.

Attribute	Renewables	Fossil fuels	Nuclear fission
Energy density	Low	Medium	Medium-high
Site specificity	Moderate	Moderate	Moderate
Fuel mass flow requirements	None (except biomass)	Very high	Low
Real-time emissions	Negligible (except biomass)	High	Negligible
Waste generation in manufacturing	Moderate	Moderate	Moderate
Waste generation in operation	Negligible	High (large volume, moderate toxicity)	Small volume, long lifetime, and toxicity
Climate impacts	Negligible	High	Negligible
Susceptibility to climate change (water availability, heat events, extreme weather)	Moderate	Moderate-high	Moderate-high
Dispatchability (ability to call up when needed)	Low	High	Medium-high
Centralized or distributed scale	Distributed (centralized utility-scale increasing)	Distributed to centralized	Centralized
Need for energy storage	High	Low	Low
Modularity	High	Medium	Low-medium (newer conceptual designs are smaller and more modular)
Cost characterization	High capital cost though declining	Moderate capital costs and high fuel costs	High capital cost with moderate operating costs

Renewable energy systems have progressed tremendously in production and cost efficiency over the last several decades (Fig. 13.2). Many published papers and books at great depth and breadth are devoted to all aspects of this maturing industry. The contribution of renewables to the energy grid has also grown at a rapid rate in the United States and globally (Fig. 13.3).

316   Chapter 13 Renewable energy systems

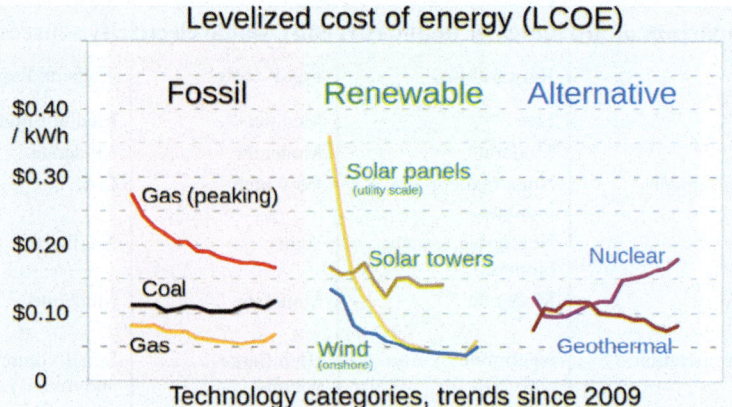

FIG. 13.2

Levelized cost of energy (LCOE) (2009–23) per kWh for major energy sources over time.

RCraig09, ShareAlike 4.0 International CC BY-SA 4.0 with data from the Lazard LCOE+ 2023 report.

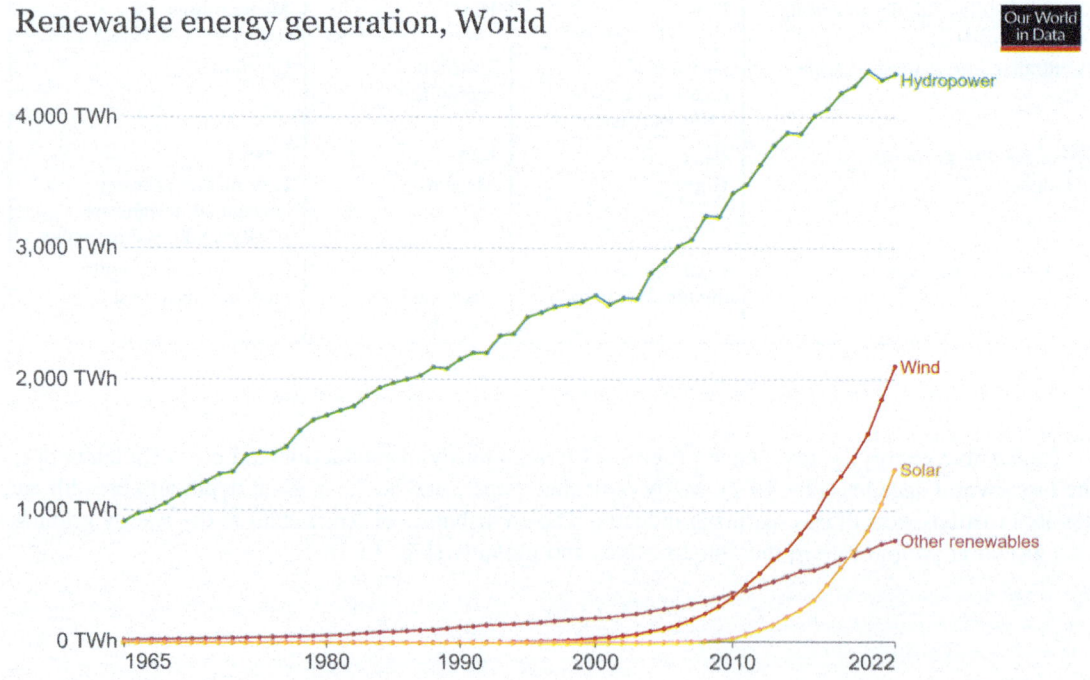

FIG. 13.3

Time history of global renewable energy generation since 1965.

Creative Commons, Our World in Data, data from the Statistical Review of World Energy, Energy Institute, 72nd edition, 2023, available at https://www.energyinst.org/statistical-review/resources-and-data-downloads.

13.2 Spectrum of solar energy and electromagnetic energy

Humans' use of solar energy in many forms—passive, solar daylighting, space and water heating, evaporative cooling, and electrical generation, among others—has been pursued for thousands of years, as detailed in the historical account "Let it Shine" (Perlin, 2013). In the United States, it dates back to the passive solar design of the Anasazi culture of the southwestern United States and their cliff dwellings and adobe structures discussed in the first chapter (Fig. 1.2). They were designed to admit the winter heat and light from the sun and exclude it in the summer. The history of solar power includes modern scientists/inventors such as Einstein and Edison and female scientists such as physicist Maria Telkes. It is worth understanding the fundamentals of solar physics to grasp the advantages and drawbacks of solar energy and other renewable sources, as many derive from the sun (wind power, biofuels, and hydropower, to name a few).

The sun is powerful, and the solar flux is meaningful, even at the distance that the Earth orbits it (Fig. 13.4). It emits radiation as a blackbody emitter with a surface temperature of ~5700 K. The flux at the top of the atmosphere (~1363 W/m^2) is moderated by atmospheric aerosols, clouds, and gas species that absorb and reflect radiation such that it is reduced by ~30% before reaching Earth's surface. Therefore, the peak surface radiative flux available is ~1000 W/m^2 on a clear day with the sun directly overhead, and this is what solar energy conversion systems have to work within the best case. Wien's Law gives the wavelength of the maximum emission (λ_{max}) from a blackbody emitter as a function of temperature (T in K) (Eq. 13.1).

$$\lambda_{max} = \frac{0.0029 \text{ nm K}}{T} \tag{13.1}$$

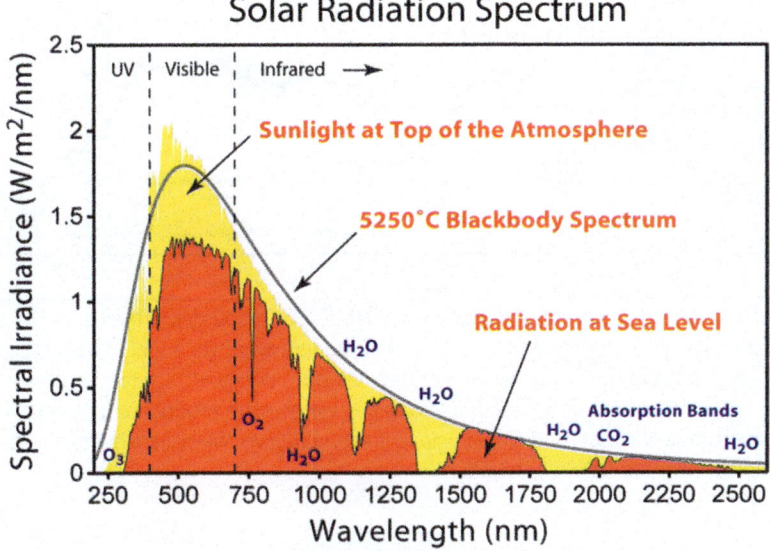

FIG. 13.4

The spectrum of solar irradiance as a function of the wavelength.

This figure was prepared by Robert A. Rohde as part of the Global Warming Art project under the Creative Commons Attribution-ShareAlike 3.0 Unported license.

The height of the sun defined by the solar zenith angle (angle down from directly overhead) can vary from 0 to 90 degrees. The result is that the direct normal solar irradiance will vary greatly depending on the time of day, latitude, altitude, and atmospheric attenuation properties (Fig. 13.5). A National Renewable Energy Laboratory (NREL) map showing the annual solar resources available as a function of location shows a variation of over a factor of 2 with the best locations in the southwestern United States. The importance of arid vs wet climate on solar viability is seen in the fact that locations in Wyoming are actually better sites for solar than Florida due to the prevalence of cloud cover in the latter (Fig. 13.5 and Example 13.1).

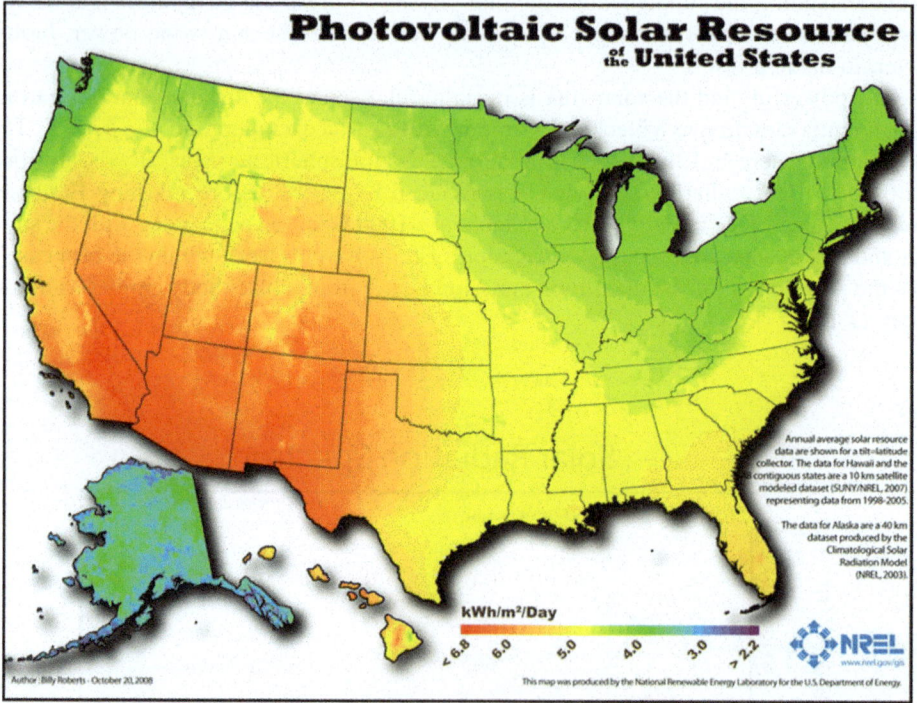

FIG. 13.5

US National Renewable Energy Laboratory map of the available solar resources in the United States expressed as kWh/m^2/day, assuming a collector facing due south with a module tilt angle equal to the latitude of the location (access data and references at the NREL website at https://www2.nrel.gov/gis/solar). The scale is 2.2 to >6.8 kWh/m^2/day. Note that states as far north as Wyoming exceed the production potential in South Florida, highlighting the importance of direct irradiance in cloud-free conditions. Map lines delineate study areas and do not necessarily depict accepted national boundaries.

Planck's Law expresses the distribution of emitted power density (B) as a function of wavelength and absolute temperature (λ and T) from a blackbody emitter, where h is Planck's constant, c is the speed of light, and k_b is the Boltzmann constant (Eq. 13.2). A plot derived from Planck's Law shows

13.2 Spectrum of solar energy and electromagnetic energy

the peak in solar radiation of ~0.5 μm and the peak in Earth's emitted radiation at ~10 μm (Fig. 13.6). This is of large consequence to the energy balance of a building and the entire Earth system!

$$B_{\lambda,T} = \frac{2hc^2}{\lambda^5} \frac{1}{\exp\left(\frac{hc}{\lambda k_B T}\right) + 1} \tag{13.2}$$

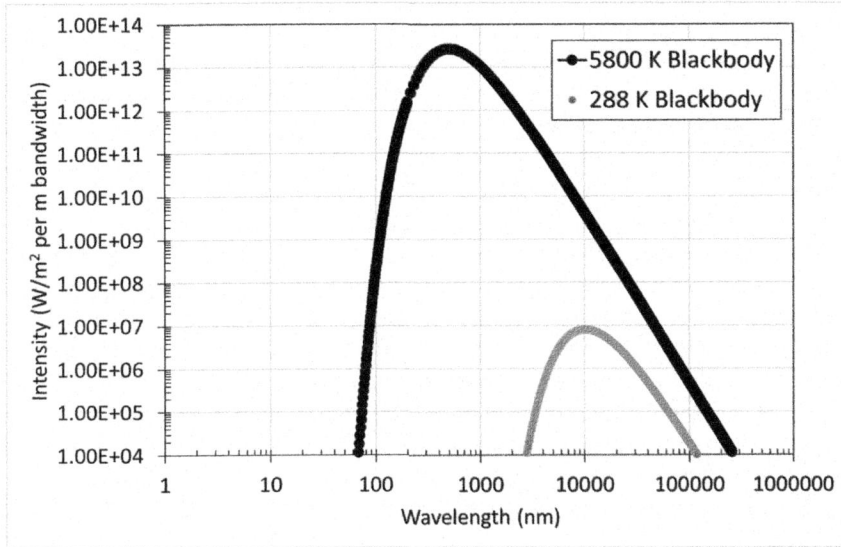

FIG. 13.6

Emission of radiation by a blackbody as a function of wavelength using Planck's Law.

EXAMPLE 13.1 Solar PV land use

Problem: SolarCity, Inc. uses a median estimate of 28 W/m² steady state for solar PV production around the clock and calendar. Estimate the land area needed for displacement of a large 1 GW$_e$ conventional power plant and compare it to that required by a coal-fired power plant. Ignore the need for energy storage, transmission, and source intermittency.

    Given: 28 W/m² PV production.
    Find: Land area for 1000 MW.
    Assume: Efficiency is incorporated in the flux and uniform flux and production over the land area.
    Solution:
    $P = 1000$ MW.
    Land area = 1000E6 W/(28 W/m²) = 35.7E6 m².
    This is equivalent to 3571 ha or 8821 acres (about 14 mile²).
    The Solar Energy Industries Association (SEIA) gives a comparable range for utility solar of 5 to 10 acres per MW of output, which brackets this estimate.
    The typical 1 GW$_e$ coal-fired power plant occupies approximately 1000 acres, not including the mining required upstream (for either case).
    As is well known, land use with solar and renewables in general is large though is not considered an ultimate limitation.

13.3 Behind the meter renewable systems

Distributed generation is one descriptor for energy resources beyond the traditional centralized power plant. Almost all renewable energy sources scale from kilowatt systems to utility scale. Here we outline the salient features of the most impactful renewable technologies focused on those viable onsite for commercial or residential buildings.

On-site grid-tied systems are often referred to as "behind the meter" renewable energy sources, as they are small scale at a given site though tied to the grid through the utility meter. Larger, utility-scale renewables are on the other side of the electric meter and include solar electric (both PV and thermal to electric), wind farms, and geothermal power stations such as the Geysers in California. In both cases (distributed vs centralized), technology, efficiency, costs, and environmental impacts are improving. On-site renewable systems have the benefit of providing energy near its end use, thus avoiding transmission losses.

The upfront costs were traditionally the rate-limiting step for scaling up on-site renewable energy systems. Compared to conventional resources, grid parity for renewables is rapidly occurring in many locations. Costs have plummeted over the past several decades, particularly for wind power and solar photovoltaics, making these alternative energy sources far more viable than in the past. Recent reports have suggested that for utility-scale developments, wind and solar are often the most cost-effective options in many markets. Though battery storage adds cost and complexity, its costs have been declining as scale-up and efficiencies of production for battery storage have accelerated. The tax incentives that are offered at the federal, state, and municipal levels are often key motivators to enabling these nascent markets. The DSIRE database (https://www.dsireusa.org/) is a useful site for state-by-state incentives, and the website maintained by Rewiring America also details the incentives available for efficiency and renewable energy systems (https://homes.rewiringamerica.org/).

Renewable systems are often easier to integrate and more cost-effective and efficient at the time of building construction. Thus, there are growing mandates specified by code (e.g., California) for adding renewables or at least making them feasible (e.g., "solar-ready" construction). Some municipalities stop short of mandating solar on new construction but will mandate that they be solar-ready homes if not fully solar. This means an appropriately sized breaker box and the wiring to add-on solar are preinstalled.

Renewable systems are a vital part of energy net-zero or carbon net-zero buildings. Net-zero energy buildings are those that produce as much energy onsite as is consumed by the operations of the building. One early example of a purpose-built zero-energy building is the Rocky Mountain Institute headquarters in Snowmass, Colorado. Carbon net-zero buildings sequester as much $CO_{2,e}$ as they emit over their lifetime.

13.4 Solar photovoltaic cells and PV systems

13.4.1 Solar cell physics and efficiency

The renewable technology of the most intense growth in the last decade and the greatest potential is solar photovoltaics (PV) for electricity production. Solar photovoltaics (PV) and associated battery storage devices are key devices reliant on advanced material science. The importance of devising lower-embodied carbon materials, improved solar photovoltaics, and improved battery chemistries (as well as novel catalysts for carbon capture and sequestration) are all reasons that material science is central to a more sustainable future (Chu et al., 2017).

PV cells are solid-state devices that directly convert sunlight into electricity, now typically at an efficiency of 20% or higher, by the photoelectric effect. Solar PV cells, depending upon the type, invariably use conductors including copper and other metals, semiconductor materials such as selenium, tellurium, gallium, and cadmium, and panel encapsulating materials including glass and aluminum. Much of the progress and cost reductions in solar PV have been tied to increasing efficiency with both energy and materials use (Gupta and Hall, 2012). The energy needs for production are substantial, and an energy input of 86 GJ/T is required to produce the metallurgical-grade silicon. Second-generation technologies such as thin-film varieties, which have lower efficiency and are less robust currently, are projected to improve the economics in part due to their lower materials and energy use. Third-generation PV includes a wide range of technologies, including nanomaterials, quantum technologies, multijunction solar cells, and others. These seek to minimize or eliminate the use of high-purity silicon and minimize production costs, all while maintaining or enhancing the efficiency found in polycrystalline silicon PV panels (Gupta and Hall, 2012).

The photoelectric effect was discovered by Becquerel in 1839, and the first (albeit ~1% efficient) PV cells were devised in 1888 by Fritz. The first useful application came out of Bell Laboratories in the 1940–50s, when silicon-based PV cells were developed primarily for space applications. Efficiency losses include internal resistance, the photons not being energetic to overcome the bandgaps, cell crystalline structure defects, and reflections off the glazing and each layer of the cell.

Solar cells use the photoelectric effect, where incoming photons of light excite electrons into the conduction band of the semiconductor material composing the PV cell. Semiconductor materials feature relatively small differences in the energy level between the valence orbitals and the next available orbitals (or conduction bands) of the semiconductor atoms. A photovoltaic cell relies upon the natural electric field setup at the junction of negative and positive or n- and p-type semiconductors. Due to the photoelectric effect, photons of sufficient energy transfer valence band electrons into the conduction band. The free electrons create a current when under a voltage potential in a complete circuit. The positively charged "hole" created by the photoelectric event migrates in the opposite direction.

Newer PV cells seek to expand the spectrum of light that they can convert to electricity. Strategic placement of reflecting and absorbing layers with different band gaps in the cross section of the cells can enhance capture efficiency. For example, cells use multiple layers, each tuned to absorbing photons of differing energies and thus collect and convert a greater range of wavelengths. Newer designs include concentrating solar cells, some of which use wavelength-shift materials that absorb one wavelength and reemit at another wavelength. An especially promising material is perovskite, which has shown greatly increased efficiency. Organic PV cells, which use layers of carbonaceous-based semiconductors much like organic LED screen technology, are another advancing front in PV cells. See Andrews and Jelley (2022) for a much more detailed review of PV technologies and advancements.

The Shockley-Queisser limit for single junction PV sets the upper limit of these single-cell crystals at 30%–40% dependent on the bandwidth gap of the PV semiconductor materials. The US NREL as of the summer of 2022 holds the record for triple-junction quantum well solar cell efficiency at 39.5% relying on bandgap optimization (France et al., 2022). Novel approaches using triple-junction cells tuned to different wavelength regimes have allowed efficiencies above this limit in research cells (Fig. 13.7).

The PV cells are built into modules or panels that are typically 300 W or thereabouts. These are then built into solar arrays. The other pieces required include an inverter for converting the DC power produced by PV into AC power used in buildings (Fig. 13.8). The PV systems can be stand-alone or off-grid but are more typically grid-tied, particularly in urban areas. Net metering allows users to be credited for electricity generated above the building's usage which is delivered to the grid with such systems. Building scale or rooftop solar is increasingly being mated with energy storage in terms of batteries. These used to be only with off-grid systems to assure electric availability during night and cloudy periods. With net metering and the need to extend the solar production day into the evening, battery storage is increasing rapidly at many scales of PV systems.

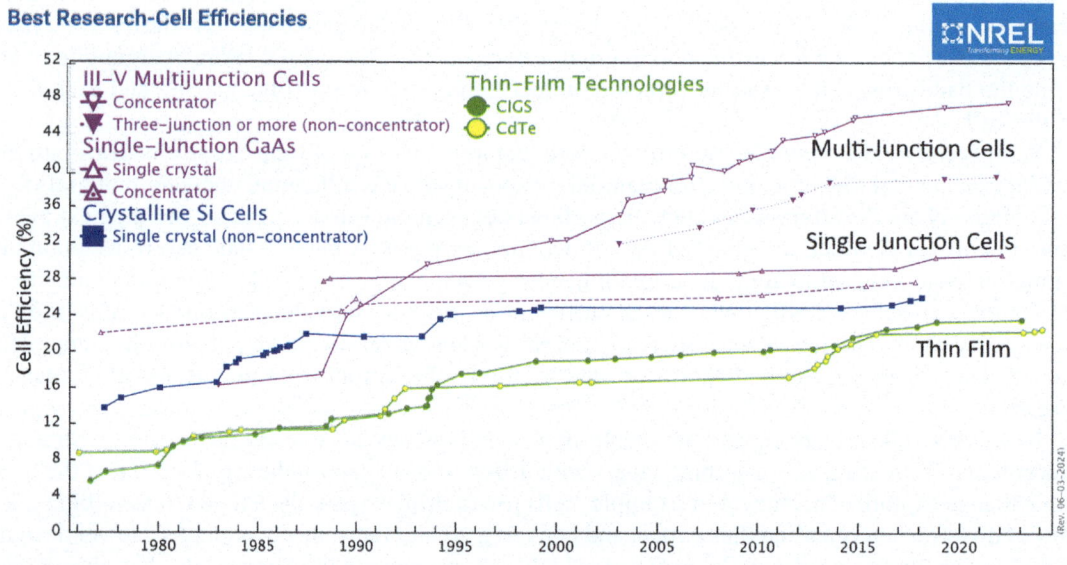

FIG. 13.7

Tracking over time of PV solar cell efficiency in the lab color-coded by the type of cell/module (US DOE National Renewable Energy Lab).

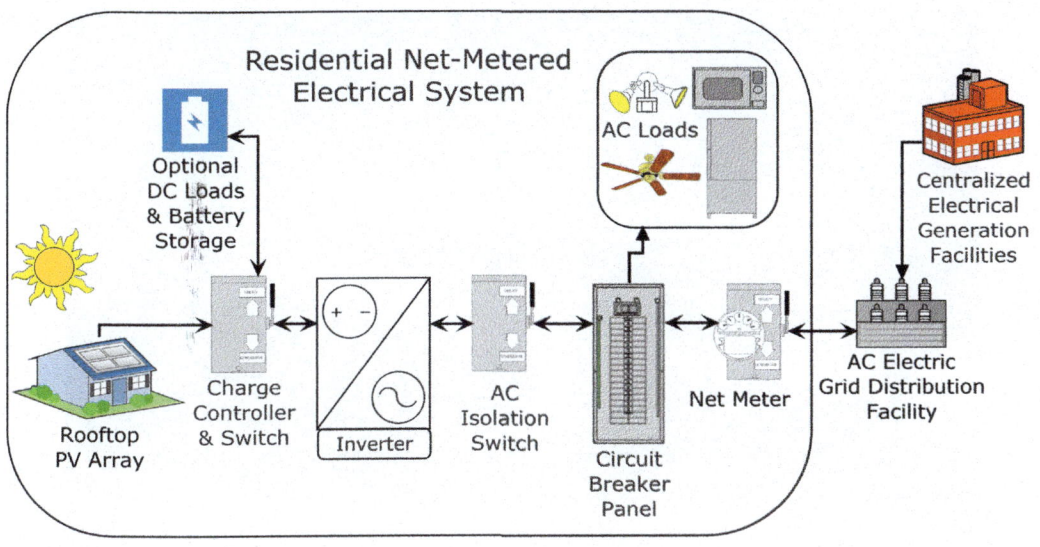

FIG. 13.8

Key components in a grid-tied solar photovoltaic system.

13.4.2 Electricity generation with solar PV panels

In the United States, the installed cumulative solar PV reached 110 GW AC capacity (at nameplate rating or maximum output) and represented 46% of grid additions in 2022, according to NREL. Solar PV has become cost-competitive with conventional sources in many markets in the United States ("grid parity"), even now including solar PV with battery backup systems in some cases.

Solar PV developments tend to be smaller scale than centralized power generators, as shown with a grid-tied system on a municipal building (Fig. 13.9). For example, in New Mexico (the desert southwest is among the best regions for solar resources), as of 2019, 69 grid-tied solar PV generators totaling 675 MW were in operation with individual nameplate capacities of <1 MW up to 70 MW, according to the US DOE. Solar PV is equally viable on much smaller scales. For example, community solar for low to moderate income can be viable.

The peak in solar PV production is close to solar noon (Fig. 13.10), whereas the peak in electrical demand is typically late afternoon. As discussed earlier (Fig. 9.19), the net demand profile or "duck curve" resembles the profile of a duck. Overall, the net capacity factor for solar PV is one of the lowest, at about 25% availability. It is not considered a very "dispatchable" power source as it must rely on the sun's output as a function of time of day, season, and location (variable though predictable) as well as local meteorological conditions (variable and less predictable). The importance of solar storage has become central to the continued scaling of PV contributions to the electrical grid.

The PV production curve is highly variable and strongly dependent on the local meteorology, time of day, season, and location, as previously seen in Fig. 13.10. As a result, solar PV has one of the lowest

324 Chapter 13 Renewable energy systems

FIG. 13.9

PV solar array designed into a parking shade structure at a municipal courtroom building in Socorro, New Mexico.

capacity factors at roughly 25%. The suitability is thus site-dependent, and often meteorological assessment of a site is prudent (Fig. 13.11), particularly for large-scale applications. Solar PVWatts and NREL Solar Advisor Model are two freely available, internet-accessible models for PV systems and general renewable applications, respectively. Helioscope by Aurora Solar is another more detailed model for designing PV systems and calculating payback periods. PVWatts will predict the month-by-month output for a given PV array, installed in a specified geometry, in a location using the nearest meteorological data for that site. An exercise using PVWatts is included at the end of the chapter.

13.4.3 Solar PV costs

The cost reductions in consumer PV panels have been both rapid and sustained for several decades. The PV panel costs dropped from $3/watt to $0.50/watt in less than a decade from 2008 to 2016, and over the longer term, the reduction has exceeded a factor of 50 since 1980 (Chu et al., 2017). A comparison of two residential systems shows the progress in both efficiency as well as cost-effectiveness of solar PV (Table 13.2 and Fig. 13.12).

One result is that the soft costs now roughly two-thirds of PV array installation costs for residential systems and one-third for utility-scale systems, according to the US DOE National Renewable Energy Laboratory (NREL) (Feldman et al., 2021). The balance of the system as well as "soft" costs of solar developments from building to utility scale are becoming more significant parts of the overall cost with the rapid decline of solar PV panel costs over the past 40 years.

13.4 Solar photovoltaic cells and PV systems

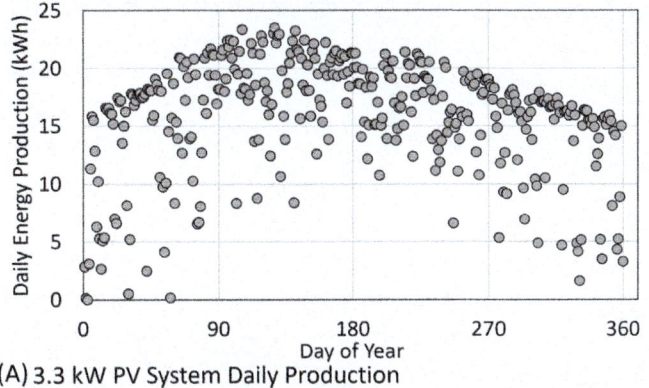
(A) 3.3 kW PV System Daily Production

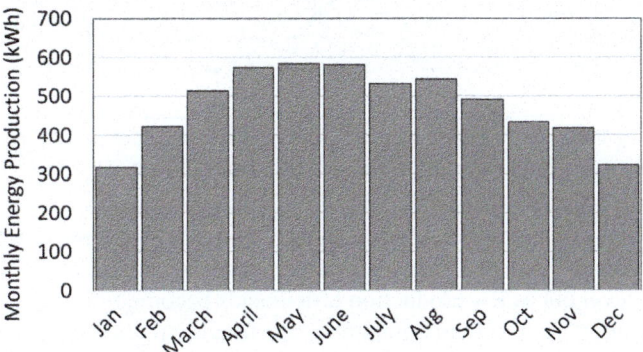
(B) 3.3 kW PV System Monthly Production

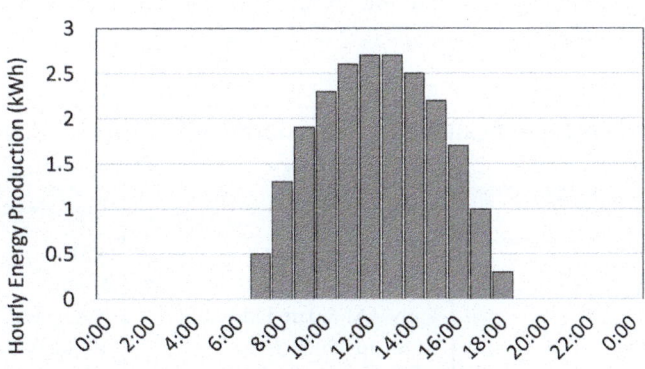
(C) 3.3 kW PV System Hourly Production (mid-May)

FIG. 13.10

Example online monitoring system for a residential PV system showing a typical (A) daily generation, (B) the monthly kWh production, and (C) hourly production of the system.

Chapter 13 Renewable energy systems

FIG. 13.11

Meteorological monitoring station for measuring the suitability of a site in California for PV development. Direct and diffuse radiation, precipitation, and temperature will all significantly impact PV production.

Increasing panel size with fewer cells and higher efficiency is one trend. Systems that allow more automated planning, permitting, and sizing systems are under development. At some point, those cost reductions will likely level out as PV production efficiencies become fully realized and as the balance of system costs (racking, wiring, mounts, inverters, electrical parts, and soft costs such as labor and permitting) becomes a larger fraction than panel costs. The materials and energy input for solar PV production are not inconsequential owing to the purity and clean manufacturing conditions associated with the semiconductor industry.

Table 13.2 Comparison of specifications for two residential PV systems installed 5 years apart by the author.

Parameter	Colorado PV system	New Mexico PV system
Array maximum capacity	2.4 kW	3.3 kW
Orientation	East	South
Shading	Early AM & late PM (20%)	Negligible
Panel details	14 Sharp 170 W 13% efficiency	10 SunPower 330 W 20% efficiency
Panel efficiency	13%	20%
Annual production	~2500 kWh	~6000 kWh
Full capital cost	$17,000	$17,000
Out-of-pocket capital cost after incentives	$8000	$7000
Retail electricity cost	$0.07/kWh	$0.13/kWh
Value of electricity per year	$200	$815
Estimated payback (simple)	40 years	9 years

13.4 Solar photovoltaic cells and PV systems

FIG. 13.12

Residential solar photovoltaic power systems that are compared in Table 13.2.

PV cost-effectiveness relates directly to the utility electric rate structure. Many utilities are required to "net meter" or give credit back to the consumer for excess generation, usually credited at the wholesale rate. Net metering allows the consumer to generate and use solar when it is available, put excess electricity onto the grid, and take power from the grid during overnight or cloudy times. Thus, the meter needs to be a smart meter or production plus a consumption meter needs to be paired together.

Another encouraging development in the last couple of decades is Building-Integrated Photovoltaics or BIPV. The first effort is the use of standard PV arrays on pergolas, carports, or other shade structures to provide a second purpose and utility of the land area underneath (Fig. 13.9). In a more integrated application, the PV cells are incorporated directly into the roofing shingles or even the windows of a building rather than installing separate panels on a roof.

Efforts are emerging to reduce the costs beyond the solar panels themselves including the increasingly important "soft costs" of solar PV. NREL offers a tool that local governmental jurisdictions can use for "automated" permitting of residential solar installations. The Solar Automated Permitting Process (SolarAPP (nrel.gov)) available from NREL at no cost speeds the local permitting process, one that can delay installations and overwhelm local permitting agencies in areas with large demand for residential rooftop solar. The SolarAPP can automate the majority of the process beyond the required on-site inspection, particularly aspects that invoke universally applied common safety and code conditions (Cook et al., 2024).

Though expensive at the outset, solar has the benefit of very low operational costs with no fuel cost, a plug-and-play nature not requiring much oversight, and very low maintenance and high reliability. Solutions to the high upfront capital cost of solar PV include such options as on-bill financing, community solar, solar leasing programs, and other creative ways to cover the system cost.

Dual-use solar is a step in the right direction to mitigate the significant land use requirements. With "agrivoltaics," a less panel-dense solar array mounted at a height that allows agricultural production underneath and between rows of panels. This can provide advantages to certain crops that thrive in less than direct sun as well as "shield" them from extreme temperatures while also reducing water losses.

13.5 Solar daylighting and passive solar space heating

Arguably the best solar energy applications are those that rely primarily on passive features. Two that are most common are passive solar space heating and solar daylighting. These have been used throughout human history, such as Mesa Verde cliff dwellings mentioned earlier (Fig. 1.2) that relied on the low winter sun to help heat the dwellings in winter while the cliffs above shadowed them during the high summer sun periods.

Solar daylighting involves using diffuse solar radiation for lighting interior spaces. Daylighting saves both electrical use for artificial lighting and provides a more pleasant natural solar spectrum that research shows promotes productivity, mood, and human health (Halliday, 2008). Diffuse natural light is almost always desirable in a building, as humans respond to the solar spectrum most favorably as opposed to more "artificial" light sources. Quality of light is important and can be quantified by correlated color temperature (CCT, from 2700 K warm white to 5000 K or greater approximating daylight), color rendering index (high CRI desirable), light stability, lack of flicker, light intensity, and the diffuse-to-direct intensity ratio.

The penetration of diffuse daylight into interior spaces can be facilitated by skylights, light trays, and other devices, some of which operate passively. Other design elements include central atriums, light shelves for blocking direct irradiance but reflecting diffuse radiation off the ceiling and into the interior (Fig. 13.13), clerestory windows with overhangs to block direct radiation, and light tunnels in interior spaces far from windows. Also shown in Fig. 13.13 is the Student Center at New Mexico Highlands University in Las Vegas, New Mexico. The LEED gold rating building uses daylighting with extensive windows, and the louvered system on the front of the building is used to control solar gain. Other features include ground-source geo heat exchange, a no-paint exterior, and a 96.8% construction waste recycling rate combined with 23% recycled materials in construction (https://www.nmhu.edu/student-union-achieves-leed-gold-for-green-building-practices/).

FIG. 13.13

(A) Light trays on south-facing windows at the New Mexico Bureau of Geology. The device helps block direct solar heat gain and glare in the summer while also reflecting indirect diffuse light up onto the ceilings and deeper into the building for daylighting. (B) The New Mexico Highlands University Student Center building was designed with an automated louvered system to control solar gain.

A novel application is a "virtual skylight" that connects a small solar panel to an LED light fixture and thus is a much simpler retrofit than a skylight or sun tunnel. These passive LED lights manufactured by Solaro, Inc. provide diffuse, natural spectrum light that turns on at sunrise and off at sunset and is particularly useful for interior rooms that lack natural light from windows (Fig. 13.14). It also mimics the spectrum and the circadian nature of sunlight by following its intensity.

FIG. 13.14

Passive solar light fixture in an interior bathroom. The fixture is interfaced to a small rooftop PV panel and turns on at sunrise and off at sunset. The adjacent ventilation fan is a Panasonic Inc. WhisperGreen model that is both energy-efficient and quiet.

Just as landscape watering running during a rainstorm is futile, the operation of artificial lighting with adequate natural lighting is similarly egregious. Integration of natural lighting and artificial lighting in an intelligent manner is important. Light level sensors can adjust artificial light levels depending on the solar daylighting available. There are numerous software tools to help design lighting systems for buildings. Among the common ones include Cymap, HyperLight, Virtual Lighting Designer, Radiance for daylighting, Lightpro, FlucsDL, and FlucsPro for lighting design (Halliday, 2008).

Depending on the location, building design, and seasonality, solar heat gain may be desirable or not. A building in Anchorage, Alaska, in winter should take advantage of passive solar heat gain as much as possible, while Phoenix, Arizona, in summer will be looking to avoid solar heat gain. Passive solar heating is primarily something that must be designed early on with respect to the building and site it occupies.

Other passive solar building designs ideally incorporate the following concepts:

- Orienting the long axis of the building on an east-west transect to allow significant southern exposure.

- In all but the hottest environments, a large window area on the sunny side (south side in the Northern Hemisphere) facilitates solar gain and daylighting.
- Additional skylights, sun tunnels, and other features allow admission of diffuse visible light for daylighting interior spaces.
- Windows and skylights are a careful balance between views, glare, ventilation, daylighting, and energy loss.
- Large overhangs on the south side to block high summer sun, less window area on east, north, and particularly the westside to minimize the solar gain in summer with the setting sun.
- Sloped roof area facing south for solar access for on-site solar photovoltaics.
- Promoting natural ventilation by excluding winter winds while enhancing summer evening cooling winds to provide a nighttime cooling effect.
- Unheated vestibules can buffer interior spaces from exterior conditions as well as noise.

The photovoltaic systems were previously discussed, but a cross-over passive stand-alone use of solar is self-operable and self-charging landscape or street lighting. A small solar cell or panel charges a rechargeable storage battery, and a brightness sensor turns on the light (typically LED) at night. These can be effective stand-alone low-intensity landscape lighting and are now commonly used even in street lighting.

A classic passive solar feature is a Trombe wall which is an absorbing wall behind a large south-facing glazed surface (Fig. 13.15). Such a wall is connected to the conditioned area by venting and through the exchange of heat directly via its thermal mass. Although these are not common in modern green buildings, they still maintain some currency in green building designs, particularly in sunny locations that require significant wintertime space heating.

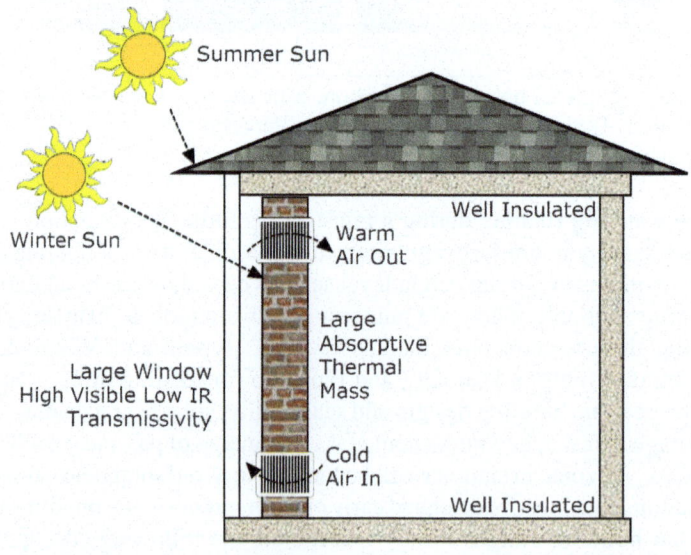

FIG. 13.15

A Trombe wall shows some of the key design features including a large thermal mass behind a large window designed for absorption and release of stored thermal energy.

13.6 Case study: CSU powerhouse research facility

The Colorado State University Powerhouse energy campus is a notable example of adaptive reuse (https://energy.colostate.edu/powerhouse/). It is a 100,000-ft^2 research facility that was LEED platinum designated in 2015. The former coal-fired power plant building was repurposed and added onto, forming the CSU Powerhouse research labs investigating solutions to our energy and climate crises as well as an incubator for private sector spinoffs. The original coal-fired power plant was built in the 1930s to service the town of Fort Collins, Colorado, and was decommissioned in 1973 due to its obsolescence and inadequate size to service the growing city and university.

The power plant was repurposed into the Engines and Energy Conversion Laboratory in the early 1990s and then expanded into the Energy Institute at the Powerhouse energy campus in 2014, when the lab was expanded, tripling in size.

The CSU Powerhouse building includes numerous additional green energy features, including rooftop smokestacks that have been converted into vertical wind turbines. The building uses natural daylighting as much as practical, including a roof-mounted solar tracker that connects to the building's interior via a sun tunnel (Fig. 13.16).

FIG. 13.16

How do you use solar daylighting in interior areas, not on the upper level of a building? This tracking solar "light collector" (left) at the Colorado State University Powerhouse features a solar light concentrator that delivers collected light through a reflective tunnel to an interior light tunnel (right).

13.7 Active solar thermal systems

Active solar thermal collectors are the predecessor of solar photovoltaics and reached a US peak in interest before 1980, while solar PV was very cost-prohibitive. In some ways, direct use of the sun's spectrum of energy is superior with collection efficiencies exceeding 50%. It turns out it is easier to take the radiant energy of the sun to heat a moving fluid than to use it to move electrons via the photoelectric effect.

The most common are solar hot water systems, but air heating systems are also still employed. A large enough solar hot water system can provide space heating as well, although this requires many more panels. Collector designs are diverse and range from parabolic troughs, concentrating dishes, flat plate collectors with a fluid loop passing through, and evacuated tube collectors.

The greater complexity of active solar systems with pumps, heat exchangers, fluids, overheating potential, freeze concerns, assorted controls and sensors, and the typical need for a backup system all make them generally less plug-and-play than PV systems. Some have even forecast the death of solar thermal, though this may be hyperbole. There are certainly applications where they make sense, such as somewhere like Hawaii with abundant sunshine, high conventional energy costs, and direct water heating potential due to no freeze potential. The solar thermal industry suffered from early poor designs and installations during a boom era in the 1970s, much higher initial installation costs compared to other countries, a general lack of demand, changing government incentives, plus other barriers to greater utilization. The market and companies in the continental United States mainly focus on commercial-scale and solar pool heaters, although there are still other residential applications as well (Example 13.2).

EXAMPLE 13.2 Solar thermal collector design

Concepts: Draw a diagram providing seven design elements for a flat plate solar thermal collector that uses a glycol heat transfer fluid. Focus on the collector itself and what attributes it should have, returning to your knowledge of heat transfer.

Solution: See Fig. 13.17.

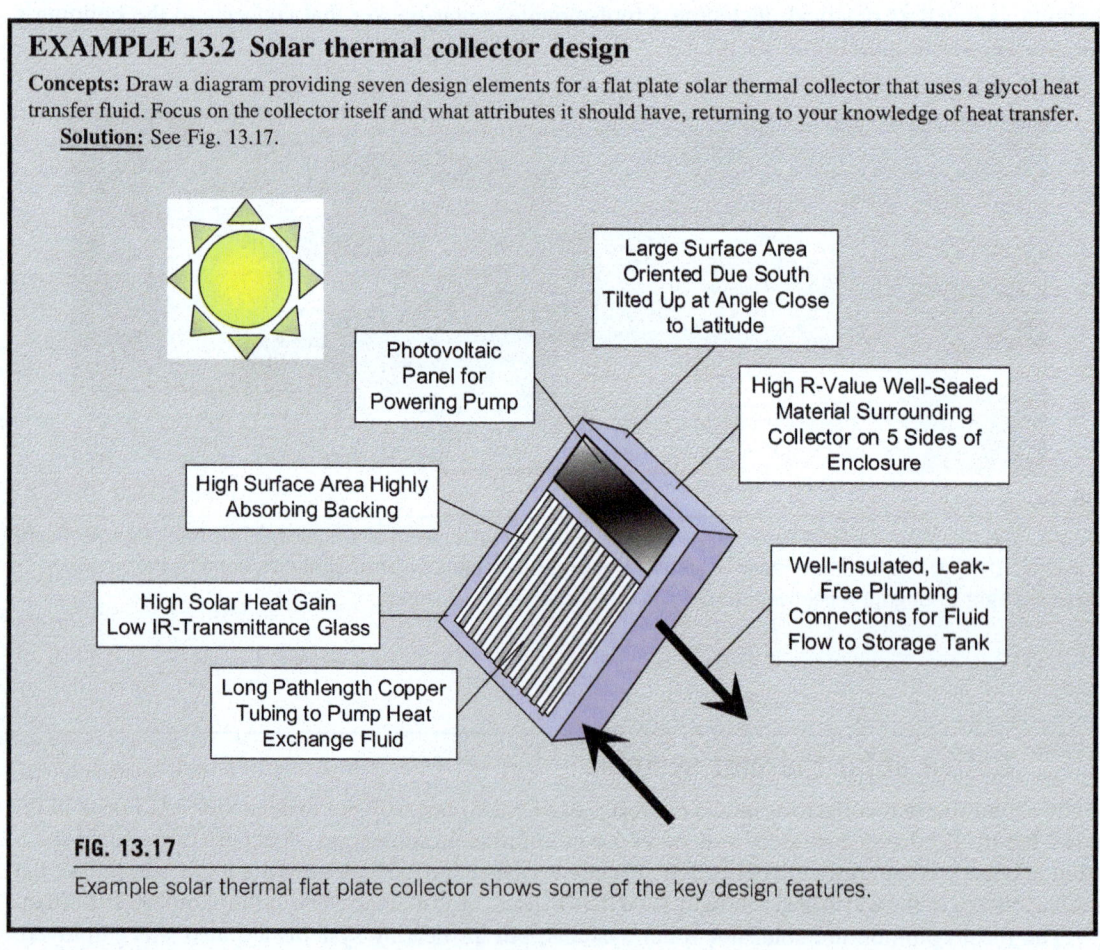

FIG. 13.17
Example solar thermal flat plate collector shows some of the key design features.

13.8 Geothermal energy systems

An often-applicable on-site renewable system is geothermal energy, which can be implemented in several ways. Geothermal can use the heated interior core of the Earth as a hot reservoir for space/water heating, for a heat engine (for geothermal electric power production), or simply as a steady temperature heat source/sink (for ground-source heat pumps).

The use of "hot rock" or steam geothermal energy systems for power generation is a topic that was discussed briefly in the chapter on energy systems and heat engines. The Earth's molten core is approximately 6000°C, as hot as the sun's surface. This 30 GW source of thermal power has been sustainable for billions of years and is driven by the decay of naturally occurring radioactive elements. The problem, of course, is that the power flux at the Earth's surface is extremely low making it difficult to harness this energy source. There also are not many sites where the near-surface temperature of the rock or the groundwater is hot, limiting the Carnot efficiency of any system. Nonetheless, select sites in California, Iceland, Italy, Kenya, and others that have these high-temperature resources and have implemented this for electricity production using a variant of the Rankine Cycle.

A simple, passive way for new construction to take advantage of geothermal energy is an Earth-bermed building. The more a building is in contact with the Earth, it helps to moderate the temperature swings in the ambient air temperature. If the air in a building only requires conditioning from the near-constant 55°F (13°C) ground temperature to room temperature, energy needs are substantially reduced. This is particularly true compared to a starting point of zero or 100°F (−18°C or 38°C), which are not uncommon for ambient air temperatures.

Geothermal or ground-source heat pumps have been an alternative building HVAC system for many decades. They take advantage of the constant temperature of the Earth's subsurface and use it as a heat source or sink depending on the season (Fig. 13.18). Geothermal heat pumps were briefly discussed in the chapter on engineered systems, and a diagram is shown in Fig. 13.18.

Like air-source heat pumps, they are reversible and run as a heating or cooling system depending on the season. They have the advantage over air-source heat pumps of not having to extract heat from cold winter ambient temperatures or dump heat into warm summer ambient temperatures.

They require a significant (~100 m or more) heat exchange loop underground, either trenched horizontally approximately 3-m deep or vertically 100+ m deep. The drilling and excavating costs are significant in either case and best done on new construction where buildings, landscaping, and other infrastructure and obstacles are minimal. Another alternative, if available, is to use a large and deep enough pond as a constant temperature heat source/sink throughout the year.

The upfront costs, system complexity, and need for drilling wells can also be limiting factors as can finding contractors to properly install and service them. Thus, it is often viable for new construction and larger commercial-scale buildings but can be a challenge for small residential and retrofit applications. However, their operating costs are generally far less than those of a traditional HVAC system with a forced air gas furnace paired with a standard compressor-based cooling system.

The improved performance of heat pumps, particularly at low ambient temperatures, has helped geothermal. It has, however, made air-source heat pumps much more viable in many locations, including cold climates, and steered many households toward these simpler and less upfront capital cost alternatives.

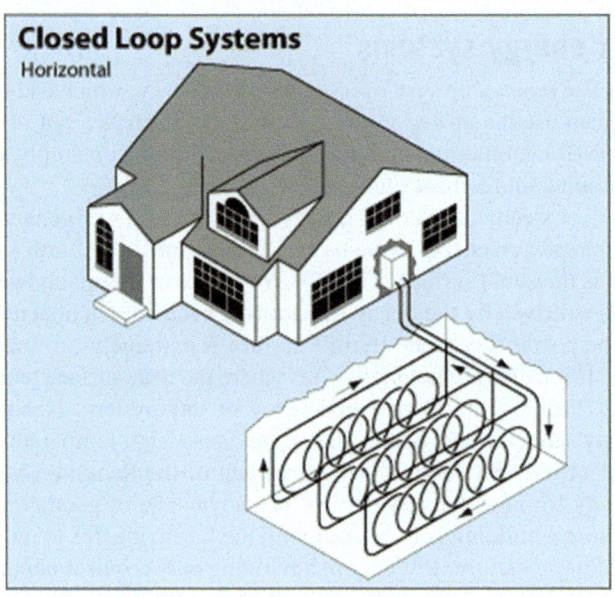

FIG. 13.18

Diagram of a closed loop geothermal system with horizontal arrays of heat exchange loops. The heat exchange can also occur through vertical wells drilled 100 m or deeper into the ground or with a large enough pond. The compressor and air handler are inside the building.

Public domain image from US Energy Information Administration.

Earth tubes are another low-tech application of geothermal systems. These are long fresh air inlet tubes that pass through a significant pathway through the Earth. This can preheat ambient air in the winter or precool it in the summer. The systems can provide a few hundred to a few thousand BTU/h and typically require mechanical ventilation to move the air (e.g., a heat recovery ventilator fan). These are essentially building inlet air conditioners that take advantage of the relatively constant underground soil temperature to preheat input fresh air in the winter and/or precool in the summer. They also require excavation of lengthy, sloped horizontal trenches 10 ft or more and then refilling. Important design aspects are the longevity of the system vs the underground environment (structural, corrosion-resistant, leak points) and the potential for indoor air quality problems (mold, mildew, soil gas, e.g., radon intrusion). There have been problems with these when condensation occurs and mold or mildew growth risk making them a not-so-fresh air inlet.

Geothermal systems are also not entirely sustainable in that the hot reservoir can be depleted locally, or the heat sink/source ground temperature can be affected, causing system performance issues. The emissions are often nonnegligible in that pumping heated water or steam often brings up other possible contaminants, including H_2S, SO_2, H_2SO_4, mixed hydrocarbons, radionuclides, and other species. The reinjection of this fluid can lead to the same sort of induced seismicity associated with oil and gas exploration.

13.9 Wind power systems

Wind power harnesses the fluid energy of the blowing wind, extracting some of its kinetic energy to turn a turbine that connects to an electrical generator. Although many applications are on the utility scale, commercial wind turbines scale down to <10 kW. It is most viable in open, rural areas with strong, steady winds, preferably blowing from a consistent direction. In such remote areas, it can be implemented on the individual building or property scale. Wind power developments range from individual residential-scale turbines that are ~1 kW or less (Fig. 13.19) to large-scale utility turbines that can exceed 10 MW apiece put in wind farms of hundreds of MW. Smaller-scale wind turbines for distributed generation at individual rural homes or ranches face a range of challenges compared to the large-scale wind turbines including high cost per kWh, low capacity factor, lower windspeed, and higher turbulence near the Earth's surface, and generally less research and development into the aerodynamics and design of small turbines (Bianchini et al., 2022).

FIG. 13.19

Small residential-scale horizontal axis wind turbine on a monopole located in Estes Park, Colorado, United States. These are feasible in off-grid and more remote areas where open areas make them viable sites. The plume from the High Park Fire in Colorado is seen in the background.

A wind rose maps the wind speed frequency as a function of direction (Fig. 13.20). The illustrated wind rose from southwestern Wyoming shows strong, consistent winds from the west-southwest at this site. This is an attractive site for wind power, and not surprisingly, the area has been subject to extensive wind power use, including utility-scale projects. Notably, one required a transmission line called Trans-West from southwestern Wyoming to Las Vegas, Nevada, over 700 miles apart. Currently, the SunZia project linking the eastern plains of New Mexico with abundant wind and solar to load centers in Arizona and California is the largest renewable energy project to date at 3 GW. It has taken nearly 2 decades from commencement to its construction, a testament to the complexity and controversy of such grid-scale projects.

336 Chapter 13 Renewable energy systems

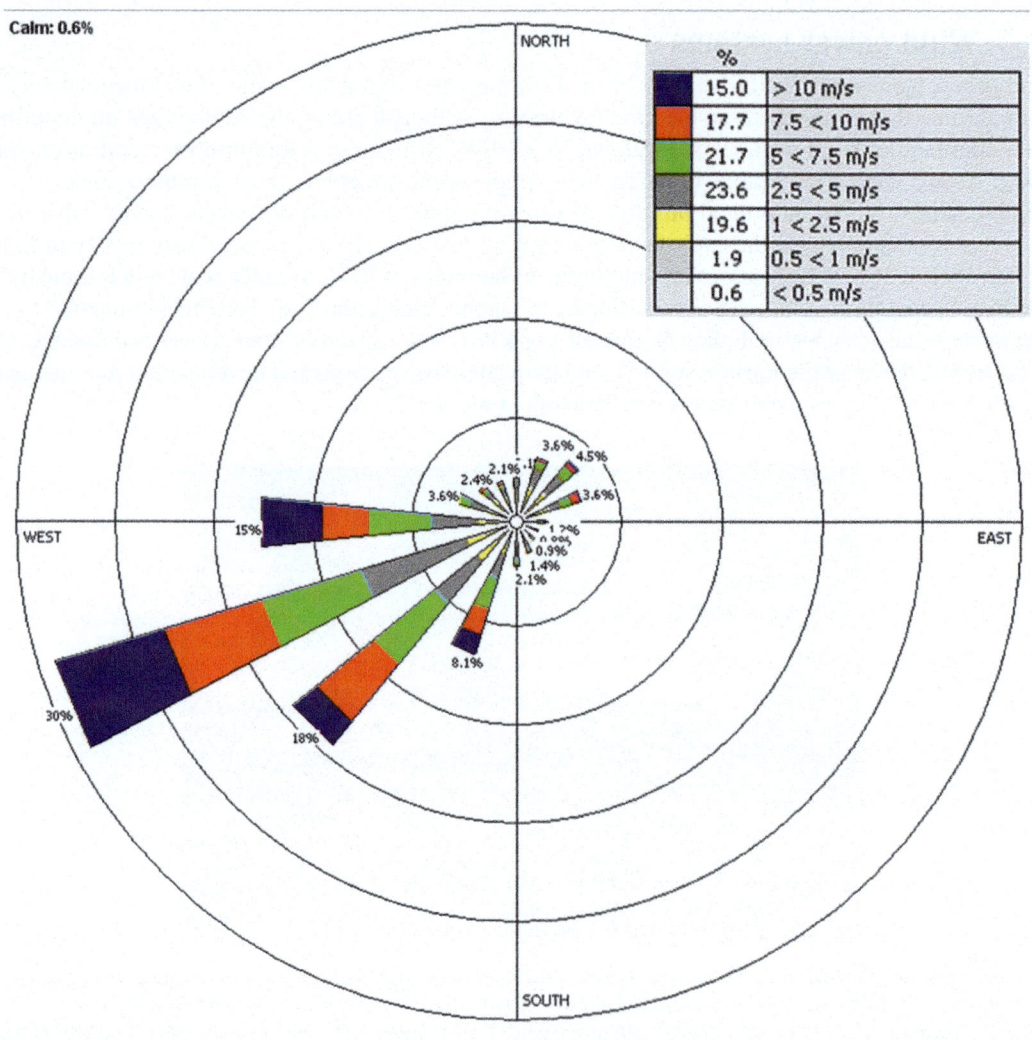

FIG. 13.20

Wind rose showing the frequency distribution of winds at a 10-m height at a site in southwestern Wyoming. Sites with strong, steady winds from a relatively narrow directional range are the best for wind power generation.

A wind turbine does not produce power under all conditions, as it has a cut-in and cut-out wind speed outside of which it does not operate (Fig. 13.21). The figure shows a typical power production profile as a function of wind velocity, where power production starts at 4 m/s and shuts off at 25 m/s, when the turbine blades are feathered to shut off the spinning turbine to avoid damage.

13.9 Wind power systems

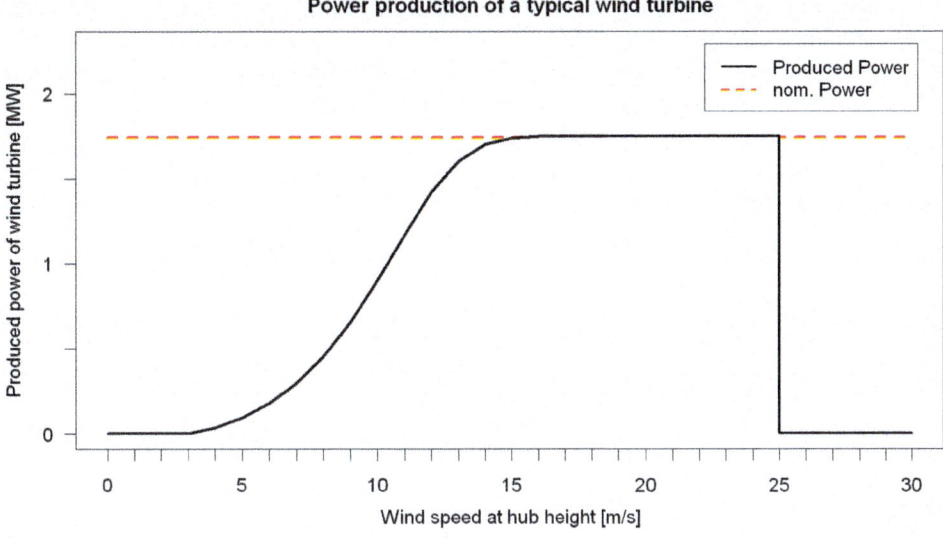

FIG. 13.21

Typical wind turbine power curve as a function of wind speed.

Source: Isjc99, CC BY-SA 3.0, via Wikimedia Commons.

The wind velocity is the most important parameter in terms of power production (Eq. 13.3). The hub height and the swept area of the rotor are both important as well. The power contained in the wind is the product of the kinetic energy of the wind and its velocity, where ρ is the fluid density, A is the cross-sectional area or "swept area" normal to the wind direction, and v is the horizontal wind speed normal to the turbine swept area (Eq. 13.3):

$$P = \frac{1}{2}\rho A v^3 \tag{13.3}$$

If the wind were stopped, the accumulating air behind it would shut off the turbine. The optimal downstream wind speed for wind generation is one-third the upstream velocity. A turbine operating ideally at this ratio is 59.3% efficient and reaches the theoretical maximum efficiency described by the Betz Limit. The real-world efficiencies of actual turbines are typically 30%–50% depending on design and size (larger turbines often also have larger efficiencies).

Small wind turbine designs include the horizontal axis turbine shown, vertical axis turbines, and numerous other designs. The dominant design for wind turbines of the last few decades is the Horizontal Axis Wind Turbine (HAWT) that is mounted on a monopole as in Fig. 13.19. The latticed towers' past designs caused wind field interference and were more visually intrusive than the monopole tower. Numerous other types have been designed, the largest category being the Vertical Axis Wind Turbine (VAWT) that looks like eggbeaters and is more viable in smaller-scale applications more relevant to the building scale. Other VAWTs include the Savonius and Darrieus turbine designs.

Wind power is also not immune from environmental impacts. The land use factor plays into most renewable power systems due to the relatively low power density of the wind. The other concerns are impacts on avian species, wind turbine hum and shadow flicker nearby large units, and general esthetics. Manufacturing also involves significant energy and materials use.

Wind is about a decade ahead of solar in its market scale. Wind turbines due to their mechanicals involve more maintenance than solar PV with moving parts, gears, frictional losses, and all the mechanical and electrical parts of the system. The wind stresses on the exterior components are also important considerations related to the turbine service life (Example 13.3).

> **EXAMPLE 13.3 Wind power generation**
>
> **Problem:** A wind power generator is designed for a site where air density is $1\,kg/m^3$, rotor radius is 30 m, and average wind speed at the site is 10 m/s.
>
> (a) What is the wind power at this site that can be generated by a wind turbine?
> (b) What is the maximum that can be generated by an ideal wind turbine?
> (c) If the turbine has an actual efficiency of 45%, what is the actual output of the turbine?
> (d) If the wind speed doubles, what is the new actual output?
> (e) How much larger would you have to make the turbine radius equal the new actual power output if the wind speed stayed constant at 10 m/s?
>
> **Given:** Turbine 30 m; $u = 10\,m/s$; $\rho = 1.0\,kg/m^3$.
> **Find:** P_{wind}; P_{max}; P_{actual}; P_{2xwind}; R_{new} for P_{2xwind}.
> **Assume:** 45% efficient; Betz limit applies for max power.
> **Solution:**
> Area $= pi\,R^2 = 3.14159 \times 30\,m^2 = 2827\,m^2$.
> The power contained in the wind is $P = \frac{1}{2}\rho A V^3$.
> Power wind: $P_{wind} = 0.5 \times 1\,kg/m^3 \times 2827\,m^2 \times (10\,m/s)^3 = 1.4\,MW$.
> Power max: $P_{max} = 0.593 \times 1.4\,MW = 0.83\,MW$.
> Actual power: $P_{actual} = 1.4\,MW \times 0.45 = 0.63\,MW$.
> Doubled wind speed: $P_{2xwind} = $ Initial power $\times (V_f/V_i)^3 = 0.63\,MW \times (20/10)^3 = 5.04\,MW$.
> $P_{2xwind} = 5.04\,MW = 0.5 \times 0.45 \times 1\,kg/m^3 \times piR^2 \times (10\,m/s)^3$.
> → $R_{new} = sqrt[(5040E3) \times 2/(0.45 \times pi \times 10^3)] = 84.4\,m$.

13.10 Biomass energy

Biomass is arguably humans' oldest energy source (along with human and animal power). It is well known and versatile in that it can be used for heating, water heating, cooking, as a transportation fuel, as a power plant feedstock, and in many other ways. Biomass combustion may offer alternatives that are closer to "drop-in" to replace fossil fuel systems for heat generation, vehicles, or even power generation. It has applications in building-scale systems as well.

Biofuels in building applications are mostly used as a source of space heat. The range of combustion appliances includes wood and pellet stoves. Modern designs produce far less emissions of biomass smoke though harmful emissions cannot be eliminated. With an impetus to move away from indoor combustion sources, the installation of new biomass combustion devices indoors is often avoided.

Biofuels more often are transportation fuels and include biodiesel and ethanol (typically corn-based in the United States) as well as various other hydrocarbon fuels produced from a multitude of feedstocks. The upside is that many of these alternative fuels are drop-in replacements for fossil fuels or can be used with minimal modifications to internal combustion engines.

Biofuels relying on the combustion of hydrocarbon fuels are subject to many of the same concerns as fossil fuel emissions. This includes primary and secondary air pollutants largely from incomplete combustion, including particulate matter, oxides of carbon, nitrogen, and sulfur, and many trace gas and particulate species. Burning wood products is marginally better in some ways and worse in others in terms of these emissions.

In concept and if done right, biomass is carbon neutral as it removes CO_2 from the atmosphere to build into plant tissues (while releasing O_2) followed by the oxidation to reform CO_2 in combusting the biomass. Many studies have assessed the major biofuels and have found our current use of corn-based ethanol is perhaps marginally better in terms of greenhouse gas emissions than gasoline. The problem lies in how biofuel is produced and distributed and how much energy is spent in doing so (Example 13.4).

EXAMPLE 13.4 Biofuels

Problem: According to the US EIA, in 2011, the United States consumed about 3.19 billion barrels of gasoline. How many acres of corn would it require to provide all of this?

Given: 3.19 G bbl/year of gasoline.
Find: Acres of Corn.
Assume: 300 gal/acre of E100.
Solution:
According to the US DOE, an acre of corn produces roughly 300 gal of ethanol.
We will use the following energy contents of fuels:
Gasoline: 125,000 BTU/gal.
Ethanol E100: 76,330 BTU/gal.

$$\text{Land area} = 3.19\text{E}9\, \text{bbl} \times \frac{42\, \text{gal}}{\text{bbl}} \times \frac{125{,}000\,\frac{\text{BTU}}{\text{gal}}}{76{,}330\,\frac{\text{BTU}}{\text{gal}}} \times \frac{\text{Acre}}{300\, \text{gal}} = 731\text{E}6\, \text{acres}$$

➤ That sounds like a lot of land! In 2000, ~72E6 acres of corn were grown in the United States (USEPA), so it would require upping our corn production by a factor of 10.
➤ 1 acre = 0.0015625 mile^2, so it is 1,143,000 mile^2.
➤ That is about one-third of the US land area or about 1.7× Alaska's area, 4× Texas's, or 6.5× California's! This shows the large-scale limits of biomass-based energy systems, though smaller individual applications may make sense.

A related limitation with all biofuels is that the upper limit on efficiency of converting sunlight into stored chemical energy in the plant's tissues is on the order of 5% (DeLucia et al., 2014) or even as low as 1% (Kleidon, 2021). Much of the energy associated with solar radiation absorbed by plants goes into driving evapotranspiration of water and transport of CO_2 involved in synthesizing carbohydrates contained in plant biomass (Kleidon, 2021). That begs the argument that, rather than annual biomass crops, why not just cover the land area with PV once every several decades and reap 20% conversion efficiency to electricity?

Considerable research is going into using less agriculturally intensive feedstocks to produce biofuels, including bamboo, switchgrass, agricultural waste products, and other cellulosic materials. They all require considerable processing, energy inputs, and the right microorganisms to ferment these into fuels. Surely biofuels will play a role in our energy future but are far from a complete or even large part of the solution.

13.11 Micro hydro applications

Although hydropower applications are more common in larger rivers and water bodies where impoundments store water behind a dam, smaller-scale run-of-river-type systems that are tens of kW can be used in rural properties with creeks or streams given sufficient flow and vertical drop (Dunlap, 2019). It is mostly applicable in rural sites, although urban areas can benefit in some cases from hydro applications as well. These are fairly niche applications since suitable sites where it makes sense are infrequent.

Their smaller scale obviously makes them more suitable for on-site power generation and use, but they have the advantage of considerably lower environmental impacts compared to impoundment systems that have a dam and reservoir with water released through turbines. The run-of-river systems have far less ecological impacts, though the suitable sites and power scales are relatively limited. These systems have considerable permitting requirements and water rights implications and are not something a rural landowner with a creek can up and decide to implement.

Another application is hydro storage systems. This has received greater attention with the increasing renewable slice of the global energy pie as a means of mitigating the intermittency of sources such as wind and solar. The use of already built and functioning dam systems for this purpose is attractive. The prospect of building new such facilities is fraught with uncertainties in ecological impacts, construction, scheduling, permitting, and public acceptance.

Various approaches for tidal power, ocean thermal, salinity gradient osmotic power generation, and other uses of water sources are actively being used and researched. Most such applications are either demonstration-level or intended for utility scale (or both) and thus are beyond building-scale systems.

There are also increasing applications of waste heat recovery systems from heated process water, cooling water, or domestic wastewater. These types of systems are applicable at the facility or building-scale level.

13.12 Energy storage systems and transmission

As critics often cite, renewable energy sources have low capacity factors (the sun does not shine, and the wind does not blow all the time argument), which makes storage a critical component as intermittent sources scale-up. At a low level, the variability of renewable sources can be compensated for by employing them as grid-tied systems. These power a building when available, put extra electrical energy onto the grid when production exceeds consumption, and take from the grid when needed. At a higher level of reliance on renewables, provisions for energy storage are needed. As a result of the time and space dependency, a growing importance of battery storage and load shifting has emerged. Microgrids can also help alleviate the issues and are discussed in more detail in the chapter on resiliency.

Energy storage needs are acute, and the scale-up for utility-scale electric storage and electric vehicle storage are both factors of ~ 100 for global decarbonization goals, according to the International Energy Agency. A large and creative range of energy storage systems have been implemented at the facility or

13.12 Energy storage systems and transmission

even utility scales on pilot scales. Among these are pumped hydro storage (using water pumped into a high lake to store energy) as discussed in the section on hydropower systems, other similar gravitational energy storage means, compressed air storage in underground caverns, flywheel storage, ice or other frozen storage means (integrated with building cooling), and molten salt storage (often integrated with solar thermal generation systems). These systems all have important site-specific applications in our energy grid.

Battery storage is one area where building-scale systems are viable. One of the most promising developing and currently viable technologies is electrical energy storage in chemical potential energy in the form of battery banks (Fig. 13.22). Battery storage systems have emerged as a key technology for energy storage from individual devices to the utility scale. This is currently pursued for shorter-term ($\sim$4–6 h) utility storage applications. This is not unlike a hybrid or electric vehicle which frequently charges and discharges the storage battery onboard. Storage is making it possible to deal with matching the peak in demand to the peak in renewable generation and generally mitigating the intermittency of renewable energy sources. A grid-tied customer can collect and store plentiful, inexpensive solar energy during the peak of the day and use this later in the evening when peak demand hits the grid. Battery storage has long been how off-grid residences power their electrical needs during night or other times of low solar input. These off-grid, building-scale applications are also becoming more viable with decreasing costs and complexity of battery storage systems.

FIG. 13.22

Students at New Mexico Tech operate an air quality collector using a small solar PV system with a battery storage pack.

Although innovative chemistries are emerging, much of electrical storage in the last decade is now lithium-ion batteries (LIBs). Their energy and power density and continuing evolution have enabled LIBs to play a dominant role in this market, particularly for portable applications, including electric vehicles. This has been driven by the consumer electronics industry in part and due to their relatively high energy density and number of lifetime charge-discharge cycles (Chu et al., 2017).

The costs of battery storage have been decreasing like the costs of solar PV. Despite improving economics, a renewable energy system with battery storage will roughly double the cost. Some of this is currently covered by municipal tax incentives including a 30% tax credit from federal sources as of 2024.

A strident competition is on to advance the storage options for the current rapid development of renewable energy as well as accompanying storage technologies. A very active area of research in chemistry and materials science, emerging technologies and improvements in battery chemistry can allow foreseeable achievement of energy storage density approaching 600 Wh/kg (Chu et al., 2017). One interesting example is Repurpose Energy which is pursuing the construction of modular energy pods on the grid that use recycled and repurposed batteries that are ramping up as electric vehicles are retired.

After a decade-plus lifetime, EV batteries are put under high strain and can only function while still retaining ~75% of their capacity; they can however be repurposed for grid or off-grid storage for renewable energy systems. Other nascent recycling and reuse opportunities are currently under intense global research. B2U is another startup company that is taking used battery packs from electrical vehicles no longer useful in the vehicles and repurposing them for the less taxing use of electrical storage with grid-scale PV applications.

The US DOT and Federal Highway Administration have launched a financial incentive program for EV charging stations called the National Electric Vehicle Infrastructure (NEVI) program. Electrical energy storage has one additional upside in that solar PV and vehicle battery charging pair well. This mating of technologies deployable on the building scale can help take the strain off the grid for an electric power-intensive and increasingly electrified vehicle fleet. Improving Vehicle to Grid (V2G) technology can make this a two-way street and turn EV vehicle batteries into on-grid electrical energy storage systems, as explored in Example 13.5.

EXAMPLE 13.5 Large-scale electrical energy storage

Problem: Examine the scale of the energy storage issue with renewables in the United States. Take the entire US-registered on-road vehicle fleet and replace it with the Tesla Model 3 long-range. Assume it can be discharged to 80% of its capacity. Should we need to use this stored electricity to power the nation's electrical grid, how long would that fleet of vehicles run the system?

Given: US vehicle fleet replaced with Model 3 LR.
Find: T to operate grid on vehicle storage.
Assume: 80% discharge.
Solution:
According to Statista.com, the United States has approximately 274 million registered vehicles on its roads.
The Tesla Model 3 long-range stores 85 kWh of electricity.
Thus, this storage capacity of a Tesla fleet is $E = 274\text{e}6 \text{ vehicles} \times 85 \text{ kWh/vehicle} \times 0.8 = 1.86\text{E}10 \text{ kWh} = 1860 \text{ GWh}$.
United States current annual electrical use is 4.1E12 kWh/year according to the US EIA.
Steady-state power use = $4.1\text{E}12 \text{ kWh/year} \times (\text{year}/8760\text{h}) = 468{,}036{,}000 \text{ kW}$ equivalent steady state or 468 GW.
Thus, this fleet of vehicles could take on the national grid load for 1860 GWh/468 GW = 4 h.

This is not a trivial addition but does show that additional battery farms beyond an electric car fleet may be necessary for a mostly renewable grid (or some other baseload & backup generation). Keep in mind that a diversity of sources and interconnected regions assures that the whole grid will not go down at once, e.g. when the sun is not shining, but gives a scale for the problem and indicates we use an immense amount of electricity.

13.13 Constraints and opportunities in a renewable energy transition

The mineral resources are significant compared to current production, though they are generally not geologically limited for most elements. With the massive increase in battery use, the scaling up of mining, processing, and production is a daunting task, as next discussed. There are bound to be conflicts regarding the land use, mining operations, and environmental liabilities of such a scale of mining that are emerging.

13.13 Constraints and opportunities in a renewable energy transition

Renewables are poised for continued growth and are pivotal for addressing climate change. For the individual building owner, the attractions are strong of a low environmental liability energy source and self-sufficiency. On-site systems, of course, can add complexity and cost considerations. The other liabilities with renewable sources are worth a sober assessment of their potential and how to mitigate drawbacks, as summarized in Table 13.3 for solar PV generation.

Table 13.3 A summary of the health, safety, and environment (HS&E) liabilities associated with solar PV.

Process	HS&E concern	Species of concern	Mitigation
Manufacturing PV panels	Energy-intensive	Embodied energy and associated emissions	More energy-efficient and cost-effective manufacturing processes are occurring continuously
Silicon processing	Waste stream (up to 80%); worker exposure	Metallurgical-grade Si	Recycling of used PV panels is seeking to reuse silicon from used panels
Silicon processing	Explosion hazard; worker exposure	Silicon tetrachloride Silane gas SiH_4	Alternative processing agents and processes
Silicon purification	Toxic exposure or release; waste stream	HCl, HF, HNO_3, H_2SO_4, $NaOH$	Less hazardous chemicals are being introduced to manufacturing
Silicon doping	Toxicity	Arsine, phosphine gas	Less toxic dopants and carrier gases are being introduced
Electrical engineering	Metals exposure	Lead, silver, aluminum	Metal recovery from used panels is advancing
Process cleaning	Climate impacts	SF_6 (global warming potential of $\sim 25{,}000\times$ CO_2 on mass basis)	Alternative electronics materials solvents are in production
Cleaning and purification	Toxic exposure or release	TCE, acetone, NH_3, IPA	Alternative electronics materials solvents are in production
PV land use	Land degradation	Ecological impacts of large solar PV farms	Dual use, including agrivoltaics, where shade-grown crops can be grown underneath panels

Despite its environmental friendliness, substantial liabilities still exist.

Wind and solar are highly variable sources; electricity generated from these assets varies by time of day and season and thus is not "dispatchable." Thus, a drawback of many renewable energy systems is often the need to have a backup conventional system or storage for downtime (when the sun is not shining, or the wind is calm). This adds to the complexity of a building's systems, a complication that some building managers or owners would rather avoid. As another viable option, many times this can be accomplished by using a grid-tied system so that the grid becomes the backup and storage system.

Another notable drawback of renewable power generation systems, such as solar photovoltaics, is the low power density requiring significant land use for large-scale production. Thus, land use is significant due to the low power density of sunlight ($\sim 1000\,W/m^2$ peak) hitting the Earth's surface. To mitigate the land use and habitat destruction associated with large-scale solar PV, emerging applications of solar PV include "agrivoltaics" and "crustivoltaics" or desert restorative PV. The semishade provided under solar PV can be ideal for some shade-tolerant crops and can also provide a suitable environment for the restoration of desert bio-crusts, which mitigates desert dust generation (Heredia-Velasquez et al., 2023). The concept of dual use or shared resources is becoming an area of interest, including agrivoltaics, where land used by solar PV can also be used for agricultural activities, including cropland or rangeland for grazing animals. Choosing the right crops—e.g., those that have some shade tolerance and are not location light limited—is critical. The reduction of direct irradiance, barrier to water vapor upward flux, and direction of rain runoff from panel surfaces can actually increase production in some crops when planned effectively. Wind farms have even more open space between turbines and transmission hardware that allows wider applications for co-uses for farming or ranching in conjunction with wind energy production.

The materials needed for producing the machines that enable a low-carbon economy are quite intensive with regard to mineral inputs. A low-carbon electrical infrastructure relies upon various minerals and metals, including copper, nickel, cobalt, lithium, and a range of rare Earth metals that are scarce, distributed over wide regions, and/or at low concentrations (IEA, 2021). This contributes to the embodied energy and embodied carbon associated with manufacturing renewable energy technologies. For example, the mineral intensity of an electrical vehicle is ~ 6 times greater than that of a conventional vehicle, while an onshore wind turbine is 9 times more mineral intensive than a gas-fired power plant (IEA, 2021).

The challenge is that the mining of these will have to scale up—and dramatically in some cases—to meet the growing needs. The mineral intensity of renewable energy sources involves mining operations that have considerable impacts on biodiversity on land (Sonter et al., 2020) as well as in the controversial practice of ocean mining, where reserves of many metals are larger (Paulikas et al., 2020). The use of lithium and cobalt in particular is expected to be dominated by renewable energy applications by mid-century; even copper and rare Earth will require over 40% of their global use devoted to renewable energy sources based on the scale-up required (IEA, 2021). The World Bank has detailed these and their efforts for climate-smart mining to minimize these impacts, as projections show increased material extraction up to a factor of 5 (e.g., lithium and graphite) by mid-century depending on the element and pathway. The additions to global production are more modest for common metals such as copper and aluminum. New efforts focused on efficiency of material extractions and use, efficient energy use in the mining sector, innovation in materials processing and waste management, and using low-carbon energy sources in mining are emerging (Hund et al., 2020). The constraints and disposal costs for PV, for example, are real and a challenge, though not a reason to delay a transition to renewable energy resources (Mirletz et al., 2023).

Also related, PV electric generation is the electronic waste created a few decades down the road at the panel end of life. There is more interest in the recycling of solar PV panels as the supply chain has been disrupted and materials constraints have come into play. In 2018, the waste management company Veolia, based near Paris, opened what it says is the first recycling line developed specifically for recycling solar panels in Rousset, France. A small number of US states have enacted requirements for PV panel recycling. However, this is a nascent industry with many of the processes needing refinement and improved economic viability. Rystad Energy is a company that is pursuing the recycling of solar panels and turning the dead solar panels into a commodity rather than a waste product requiring disposal. The glass can be crushed and used for new glass products. The metals (particularly silver and copper) and silicon are also recoverable, though this has been a difficult and costly process to this point. New machines for processing the end-of-life panels into their recycled fractions are becoming available in the marketplace.

The transition from a fossil to a renewable energy economy will trade massive consumption of fossil fuels for much-increased extraction and use of other minerals for renewable manufacturing. Life cycle total material needs are less for renewables than conventional sources due to little input beyond manufacturing. The recycling of these materials is obviously paramount to making the production loop more sustainable. A comprehensive assessment of powering our entire economy on a combination of Wind, Water and Solar (WWS) and efficiency provides state-by-state plans (Jacobson, 2020). The approach has been critiqued as overly optimistic in its assumptions (Clack et al., 2017). If technically feasible, there are serious questions about the grid stability and the liabilities of a renewables-only system without the contributions of "firm" noncarbon sources including nuclear, bioenergy, geothermal, and natural gas with CCS (Sepulveda et al., 2018). The question is open as to how our postcarbon energy systems will evolve.

13.14 Case study part I—Mid-century schoolhouse/church repurposed into a home

I will end this chapter with my own albeit imperfect and incomplete efforts to incorporate the preceding lessons into a lower-impact and more resilient home (Fig. 13.23). There have been several mentions and figures from this building related to my efforts to "walk the walk" (Figs. 6.4, 7.9, 8.1, 9.8, 11.6–11.9, and 13.12), and discussed here in more detail.

I started a faculty position at New Mexico Tech in 2014, and my partner and I (and several animals) sought an in-town house near campus. Commuting without a car was critical, as biking and walking are viable here year-round. The property is in Socorro, New Mexico, a small, historic former mining town (~9000 residents and dating to 1598). The location is in central New Mexico in the Rio Grande Valley at about 4700 ft above sea level. The climate is mixed heating-cooling climate zone 3B, and it is very dry at <25 cm (10 in.) of precipitation per year.

It has been often said that the greenest building is the one already built (attribution unknown), particularly if it can be renovated. The building we purchased is in the "urban core" within about a mile of campus and a similar distance to the grocery and other stores and restaurants. It was built in 1953 as a 2700 ft^2 single-classroom kindergarten schoolhouse that later served as a Spanish-Baptist micro-church (among other varied subsequent uses such as a Girl Scouts meeting venue). The building was converted into a residence by the last owner who used the space as a live-work painting studio. The building has reasonably "good bones" including cinder block construction resulting in a large thermal mass. A standard pitched roof featuring roof trusses meant there was an attic for easy insulation as compared with many "flat-roofed" buildings in this region. With a large unobstructed roof fetch to the south, it also meant solar PV is attractive.

The greatest risks and threats are extreme heat and water availability. What have we done in terms of green building, sustainability, and resilience? The following are some of the key elements:

- We "out"sulated (insulated on the outside) of the cinder block walls and re-stucco-ed (Fig. 13.23). In two prior home renovations in Colorado, we had insulated the interior and in the wall cavities which both were thermally less ideal than an outside layer of insulation which provides more continuity and brings in the thermal mass inside the thermal envelope. This was a fairly continuous insulation layer, but we did not alter the buttresses, although they act somewhat as "heat exchange fins."
- We replaced single-pane, metal frame nonoperable windows having assorted glass and sealed shut with dual-pane, low-e coated, argon-filled windows with U-values of 0.31 effectively reducing window conduction losses by ~75%.
- We found ~R12 in the attic and feared asbestos in ceiling tiles and/or attic insulation; however, this was negative, so we air-sealed and added insulation to ~R50 in the attic. The added insulation may have been a bit much as ceiling tiles dropped due to added weight!
- Can New Mexico be tolerable without air conditioning? We say yes with thermal mass and reasonable insulation and infiltration. We use two zoned whole house fans for cool air flushing during summer nights and an evaporative cooler on the hottest of afternoons when indoor temperatures approach 80°F.
- Solar PV was not the first priority on arrival, but at the time the building's 50A service needed to be upgraded, as did the roofing. Thus, with tax incentives, we decided to install a 3.3 kW PV system which provides nearly 200% of our electric use (oversized for future heat pumps and/or electric vehicles). The system was oversized to accommodate future heat pumps and electric vehicles.
- All lighting is LED, appliances are Energy Star if available, and many other small efforts to limit consumption are made.
- Most of the yard has been converted to xeriscape with only a small patch of native grasses that requires a few mowings a year and is only watered incidentally with tree watering.
- Gutters are somewhat optional in this dry climate, but we added them draining into two 200-gal rain barrels that we use for some of the landscape watering.
- Although a swimming pool in the desert seemed outrageously consumptive, a "cowboy pool" for two (170-gal stock tank) helps cool off on the hottest afternoons. The water is changed out periodically and used to water landscape plantings.

Old buildings are always a work in progress, and compromises and mistakes were made in this renovation still in progress. For example, for the replacement windows, I might have gone higher end and possibly triple pane. Doing it again, I would have given more consideration to the embodied energy of foam panel insulation and possibly used rock wool for rigid insulation and cellulose for loose-fill insulation in the attic. The leakiness is still higher than we would like to improve. The attic was air-sealed and additional insulation to ~R50 was added. However, the air changes per hour at 50 Pa pressure difference (ACH50) for the building is approximately 4. One of the drawbacks of the whole house fan as well as a number of blower door tests is that the much-worn cellulose tile ceiling has started to drop. It is currently being renovated (leaving the existing tiles in place), including better air sealing and a radiant barrier. Improvements are planned by phasing out the atmospherically vented combustion appliances as well as renovating or replacing four exterior doors, which should reduce infiltration.

Two gas-fired wall heaters installed before our occupancy are entirely functional and presently used, along with some spot-use electric resistance heaters. The annual gas usage for the building is ~500 therms per year, about 400 therms of which are for space heating and the rest for gas water heat and a gas stove (we have an induction cook plate to experiment with electric cooktop). According to the US Energy Information Administration (EIA), the average monthly residential natural gas usage in the United States is between 70 and 90 therms per month or 800–1100 therms per year (in New Mexico it is at the lower end of this range for gas-heated houses). Though we are on the low end, I still seek to reduce or eliminate gas use. It is a continued work in progress, and plans are to move the space heating to electric heat pumps of some variety. A small kit greenhouse for growing greens, starts, will be an "off-grid" experiment. Other future projects include heat pumps for space conditioning and water heating and additional water storage.

FIG. 13.23

Home renovations including "outsulation" of a cinder block building. Infrared picture showing the areas of heat loss on double front doors in a New Mexico building where the hotter colors indicate larger heat loss. Note the contrast in heat loss in the area immediately surrounding the doors with the area further above the doors which was retrofitted with 2 in. of foam board insulation.

13.15 Chapter summary and conclusions

Renewable energy applications at the building scale are becoming common and dramatically more cost-effective. Distributed power generation (as opposed to centralized stations) is fundamental to integrating on-site renewable energy systems into green buildings.

Remarkable improvements in efficiency, manufacturing, and cost-effectiveness have occurred with key renewables over the last few decades. Small wind power systems are deployable at the building scale in rural locations. Solar photovoltaics (PV) and wind generation systems that produce electricity have witnessed the most significant improvements in their cost-effectiveness over the last few decades and are scalable from residential to utility-scale production. Efficiencies of solar cells continue to improve in the commercial market (exceeding 20% now is common) and in the laboratory (around 40% presently). The vast majority of PV panels are silicon-based, while the remaining are dominated by thin film, a newer technology. Though currently more expensive and less efficient, thin film has the potential to be more cost-effective due to lower materials use. Systems range from grid-tied systems, with or without battery backup, to completely off-grid systems.

Solar thermal systems have efficiency advantages over PV but are less "plug and play" than current PV technology. The downsides of solar thermal include that (a) typically a backup conventional system is needed for heating when there is a long spell with little solar input, and (b) these systems usually involve a fair number of pumps, controllers, fluids, and plumbing which are all prone to maintenance and inevitable failures.

Other renewable systems ranging from small-scale hydroelectric to fuel cells to geothermal, often very site-specific, will play increasing roles in generating on-site energy while minimizing greenhouse gas and other emissions. The most viable building-scale application is the geothermal-based heat pump, which has a long heat exchanger in the ground. The solid Earth serves as a relatively constant temperature heat source in the winter and a heat sink in the summer. The cycle can be reversed to provide cooling in the summer and space heating in the winter. Geothermal systems are often limited due to the capital cost of installation as well as the disruption of drilling vertical wells or horizontal trenches.

The use of energy storage techniques is valuable with renewable power generation. The dominant chemistry for battery storage in the last decade has become lithium-ion batteries, and costs have dropped significantly for storage systems as well. New chemistries and solid-state batteries under intensive R&D will continue progressing to marketable products as well.

Technology has improved the reliability of renewable systems, but the intermittency of sources such as solar and wind often necessitates backup dispatchable systems, energy storage systems, grid-backed-up systems, or clever coordination of loads with source availability. The net electrical demand curve over the course of a day often called the "duck curve" resembles the shape of duck from a side view, where net demand drops at midday due to solar generation and returns rapidly in the late afternoon.

We are evolving from a fuel-intensive economy to a material-intensive economy, and renewable energy systems will be a major part of this. Renewable energy sources have other concomitant environmental liabilities as well, mostly in their production and the materials required, as well as in their land use. The extraction of materials for electrification focused on renewable energy sources, though not thought to be geologically limited, is a logistical and environmental challenge related to the scale-up in mining. The transition to a renewable energy-based economy will require the scale-up of extraction for such minerals severalfold to an order of magnitude or more to reach net-zero carbon emissions by mid-century. Efforts to reduce the land use impacts with solar PV continue to develop as well.

13.16 End of chapter exercises

(1) **Concepts:** Which of the following is true regarding renewable energy resources?
   a. They have negligible environmental impacts.
   b. They provide the majority of US energy.
   c. They can be replaced in a relatively brief period.
   d. None of them require a source of fuel.
(2) **Concepts:** Discuss three advantages of taller wind turbines. What three engineering challenges are involved with supersizing wind turbines?
(3) **Concepts:** Which is the best direction (north, south, east, or west) to maximize window area in a cold climate in the Southern Hemisphere?
(4) **Concepts:** Discuss 5 factors that will dictate the solar production of a PV array.
(5) **Concepts:** Discuss approaches that can be taken to mitigate the intermittent nature of renewables such as solar.
(6) **Problem:** Make a plot of the increase in power production vs the relative increase in (a) average wind velocity, (b) turbine radius, and (c) air density. Start from 1 and show what effect increasing that parameter by a factor of 1 to 5 has on power. Which is the most important?
(7) **Problem:** Calculate the power flux at the surface of the Earth from its interior generation of 44 TW.
   a. We want to use this flux to run the average US house which can be approximated as 1.3 kW steady-state power consumption. Assume no losses. What land area is needed?
   b. Suppose we wanted to harness this flux to convert to electricity using a Rankine cycle power plant at typical efficiency (of course the efficiency will be far lower due to such low-temperature hot reservoir). What area in hectares and acres would be needed? Compare to a comparable land area state.
(8) **Problem:** Estimate the power output and maximum yearly energy production from the turbine shown in Fig. 13.19. Assume average properties in Estes Park, Colorado, and make any other necessary and reasonable assumptions for turbine efficiency.
(9) **Problem:** A parabolic trough system collects solar and covers an area of 1 km × 2.58 km. Estimate the annual kWh electrical production near Las Vegas. Use the map of kWh/m^2/day for PV (assume it is collected as heat rather than PV electric production) in the notes at a location near Las Vegas with a solar thermal collection efficiency of 55% and a power cycle efficiency of 28%. How many average US homes' electrical use does this supply?

350 Chapter 13 Renewable energy systems

(10) Problem: The approximate land area on the planet is 130 million km^2 of land, while the roof area of buildings on the planet is approximately 0.2 million km^2 of rooftops presently (an area roughly the same size as the United Kingdom) (Joshi et al., 2021). How much of that roof area would be needed to supply the planet's steady-state power use (all energy use, approximately 18 TW)? You can presume ideal solar conditions (facing the equator at an angle close to the latitude, direct sun access, no storage or transmission issues, no shading, or other roof nonidealities). You can assume the use of SunPower E20-327 modules (use nameplate-rated output) with no interspacing required and a capacity factor of 0.25 for solar PV. Comment on the result.

(11) Problem: Use your knowledge of radiant energy. Assume the sun is 865,370 miles in diameter and has an effective surface temperature of 5800 K. Assume a perfect blackbody radiator.
  a. Calculate the radiant power output of the sun in watts.
  b. The Earth is 12,756 km in diameter and is 92.96 million miles from the sun. Calculate the average solar flux at the top of the atmosphere.
  c. What fraction of the sun's output does the Earth intercept?
  d. How much power does the Earth receive from the sun?
  e. Compare to the total humankind energy use.

(12) Problem: A typical natural gas home heating system generates 102,000 BTU/h output. What is this in kW? If you used a solar air heating device that has a ΔT of 54°F, what flow rate of air would be required (volumetric and mass flow rate) for the same heat delivery rate? If the pressure drop through this collector is 50 mBar, what size (kW) pump (assuming 75% efficiency) would be required to drive this flow? You can assume the air exits the system at STP298.

(13) Problem: Consider the water pumping requirements for a 100 MW OTEC plant, a utility-scale size though fairly modest in capacity compared to conventional power stations. Consider the best OTEC sites (highest ΔT), Carnot efficiency, and the water heat capacity difference between hot and cold flows (assume 4C and 29C, fully extractable with Carnot efficiency).
  a. Find the Carnot efficiency.
  b. Find the mass flow rate of water required. Compare this flow rate to Chicago's Stickley wastewater treatment plant's typical flow rate, one of the largest WWTPs in the world.
  c. What size pipe would this require assuming this entire flow rate is pumped through the cold-water intake? Check the engineering toolbox for an approximate limit for water velocity in pipe flow to avoid damage. Comment and compare to the mega-engineered pipe discussed below.
  d. Estimate the pumping cost (in kW) by using the volumetric flow rate multiplied by the pressure drop and assuming a 70% efficient water pump. Use the linked pressure drop calculation tool and assume 100 5-m sections of pipe to reach 500 m below the ocean surface. Consider the inlet loss (assume rounded inlet), conical grid as well as pipe frictional loss with a plastic pipe (surface roughness thickness of 0.0015 m) of the diameter calculated. Include a screenshot of your results.
  Largest diameter solid wall HDPE pressure pipe project in North America completed | WaterWorld Pressure-Drop.online

(14) Problem: Although farcically simple, an energy storage approach (gravity) has been proposed to jack up a house to store energy via gravitational potential energy storage. Assuming one could jack up a 50,000 kg house by 10 m maximum, how long would that stored energy be able to supply an average home that uses 800 kWh/month?

(15) Problem: Current lithium storage density for state-of-the-art batteries is about 250 Wh/kg (yes, a weird unit). How much is this in MJ/kg? It has been said that a lithium-ion battery has a lower energy density than a ham sandwich. Compare with the nutrition label found from online research and confirm or deny. What if you add a tablespoon of mayo and a slice of cheddar?

(16) Problem: As of the early 2020s, among the largest wind turbines is the GE 12 MW Haliade. Holy cow, its rotor diameter is 722 ft! Estimate its efficiency from the data below and assume a design windspeed of 10 m/s. If the capacity factor is 63%, calculate its annual production in GWh and compare it to the stats given below. If the average household in the United States consumes 897 kWh per month, how many households could this wind turbine power (ignore storage issues)?

(17) Problem: Heliogen is a startup that is pursuing the heliostat approach to generating heat with many enhancements to the traditional process. In a breakthrough, it announced its solar technology can generate fluid that exceeds temperatures of 1000°C strictly from concentrating solar energy. This can be very useful for cement and steel production which are high-temperature processes that historically relied upon fuel combustion; it can also boost the efficiency of using solar power towers for electricity generation. Currently, commercial solar thermal systems can only reach temperatures of up to 565 degrees Celsius (which can be used for power generation only).

Assume a cooling water temperature of 80°C. All else being equal, which would be a better improvement from the standpoint of efficiency, adopting the Heliogen technology or dropping the cooling water temperature to 25°C?

(18) Problem: Use the annual solar PV production potential map from NREL. Estimate the size (acres) of the solar PV generation facility to equal the output of a large 1000 MW$_e$ power facility (assume 85% capacity factor for the latter). Assume 100% of the land area can be covered by PV panels. Compare this to the size of Socorro County, New Mexico.

(19) Problem: Calculate the current energy density of gasoline in kWh/gal. Compare the energy density of lithium-ion batteries to gasoline energy density per kg and per m^3. 33.7 kWh/gal

(20) Problem: All else equal, which of the following would you prefer to use to run a heat engine? A geothermal spring at 90°C and using the atmosphere at 15°C as your thermal sink or an arctic ice at 253 K as heat sink with the tropical atmosphere ($T = 303$ K) as the hot reservoir?

(21) Problem: 1 L of water is poured off a 50-m-high tower every second. If the change in gravitational potential energy is converted into electricity with an efficiency of 85%, how many 60-W light bulbs can be illuminated?

(22) Problem: A grid-tied solar PV array composed of 14 panels in series is rated at a total array voltage of $V = 320$ VDC, and the current (I) measured over time is shown in the graph below.

Chapter 13 Renewable energy systems

a. Calculate the power generated in each of the six time periods below in watts if the following currents are measured: 9–10 am 1 A, 10–11 am 3 A, 11–12 pm 0.5 A, 12–1 pm 7 A, 1–2 pm 4 A, 2–3 pm 1 A.
b. For the entire day, how much energy did the array generate in kWh?
c. The owner had no electrical consumption on this day, so all the energy generated went back to the grid. If the utility rebates $0.10/(kWh) for solar electricity put back on the grid, how much income did the solar array produce for the owner on this day?

(23) **Problem:** Use the PVWatts online software (http://pvwatts.nrel.gov/). Calculate the estimated annual output (kWh) for the following solar array. Attach the PVWatts report to this assignment. Use the following (and otherwise default values):

> Location: Socorro, NM
> Weather data source: Albuquerque
> System size 3.27 kW
> Module type: premium
> Fixed mount (roof), 20 degrees angle, array azimuth 180 degrees (due south)
> System losses: 14%; inverter efficiency: 96%
> Cost of electricity from utility: $0.125/kWh
> Capital cost: $5.22/watt (all costs)
> Incentives: 30% federal, 10% state

The actual monthly production in 2015 is given below. Plot a time series of the monthly production. Plot the actual vs predicted output and find a best-fit linear regression to the actual (y) vs predicted (x) monthly kWh production. Describe the trends and comment on any differences or observations.
Actual (kWh from Jan. to Dec. 2015):
317.28, 421.49, 513.47, 573.55, 582.94, 580.91, 531.63, 543.71, 490.71, 433.3, 417.87, 323.02

(24) **Problem:** Investigate the viability of a solar jumbo jet. Research the Boeing 747 and find the power needed to run this aircraft (go with the max output). How much solar panel area would be needed? Perhaps you can cover the wings and use the cross-sectional area of the fuselage. Assume 250 W/m^2 as the daily average solar flux. Does it seem viable? Using Tesla Powerwall specs, how much battery mass would be needed to provide 1 h of energy needed to fly the aircraft to bring it back to an airport if things are not going well?

(25) **Problem:** A large solar power tower application is Ivanpah in California's Mojave Desert. The facility contains 173,500 heliostats, each with 2 garage door-sized mirrors. Each heliostat is approximately 8' × 10'. You can assume 55% of the input solar energy is collected (the rest is lost to reflection, dirty mirrors, and heat loss in collection system), and assume a 28.7% efficient steam power plant downstream. Estimate the maximum electrical power output.

(26) **Problem:** A parabolic trough system collects solar and covers an area of 1 km × 2.58 km. Estimate the annual kWh electrical production near Las Vegas. Use the map of kWh/m^2/day in Fig. 13.5 at a location near Las Vegas with a solar thermal collection efficiency of 55% and a power cycle efficiency of 28%. How many average US homes' electrical use does this supply?

(27) **Problem:** Calculate the energy (in eV) of the photons at the peak of the solar spectrum. Thermoelectric PV cells can absorb IR energy. Find the eV of a PV cell optimized for a radiative emitter at 1000°C.

CHAPTER 14

Socioeconomic context and equity in green building: Policy, tools, codes, and certifications

Learning objectives

(1) Establish the importance of policy and codes in directing green building projects.
(2) Understand the organizations, certifications, and standards related to green building.
(3) Evaluate the economics and incentives in green building projects.

Engineers and scientists focus on the technical and quantitative aspects of societal imperatives such as green building. However, the nature of green building merits placing these problems within the social, economic, political, cultural, and overall human factors context. Though only a snapshot, this chapter explores these facets of green buildings to contextualize the broader issues.

14.1 Economic factors, financing, and payback times

The notion that green building up-front costs are significantly higher (a "luxury" good and focused on such visual elements as green roofs) than conventional buildings is often not the case when examining the data (Hu, 2024). The up-front capital costs of sustainable buildings can be less than, equivalent to, or within a modest extra cost of 5% or 10% of conventional builds (Hu, 2024). Some key innovations and more sustainable materials have greater up-front costs. Many green technologies, though, have a reasonable payback time with cost savings due to lower energy consumption or greater longevity. They can also lead to better building performance plus improved occupant comfort and productivity. Research also points to the project team's expertise and effective design process rather than the system's capital costs as a key driver of how the total project costs compare (Hu, 2024).

Owner-occupied versus rental facilities have their own unique challenges related to green building. One constraint for green building is termed the split-incentive problem that applies primarily to rental properties. Split incentives with renovation projects involve the priorities that lessors vs lessees have with rental properties. The owner who pays for the updates and repairs seeks to minimize their capital costs, while the renter seeks to minimize their monthly expenditures on rent and utilities. In other words, there is little incentive for the owner to spend more to minimize renters' utility costs, while renters have little incentive to invest capital to improve the efficiency of buildings they are renting. This has improved somewhat as owners can market more efficient buildings to energy-savvy renters.

One of the major impediments to the integration of renewables and the use of lower-carbon solutions is the up-front capital costs. New directions include the use of community-based solar systems, solar leasing programs, and such programs as electric heat pump installations financed and leased by companies like BlocPower for multiunit residential buildings (BlocPower—Smarter, Greener and Healthier Buildings). The approach of many energy services companies, such as Ameresco, Inc., is to offer performance-based contracts. The contractor will improve the energy performance of a facility and will be paid for the energy savings from the upgrades.

Straight payback is often the simplest approach to examining the cost-effectiveness of a given green building effort. The payback time is found by taking the additional up-front cost of a green building feature divided by the annual cost savings associated with its operations, as shown in Example 14.1. The costs for the more efficient technologies have dropped considerably since this problem was first written, and this is left as an exercise at the end of the chapter.

EXAMPLE 14.1 Straight payback times

Problem: Based on use for 4h per day and an electricity cost of $0.11/kWh, calculate the payback period for (a) CFL bulbs and (b) LED bulbs in comparison with incandescent bulbs, using the following costs: four 60 W incandescent bulbs for $1.12; six 13 W CFL bulbs for $9.97; one 9 W LED bulb for $2 (when I first used this problem the LED cost was twelve times this!). Consider both capital (purchase) and operating costs; you can do a straight payback rather than discounting future cost savings.

Given: Incand=$1.12/4; Fluor=$9.97/6; LED=$2; Elec=$0.13/kWh
Find: simple payback
Assume: Only costs are capital and electric
Solution:
Set up a relationship to equate the costs of the more efficient bulbs to that of the incandescent
Total Cost Incandescent = Capital + Operating = $1.12/4 + Time × $0.13/kWh × 0.06 kW
Total Cost Fluorescent = Capital + Operating = $9.96/6 + Time × $0.13/kWh × 0.013 kW
Total Cost LED = Capital + Operating = $2 + Time × $0.13/kWh × 0.009 kW
Equate total costs to find payback time for more efficient bulbs:
Fluorescent: Assume payback in less than 1 incandescent lifetime:
$1.12/4 + T × 0.13/kWh × 0.06 kW = $9.97/6 + T × 0.13/kWh × 0.013 kW
T × (0.13 × (0.06−0.013)) = 9.97/6 − 1.12/4
→ T = 226 h or 56 days @ 4 h/day (payback for fluorescent)
LED: Assume payback in less than 1 incandescent lifetime:
$1.12/4 + T × 0.13/kWh × 0.06 kW = $2 + T × 0.13/kWh × 0.009 kW
T × (0.13 × (0.06−0.009)) = 2 − 1.12/4
→ T = 268 h or 67 days @ 4 h/day (payback for LED)

The payback on more efficient lighting is well less than a year. Notably, the purchase price for more efficient lighting has decreased significantly over the last decade, particularly LEDs, which are overtaking the market.

Present value analysis offers a way to calculate the cost payback of efficiency projects, taking into account the time value of money. The concept is that $1000 today is worth more than $1000 a year from today since it can be invested or used otherwise. The net present value (NPV) of future costs can be calculated by discounting the future dollars to the current value. The net present value of a future recurring cash flow (either positive or negative) is a future cash flow (R_t), divided by 1+the fractional discount rate raised (i) to the power of the project lifetime (t) (Eq. 14.1) (Example 14.1). Using the same terms, the analogous relationship for a set of recurring fixed payments occurring each time increment over the project lifetime t follows (Eq. 14.2) (Example 14.2). The choice of an appropriate discounting rate (e.g., prevailing interest rates, mortgage rates, or inflation rate) is an ongoing debate. For many purposes, a discount rate of 4%–6% is often used.

$$\text{NPV} = \frac{R_t}{(1+i)^t} \tag{14.1}$$

$$\text{NPV} = \frac{R_t}{i}\left(1 - \frac{1}{(1+i)^t}\right) \tag{14.2}$$

EXAMPLE 14.2 NPV for recurring payments

Problem: Based on an annual recurring electricity savings of $900, does a 30-year solar PV project make financial sense? Assume an initial cost of $7000 with a discount rate of 4% annually.
  Given: CapCost = $7000; Annual Elec Savings = $900/year
  Find: NPV cost analysis
  Assume: Only costs are capital and electric, which are fixed
  Solution:

$$\text{NPV} = \frac{R_t}{i}\left(1 - \frac{1}{(1+i)^t}\right)$$

$$\text{NPV} = \frac{\}24900}{0.04}\left(1 - \frac{1}{(1+0.04)^{30}}\right) = \}2415{,}563$$

The net present value of $15.6 K exceeds the installation cost of $7 K, and thus the project is favorable from a financial perspective at the discount rate used.

14.2 Federal policy and financial incentives

The policy tools to promote green building extend from local municipalities to the federal level, the latter of which can set the tone throughout the political economy. The efforts have been led by the US states facing the highest energy costs such as California, Hawaii, the Pacific Northwest, and the Northeast United States. At the federal level, the Inflation Reduction Act and Bipartisan Infrastructure Act contain multiple provisions promoting energy efficiency and decarbonization. They represent one of the largest single policy efforts to address climate change undertaken in the United States and followed past energy efficiency incentives contained in previous legislation.

The financial tax incentives for efficiency and decarbonization of energy systems are diverse and target many parts of the supply chain, from manufacturers to end-users. The cost-effectiveness of greener building options is often facilitated by financial and tax incentives offered by the government. These include 30% tax credits for individuals for energy efficiency measures, heat pumps, and renewable energy systems such as wind, solar, and geothermal energy systems. These are often in conjunction with utility incentives for such measures. Other financial incentives target home builders, mortgages, and grants or loans for rural and low-income households. The Database for State Incentives for Renewables and Efficiency (DSIRE) compiles, as a function of location, the energy tax incentives both at the federal, state, and local levels (https://www.dsireusa.org/).

The federal government has recently invoked the Defense Production Act to incentivize domestic heat pump production. This has been largely in response to the threat from the Russia-Ukraine war and the West's dependence on Russian oil and gas, though climate concerns are also a factor. The effort resulted from a proposal by environmental activist and writer Bill McKibben who proposed the effort as Heat Pumps for Peace and Freedom. The threat of climate change has also played into the use of federal grants and loans to develop the domestic heat pump industry.

Several federal-level programs exist mainly through the US Department of Energy. The Energy Star program, as discussed later in the certification program section, is a long-standing program that focuses on appliance efficiency. The Government Services Agency (GSA) also discussed later administering government purchasing efforts to improve the sustainability of government operations.

The Federal Emergency Management Agency has embarked upon improving the resilience of buildings and communities, now offering a Building Resilient Infrastructure and Communities (BRIC) grant program (https://www.fema.gov/grants/mitigation/building-resilient-infrastructure-communities).

14.3 Zoning and building codes

Local construction and land use codes dictate permitted land uses for assorted land use categories (e.g., residential versus commercial versus industrial). These include urban density, parking requirements, and affordable housing, among many other aspects that connect to sustainability. Building codes are the minimums specified for such systems as mechanical, electrical, and plumbing (MEP), structural integrity, fire risk, egress, and notably energy efficiency, among other provisions.

The first focus of code requirements is buildings that function properly without undue hazards to occupants or the public. Thus, codes are by nature conservative and slow to change concerning new technology, techniques, and materials. Recently, codes have had to keep pace with the rapid evolution of on-site renewable energy sources across all sectors, including residential.

Code drives the minimum energy efficiency of new structures as well. Codes are very locally dependent, though there are overarching international standards often adopted. The International Building Code (IBC) and International Residential Code (IRC) are the two most prevalent codes at the global level. The International Energy Efficiency Code (IECC) dictates the minimum standards for building energy efficiency.

Though some builders strive for high performance well beyond code, many will only meet code requirements out of concern for budget. Sometimes these issues for new construction or renovations can be alleviated by "on-bill" financing. This means the up-front costs can be paid incrementally on the utility bills resulting in overall similar or lower monthly utility costs than with conventional builds.

An example of recent building code amendments related to sustainability concerns the requirements for a minimum number of parking spots in urban areas for a given square footage and property type. This is being adopted to drive the use of alternate modes of transit as well as to alleviate the shortage and cost of housing in dense urban areas. As of 2023, Austin, Texas, is one of the earliest big cities to relax the code requirements for a mandatory number of parking spaces per square footage of building space. Building and land use codes will continue to evolve and intersect with green building applications.

14.4 Utilities and rate structures

Energy costs are directly related to the viability of energy efficiency measures in a given location. Utility services, such as electrical providers can be municipally owned, investor-owned utilities (IOUs), or cooperatives that are member-owned, the latter particularly in rural areas. No matter the provider, the rates that utilities charge are heavily regulated by a public utilities commission that must approve rate increases with convincing evidence provided for justification. Notably, this does not always assure utility best practices for protecting ratepayers.

The typical utility billing approach for electrical utilities includes the following components:

1. A customer charge, which is typically a flat fee for covering the infrastructure cost, maintenance, and other overhead costs.
2. An energy usage charge to cover the usage of each customer, typically billed based on electrical consumption (kWh).
3. Typically, for larger commercial customers, a peak demand charge based on the kW demanded by the customer at the time of peak utility demand.
4. Taxes and other specific fees, sometimes related to bond issues for construction projects.

Old-school rotating disk electric meters were used for decades for recording the electrical consumption of a building (Fig. 14.1). These now make a great application as a Junior Achievement product to sell as meter lamps (this one is still functional 40 years later recording the consumption of this lamp). The analog meters did spin backward when solar production exceeded household consumption. Newer digital approaches to electrical metering include "net metering" that records the building consumption as well as the on-site solar production. Here a meter showing the building's consumption and production are shown (sometimes a single meter will record both data streams). This system has produced nearly twice what the household consumes, and the excess is delivered back to the electrical grid under a net metering program. The solar REC meter records the production for paying the customer solar "Renewable Energy Credits" to meet their renewable energy goals or mandates.

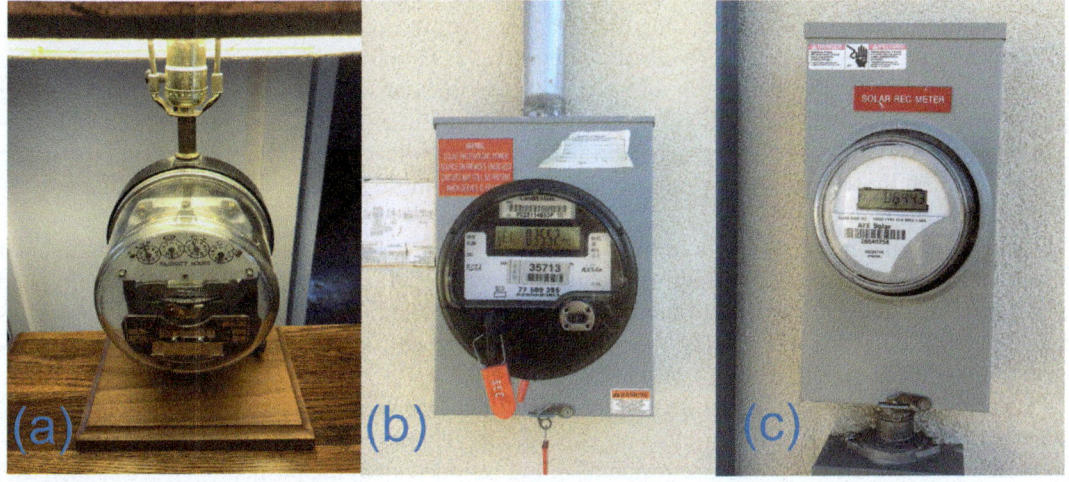

FIG. 14.1

(A) Traditional spinning disk mechanical meter for recording electric consumption and newer digital meters showing (B) the household consumption meter on the left and (C) the solar generation on the right meter.

Advanced metering technologies combined with more real-time data and feedback have allowed more accurate pricing of energy resources, including time-of-day pricing. These tools and approaches become more important with a utility grid with more variable generators. The net energy demand curve or "duck curve" was described previously (Fig. 9.19) in discussing renewable energy resources. It characterizes regions with high renewable energy penetration, particularly solar PV, where net demand drops during the daylight hours due to abundant solar resources. In the evening, when electric use surges and solar generation drops off, the net demand increases abruptly. Although tools and approaches have improved, this is still problematic for utilities, particularly as more intermittent energy sources are integrated into the grid.

Advanced metering and other demand-side management approaches can mitigate the duck curve problem. Advanced utility metering/billing approaches include:

- Tiered rates or variable peak demand charges
- Time of day (TOD) electric rates
- Smart meters that allow real-time feedback and two-way net metering (Fig. 14.1)
- Utility capability to modulate residential loads
- Other advanced metering applications, including prescribed load shedding by the utilities

14.5 Environmental justice issues and the building industry

Economics are also embedded in the political, social, and environment of a given region. Indicators almost uniformly show the socioeconomic progress of the global population over the last century, but persistent and severe issues including chronic poverty (~10% of the global population), illiteracy, and inequality need to be addressed.

Environmental justice issues center around the fact that economically and socially disadvantaged groups, particularly communities of color, often suffer disproportionately from the environmental impacts of a given issue, facility, or project. At the same time, these marginalized communities are often minor to negligible contributors to the problems.

For example, related to green building, those on the margins in terms of ability to relocate, adapt, and recover from increasingly severe climate change impacts will face the highest impacts. In most cases, they also contributed far less in terms of their lifetime emissions of greenhouse gases, raising equity concerns. Marginalized neighborhoods often suffer liabilities (e.g., forced relocations) with public infrastructure projects due to depressed property values and lack of political representation in the process. In short, those that bear the most responsibility and those that feel the most effects are quite different populations.

Furthermore, marginalized communities have fewer resources to mitigate or oppose environmental impacts in their communities. Combining this with lower property values, less stringent zoning, and less political resistance, it can lead to the location of undesirable facilities in marginalized communities. These issues will continue to play central roles in implementing solutions to environmental problems including in the realm of green building.

One of the downsides of green redevelopment and urban renewal projects is that they can occur at the expense of established working-class neighborhoods. The term gentrification denotes neighborhood renewal where the lower-income residents of redeveloped areas are pushed out as costs increase. Large influxes of new residents to an area cause friction in the best of circumstances. The economic aspect can be an additional strain on top of it. Gentrification of such areas can change the fundamental character and nature of the businesses and population of these areas in not always positive ways. The maintenance of living and working options for lower-income wage earners is the key concern. Concerns over affordable housing options have emerged in many major markets, particularly those that have experienced appreciating real estate.

Climate change is noted as a global problem, a "tragedy of the commons." It is not lost that those who contributed the least to the climate crisis (e.g., the global poor) are those who will suffer the most. The lowest-income nations are the ones who have contributed the least to our atmospheric carbon burden and are least prepared to deal with its consequences. An international-scale example of climate justice revolves around the climate impacts suffered by the residents of small island developing states, several of which are facing existential crises related to rising sea levels. Island nations such as the Maldives, Tuvalu, Kiribati, and the Marshall Islands are among those that are particularly at risk due to sea level rise and exacerbated tropical storms. Developing nations, particularly coastal, island, and low-lying nations, will undoubtedly suffer the greatest costs. Thus, it is vital that the wealthier nations, those that benefited from the prodigious use of fossil fuels, lead and finance the global decarbonization effort.

Even in the United States, in part due to redlining practices, the lowest-income districts of US cities are typically those that have a magnified urban heat island effect due to a higher proportion of asphalt and concrete compared to vegetated lots (Fig. 14.2). Solutions are numerous including locating cooling

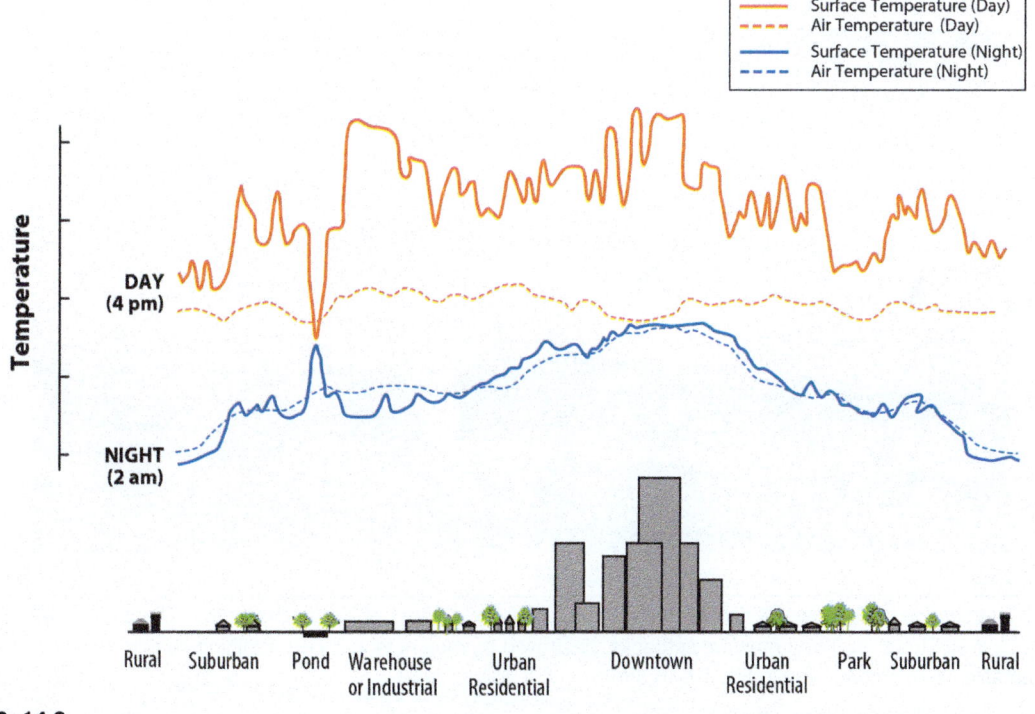

FIG. 14.2

Urban Heat Island effect (schematic courtesy of USEPA), where urban core temperatures can exceed surrounding rural temperatures by up to five degrees centigrade. Daytime temperatures are higher in urban cores due to materials, namely concrete and asphalt, and lack of shading (e.g., vegetation). This is often exacerbated in urban core areas with significant fractions of marginalized groups.

centers, assistance for energy efficiency and residential thermal insulation, and increasing shading in such districts to reduce the extreme heat burden. Again, just as with health-related air pollution issues, vulnerable populations include lower-income people, people of color, the very young and old, and those with underlying health issues. Even today, the formerly red-lined districts in many cities have across the board lower property values, funding for school systems, internet access, and health and well-being indicators.

A distinct example of an environmental justice issue in the United States is that of the Flint water crisis. A predominantly African American community suffered through lead poisoning and degraded water quality for years. Another specific environmental justice issue centers around the 20th-century African American community in the Greenwood district of Tulsa, Oklahoma. The area suffered one of the most violent race-based massacres in 1921, in an area described as "Black Wall Street" due to its emerging business community. After the near destruction of this Black business and residential neighborhood, the area rebuilt and became thriving again though subject to the redlining policies common in the post-Civil War United States. The region suffered further setbacks with the interstate project which cleaved north and south Tulsa and intersected the Greenwood district directly (Fig. 14.3). The effects of policy, zoning, and efforts at "urban renewal" associated with the interstate highway system fractured a recovering Black business community in the Greenwood district. The project is emblematic of other federal highway projects in the 1950s–70s that imposed much of the costs of new interstates on already marginalized communities.

FIG. 14.3

The intersection of Greenwood and Archer in the Greenwood District of Tulsa, Oklahoma. The area known as "Black Wall Street" suffered massive setbacks, destroyed by race riots in 1919, redlining zoning practices, and bifurcated by interstate built in the 1960s shown in the background.

The issue of socioeconomic equity of energy use and efficiency is one being addressed by both government agencies as well as private entities, both nonprofit and for-profit. An example of the latter,

BlocPower is a startup in New York that focuses on energy upgrades in multiunit buildings, particularly lower-income neighborhoods (BlocPower—Smarter, Greener, and Healthier Buildings). These dwellings often face the split-incentive problem of landlord vs tenant needs. They also often have economic barriers to raising the capital for energy upgrades. On the building to the neighborhood level, BlocPower leverages building electrification and decarbonization projects while trying to reduce the barriers to up-front costs via grants, loans, and tax incentives.

14.6 Incorporating social, environmental, and economic costs over the life cycle

A more holistic business model that values more than dollars has developed in the last several decades. The triple bottom line economy incorporates monetary costs but also values environmental preservation and social equity. Performing triple bottom-line analysis attaches value to these other parameters including minimizing environmental impacts and considering the welfare of those involved in the process from workers to customers.

The human development indicator (HDI) is a metric that incorporates economic, environmental, and social indicators such as education levels. Rather than strictly an economic indicator like the gross domestic product (GDP) per capita, HDI includes indicators such as life expectancy, education levels, infant mortality rate, and other social and environmental indicators to gauge the quality of life for a population. The HDI scales directly with energy use up until mid-range per capita energy use, at which point the curve flattens out (Fig. 14.4). Above about 5000 kWh/capita annually, further gains in HDI are negligible.

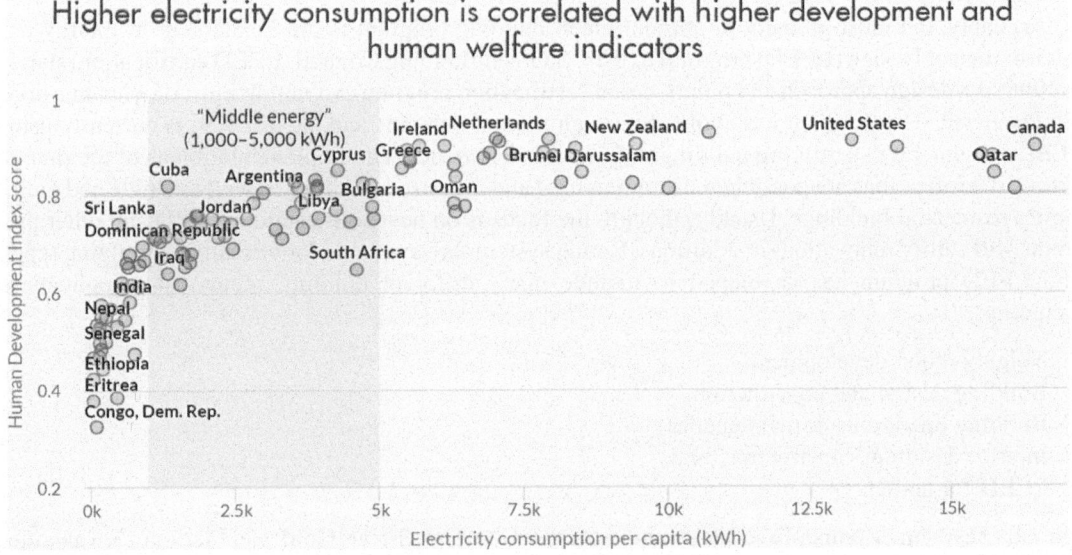

FIG. 14.4

The relationship between the human development index (HDI) and electricity consumption (annual kWh/capita) as a function of nation (Center for Global Development using UN and World Bank data, used with permission) (CGD, 2016).

A related environmental priority is what is termed the circular economy. Rather than a straight-through linear production model from mining to disposal, the circular economy concept returns materials to the production cycle at many points in the process. Such "cradle to grave" (or cradle to cradle) analyses have become vital to exploring the entire life cycle costs, economic and otherwise.

Life cycle analysis (LCA) takes a long-term, holistic approach to determining the costs of a given technology. The approach looks at the process from extraction to disposal to quantify the economic and environmental costs of the product or project. Whole-building LCA takes a comprehensive view of the building, its components, the materials, and its energy use in assessing these from cradle to grave (or more desirably, cradle to cradle). Much of the life cycle analyses focus has been on urban, new construction, more energy-conscious, and sustainable buildings (Cabeza et al., 2014). Life cycle analysis includes such facets as energy consumption, water consumption, use of recycled materials, and recyclability of the product. The International Organization for Standardization (ISO) in ISO 14000 specifies the protocols for documenting the LCA. The number of tools available for green building design and planning is constantly expanding. Building information modeling is a building design, drawing, costing, and modeling approach using software tools, and numerous software packages exist for such applications.

14.7 LEED: Leadership in energy and environmental design

Numerous sustainable and green building organizations and certification programs have developed over the past several decades. Here we will discuss several of the most prominent certification programs. Various programs focus on certifying the design and performance characteristics of devices and materials, the green building professionals, and the buildings as a whole. Some programs certify or not, whereas others have scoring and several certification levels.

Arguably, the most globally prominent green building program is the Leadership in Energy and Environmental Design (LEED) program of the US Green Building Council. LEED certification, started in 1998, is a design approach and points-based certification program evaluating most of what this book has discussed. The program has evolved through several versions and, as of 2024, is currently using LEED version 4.0 in certifying buildings. Version 5 is in review and implementation as of the time of writing. Certifications are possible with commercial and industrial facilities as well as single and multifamily residential buildings. Usually, though, the focus is on new construction; LEED and other programs will certify renovations or additions. Rating systems are available for virtually all building types. The LEED program has developed to include many different building sectors and applications including:

- Neighborhood development,
- Building design and construction,
- Building operations & maintenance,
- Interior design & construction, and
- LEED for homes

The LEED system is points-based, and rankings from Certified, Silver, Gold, and Platinum are awarded to buildings after an extensive review and documentation process. The areas for which points are awarded are summarized in Table 14.1. LEED Green Associate (GA) and LEED Associated

Professional (AP) are individual professional credentials that are also valued in the LEED process. LEED-GA or AP are valuable credentials in the marketplace and a way for an engineer to learn about the green building arena.

Table 14.1 LEED v4 program details as of 2024.

Categories evaluated	Description	Points awarded
Location and transportation	Site location (e.g., urban infill) and transportation options	16
Sustainable sites	Site-related outdoor management, heat island, and light reduction	10
Water efficiency	Indoor and outdoor water use reductions, plumbing system's efficiency, rainwater collection	11
Energy and atmosphere	On-site energy systems, efficiency efforts, refrigerant management	33
Materials and resources	Construction materials, recycling programs	13
Indoor environmental quality	Indoor air quality, thermal comfort, light, and noise	16
Innovation	Innovative systems and materials and involvement of a LEED-accredited professional	6
Regional priority	Dependent on the local to regional priorities	4
Total points	Certified 40–49; Silver 50–59; Gold 60–79; Platinum 80–110	110

No certification program is perfect, and all have their limitations. Some studies have shown that in practice LEED buildings are not outperforming conventional (at least at lower certifications) (Schendler and Udall, 2005). An overfocus on points means the accreditation can be gamed or turned into a "greenwashing" marketing tool with a checkbox-type system. For example, under LEED 4.0, remediating a brownfield site and installing bike racks are rated as equivalent efforts at 1 point apiece. With its design focus, LEED prioritizes the design of green buildings and modeling tools over real-world performance. It has also received criticism for its cost of certification, complexity, and bureaucratic level of documentation. However, LEED remains the gold standard for green building certifications and is constantly improving (Fig. 14.5).

Nonetheless, LEED is constantly evolving and will continue to be the global standard going forward. LEED has incorporated the many improvements that practitioners have recommended. One new credit is social equity. LEED is currently rolling out v5 of its certification program with a strong emphasis on climate action/decarbonization, health and well-being associated with buildings, and restoration of ecological systems. Over time, the LEED system has moved from primarily focused on energy efficiency to a broader sustainability context. LEED v5 is even more holistic and emphasizes decarbonization, equity, biodiversity, and resilience, among other aspects. It is also making more actions mandatory as prerequisites for LEED accreditation.

The Greensburg, Kansas, recovery from a tornado that destroyed the town in 2007 is an example of a community that went in big on LEED certification and sustainability in general. After a devastating tornado with 200 mph winds flattened the town, Greensburg, Kansas, decided its rebuilding effort

FIG. 14.5
The Lopez Chemistry Building at the New Mexico Institute of Mining and Technology is a LEED silver building. Many states require that municipal buildings meeting certain thresholds be built to meet LEED accreditation.

would lean heavily into sustainability and LEED certification, with many buildings achieving platinum status (NREL, 2012). A few of the highlights include the use of geothermal heat pumps, exceptionally tight thermal envelopes, reducing energy consumption by two-thirds to three-quarters, and producing more energy than consumed with nearby wind turbines. Many of the rebuilt buildings reduced energy consumption by over 50%. Unfortunately, the increasing incidence of severe weather will guarantee that we will have an increasing opportunity to rebuild better. It merits approaching it more sustainably!

14.8 ISI & ASCE ENVISION sustainability program

The Institute for Sustainable Infrastructure and the American Society of Civil Engineers (ASCE), a venerated leading professional organization in the civil engineering field, have integrated sustainability throughout their programs and future planning of their operations. ASCE is an organization that focuses on infrastructure and construction projects. ISI is an organization founded under the auspices of ASCE and several other organizations to implement a sustainability program.

Together, ISI and ASCE have established a program called Envision that offers accreditation for projects and for students and practitioners in civil engineering (https://sustainableinfrastructure.org/). Through the Envision program, ISI and ASCE provide technical resources such as reports, journals, books, case studies, and short courses related to the sustainability of infrastructure.

ASCE describes sustainability consistent with other organizations' and fields' definitions: "ASCE defines sustainability as a set of environmental, economic, and social conditions—the 'Triple Bottom Line'—in which all of society has the capacity and opportunity to maintain and improve its quality of life indefinitely, without degrading the quantity, quality, or the availability of natural, economic, and social resources." ASCE offers short courses addressing the triple bottom line approach and life cycle analysis approaches as well as a professional certificate program in sustainable infrastructure.

The Envision program, also including the participation of the American Public Works Association, and the American Council of Engineering Companies, is focused on improving the sustainability of infrastructure-related projects using a project rating system evaluating 60 criteria related to environmental, social, and economic aspects. These are broken down into the categories of Quality of Life, Leadership, Resource Allocation, Natural World, and Climate and Risk. The projects highlighted under Envision include such facilities as airports, waterworks, wastewater treatment plants, managed wetland restoration projects, and community parks. As of summer 2024, Envision has accredited 344 infrastructure projects. A professional accreditation, the Envision Sustainability Professional (ENV SP), is offered through the Envision program via the Institute for Sustainable Infrastructure.

14.9 Indoor environmental quality certification programs

The WELL Building standard from the International Well Building Institute is a certification program similar to the LEED standards but focused on the interactions of people with the built environment and specifically the occupants' health and well-being. It focuses on the following elements of the indoor environment: air, water, nourishment, light, fitness, comfort, and mind (WELL Building Standard® | WELL Standard (wellcertified.com)). It features required aspects as well as elective aspects plus a category for innovation. It has been developed using evidence-based research and relies upon over 7000 peer-reviewed articles from the literature on indoor environmental health. Like LEED, it has mandated technologies or efforts that must be taken as well as measurable performance-based standards. The standard is in its v2.0 stage for commercial buildings as of 2024.

14.10 Wider World of green certification programs

The number of certifications and accreditations regarding green building has proliferated over the past several decades. This is good news for the field but can be daunting for practitioners and confusing to the public (Doan et al., 2017). Programs span from local municipalities (e.g., Austin Energy Green Building) to international (e.g., Green Globes). We give some highlights here with the caveat that this list of green building programs is incomplete and a moving target. Austin Energy Green Building is one of the most recognized and oldest municipal green building programs. Green Globes is an international program that provides flexible assessments for energy, water, materials, and more, awarding One to Four Globe ratings. Similar to the Austin Energy Green Building program, numerous programs have been launched at the local level that offer localized solutions to green building needs.

Architecture 2030 was mentioned earlier in this book, and it is a program that focuses on environmental design. Architects have many venues for green building, including the Architecture 2030 organization, which was a pioneer in both quantifying and working to minimize the contributions of the built sector to climate change (https://architecture2030.org/).

The Energy Star program began in 1992 and is administered by the US EPA and Department of Energy. It focuses on the energy efficiency of appliances and other energy-consuming devices. The Energy Star labels give the annual energy consumption and approximate costs of operation of the rated appliances. Depending on appliance type, an Energy Star rating typically means 20%–40% lower energy consumption than the average appliance. A related program for homes, the Home Performance with Energy Star (HPwES), certifies buildings and contractors meeting performance criteria related to energy use.

The American Society for Heating, Refrigeration, and Air Conditioning Engineers (ASHRAE) publishes performance and efficiency standards for various HVAC systems and related building operations. One of the most relevant is ASHRAE Standard 90.1, most recently updated in 2022 (Standard 90.1 (ashrae.org)). ASHRAE Standard 90.1 covers energy performance for commercial buildings except low-rise residential buildings. It offers two pathways for compliance: a prescriptive pathway and a performance pathway. The prescriptive pathway provides detailed specifications that must be met for power use, lighting, the building envelope, HVAC, hot water, and assorted other equipment, while the performance pathway relies upon modeling of the building's performance.

Under Title 24, California is now the first state that has mandated embodied carbon reductions in new construction starting in 2024. The limits are specified for steel, glass, mineral wool, and concrete for nonresidential buildings of 100,000 ft^2 or larger, schools of 50,000 ft^2 or larger, and all nonresidential buildings in 2026.

Certification is voluntary though important in the solar industry with the Solar Rating and Certification Corporation (SRCC) certifying solar thermal equipment and the North American Board of Certified Energy Practitioners (NABCEP) certifying installers of solar PV panels.

The Living Building Challenge sets stringent sustainability standards. The NAHB Green Building Standard guides sustainable residential construction. Resnet with its Home Energy Rating System (HERS) rates the energy efficiency of residential buildings. A summary of programs relevant to green building is provided in Table 14.2, although incomplete, and many other programs exist internationally.

Table 14.2 An incomplete list of green building relevant certification programs.

Green building program	Organization	Description and attributes	More information
LEED (Leadership in Energy and Environmental Design)	US Green Building Council	Points-based green building design rating system to certify buildings and professional accreditation	https://www.usgbc.org/leed
HERS Index (Home Energy Rating System)	RESNET (Residential Energy Service Network)	Rating system for homes focused on energy efficiency	https://www.resnet.us/
Energy Star	DOE (US Department of Energy)	Federal energy efficiency rating program for appliances and energy-consuming devices	https://www.energystar.gov/
WELL-Certified	IWBI (International Well Building Institute)	Certification program for buildings for healthy and safe indoor environments and professional accreditation	https://www.wellcertified.com/

14.10 Wider world of green certification programs

Table 14.2 An incomplete list of green building relevant certification programs—cont'd

Green building program	Organization	Description and attributes	More information
BIFMA-compliant	BIFMA (Business and Institutional Manufacturers' Association)	Standards for safety and durability of indoor furnishings	https://www.bifma.org/page/bifma-compliant
EPP (Environmentally Preferable Products)	GSA (US General Services Administration)	Government procurement program focused on reducing environmental impacts of government purchasing of materials	https://www.gsa.gov/climate-action-and-sustainability/buy-green-products-services-and-vehicles/buy-green-products/environmentally-preferable-products
Architecture 2030	Architecture 2030	Architectural design program focused on reducing the greenhouse gas emissions from the built sector and making it a solution to the climate crisis	https://www.architecture2030.org/
BREEAM (Building Research Establishment Environmental Assessment Methodology)	BRE Group	Validation and certification for a sustainable built environment from design, construction, operations, and repurposing	https://bregroup.com/products/breeam/
ASHRAE standard 90.1	ASHRAE (American Association of Heating, Refrigeration, and Air Conditioning Engineers)	Performance standards for HVAC systems and equipment	https://www.ashrae.org/technical-resources/bookstore/standard-90-1
ENVISION Sustainable Infrastructure Rating System	ISI (Institute for Sustainable Infrastructure) and ASCE (American Society of Civil Engineers)	Certification and professional accreditation of the social, economic, and environmental performance of infrastructure-related projects with a focus on sustainability and resilience	https://sustainableinfrastructure.org/envision/
FSC-Certified	FSC (Forest Stewardship Council)	Certification program for responsibly grown lumber and wood products	https://fsc.org/en
BPI-Certified	BPI (Building Performance Institute)	Certification program for buildings as well as professional accreditation for home performance contractors	https://www.bpi.org/
SRCC-Certified	SRCC (Solar Rating and Certification Corporation)	Certification of meeting performance standards for solar products	https://solar-rating.org/

Continued

Table 14.2 An incomplete list of green building relevant certification programs—cont'd

Green building program	Organization	Description and attributes	More information
Passive House Enerphit	Passivhaus Institute	International certification program for new construction and retrofit focused program emphasizing energy efficiency, comfort, and affordability	https://passivehouse.com/
Austin Energy Green Building	City of Austin, Texas	Local green building program credited as the first local municipality-administered green building program	https://austinenergy.com/energy-efficiency/green-building

New certifications are emerging for net-zero energy and net-zero carbon buildings from LEED, Net-Zero Energy NZE 1.0 standard from the International Living Future Institute (ILFI), ZeroCode, Zero Carbon Certification, and the Living Building Challenge, among other organizations. The details of how this is calculated are under intensive discussion, including how to define boundaries, whether or not to allow indoor combustion, and other green building issues. US DOE has a designation of zero-energy-ready homes that are built to a standard of efficiency such that adding a renewable energy system such as solar PV could provide most or all the energy needs of the building. The evolution of all these programs will continue to define the green building field.

14.11 Career directions in green building: Energy systems engineer

One example of key career pathways related to green building is energy engineering, represented by the AEE (Association of Energy Engineers | (aeecenter.org)). Jobs in this area exist in all sectors: industrial production, utilities, government building operations, academic institutions, consulting firms, and others. Many of the practitioners follow this career pathway with backgrounds in mechanical or electrical engineering with a focus on energy systems. Advanced degree programs in energy systems engineering and related fields also serve as an entry point to the profession.

Overall, energy engineers design, operate, and maintain the complex systems that generate, transmit, and use energy. Their work may involve both conventional energy generation as well as renewable and on-site distributed sources. The job often involves design through operations of buildings with a particular focus on energy systems, HVAC, lighting, and refrigeration. The goals include reducing energy costs, minimizing energy-related emissions and other environmental liabilities, and integrating on-site energy production in some cases, all while maximizing the facility's function and comfort. Depending upon the job details, facilities management, overseeing retrofit projects, emissions tracking, and environmental compliance efforts may also be part of the job. Operating, measuring, and troubleshooting mechanical and electrical systems and analyzing data related to their operation are important skills. Control systems for such operations are evolving to higher levels of automation, and having engineers trained in these details is important.

14.11 Career directions in green building: Energy systems engineer

The AEE offers many professional resources for energy engineers including several accreditation programs. Among others, the Certified Energy Manager (CEM) or Certified Energy Auditor (CEA) certifications. To become certified, they require a combination of educational degree and practical, relevant work experience, training seminars, and a body of knowledge test covering many of the topics discussed in this textbook (Table 14.3). The CEM targets energy engineers in energy management

Table 14.3 Body of knowledge tested in the Certified Energy Manager certification test (Association of Energy Engineers, AAE).

Body of knowledge	Exam coverage	Description and example subtopics
Codes and Standards	3%–5%	Covers ASHRAE, LEED, ISO, Energy Star, Green Globes, and others
Energy Accounting and Economics	6%–10%	Simple payback, net present value, depreciation, life cycle cost analysis
Energy Audits and Instrumentation	8%–12%	Combustion analysis, duct air velocity, light intensity, electric metering, T, P, & RH, heat, infrared equipment
Electrical Power Systems and Motors	9%–13%	Power factor, peak demand, AC and DC motors, fans and pumps, variable flow systems, variable frequency drives
HVAC Systems	9%–13%	Degree days, heat transfer, vapor compression cycles, refrigerant systems, chillers, economizers, distribution systems
Industrial Systems	6%–8%	Waste heat recovery, heat exchangers, turbines, compressors, pumps, boilers, steam cycles, compressed air systems
Building Envelope	3%–5%	Insulation, solar heat gain, thermal control, heat transfer, vapor barriers
Combined Heat and Power (CHP) Systems and Renewable Energy	4%–6%	Combined cycles, solar, wind, biomass, hydropower, distributed generation, microgrids, battery storage
Fuel Supply and Pricing	2%–4%	Natural gas, fuel oil, electricity costs, regulation, and deregulation
Building Automation and Control Systems	7%–11%	Control systems and strategies, analog, digital, PID controls, web, cloud, and Internet of Things
Thermal Energy Storage Systems	2%–4%	Chilled water, storage media, thermal storage, ice systems, phase-change materials, sizing, and operations
Lighting Systems	6%–8%	Lumens, color temperature, color rendering index, retrofits, efficacy, controls, and dimmers
Boiler and Steam Systems	3%–5%	Combustion efficiency, air-to-fuel ratios, condensing boilers, waste heat recovery, scaling and fouling
Maintenance and Commissioning	8%–12%	Maintenance of controls, leaks, insulation, relamping, scheduled and preventative, water treatment, and commissioning phases
Energy Savings Performance Contracting and Measurement & Verification	3%–5%	Measurement and verification, financing, risk assessment, demand-side management, energy service companies, and contracts

positions, whereas the CEA is geared toward third-party energy auditors who come into facilities to audit for energy efficiency. Additionally, the Certified Green Building Engineer designation is available to engineers who are CEM certified as well as Professional Engineer certified.

Energy systems engineers have diverse pathways and applications. The operations and maintenance of the physical facilities of a university is one such job. As an example of a problem that arose, NMT energy management engineers identified a building installation problem related to the chiller-based HVAC system that was discussed earlier in the book (Fig. 9.13). The use of real-time measurements and controls allowed the determination that the inlet and outlet to the system in a building on campus had been reversed during construction. The problem was readily corrected, and the performance of the building climate control immediately improved.

14.12 Chapter summary and conclusions

A grand challenge of our building sector is how to make inspiring buildings that are functional, aesthetic, affordable, accessible, equitable, well-located, and enhance the planet rather than degrading it. Though this is not a small task, it is achievable if we prioritize it. In this chapter, we looked at how these buildings must meet socioeconomic goals that are equally important to technical performance.

The economics of any green building application will always intrude upon the ideal green building solution. Green building is often associated with high-end and luxury buildings, and the perception, often mistaken, is that building green costs significantly more than conventional. The up-front cost differential between green and conventional ranges from significant to minimal to negative. Even with greater capital costs in some cases, green buildings often pay back over time in lower operational costs. Paybacks can be calculated by simple payback or by discounting the future savings using an appropriate discount rate. A life cycle cost analysis looks at these costs, economic and/or environmental and social, from cradle to grave in evaluating the supply chain associated with a project or product.

Building codes in a local jurisdiction are a top-line determinant of green building options and define the up-front capital costs of a project. The ongoing operational costs of a building are heavily dependent on the utility rate structures and equipment efficiency. With new "smart meters," the use of time-of-day rates and other tools can help to more accurately price energy.

Life cycle assessment (LCA) takes the holistic view through a product's lifetime to consider the extraction to manufacturing to use to disposal costs. Such a cradle to grave cost analysis looks thoroughly throughout the supply chain of a product. This approach, along with other sophisticated tools for planning, including building information modeling, helps to minimize impacts.

Green building programs and tools are now available to assist the practitioner with greening their building. Green building certification programs may focus on certifying buildings themselves, the professionals who work on the buildings, and the materials and components that go into buildings. The Leadership in Energy and Environmental Design (LEED) program from the US Green Building Council is a valuable tool for promoting green building. LEED is a dominant certification process and credentialing in the field of green building. LEED, among other rating systems, provides a framework for healthy, efficient, and economic green buildings. The green building world requires many professionals to implement the programs. These include energy systems engineers, environmental compliance professionals, and sustainability consultants, among others. A range of programs, including ASCE's Envision and the LEED AP designation, offer professional accreditation to individuals in this pursuit.

14.13 End of chapter exercises

(1) **Concepts:** Provide a one-page summary of one of the major green building certification programs.
(2) **Concepts:** Make a prediction for the price of WTI crude oil, natural gas, and polysilicon for the end of the semester. Provide reasoning for your predictions.
(3) **Problem:** Take the most recent LEED scorecard and perform a mock evaluation of the building you are taking this class in or another relevant building. Some attributes will have to be estimated. Compare the results among all the students to see how consistent the ratings are.
(4) **Problem:** Examine the Empire State Building's green renovation. What were some of the key features? Challenges? How do skyscraper renovations differ from smaller structures? Do some research and find the cost of renovation vs the initial cost to build the building in 2011 dollars (using a discount rate of 4% annually and exponential growth). Do some research and estimate the cost to rebuild a comparable building from the ground up in New York City present day. http://www.esbnyc.com/esb-sustainability
(5) **Problem:** Based on use for 4 h per day and an electricity cost of $0.13/kWh, calculate the payback period for (a) CFL bulbs and (b) LED bulbs in comparison with incandescent bulbs, using the following costs: 60 W incandescent bulbs for $1.00 (2000 h life); 14 W CFL bulbs for $1.50 (lifetime of 10,000 h); one 9 W LED bulb for $2 (15,000 h life). Consider both capital (purchase) and operating costs; you can do a straight payback rather than discounting future cost savings.
(6) **Problem:** Public restrooms used to have hand dryers that were like hair dryers (blowers with heating elements). Newer dryers are high-velocity air jet blowers, typically without heat elements. The old-school dryer takes 45 s to dry one's hands and consumes 2500 W while doing so. The newer dryer takes 1000 W and 12 s to accomplish the same task. Other restrooms still use paper towels. Let us compare the costs.
   (a) An ever-open truck stop averages 20 uses per hour. Electricity cost is $0.14/kWh. Assume an installed cost of $1500 for the newer air jet dryer. Calculate the simple payback time of replacing the older electrical resistance heater hand dryer with the newer air jet dryer.
   (b) Assume a roll of commercial dispenser paper towels costs $12 delivered for a 600-ft roll weighing 15 lbs. Assume each average use involves two paper towels at 12 in. length apiece to dry one's hands. Calculate the payback time and annual solid waste savings of replacing the paper towel dispenser with the air jet dryer.
   (c) Describe the heat transfer and thermodynamic principles involved in each drying process. Which would you evaluate as more energy efficient?
(7) **Problem:** Compare the carbon emissions with the options examined in the preceding problem. Compare in terms of the CO_2 footprint of paper towels vs pure carbon coal-generated electricity with a typical power plant efficiency of 33%.

Chapter 14 Socioeconomic context and equity in green building

(8) **Problem:** Using Example 14.2, construct a spreadsheet to calculate the NPV of the recurrent payments example. Does it give an equivalent answer?

(9) **Problem**: Using Example 14.2, in what year does the project just achieve financial payback? Use a spreadsheet or solve analytically.

(10) **Problem:** Using Example 14.2, at what discount rate does the project just achieve financial payback? Use a spreadsheet or solve analytically.

(11) **Problem:** The average office building in 2000 used 100,000 BTU/ft^2/year. Convert to kWh/m^2/year.

CHAPTER 15

Climate adaptation, resiliency, and the built sector: Designing for future disruptions

Learning objectives

(1) Recognize our current meteorological extremes are becoming more extreme as we continue to add energy to the climate system motivating the need for more resilient buildings.

(2) Investigate that, in a world of accelerating changes in climate, complex and regionally dependent effects are occurring that involve the built sector.

(3) Define resilient system design that offers improvements in making systems that will bend and recover rather than break.

15.1 Overview of global change effects on the built environment and adaptation

Our closing chapter seeks to pull this all together with an eye toward the future of building construction. The area is a very emerging area of research in the field of civil and environmental engineering. It is likely that it will have changed considerably by the time this book is published, and thus please take this as only a sampling of this field. It is one that uniquely combines the physical sciences, social sciences, and economics. It also, like geoengineering, offers a moral hazard in that it cannot be relied upon as the sole or primary solution to climate change.

The changes in our earth system have been accelerating and are quickly becoming staggeringly disruptive to both the natural and built environments. One aspect of climate change that is not universally appreciated is the cumulative nature of the problem. This results from the century-plus lifetime of CO_2, the decades-plus time lags in the climate system, and the fact that energy continues to accumulate in the earth system due to legacy emissions (i.e., the "excess" CO_2 currently in the atmosphere). The "in the pipeline" warming of our planetary system insures that some further change is now unavoidable, and the built sector must plan for this reality (Hansen et al., 2023). Notably, this is not to say that mitigation by rapidly reducing our emissions is not the first, best, and most urgent action we can take! The movement to electrify and make more efficient most energy end uses and provide that electricity by low- to zero-carbon energy sources is the first key to addressing climate change (Griffith, 2021).

However, our civilization needs to prepare for the increasing frequency, severity, and cascading nature of multiple extreme events in vulnerable areas as climate change accelerates (NASEM, 2022).

The desired goal is that infrastructure dampens rather than amplifies the effects of disruptive events. The challenge will be integrating the pace of change and calibrating an appropriate response to these events. For example, Houston may be able to recover from the 500-year flooding event associated with the 2017 Hurricane Harvey. However, when 500-year flooding events happened on an annual basis from 2015 to 2019, it strains the response capability and brings into question future habitability. Furthermore, extreme events can be disparate. Amidst the flooding events in Houston, Winter Storm Uri in 2021 resulted in hundreds of deaths and an estimated $100 billion cost to Texas as a whole. Additionally, an intense windstorm was associated with a derecho event on 16-17 May 2024, with early estimates of $7 billion in damages. This was followed shortly by Hurricane Beryl landfalling in July 2024, also causing extensive flooding. As another example on a smaller scale, the June 2024 South Fork and Salt wildfires near Ruidoso, New Mexico evacuated a town of 8000 residents and destroyed 1400 structures. It was followed within a day by intense flooding events and mudslides from remnants of a tropical storm, which then created a 200-mile-long haboob (wall of blowing dust from intense thunderstorm outflow). Such iterative and recurrent hazards will strain emergency response as there will be populations who have not recovered from an earlier event when the next disaster arrives.

Resiliency is an area that connects the building, site, and surrounding infrastructure/environment in a way that merits more holistic consideration. Discussion of extreme events is a starting point for understanding how this will impact the built environment. A changing climate is not only a motivation to build greener (the mitigation approach) but will also alter buildings to a new and accelerating climate reality (the adaptation approach) and could also delve into intentional climate modifications related to buildings (the geoengineering approach, e.g., painting all roofs white).

15.2 Extreme weather events & the built environment

Climate disruption has already altered the natural world and human-built environment on a global scale. More urgent terms such as climate crisis or climate disruption have emerged as symptoms have become further manifest and more extreme. Extreme weather events include those associated with extreme heat, water, and winds. They also include winter storm events that can deliver enhanced precipitation and wind speeds, including Arctic outbreaks, nor'easters, ice storms, and blizzards, including lake effect snow (Halverson, 2024).

Two areas of active research ask the questions: (a) is weather becoming more extreme because of climate change, and (b) can attribution (or a contribution) be made of extreme weather events to anthropogenic climate change? With extreme events due to being on the tail of the probability distribution and thus infrequent, it is often difficult to get statistics with a meaningful sample size (e.g., hurricanes). Nonetheless, climate change impacts beyond increasing global average temperature rise are beginning to emerge in the occurrence of extreme events. More specifically, the climate change signal is often seen as intensifying extreme weather events by increasing the precipitation amounts or maximum wind speeds (Reed et al., 2022).

Increasingly extreme weather is clear with the number of high vs low-temperature records. This has profound implications for resilient building approaches for pavement surfaces, road buckling, worker health and safety, and numerous other aspects related to green building. As of the early 21st century, this ratio, obviously expected to be ~1, is approximately 2 with projected continued increases (Meehl et al., 2016). This is also observed in the range and seasonality of plant cycles (earlier spring and later fall). Data is increasingly showing more variability in precipitation over landmasses as well (Zhang et al., 2024).

Almost globally, sea ice and glacial extent and mass are declining, while permafrost area is declining. The cryosphere is the "air conditioner" of the planet, and this has led to feedback where warming has been amplified at the poles (IPCC, 2021b). The mean sea level globally has increased by about 0.25 m and continues to accelerate. The "mega-drought" in the western United States over the past 20 years is the worst in the last 1200 years (Williams et al., 2022). These are sobering indicators of a planet in flux. This underscores the longevity of these changes as a century- to millennial-scale for enhanced greenhouse warming. The extreme monsoon conditions in Pakistan in the summer of 2022 submerged one-third of the nation, with over 1000 deaths and 1 million homes destroyed, along with extreme heatwaves in many locations, including the western United States and Europe.

Some key threats in the natural world associated with the climate crisis and of relevance to the built environment include:

- Longer, hotter, drier summers, particularly in the western United States
- Longer, hotter, wetter summers, particularly in the southeastern United States
- More frequent and severe extreme heat (e.g., heat dome events)
- More frequent and severe wind events damage both those associated with cyclones and straight-line winds
- More persistent and pervasive drought conditions
- Freshwater scarcity issues and freshwater depletion in arid regions
- Greater intensity, duration, and destruction of wildfires and expanded fire-prone regions
- Heavier and more frequent extreme precipitation events
- Flooding events of higher frequency and severity

Extreme events are causing increasingly costly and frequent losses (Fig. 15.1). Annual "billion-dollar" weather events in the United States are often numbering a couple dozen per year with summed losses often reaching into the hundreds of billions in USD. The analysis of billion-dollar events accounts for 80% of all storm losses and includes public and private, insured, and uninsured losses (Smith and Katz, 2013; Smith and Matthews, 2015). The data relies on insured loss reports including the National Flood Insurance Program and then uses a best available multiplier factor to convert insured losses into total losses for each event.

By count, the largest number of events are typically severe storm events (wind, hail, severe thunderstorms), while the most damaging in terms of dollar costs and deaths are tropical cyclones (hurricanes). Heat waves cause the most human deaths, typically. Wind events are included in "Severe Storms" and can be straight-line winds not associated with hurricanes or tornadoes. With each degree Celsius of temperature increase adding 7% to the water vapor concentration (all else being equal), extreme precipitation events dubbed "rain bombs" are becoming more commonplace. The "atmospheric rivers" multiple-day events that are impacting the West Coast of the United States are another heavy precipitation event of notable increase. As we continue to add energy to the earth's system via increasing greenhouse gases, these losses will multiply.

Chapter 15 Climate adaptation

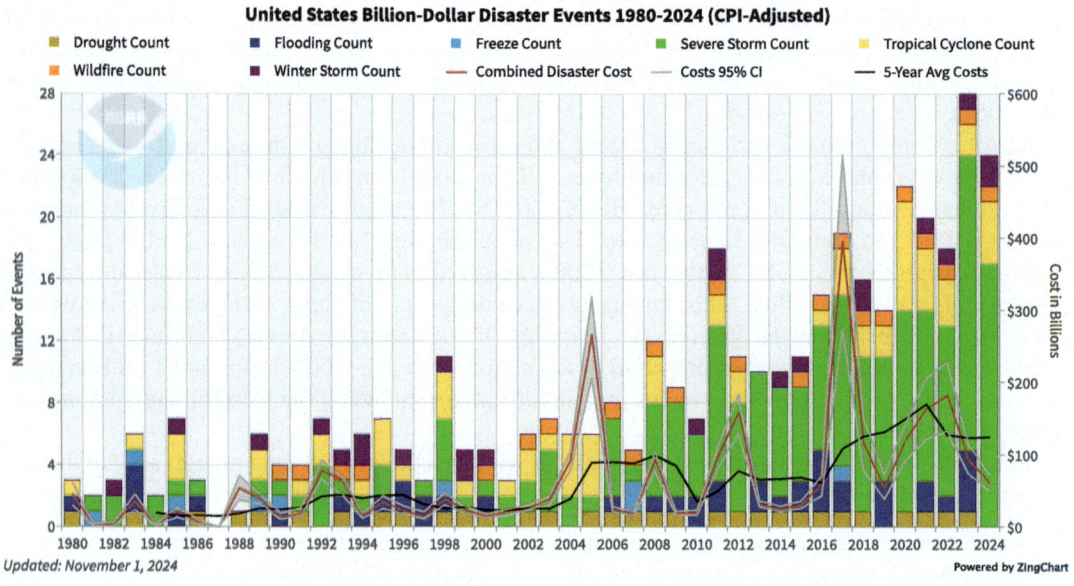

FIG. 15.1

US National Oceanic and Atmospheric Administration "Billion-dollar disasters" (in adjusted dollars using the Consumer Price Index) from 1980 through 2024 (NCEP, 2024). The number count is shown by the stacked bars for each type of event, and total costs and 5-year averages are shown as linear traces.

The impacts on the built sector with extreme weather episodes linked to climate change are manifold, some well-known, while others still uncertain (Table 15.1). Emerging evidence shows that the extremes are becoming more extreme with maximum records eclipsed by wide margins for precipitation and high temperature in particular (Fischer et al., 2021; NASEM, 2024). Fortunately, despite the increasing severity and cost of such events, our ability to plan for, forecast, and mitigate these events has improved over the last decades, typically leading to less loss of life even as the event severity increases.

Table 15.1 Extreme weather risks to buildings and infrastructure.

Extreme weather threat	Built environment impacts	Mitigation
Extreme heat waves	Buckling of structural members, roads, and other damages	Increased expansion joints, decreased thermal expansion
Extreme flooding events	Damage to building structure and interior at, near, or below grade	Lifting critical building infrastructure well above grade
Extreme wind events	Roof and structural damage	Improved fastening of structure, particularly the roof to the walls
Extreme winter storms	Freeze-up of infrastructure not designed for such events	Improved winterization of critical assets

Climate attribution studies link changes observed in the natural world to anthropogenic climate change (Shindell et al., 2009). The area of extreme weather events and their attribution to climate change is an emerging research area (Stott et al., 2016; Trenberth et al., 2015). Rather than asking if climate change caused this event, the relevant question is what the probability (and severity) of the event is with and without the climate change's impacts. Recent studies have examined the climate change contribution to extreme meteorological events (Fischer and Knutti, 2015), drought (Barnett et al., 2008), flooding (Pall et al., 2011), wildland fire occurrences (Abatzoglou and Williams, 2016), as well as changes observed in the biosphere (Parmesan and Yohe, 2003). Attribution science is becoming much more definitive in quantitatively linking given extreme weather events to climate change with higher confidence levels.

Climate-related risks can be amplified by other societal structural changes. One well-understood synergism between climate change effects and the built world relates to the continuing global urbanization (Zhao, 2018). The urban heat island effect (UHI), discussed earlier, adds a few °C in urban areas on top of the ΔT associated with a warming planet and may combine with nonlinear effects (Fig. 14.2). This is not unlike the combined effects of rising sea levels and increased storm surges, exacerbated by subsiding or eroding landmasses. UHI has a complex dependence on season and climatic zone driven by differential moisture and energy balance (Manoli et al., 2020), the result of which tropical cities have fewer options than drier cities for using albedo (reflectivity) management and vegetation planting for mitigation (Manoli et al., 2019). Urban planning and mitigation measures are important adaptation approaches for increased warming in growing urban regions (Georgescu et al., 2014). "Cool roofs" seek to reduce the UHI effect as well as unwanted summer heat gain in individual buildings. This, however, faces constraints, particularly during nights, which are warming faster than days, and other situations where cool roofs are less effective (Krayenhoff et al., 2018).

As a larger-scale example, one area of continuing climate investigation is the Arctic amplification on climate change. This has effects on meridional temperature and pressure gradients affecting meteorology. The reduced temperature gradient from the equator to the poles impacts the circumpolar circulation (polar vortex) that is contained by the upper-level jet stream (the high-altitude wind speed maximum that circles the Northern Hemisphere). A weakening of the circulation, or greater "waviness" of the jet stream and its normal meanderings, allows blobs of cold Arctic air to spin off more easily, causing cold wintertime outbreaks as well as blocking patterns that can slow the mid-latitude circulation (increasing storm persistency) (Francis et al., 2017, 2018). Winter Storm Uri, which brought Texas to a multiday standstill in February 2021 is an example of a winter event coupled with the dynamics of a southerly shift of the polar vortex. The hypothesis is fairly straightforward and logical, and this is a continuing area of research and somewhat of an open question. Emergent evidence is clearly showing this contribution (Francis and Vavrus, 2015), while modeled understanding of the behavior is still nascent (Cohen et al., 2020).

Furthermore, a weakened, loopy jet stream resulting from the reduced equator-to-pole temperature gradient also sets the stage for temperature records. Elevated temperatures with "heat domes" formed where an omega-shaped jet stream creates a persistent high-pressure center that can stall over a region. Extreme heat is leading to increasing hospitalizations and deaths due to heat exhaustion, heat stroke, dehydration, and even asphalt burns, especially in areas unaccustomed to such heat extremes. This happened over the Pacific Northwest United States and southwest British Columbia, Canada, in the Western North American heat wave in late June 2021 (Fig. 15.2). Many all-time high temperature records, including Portland, Oregon, and Seattle, Washington, were shattered by 2–3°C or more. Concerns are emerging that some locations in even hotter and more humid conditions will test the survivability of humans in these environments (Vecellio et al., n.d.). The nearby Canadian town of Lytton, British Columbia, recorded the highest temperature on record in Canada at 49.6°C during the event, and the following day burned down almost entirely due to wildfires exacerbated by the heat dome event. As an example of adaptation, the rebuilding of Lytton, Canada, following these fires is proceeding with much more fire-resilient construction. Using the approach of Nexii Building Solutions Inc., manufactured wall and roof panels are being used that also create little waste plus an airtight barrier. These win-win scenarios for greater sustainability and more resilience are important recovery opportunities.

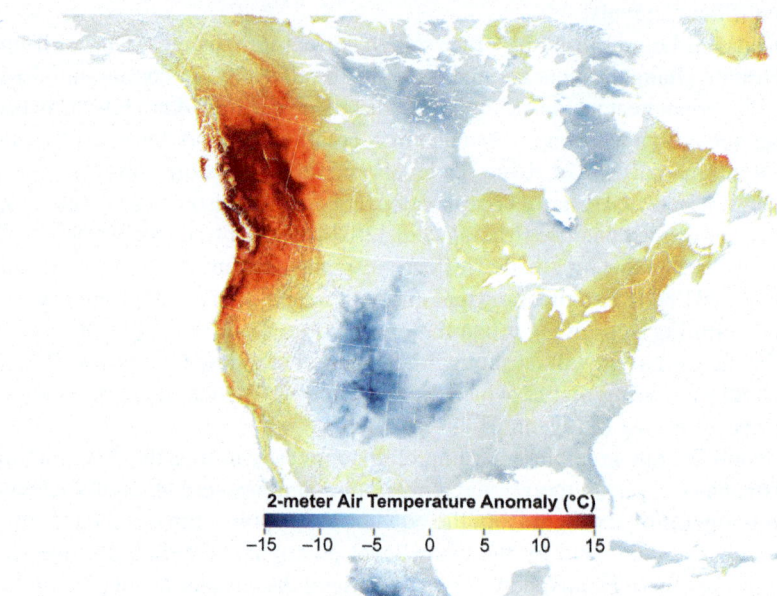

FIG. 15.2

NASA Earth Observatory image of heat dome event in the Pacific Northwest showing the perturbation in temperature from the norm on 27 June 2021 (NASA Earth Observatory image by Joshua Stevens, using GEOS-5 data from the Global Modeling and Assimilation Office at NASA GSFC with data courtesy of Joalda Morancy/NASA/JPL-Caltech and the ECOSTRESS science team).

Climate mitigation and adaptation will be more challenging with the havoc that climate extremes cause. The susceptibility of our energy infrastructure including central power stations, distributed resources, and the electrical transmission grid is a case in point. Although designed for ambient conditions, as "outdoor" generators, wind and solar sources are exposed to extremes of weather. The upshot is that weather extremes often lead to periods of greater carbon emissions and heavier reliance on conventional power generators (Zhao et al., 2023). A case in point where the initial damages from a vehicle collision were followed by intense winds caused a partial panel collapse of a parking canopy solar array (Fig. 15.3). Accelerating extreme weather events are a risk factor for both conventional and renewable energy systems, particularly winds, ice, hail, and extreme cold, as happened during Winter Storm Uri.

FIG. 15.3

Structural damage followed by intense wind damage caused the collapse of a section of a parking canopy solar array. This caused an extended outage while the damaged section and its structure had to be replaced.

Planning for more extreme and variable weather events is prudent planning. Where does one make the distinction between prudent engineering vs overengineering? The 20-, 50-, 100-, and 500-year past storms are likely arriving more frequently in a climate system juiced by increased water vapor concentration and warming from greenhouse gases. Tough decisions under budget and time constraints are certainly ahead.

The climate hazards to the built sector extend beyond the physical to socio-economic factors, public policy, and land use planning. The requirement for homeowners' insurance for conventional mortgage loans is a key driver reshaping the housing market in such risky locations. For example, recent years have seen increasing refusal to write new policies or even the exodus of private insurers from some of the highest-risk locations, such as Florida and California. In some cases, the increasing costs of insurance have driven existing homeowners out of hazard zones. Some states restrict cost or cost increases that insurers can charge. This can protect the consumer from price shocks; however, it can also distort the market by discounting the real risks due to increasing hazards. In other cases, the states have more often become the "insurer of last resort" by filling gaps in the private sector or insuring particularly risky properties. Cost and availability of insurance will be an increasing constraint on what areas remain habitable.

15.3 Adaptation and resiliency: Planning for current and anticipated changes

Adaptation is the adjustment of human and, in some cases, natural systems to current and expected changes (IPCC, 2021a). It is recognized as one of the key approaches required in our response to climate change. Adaptation is changing physical infrastructure or human systems (plans, processes, and procedures) to respond to known and reasonably anticipated risks. Adaptation usually refers to human adaptations though human interventions in natural systems' adaptation are possible (a constructed wetland to mitigate flooding would be an example). The general adaptation process for a municipality can be described in five stages: awareness, assessment, policy and planning, implementation, and monitoring and evaluation (Wasley et al., 2023). Adaptation happens before, during, or after an event, though it usually is more effective with advanced planning. Also, there is a distinction between proactive and reactive adaptation, the former being a more planned response and generally more successful (Wuebbles, 2013).

There is some unavoidable change baked into the system concerning our legacy emissions of greenhouse gases and inertia within the system; resilience engineering plans for these associated changes in our earth system as best as possible. Climate resilient development as described by the IPCC marries together sustainable emissions reductions with rational adaptation to climate changes in the pipeline (IPCC, 2022). Resilience is often described as the ability to bend but not break and has been applied in fields as diverse as psychology, ecological systems, business, computer science, and engineering. Resiliency incorporates elements of risk management as well as safety engineering but transcends simply keeping the occupants safe while keeping systems operational (IPCC, 2021a). Resilience connotes a stable equilibrium in chemistry or a reversible process in thermodynamics where a system can return to a previous state after a disturbance. It involves maintaining the infrastructure's essential function, identity, and structure, but also the capacity for transformation in the face of new circumstances (IPCC, 2021a). One description is rather than making infrastructure or landscapes "fail-safe" to natural threats and extreme conditions, making them "safe to fail" due to their self-regeneration and ability to bounce back (Ahern, 2011). A resilience approach applied to green building focuses on up-front preventative measures, emphasizing a "precovery" as opposed to the recovery after disaster strikes.

As an example, one of the adaptations to warmer average temperatures and more extreme heat events is the use of urban "green infrastructure," including green roofs and walls, green open space in parks, tree planting, and other means to modulate these extremes as well as alleviate urban heat island effects (Norton et al., 2015). The bulk of studies related to urban green infrastructure show a cooling effect locally of $\sim 1°C$ in the daytime with this infrastructure in comparison to the surrounding urban region (Braham and Casillas, 2021). This up-front effort mitigates the effects of naturally occurring heat waves compounded by continued greenhouse warming.

The increasing frequency of extreme weather events will require construction modifications. Adaptation to extreme heat is a challenge, particularly in already heat-stressed regions. In the United States, Phoenix is well-versed in mitigating extreme heat risks, and even there, the number of deaths linked to heat exposure is significant. Construction work and other intensive outdoor activities in some regions may have to shift from daytime to something like a 4 a.m. to noon timeframe in the summertime.

An adaptation example is given of the city of Chicago and its planning efforts for future heat waves. This includes urban tree planting, green roofs on urban buildings, neighborhood cooling centers, efforts to get portable air conditioning units to the most vulnerable (old, young, and poorest populations), and extra efforts to reduce the air pollution effects (e.g., ozone) during the most extreme heat waves (Wuebbles, 2013). These measures are equally applicable to reducing the urban heat island effect discussed earlier, irrespective of a changing climate.

Key categories of risk include those from extreme weather events and abrupt, irreversible climate shifts (Table 15.2). The risk areas of most concern include coastal infrastructure with sea level rise, infrastructure related to hydro-systems (e.g., dams, levees, water and wastewater treatment),

15.3 Adaptation and resiliency: Planning for current and anticipated changes

Table 15.2 Climate change risk categories or reasons for concern, specific risk examples, and example adaptation planning responses (IPCC, 2021a).

Risk category or reasons for concern	Risk examples	Description	Adaptation planning response examples
Extreme weather events	Increased hurricane intensity and damage	The increased temperature gradients between sea-surface temperatures and space accelerate the engine driving hurricanes	Stricter coastal code requirements for the roof structure, upgraded structural attachment hardware, improved window strength, and glazing integrity
Increased extreme precipitation events and associated flooding	Flooding from runoff due to events such as Hurricane Harvey	Extreme precipitation events stress infrastructure with contributions from sea level rise in some cases	Floodplain enhancement, pervious surfaces, water detention features, land use planning, and codes
Aggregate impacts on global biodiversity & economics	Increased rate of species loss such as in the Amazon rainforest	The increased rate of species loss due to habitat loss and fragmentation is acute in regions such as the Amazon	Incentives for conservation tracts to preserve at-risk species
Abrupt climate shifts	Perturbation of Gulf Stream conveyor belt	Slowdown or cessation of the Atlantic Meridional Overturning Circulation (AMOC) or Gulf Stream	Robust emergency response system, but otherwise, due to uncertainty, limited options outside of mitigation of climate change
Heterogeneous distribution of impacts	Agriculture-dependent developing countries	In many cases, those most at risk (and least to blame for climate change) are the poorer nations more dependent on agriculture	Crop shifting, implementing more drought-tolerant varieties and species
Unique and threatened systems	Arctic sea ice loss; coral reef collapse	Particularly affected regions such as the Arctic due to amplified warming in polar regions or coral reefs sensitivity to temperature and acidity amplify risks	Mitigating other cothreats such as habitat reduction plus species-specific conservation efforts; assisted species migration

infrastructure related to the cryosphere, ecosystem services associated with sensitive marine ecosystems, including coral reefs (aquaculture, fishing, tourism), and any systems sensitive to temperature extremes (agriculture, aviation, construction). This is hardly an exhaustive list.

Adaptation efforts have begun to permeate government planning processes from the local level (coastal cities and sea level rise) to the international level (e.g., the policy implications of an ice-free Arctic summer) (IPCC, 2021a). This can be approached incrementally and iteratively with learning occurring along the way. In many cases, adaptation parallels the steps that could be taken to reduce the risk of extreme weather in general. In some cases, the adaptation may reduce other complementary threats such as minimizing human encroachment in wildlands as a means of reducing wildland fire risk. An emerging area of research, examples of some of the major risks and adaptation strategies are given

Table 15.3 Examples of problems of climate extremes with incremental solutions and more transformative solutions.

Problem	Incremental solution	Transformative solution
Extreme heat	Open cooling centers in vulnerable areas	Reduce materials contributing to urban heat island effect while tree planting in under-vegetated areas
Coastal flooding	Elevate houses or key infrastructure above ground level on stilts	Restore mangroves and estuarine ecosystems to provide natural defenses to coastal flooding
Inland flooding	Raise infrastructure such as roads above grade	Vacate floodplains and flood-prone areas
Tropical cyclones	Bolster codes to mandate improved roof connecting hardware and hurricane-rated glazing	Build communities with setbacks from shore and elevations well beyond coastal sea level

in Table 15.2. The distinction between incremental and more transformative solutions is an important one. Although both can contribute, the transformative solutions offer the best hope for long-term, permanent solutions acknowledging a new reality (Table 15.3).

Materials and construction methods will evolve to build more resilience in buildings. The imperative has multiple goals: lower building energy use, lower embodied energy and carbon, or better yet, carbon-sequestering materials, and making buildings more resilient to expected and unexpected changes. Concrete was discussed at length in Chapter 7 on building materials. There are new efforts to make concrete much lower embodied energy and carbon, a better insulating material, increase its longevity, and mitigate its contribution to the urban heat island effect. Some of these take inspiration from ancient materials such as concrete that can self-repair cracks following ancient Roman approaches to concrete with longevity (Seymour et al., 2023). Some of these concrete mixes that have persisted for a couple of millennia were thought to purposely maintain unmixed lime clasts into the mix to heal cracks with water entry that produces hydrated lime. With the right concrete mix in the right application, one can reduce or eliminate the emissions associated with Portland cement, increase the longevity of concrete, and provide greater thermal insulating properties.

Another example: the company Renco USA has developed a Lego-block-like system for commercial buildings constructed of recycled glass, fibrous material, calcium concrete-like material, and a binder holding the blocks together. The system is designed to withstand fire and the hurricane-force winds required by code in Florida and other risky locations. A similar system by Belgian company Gablok uses compressed wood chips and other reuse materials in blocks designed for residential homes. There are other similar systems under development, and such systems have encouraging tests for longevity, low impacts, fire resistance, structural strength, and sustainability. Some of these materials are also carbon sequestering. Meeting such diverse performance goals is challenging.

Worth underscoring is that resilience engineering and adaptation efforts do not minimize the need for mitigation, the arduous work of reducing our emissions. Rather, it complements mitigation and recognizes that though preventing extreme events in the first place is the best approach. Some "baked in" changes to climate can only be adapted to rather than prevented. These events, however, can be mitigated, slowed down, and made less likely and less damaging by continuing to reduce our emissions, all while building greater resilience into the built environment.

15.4 Resilience organizations and tools

Several organizations—public, private, and nonprofit—are pursuing greater resilience in construction. According to the Resilience Engineering Association, design for resiliency allows for the infrastructure to sustain its functioning during disturbances, both expected and unexpected. The US federal government has provided resources to communities to better incorporate resilience into their community planning through the US Climate Resilience Toolkit (https://toolkit.climate.gov/).

As quantified by FEMA, the risk of impacts of climate change is the product of the physical hazard, the exposure of a population or asset to the hazard, and the vulnerability of the population or asset to the hazard (Fig. 15.4). The approach involves anticipating the future risks to the built environment and building to minimize those risks with an adequate margin of safety, recognizing the significant uncertainties.

Many details go into quantifying each term, and a National Risk Index (NRI) for 18 natural hazards (some unrelated to climate such as tsunamis) is calculated according to Eq. (15.1). The expected losses include human casualties, infrastructure costs, and agricultural losses. This is modified by the social vulnerability and community resilience assessed for the location. A map showing the general climate NRI as a function of the county is given in Fig. 15.4.

$$\text{NRI} = \text{Expected losses} \times \text{social vunerability} \times \frac{1}{\text{Community resilience}} \quad (15.1)$$

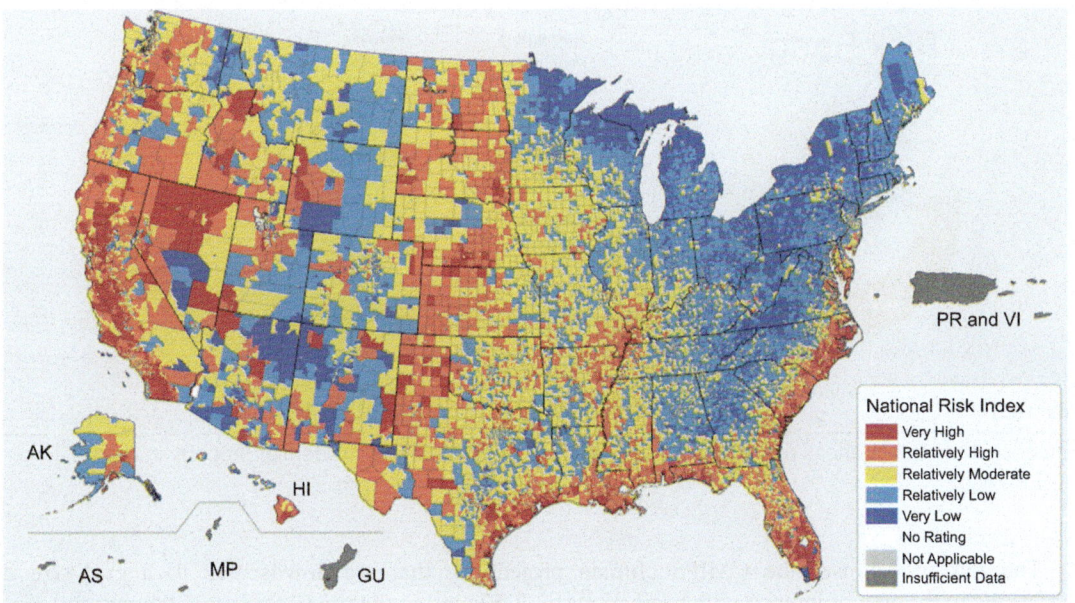

FIG. 15.4

National Risk Index (NRI) for the continental United States for a combined 18 natural hazards as calculated by (public domain image from FEMA) (FEMA, 2020).

386 Chapter 15 Climate adaptation

The US Resiliency Council (USRC) is a professional organization that rates buildings on their safety to occupants during a disaster, the projected percentage damage to the building, and how well-positioned the building is for rapid recovery from disruptive events (https://www.usrc.org/). It currently rates buildings on earthquake preparedness and windstorm resistance though it is ramping up its activities.

The USEPA has models available to assist communities with resilience planning. The Locating and Selecting Scenarios Online (LASSO) tool allows users to identify and use likely climate change scenarios of relevance for the location. The output of the model includes recommendations for adaptation, maps, figures, and GIS data. EPA's Integrated Climate and Land-Use Scenarios (ICLUS) project allows users to use what-if scenarios of likely regional growth, including human population, housing density, and impervious surfaces. The models link to information of use for the adaptation planners and the challenges of their regional climate.

The risks associated with extreme events have led to the development of new, publicly available models to assist homeowners and real estate professionals with assessing these risks. The models include Flood Factor (www.floodfactor.com) from the First Street Foundation and Climate Check (www.climatecheck.com). With the input of an address, both models can give a relative assessment of flood risks for the property. In the case of Climate Check, a broader assessment also includes heat risks, drought risks, wildfire risks, and heavy precipitation event risks. An example of the estimated risks to a residential property in central New Mexico shows a substantial risk of high temperatures in the desert Southwest United States (Fig. 15.5) (www.climatecheck.org).

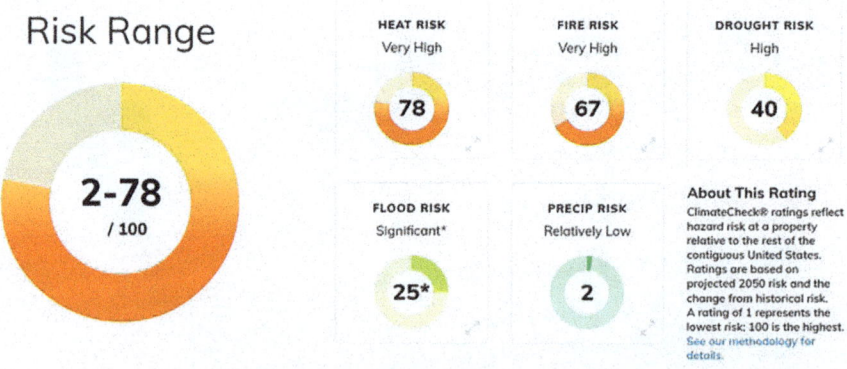

FIG. 15.5
Screenshot of climate check risk analysis tool performed on the author's residential property in central New Mexico.

The risk model uses the CMIP5 climate projections that are downscaled to a grid size of 3.125 km^2 to the area of interest with projections to 2050, within the timeframe of a 30-year mortgage. The model uses the RCP8.5 emissions scenario that invokes 8.5 W/m^2 radiative forcing by 2100, which is a low-probability case but a plausible worst-case scenario. Though the uncertainties are considerable, such assessment tools will become more important in the real estate industry, much like third-party home inspections and seller disclosures starting in the 1970s.

Addressing climate change and the natural hazards associated with it will invoke required changes in construction and building codes. Municipalities have often sought to mitigate their greatest regional risks through building codes. Examples are Florida building codes that mandate the hurricane resistance of new construction and fire resistance codes in fire-prone areas of the WUI.

15.5 Coastal construction and sea level rise

Though this is not an exhaustive review, a few of the threats and resilience-oriented adaptations are discussed here in more detail. According to the NOAA Office for Coastal Management, roughly 40% of the US population lives in a coastal county as of 2020. Resultingly, coastal population density is about four times the density of the continental United States as a whole. Generally, coastal areas and infrastructure are particularly vulnerable to climate change-driven risks for numerous reasons (Fig. 15.6). A good illustration of those regions at most risk was shown earlier in the surface elevations in Fig. 4.13.

Sea level rise is an existential threat to low-lying islands and other settlements at, below, or near sea level civilization. Global average sea levels have risen approximately 0.25 m over the Industrial Age and are currently rising by a few millimeters per year on average. Sea level rise in some regions, such as the Northeast United States, is larger due to various regional factors, such as sinking land masses and declining ice masses, such as Greenland, providing less gravitational pull. The best estimate of mean sea level rise by 2100 is 0.3–1 m (IPCC, 2013, 2021a). Notably, depending on our emissions trajectory, it will not cease there. Added on top of this is the storm surge associated with cyclonic-driven storm systems and extreme events such as hurricanes and tropical storms.

FIG. 15.6

Buildings and infrastructure located near sea level, the island of Curaçao shown here, are exposed to increasing risks from sea level rise, erosion, extreme weather, more intensive tropical storms, and associated storm surges.

Coastal adaptation and climate mitigation work best when pursued hand in hand. The example of incremental adaptation in the fifth National Climate Assessment is raising the elevation of housing in flood-prone areas. The example of a transformative adaptation is restoring coastal ecosystems that mitigate the impacts of flooding events while simultaneously sequestering carbon in these ecosystems (Table 15.3) (Wasley et al., 2023).

Among other specific local risks, hazards of concern with respect to sea level rise include storm surge and intensity, coastal erosion, saltwater intrusion, degraded estuaries, river deltas, and other coastal ecosystems (Mehta et al., 2013). This has profound implications for the built environment along coastlines in terms of flooding, erosion, and general habitability of these locations. In the built sector, extreme impacts on coastal infrastructure including roads, bridges, and ports are an accelerating concern. During extreme events, heavy precipitation on less permeable urban areas combined with increased storm surges on top of rising sea levels all synergize to cause inundation. This happened with the breaching of flood walls and levees in New Orleans during Hurricane Katrina in 2005 and the flooding in Houston in 2017 during Hurricane Harvey. Also, a concern, the intrusion of saline water into freshwater bodies including rivers, lakes, and aquifers has vital implications for ecosystem health as well as potable water resources. For example, the record low flows in the Mississippi River in summer 2023 resulted in creeping saltwater intrusion further upstream, affecting municipal water supplies.

The first-order transformation is to reinvigorate the natural barriers and defenses that have been degraded in many coastal areas, such as mangroves, coral reefs, and other natural features that blunt the effects of sea level rise and extreme events. Soils and water management techniques, if applied effectively, can also offer nature-based CO_2 reductions that can help with climate mitigation (Griscom et al., 2017). These adaptation efforts are complex, often on a decadal scale, and they involve coordination of land management at the local to federal levels making them challenging (Seddon et al., 2020). However, such transformations offer the best potential to simultaneously (1) reduce our atmospheric greenhouse gas burden, (2) adapt to a more chaotic climate system, (3) reinvigorate degraded ecosystem services, and (4) promote enhanced biodiversity.

Numerous low-lying communities have vast experience in living with water worth exploring. These range from island nations such as Kiribati or the Maldives to cities such as Venice to nations like the Netherlands built on coastal plains near sea level. Engineering approaches in such locations including floating infrastructure responsive to changing water levels are underway (Fig. 15.7). These structures can respond to changes in sea level, storm surge events, king tides, and extreme precipitation events by moving vertically upward.

Engineered responses can also extend the viability of limited marginal areas near sea level such as Venice, Italy, and numerous locations in the Netherlands. The Netherlands provides a case study of a nation that has battled the threats of sea level rise and flooding events for millennia due to its ~sea level location. It is well known for the dikes, sea walls, and engineered barriers that have allowed it to exist near sea level. The engineering of coastal structures to resist intensified storms and storm surges includes elevated infrastructure, sea gates that can be closed to reduce storm surge, deep structural piles to resist storm surge and erosion, upgraded connections between structural elements (e.g., roofs and walls), and upgrading cladding and structural elements to resist flood waters and wind damages (Mehta et al., 2013). The selection of engineered responses will be an intricate problem as some barriers, such as sea walls, can amplify the effects on neighbors. This can have ongoing negative effects, such as scour of the wall-ocean boundary weakening coastal defenses. Much of recent adaptation work with buildings in the low-lying Netherlands is moving toward houseboats, floating buildings, and other adaptations that can passively resist rising sea levels, storm surges, and flooding events.

An example of a current engineering project moving forward is the "Big U" storm surge barrier in the works to protect lower Manhattan in New York City from Hurricane Sandy-scale events. It includes a broad range of defensive measures including enhanced nature-based defenses, elevating infrastructure, permanent and deployable sea walls, and other barriers. The project is also enhancing the green

FIG. 15.7

Floating residences in the Netherlands.

Image by www.hollandfoto.net/Shutterstock.

space along the coast in conjunction with waterfront parks doubling as defense and drainage. It is an ambitious project with a matching price tag exceeding $100 billion, and its "field testing" still remains. Though this response may be merited in small areas with vast infrastructure, this scale of adaptation can hardly be applied throughout the world's coastal infrastructure.

On the wet end of the spectrum, constructed barriers are an option for urban areas at risk of flooding and sea level rise. The conversations will be difficult in terms of what at-risk locations are salvageable and where a managed retreat is the best or only option. One of the most difficult adaptations will be to an increasing mean sea level, something that has increased about 0.3 m over the Industrial Era due to ocean thermal expansion and melting of land-based ice. Already, the first climate refugees have been recorded including the Isle de Jean Charles, Louisiana, and Inuit coastal communities in Alaska located near sea level. In some major urban areas, such as Norfolk, Virginia, this problem is compounded by land subsidence which occurs from several factors, most notably unsustainable pumping of groundwater. The primary response methods may be summarized as (1) protect, using dikes, canals, detention ponds, and seawalls; (2) elevate infrastructure; (3) accommodation where some level of living in a wet environment is accepted; and (4) retreat from areas too at risk (Erten-Unal and Andrews, 2018). A large number of adaptations including enhanced water diversion, retention and storage, check valves on drainage systems, permeable surfaces, and others can help protect during storm events (Erten-Unal and Andrews, 2018). Many solutions will need to be pursued in tandem to give time for permanent adaptation and relocation.

15.6 Inland flooding and mitigation of its effects

A long-term risk for human infrastructure, flooding associated with extreme precipitation events also affects lowland areas away from the coasts. For example, the Great Vermont Flood of 2023 in the northeastern United States, primarily Vermont and upstate New York, in July 2023 was of historic scale.

Urban areas such as Montpelier and Burlington suffered extensive flooding. According to the National Weather Service, the precipitation amounts were as large as 5–10 in. over the several-day event and exceeded some of the recent tropical storm/hurricane events (The Great Vermont Flood of 10-11 July 2023: Preliminary Meteorological Summary (weather.gov)). Flooding in these same areas recurred in summer 2024. Vermont is often considered a "climate change refuge" to escape the worst of the effects. Flooding has been flagged as the key driver of local-scale migrations with respect to what is being termed "climate abandonment areas" as assessed by the First Street Foundation (Shu et al., 2023). Many other at-risk flood areas surround the Great Lakes which has also become a region in the United States designated a climate "haven." A more prudent approach, along with our mitigation efforts, may be for most populations to "shelter in place" rather than migrate.

Robust land use and building codes to mitigate construction in flood plains and other areas susceptible to flooding have become more urgent in the last decade or two. This has been accelerated by insurers and federal flood insurance programs less willing to provide coverage in at-risk areas. Current and future insurance availability as well as the National Flood Insurance Program play a significant role in whether a given location remains habitable. When flood maps are redrawn due to new data, it can make further development in a flood zone difficult to impossible and thus negatively impact property values.

Since most homeowners' policies do not cover flood risks, the National Flood Insurance Program (NFIP) was established in 1968 to provide consumers with a flood insurance option as well as a means to restrict development in flood plains (CRS, 2024). The latter typically focuses on the 100-year floodplain determination. This is a historical determination based on past data, and what happens when the 100-year event occurs every decade or two? The program has amassed a substantial debt after claims paid during major hurricanes of the past 20 years and is no longer supported strictly via premiums. A high priority for NFIP in its 2023 Risk 2.0 model is incorporating the real risks for individual properties into flood insurance rates. Also, NFIP and other insurance programs seek to break the cycle of repetitive losses followed by rebuilding for particularly risky locations.

Fortification of areas prone to flooding can at times be an appropriate adaptation. Open, permeable natural spaces should be created to allow infiltration of floodwaters as the first and best step. Constructed wetlands and stormwater detention areas are the next steps to reduce flood risks. On the building scale, adaptations include shifting critical infrastructure to higher levels in a structure. This, coupled with minimal, low-impact use of any spaces at or below grade, and potentially appropriate engineered barriers, as discussed with coastal adaptation previously, can work together to reduce the flood risk (Hammond et al., 2015). A variety of deployable flood barriers, both manual and automatic, have been devised to protect a building temporarily until floodwaters recede. The American Society of Civil Engineers (ASCE) details appropriate design elements for structures in Flood Resistant Design and Construction, ASCE 24.

15.7 Case study: Babcock Ranch Community in Florida

A US state with the high vulnerability to climate change and weather extremes in general is Florida with respect to landfalling hurricanes, wind damages, flooding, tornadoes, extreme heat events, and sea level rise (Fig. 15.4). A near-coast, sustainably planned community powered by on-site solar began in 2018. Babcock Ranch in Florida (Fig. 15.8) is a prime example of a planned community focused on sustainability and resilience. Its history was not an easy path as it traces back to the early 2000s, and the project survived the housing bubble and the Great Recession of 2008–9.

15.7 Case study: Babcock Ranch Community in Florida

During Hurricane Ian, Babcock Ranch was the only community in southwest Florida to not suffer power, water, or internet outages. It also fared much better than most of the surrounding towns considering wind damage and flooding damage. Damages were limited to roof shingles, uprooted trees, and some pool and recreational damages, but the building structures survived generally unscathed.

The town was purposely located inland away from the shore and ~30 ft above sea level and buffered against storm surges associated with landfalling hurricanes. The community also features more solar PV production than the community consumes, battery storage, and underground utilities. The electricity is supplied by solar PV production, and it maintains its wastewater treatment and potable water supply from deep aquifers. The streets were designed for flooding so that structures avoided the worst of the inundation. It exceeds the increasingly stringent building codes in Florida related to hurricanes and winds. Eighty percent of the land of the development was transferred to the state and devoted to natural preservation, while 50% of the in-town land is parks, greenspace, and water.

What critiques can be made? Although it would not be described as affordable housing, as of summer 2024, there were numerous condos as well as single-family detached houses at Babcock Ranch listed below the United States and Florida median prices of ~$420,000 and ~$400,000, respectively. Homeowner's Association (HOA) fees of ~$250/month for detached homes and $1000/month for condos add significantly to these costs but are not unheard of for HOA fees. Though it is targeted for retirees, isolation and a likely need for commuting to many local employers are a reality. Some criticisms of the community as too "planned" and homogeneous have been heard though of course subjective. One can argue that establishing a new development in Florida, where we arguably should be retreating from, is a fool's errand. However, this argument could be made about many locations, notably those in red in Fig. 15.4. They have consciously made the effort to locate away from the shore ~30 miles and at 30 ft above sea level.

FIG. 15.8

The Babcock Ranch development in Florida, designed from the ground up for resiliency, took a direct hit from Hurricane Ian and survived largely unscathed.

Image by Bilanol/Shutterstock.

15.8 Drought risks

On the other end of the spectrum is the risk of too little water. Water constraints will also be paramount in a world of greater climate instabilities. Reduced snowpack and freshwater constraints are accelerating in the western United States and other arid regions. Already cities, particularly in the southwestern United States, are restricting landscapes and watering. The current situation in the Southwest US is thought to be the worst mega-drought in 1200 years and roughly half attributable to anthropogenic climate change effects (Williams et al., 2022). The situation on the Colorado River which supplies much of the surface water for the region is dire with Lake Powell and Lake Mead reaching dramatically low levels (as of 2021, approaching 200 ft below their standard level) in the current mega-drought. The water stresses in the west will continue to grow and cause strains on the ecology and economy of this region (Wuebbles, 2013). These will greatly impact urban development as well as agricultural viability in these regions. Drought risks also exacerbate the risk of wildland fires, as discussed next.

15.9 Wildland fire and the built environment

The interconnections between climate change, wildland fire, and air quality were discussed previously in Chapter 4 related to climate change. Here we will relate this extreme behavior to the built environment. The multiple hazards associated with wildland fire include the structural hazards of the fire, the postfire hazards of floods and mudslides, and the occupant hazards associated with indoor air quality effects of the smoke emissions as previously discussed as well (Fig. 15.9).

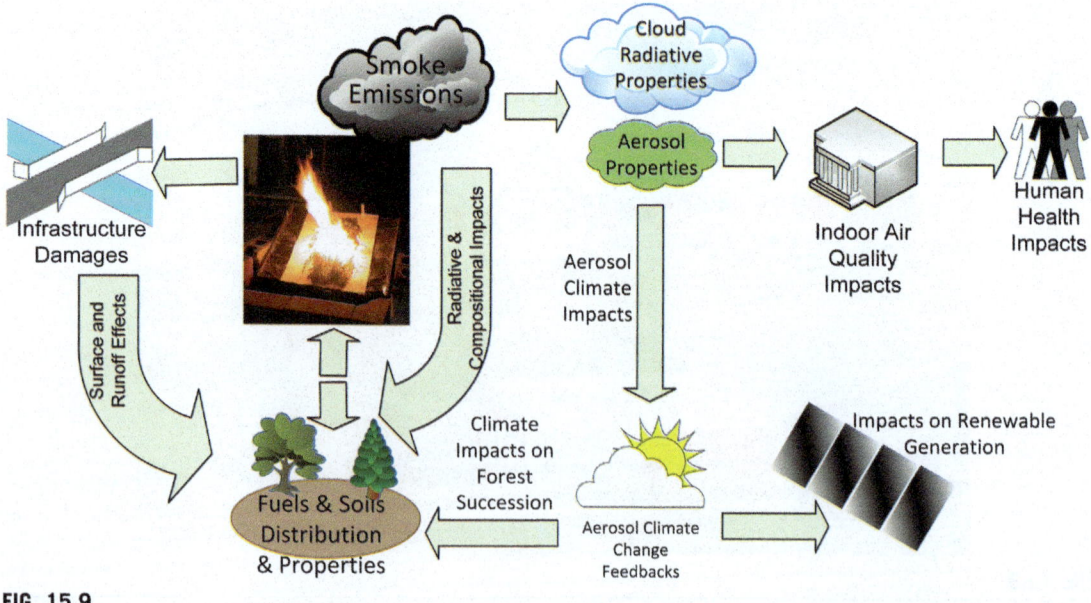

FIG. 15.9

Wildfire impacts on the built sector.

The threat of massive wildfires has grown in recent decades as a result of (a) a longer fire season with warming temperatures, (b) lack of or poor land management practices, and (c) the greater interaction of civilization with fire in the expanding WUI (UNEP, 2022). The first response when wildland fire threatens the built environment is fire suppression. This was the policy for a century with the mantra of "all fires out by 10 a.m." However, evidence shows this has exacerbated the fire problem and made the inevitable fire episodes more severe (Kreider et al., 2024). Fire incidence and severity has profound impacts on the development of new buildings and infrastructure in what is termed the Wildland-Urban Interface, or "WUI." The WUI occupies only ~5% of the land area globally but is home to nearly half of the human population (Schug et al., 2023). The occurrence of wildland fires is becoming more prominent in grasslands as well making fires like the Lahaina (Hawaii) and Marshall (Colorado) fires higher-risk events (Radeloff et al., 2023).

The increasing prevalence of wildland fire also affects solar production due to the attenuation of solar radiation by biomass smoke aerosols. Early analyses show this can be significant (10s of %) for discrete-time periods for solar production quite near wildland fires, but on a larger scale, the reduction of direct radiation is somewhat offset by the increase in diffuse radiation (Duncombe, 2023). These perturbations may affect forecasting electricity production and grid balancing.

The engineering response to the fire threat will undoubtedly take a multipronged approach of mitigation, adaptation, planned retreat, and generally designing new ways to live in the age called the "Pyrocene" (Schlickman and Milligan, 2023). The acres burned and acres per fire have dramatically increased in the western United States, while many more structures are being built in the WUI. The first-order change is to reduce the susceptibility of human settlements to the threat by disincentivizing construction in the WUI. Firewise construction for buildings in the WUI-relates to sustainable forestry and -appropriate timber use in buildings as well. This has prompted a strong desire (as well as evolving building codes) to build "firewise" properties in these locations. This adds another goal for resiliency and constraint on budget to new construction in such areas.

Australia may have one of the best adaptation policies called the "stay or go" (less succinctly, prepare, stay and defend, or leave early) approach which emphasizes preparation, hardening, and survival in fires or an early evacuation for even moderately risky events. This contrasts with the often-used approach of "wait and see" how bad it gets and then evacuate if all else fails.

Many of the efforts to reduce fire risk are prudent, irrespective of the contribution from a changing climate (Fig. 15.10). Efforts to mitigate fire risk in buildings include (www.firewise.org):

- Building codes that address fire risk and avoid the most susceptible areas
- Lower flammability and nonflammable building materials (e.g., roofs, decks, sheds)
- Fire-rated doors and windows with multiple panes, tempered glass, and low-emissivity coatings
- Soffit and roof vents and chimneys that are screened and/or closed to prevent firebrands from igniting the building from the inside
- Choosing appropriate landscaping/vegetation and its spacing
- Maintaining landscape by trimming, removing debris from roofs, gutters, and near the house, chipping and shredding, and composting landscape waste away from the building
- A zoned approach to vegetation and combustibles around a firewise property with minimal flammable materials near the structure

FIG. 15.10

Firewise program specifies the measures that homeowners can take in fire-prone areas (public domain image from the NIFC website at https://www.nifc.gov/fire-information/fire-prevention-education-mitigation/wildfire-mitigation/home).

The town of Paradise is rebuilding after it was nearly completely destroyed by the Camp Fire in 2018. According to FEMA, the Camp Fire in November 2018 displaced 50,000 Paradise and nearby residents, destroyed 19,000 structures, and resulted in 85 fatalities. The town began rebuilding—a process that will take decades to complete—under the guidance of a local building resilience center using a disaster recovery plan enacted 7 months after the tragedy. The process for a community with many elderly, disabled, and/or low-income residents has faced early challenges such as hazards with dead trees needing clearing and reestablishment of water and wastewater services. The reconstruction is focused on fire resilience from the planning level to the individual home construction, such as in Fig. 15.10. Building techniques employed included autoclaved aerated concrete, insulated concrete forms, metal fire-resistant roofing, and kits homes preapproved to meet more stringent fire and earthquake codes (FEMA, 2022).

A lack of trained contractors, high construction costs, and new requirements for fire and earthquake resistance made rebuilding impossible for some disadvantaged populations who were displaced. Significant numbers of residents slipped through the cracks between federal, state, nonprofit, and insurance claims related to the recovery. The city of Paradise established a Building Resiliency Center to help homeowners navigate the rebuilding process including financing, insurance, permitting and fees, site hazards, code requirements, and other facets of the complex process of rebuilding. Despite noteworthy progress, it remains to be seen the effectiveness and permanence of the effort to rebuild this town.

Responding to wildfire risks offers one potential climate win as well by removing dead, burnt, and overgrowth trees and burying them underground to sequester the carbon in a process called "wood vaulting." It must be done near the carbon source to minimize transport costs, in suitable soils, and where land use issues do not complicate it. The other keys are the permanence, monitoring, and prevention of anaerobic conditions leading to methane formation.

The increasing destructiveness of wildfires has bolstered the argument for microgrids capable of "island mode" when, for example, high-voltage power interconnections must be shut off due to fire or other hazards. The fires in remote areas of California and elsewhere have been caused by downed power lines (and have caused them), both leading to sustained power outages. Microgrids are discussed next as an adaptation measure.

15.10 Microgrids, smart grid, and grid resiliency

Many adaptation measures need better implementation, and many are policy reforms strongly connected to economic, political, social, and regulatory factors. An area of focus related to engineering and the built sector is the need for upgrades and new models for the national electrical transmission grid. The grid which connects power production to hundreds of millions of end users has been described as the nation's largest engineered system (Schewe, 2007). The electrical transmission system is antiquated and prone to breakdowns such as happened in Texas with Winter Storm Uri. Grid modernization is needed irrespective of climate change.

Moreover, climate change symptoms—extreme heat, more frequent storms, wildfires—threaten damage to the nation's electric grid, much of which was built in the mid-20th century with a 50-year life expectancy. The grid in the United States encompasses 600,000 miles of transmission lines and 5.5 million miles of local distribution lines, according to the American Society of Civil Engineers. Over 300 transmission outage events from 2014 to 2018 were caused by severe weather, according to ASCE.

Electrical grid failures are key susceptibilities to climate disturbances. The traditional grid model is a one-way street where centralized power generation facilities (electric, gas, and water utilities) provide electrical (or other utility) services to customers in a broad US region. Newer approaches from the building to the community level invoke microgrids (Fig. 15.11). Microgrids allow more distributed resources to be integrated with the grid as a whole. These are advantageous for buildings in the mountains, islands, or other remote locations, particularly those susceptible to natural disasters, and with a single linkage to the national electrical grid.

Solutions are difficult, expensive, and at multiple levels from the national grid scale to the individual consumer scale. Can we get to a place where clean energy resources with appropriate storage, serve as the backbone, and the more dispatchable conventional generators as the backstop?

The National Academies have been tasked with reporting on the resiliency and future pathways for the national electrical grid (NASEM, 2021). Grid failures are a growing risk factor with extreme weather events of the past few decades. The extent of hurricane damage (and other natural disasters) and the prospect of extended power outages have made microgrids more attractive, such as those employed at Babcock Ranch discussed earlier. Again, a case study is the failure of the electrical grid in much of Texas during the mid-February 2021 Winter Storm Uri. The event was a massive electric grid and gas distribution failure due to extreme cold and frozen precipitation. It led to failures to start, system deratings, and outages from over 1045 individual systems and 4124 individual outages including all major sources with 58% natural gas systems (FERC, 2021). By power generation, the outage affected 55% of natural gas generators, 22% of wind generation, 18% of coal generators, and 2% or less of nuclear, solar, and other generators. The largest (75%) proximate causes were frozen instrumentation and fuel delivery issues associated with the extreme cold and snow/ice and lack of winterization (FERC, 2021). These extreme events tend to have a convergence of factors both maximizing demand due to space heating during Uri as well as constraining supply due to weather-related outages from grid failures and downtime for both conventional and renewable sources. No matter the driver, the increase in extreme meteorological events related to temperature, precipitation, and winds is a particular concern for our built environment from the grand scale of our national electrical grid to the individual building scale.

Decentralization of power production by nature involves more distributed generation sources, as is the case with renewable energy sources. This goes hand in hand with smart grid efforts to make communications more real-time and two-way. Control systems, hardware, software, energy sources, energy storage, communications, and data flow are all part of this. Rather than a static pricing structure, advanced metering infrastructure and time of day (TOD) rates may be integrated with the utility end of the microgrid to more accurately price power resources. This can help with load shifting to times of lower demand including the use of energy storage during grid-peak periods to help alleviate the problem. Even when microgrids are connected to the national grid, they offer an "Island mode" when the larger grid goes down. Microgrids can also avoid the problem of residential on-site solar PV that is often offline when grid power is unavailable for safety purposes. Overall, they offer increasing resilience options for hospitals, industrial facilities, campuses, and the like.

Grid stability is enhanced by having energy resources that are baseload (i.e., always on), load-following, and those variable in nature. Advanced grid integration offers hope to mitigate these issues by such approaches as (a) higher time resolution load balancing, (b) greater geographic sharing of reserve capacity, (c) diversity of renewable sources and resource sharing, coupled with enhanced transmission, and (d) smarter inverters and other technologies and procedures associated with integration (Cole et al., 2021). A case in point is the California utility-led Western Imbalance Energy Market that is drawing an increasing fraction of all the electrical generators in the western United States to help cost-effectively alleviate supply and demand imbalances particularly associated with renewable integration (https://www.westerneim.com/).

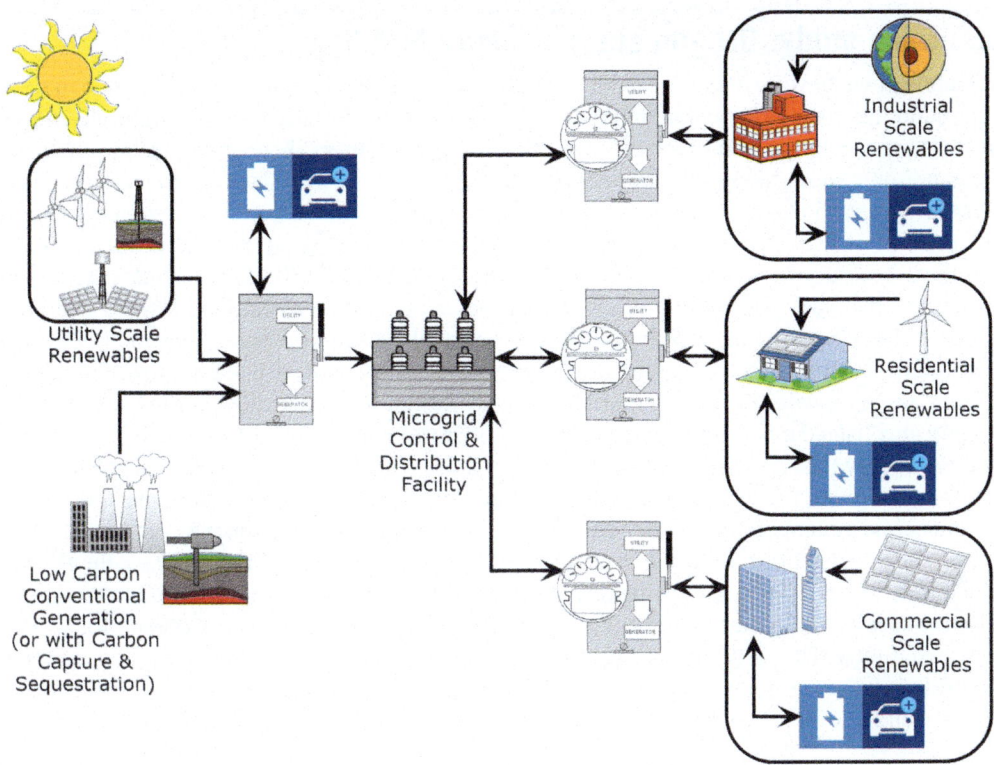

FIG. 15.11

Example architecture of a microgrid system. The primary upgrades compared to a traditional centralized model are connectivity to but independence from the national grid, decentralization of energy sources, integration of renewable energy sources plus storage, and multiple flow directions of energy and information in the system.

Grid interconnectivity on the larger scale is woefully limited when looking at integrating renewable energy sources, both distributed and centralized. The SunZia transmission project, the largest US renewable-grid integration project in construction as of 2024, is linking the wind and solar resources on the eastern plains of New Mexico to urban centers in California. The project has taken 20 years to get to this stage. It is a testament to the complexity and difficulty of implementing new transmissions.

Rather than replacing or adding new electrical grid distribution, a recent simpler approach is using existing rights-of-way, transmission towers, and other infrastructure and "reconductoring" the long-range high-voltage transmission lines with more advanced and higher capacity transmission lines. This offers hope for integrating greater utility-scale renewable energy contributions to the grid. A great resource for the latest developments in microgrids is the Microgrid Knowledge website, https://microgridknowledge.com/.

15.11 Case study: Oregon state treasury building

The disruptions we have experienced with climate change to date have deepened interest in disaster-resistant buildings. This is a movement that has always been an esoteric mix of renewable energy advocates, disaster preppers, and other off-grid thinkers and dreamers. A resilient design for local risk factors is not only more likely to endure climate extremes, but it will also fare much better in other disasters such as earthquakes, civil unrest, military conflicts, and global pandemics, among other risks.

The concept of resilience was embraced wholeheartedly in the design of the headquarters of the Oregon State Treasury in Salem, Oregon (Fig. 15.12) (Fuller, 2022). Its critical financial role in the functioning of the state government made it a prime candidate for a resilient design. The building is one of the first US platinum-rated buildings for earthquake resistance as rated by the US Resiliency Council (USRC). The building is on a suspension system composed of isolators that can reduce the sway from an earthquake by 75%. The building was designed to minimize energy use and can go completely into "island mode," where it can independently operate its infrastructure. The building generates more energy than it consumes using solar PV. It has a backup battery bank plus a backup diesel generator for failure of the former resources. The building has backup water and wastewater treatment in case municipal systems go down and redundant systems for internet and phone communications that can use satellite connections. In case of a nearby wildfire or indoor air quality problems, the HVAC can introduce fresh filtered air from outside at 2 Air Changes per Hour (ACH). Its windows are impenetrable in the case of civil unrest and rioting. It has hundreds of preserved meals on-site for workers in these essential jobs. This is a building type that will become more commonplace in an increasingly risky world, and it serves to minimize those risks and maximize recovery from threats beyond climate change as well.

FIG. 15.12

Oregon State Treasury Building in Salem, Oregon.

Image by Victoria Ditkovsky/Shutterstock.

15.12 Managed retreat: The ultimate adaptation

It is unrealistic to expect that we can maintain all current human settlements in the face of accelerating climate change. Already, the average range and distribution of plant and animal species have invariably shifted poleward to higher elevations and deeper in water bodies though with considerable species variation as a function of their climate sensitivity (Pecl et al., 2017). Congruent adaptations, including relocation, will likely be necessary for humans and our civilization. Some settlements arguably began as marginal to ill-fated, and the added stressor of an atmosphere on steroids is not helping matters. The individual decisions to rebuild vs abandon after disruptions will be excruciatingly painful for the communities and individuals who will face them.

Assisted migration or managed retreat is the relocation of susceptible homes, neighborhoods, or entire towns due to the riskiness of extreme events. The federal government or other municipalities help in this effort, often buying affected properties so homeowners can relocate.

History offers examples of the relocation of settlements from unfortunate locations to more suitable ones even before climate change-driven displacements. Valmeyer, Illinois, was a town of 900 on the Mississippi River that was flooded out multiple times and successfully relocated 2 miles away and to a bluff 400 ft higher in 1993. The Valmeyer relocation moved ~700 people and, although successful, was far more complex, time-consuming (~4 years), and expensive than anticipated (approaching $50 million according to NPR). The relocation occurred during the Great Flood of 1993 along the Missouri and Mississippi Rivers that inundated 75 towns and 20 million acres of farmland. Over several months throughout the Midwest, floods destroyed 50,000 homes and cost upwards of $20 billion, according to the USGS (The Great Flood of 1993 | U.S. Geological Survey (usgs.gov)).

The flipside approach is for residents to repair, fortify, and return to cities that are flooded. Princeville, North Carolina, is an example of a town that has been evacuated and many homes lost due to flooding associated with hurricanes and other extreme events. Princeville began near the banks of the Tar River as an African American community founded by freed slaves. It was considered marginal land for agriculture due to often saturated soil and floodplain issues, but it was the only land made available and economically attainable. The town has been caught in a limbo-like cycle of destructive events, followed by evacuations, returns, and partial rebuilds with elevated homes, increasing levee heights, and other fortifications to mitigate flood damage. Recently, FEMA has begun to build infrastructure on higher ground to relocate the town. Some homeowners have voluntarily relocated already. It is still being sorted out, and there are increasing numbers of related stories that will play out. These early affected community responses offer case studies on how to manage this process as well as its pitfalls.

Climate-related migrations are already occurring from island/coastal regions (e.g., Puerto Rico and US Gulf Coast) to more northerly locations and away from wildfire-prone locations in the wildland-urban interface (e.g., Paradise, California) to more urban areas. This will likely accelerate. Isle de San Charles in Louisiana has largely been abandoned due to encroaching seawater, while the same is happening to the Inupiaq Island village of Kivalina in Alaska due to erosion by storms exacerbated by declining sea ice. Some have characterized the Great Lakes area as the climate haven of the United States. It will have its own issues with precipitation, winds, flooding risks, and high heat indices, but it does offer abundant freshwater, more moderate temperatures, and fewer risks than coastal locations.

The US federal agencies associated with emergency management including FEMA are more seriously considering relocation of developed areas susceptible to climate hazards. Rather than paying multiple times to rebuild infrastructure increasingly in harm's way, such as that near sea level, the less risky and more cost-effective approach can be relocating communities. The Federal Flood Risk Management Standard proposed in 2016—approved though not implemented—seems to break the cycle of disaster rebuilding. The cost of rebuilding is not limited to just dollars, as each rebuild, perhaps greener than before, has a huge embodied carbon footprint.

The data show that climate migration in the United States and globally has begun and that it is a very localized phenomenon (from neighborhood to neighborhood versus in between nations). These localized areas are often due to flood risk and are termed climate abandonment areas (Shu et al., 2023). Evidence is emerging that market shifts are beginning to influence the response to climate change risks as well. The most vulnerable markets, such as South Florida, face the dual risks of sea level rise and storm surges with increasing storm intensity. For example, real estate sales volumes and prices in the greater Miami residential market have responded to risk factors such as height above sea level as buyers are increasingly concerned about climate risks (Keys and Mulder, 2020). These concerns, if not already shaping markets, will undoubtedly spread to other near-sea-level locales as well as those facing increasing inland flood and fire risks.

However, the economics and politics of managed retreat are a huge uncertainty. Both the concentrated wealth as well as abject poverty in some of the most at-risk areas will undoubtedly raise equity issues. It will be a difficult and contentious road ahead for areas and populations on the margins of survivability.

Beyond human institutions, we must also have ecosystem adaptations that can be more problematic given the speed of change. Earth's rate of species extinction has accelerated over the geological period sometimes referred to as the Anthropocene. Human efforts can only facilitate adaptations of ecosystems while acknowledging the crucial services they provide to humans. The critical role of first decelerating and then mitigating climate change is the first and best approach for the rest of the natural world as well.

It should be stated that the concept of climate "safe havens" may prove illusory. Certainly, there are more risky locations, such as hurricane- and sea level rise-prone coastal Florida, the fire risk of much of alpine California and the western United States, and the extreme heat and lack of freshwater resources faced by desert cities such as Phoenix, Arizona. Climate havens are areas often cited as regions to escape to with little risk from sea level rise, hurricanes, forest fires, and adequate freshwater availability. However, what was once deemed climate-safe havens such as the Pacific Northwest United States have experienced recent extreme heat dome events. Both in 2023, Vermont faced the recent devastating floods while Canadian wildfire smoke plagued the Northeast and upper Midwest United States. Moreover, the Great Lakes region, considered a climate haven, is projected to have much more significant flood risks (Shu et al., 2023). The notion of sheltering in place and increasing the resilience of existing infrastructure is becoming a competing narrative with climate migration. Both will likely be needed.

15.13 Case study part II—Resilience in mid-century schoolhouse/church repurposed into a home

As a follow-up, with the case study discussed in Section 13.14, we have also shifted to greater implementation of resilience efforts in our own residence. The projects never end with a building, but the focus has shifted toward more adaptation and resilience efforts. Besides alpine and riparian areas,

New Mexico is to some extent fuel-limited regarding wildfire risks. Nonetheless, the experience from fires such as the Marshall, Lahaina, and Texas panhandle fires of the last few years shows that devastating wildfires are not limited to forested areas and even built areas are subject to "urban megafires." To reduce this risk, we now have roof-mount sprinklers for potential fire suppression. These could source from municipal water and power but could also run from stored rainwater and battery-powered pumps. The previously discussed concrete-stone wall (Fig. 11.7) has an additional purpose as this is the side of the property where stormwater flows are deepest. The wall serves as a barrier to water movement into a low area on the lot. Recently, a 3-kWh battery storage system was added to provide some capacity to keep the building going during outages. This has proven useful for keeping a home business operational and has been used to run the evaporative cooler during an ~8-h power outage. The small current-draw evaporative cooler (~250 W) mates well with a small battery storage system and shows the utility of reducing loads first before attempting to self-generate electricity. These and other projects will undoubtedly continue.

15.14 Chapter summary and conclusions

In short, the growing human fingerprint is quite prevalent across natural systems as well as the built sector. Our first and best response to climate threats is to rapidly decrease our greenhouse gas emissions. Under a sober analysis, some level of climate disruption and population dislocation is already "baked in" to the system, if not due to physics, then due to human inertia.

Thus, adaptation to a changing climate is a mandatory part of our response. The potential for increasing and accelerating extreme events is a threat. Planning for unprecedented weather events to become much more precedented is prudent and not just concerning climate change. Evidence is emerging that climate change is exacerbating extreme weather events, most prominently heat waves, heavy precipitation events, damaging winds, and amplified winter storms in different regions.

This means planning for the inevitability of more heat, water, fire, winds, and other natural hazards that have been amplified by anthropogenic climate change. The dry regions are becoming drier while wet regions are wetter. Many regions are experiencing more potent windstorms. Depending on their location, islands and other low-lying locations are among the most vulnerable to climate change impacts ranging from susceptibility to sea level rise impacts, extreme events such as hurricanes and tropical storms as well as the logistical challenges of disaster response due to isolation.

The built sector has a double challenge in both reducing emissions and fortifying our systems against known and unknown changes. New tools to allow estimation of climate risks by downscaling climate predictions are emerging, for example, FloodFactor and ClimateCheck. Resilience engineering is the approach allowing our systems and infrastructure "to bend rather than break." Several broadly anticipated changes to the physical environment related to our changing climate extend to the building sector. As discussed regarding materials and some of the startup companies pursuing improved climate resilience, modular construction with reinvented construction materials can contribute to more resilient structures.

The insurance industry and what regions and homes it will insure have been an important constraint in this space. Over the past several decades, the trend has been that insurers are moving out of some of the highest-risk markets including mountainous regions of California susceptible to wildfire and coastal regions of Florida susceptible to hurricanes, tropical cyclones, and continued sea level rise. The importance of what is insurable will only grow as a climate adaptation constraint.

Wildland fire has become a critical issue in many (and expanding) pyrogenous zones including much of the western United States whether it is montane or grasslands. The localized effects are loss of life and property, while the broader effects can be seen with long-range smoke transport. The air quality gains due to the Clean Air Act are becoming increasingly counteracted by the increase in $PM_{2.5}$ due to wildland fires. For example, the Canadian fires in Quebec in summer 2023 caused huge air quality impacts in the North Central and Northeastern United States for weeks. Consistent with continued climate change, such trends are strongly linked to burn season length and regional drought conditions. Some locations may become too hazardous to wildfire occurrence and survivability.

Grid failures associated with extreme events are a key motivator for developing robust microgrids. Microgrids take the net-metered solar home to a new level. Electrical reliability and grid stability have been affected by both extreme weather in the summer (California and summer outages with heat waves and fires in summer 2020) and winter storms (Texas and snowstorm Uri in February 2021). Natural gas generators were brought down with record-setting freezing temperatures for which facilities were not prepared, while at the same time, some renewable assets experienced freeze-up of wind and the lower solar generation in winter. One effective adaptation is using microgrids. All energy resources are susceptible to varying degrees of various extreme events. Drought and water scarcity for steam power plants, plus the susceptibility of electrical generation and transmission to extreme weather, add vulnerabilities.

Not every current habitable place will be viable in the decades to come. The alternative and ultimate adaptation to retreat from some of the most at-risk locations is becoming more of a reality. Already a number of existing towns have been relocated or abandoned such as Isle de Jean Charles, located in the Mississippi River delta near sea level in Louisiana. Often multiple factors combine to make such places unhabitable—location combined with rising sea levels, sinking land masses, intensifying tropical cyclones and extreme heat, and reduced buffering from mangroves, all experienced by a population already economically marginalized. It is difficult to predict which other settlements will have to be abandoned this century. However, it is all but certain a nonzero number, and these stay or go decisions will be agonizing.

Municipalities and industries have begun to evolve to become more sustainable, whether mandated or elective. The green building approach is becoming more the norm than the exception, becoming more sophisticated and knowledgeable along the way (Fig. 15.13). Motivation stems from ethics, global concerns, reduced costs, improved comfort and functionality, market demand, and public relations. Educating students in engineering and throughout the building industry is imperative to making this happen.

Climate change is the greatest environmental threat we face and arguably the greatest challenge humanity has faced, certainly of those of our own making. It is a difficult but soluble problem, and we have the technical means to fix this. We have the solutions at hand, albeit imperfect, but only lack the motivation and will to fix the problem. The area of green building offers one of the key tools to address climate change and, in doing so, a range of other ills associated with our built environment. Fixing our built environment is thus key to fixing the perturbations in the natural environment that we have created.

15.15 End of chapter and end of book summation exercises

(1) **Group Project:** Perform an energy audit on a building or home of interest. See Appendix A for an example of specifications for the project.
(2) **Group Project:** Build a "green doghouse," solar shed, greenhouse, or tiny house (depending on the budget available) using the principles throughout the book. See Appendix B for an example of specifications for the project.
(3) **Group Project:** Have a design-build eco-concrete competition. See Appendix C for an example specification for the project.
(4) **Group Project:** Propose a university sustainability upgrade project (e.g., paper towel replacement with air jet dryers, composting of cafeteria food waste, campus landscape watering audit).
(5) **Group Project:** Build a solar oven and compete to see which oven attains the highest temperature and which maintains the steadiest temperature as solar flux diminishes.
(6) **Concepts:** What are the climate vulnerabilities of the region you live in? What steps can you recommend to reduce these risks going forward?
(7) **Concepts:** Research your community's and/or state's resilience plan and provide a summary.
(8) **Problem:** Diagram the process of a home renovation focused on energy efficiency, sustainability, and resilience.
(9) **Problem:** Investigate your local municipalities (state, county, and city) and summarize the existing status regarding climate resilience plans.
(10) **Problem:** Take the property you live in or another property of interest and discuss the ways that it can be made more firewise. Make a prioritized ranking of the most cost-effective measures that could be taken.
(11) **Problem:** Run the Floodfactor and ClimateCheck reports for an address of interest to you. Discuss the risks enumerated. Do they seem likely? What are some of the limitations of using such a model for assigning specific risks to a property?
(12) **Problem:** In 2022, total US primary energy consumption per person (or per capita consumption) was about 301 million British thermal units (MMBtu), according to US EIA. Assume you use energy over a lifetime equivalent to the average American. Determine the individual quantity of coal, gasoline, gas (therms), uranium (kg), number of solar panels, or size of a wind turbine to supply all of this. Assume continuous operation with capacity factors of 40% and 20% for wind and solar.
(13) **Problem:** Assess the vulnerability of your local community to climate change risks. Compare to the newly developed tools including Climate Check and Flood Factor.
(14) **Problem:** The following green office building design (likely the greatest ever) has been presented to you for construction in the desert southwest. Find 10 design errors or nonidealities (there are at least 15!) and corrective action. Focus on "sins of commission" related to green building and not "sins of omission" (like it does not have a front door or thermostat as not every detail is shown!). The HVAC components are all properly designed, off-the-shelf pieces of equipment. You can assume the front and sides of the building look similar. What are some resilience aspects that can be included?

Chapter 15 Climate adaptation

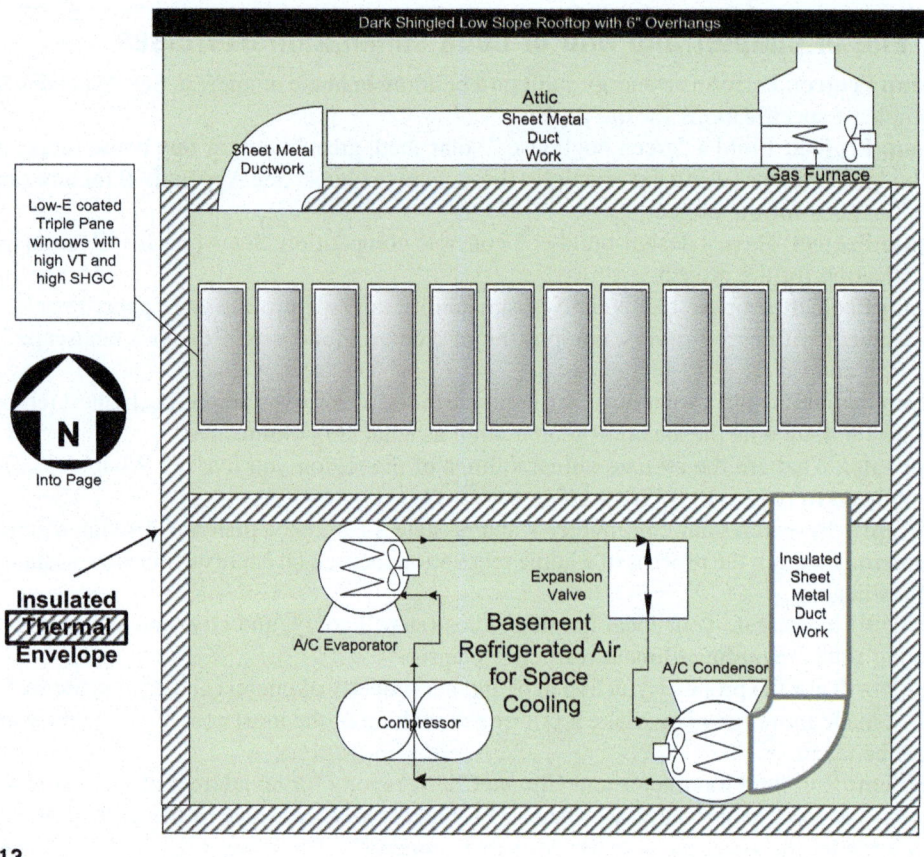

FIG. 15.13

Proposed green building for critical review.

(15) Problem: In our warming world, wildland fire is an increasing risk factor in the western United States. The US wildfire modern record is shown later, where total US acreage burned and average acres per fire are shown for wildfires.

  i. Find the acreage of the largest Alaskan fire on record in the last 50 years. The National Interagency Fire Center might prove useful. Find the total wildland fire acreage in 2015; the graph above may be useful.
  ii. Refer to the following for an estimate of fuel loading for the Northern Rocky Mountain region. Find the mass of biomass consumed by the Alaska fire and the total acreage burned in 2015.
      http://www3.epa.gov/ttnchie1/ap42/ch13/final/c13s01.pdf
  iii. For both, using this mass of biomass and the pollutant yield for particulate matter, calculate the mass of total particulate matter (PM) emitted.

iv. Use a CO_2 mass emission factor of 1.5 kg CO_2 per kg of fuel combusted. Calculate the mass of CO_2 released for both.

v. Use a biomass fuel energy content of 8000 BTU/lb of fuel. Find the energy released in PetaJoules (PJ $= 1 \times 10^{15}$ J) for both cases.

vi. In both cases, compare this energy release to the annual energy output of a typical 1000 MW electric power plant, assuming its outputs are all eventually converted into heat. Do not forget to factor in the thermal efficiency!

Appendix A. Home energy use analysis project

This project will help you perform a home energy audit on a home or apartment of your choice. You will be working as a team to perform the analysis for ultimately creating a report for recommendations for efficiency improvements. You will need to select a home (preferably) or an apartment where you have access to your utility bills, as they will be used in the analysis.

We will have access to IR thermometers as well as an IR camera which may be helpful in determining sources of heat loss. We also will have electricity consumption meters (kilowatts) that will be useful for measuring plug-in appliance electrical consumption. A window profiling instrument will be used to measure visible transmittance, IR transmittance, and UV transmittance.

A "Manual J" software package for estimating the heat loss/gain for sizing HVAC systems will be used by the group to analyze the home under study. This will allow you to take data from the home you are investigating and put it into a heat loss calculation program. There are a few spots we will discuss that need some clarity with the Manual J.

Net Zero Home Energy Efficiency Upgrades For All—ZWELL (zwellhome.com)

A few elements that this project incorporates:

(1) Home assessment
- Draw a floor plan, identify major structural components, construction details and age, and other unique features that may impact energy use (hot tub, etc.)
- Characterize sources of energy for structure (heating, cooling, water heating, other major consumptive activities)
- Look up consumption of models of HVAC and water heater, and other major appliances as available on manufacturer or energy star websites
- Assess current HVAC system: type, approximate age, fuel, efficiency if listed or search online, size. Does this unit seem appropriately sized for the space compared to the Manual J?
- Find energy star ratings for these devices (if they exist for a given appliance); compare to kilowatt measurements (if available for this appliance)
- Compile this into a table and compare with current best available efficiency (e.g., a current mid-sized energy star refrigerator is ~ 500 kW-h per year)
- Calculate the cost and payback time of the top three recommendations
- Compare your home usage to metrics for average New Mexico and US homes. Discuss why it might be different.

(2) Utility assessment
- Obtain 1-year (ideally) of electric, water, and gas bills
- Plot the data in time series, compute annual totals and monthly averages, identify any outliers, and speculate on the cause (leaking sprinkler in March cause an increase of ~5000 gallons of water consumption)
- What does the data tell you about the consumption of these resources in the building and property? What is the seasonal cycle? Apportion use categories as possible (e.g., summertime excess in electric use can be attributed to A/C, wintertime excess to home heating, summer gas usage can possibly be attributed to hot water heater)
- Utility bill analysis (compare annual use to "average" home in New Mexico and United States)
- If one of the utilities not available find suitable replacement data

(3) kilowatt measurements
- Measure plug-in device consumption (real-time and for extended period for eight or more devices)
- Preferentially choose among the most consumptive devices (e.g., refrigerator, heating system if it is electric and plugs into the wall, any large plug-in devices)
- What is the maximum power usage for these devices? What is the extended 1–2 week energy use of the devices (e.g., the power of a hair dryer is quite large at 1.5 kW but its energy use is small since it is only used briefly)
- Which are big power uses? Which are big energy uses? They may not be the same, why?
- Compare with faceplate ratings for five devices (including refrigerator). Search for the refrigerator Energy Star label online (yellow card comparing to average consumption). Does this seem accurate from your measurements?

(4) Window profiler
- Measure a minimum of five windows in your structure based on feasibility and getting a reasonable assessment of your window types. How repeatable or variable are they for the same type window?
- Compile the VIS, IR, and UV transmittance data into a table
- What does this information tell you? What would the ideal window characteristics be?

(5) IR Camera pictures
- Include 6–12 pictures highlighting areas of heat loss/gain or other interesting features of your building shell.
- Infrared pictures inside and out (6–12 to include with report) of structural components (walls, windows, doors, roofs, etc.).
- Best time: large delta T inside to out, no solar input, cold morning. Turn off lights and any other heat sources inside if possible.
- Think: leaky windows and doors, poorly installed or settled insulation, problem areas like knee walls, cathedral ceilings, and cantilevered floors.

Interpreting Thermal Images—GreenBuildingAdvisor

(6) Site solar assessment
- Include a profile of the solar site assessment tool (shading analysis) from a favorable location for solar on the site
- Do not go onto a rooftop-assume a ground-mounted solar application
- Discuss the viability of solar applications at this property

(7) Blower door test
- Blower door test will be conducted on the test property
- ACH50-set pressure drop to 50 Pa and measure the flow rate
- Documentation of "leakiness" for air infiltration of the building using a common standard
- Fogger will allow the illustration of leak points in the structure

(8) Indoor and outdoor air quality monitoring
- Ambient PM2.5 concentrations will be measured near Jones Annex using a PurpleAir sensor
- Indoor air quality will be monitored with an equivalent device at the home under test
- Comparison of indoor and outdoor measurements
- Determine what are any contributors to poor indoor air quality

Suggested project roles include but are not limited to:

(1) project manager (grad student unless none on team) edits documents, makes sure subtasks are accomplished on schedule, makes sure deliverables are delivered on schedule, interfaces with client to provide updates;
(2) laboratory technician-examines documentation and parts, assesses completeness, and researches the needed parts of the system with examining the documentation; and
(3) field technician operates IR camera, kilowatt measurements in the field assuring data collection and proper operations;
(4) data analyst plots data, looks for anomalies, assembles into appropriate tables and figures;
(5) report writer provides interpretation of data and writes the text of the report;
(6) presenter-prepares and delivers a 15-min PowerPoint presentation on the project to the client;
(7) modeler-inputs data to Manual J software; and
(8) background researcher: does literature source search and compiles relevant information on energy audits.

Schedule and deliverables:

Task (25 points are team member assessments)	Date	Deliverable (Wednesdays)
Determine group member roles (5)	Canvas	List of roles for group members and paragraph description of your game plan for the project
Identification of candidate structure, pictures, and obtaining of utility bills (5)	Canvas	2-Page description of the structure, floor plan, diagram, and tabular data on structure
Online Manual J calculations (5)	Canvas	Submit a printout of Manual J calculations for building
Solar site assessment (5)	Canvas	Image and bullet point analysis
Utility records analysis Registration deadline for groups for SRS poster presentation (10)	Canvas	Provide graphical and tabular data for utility usage and provide analysis of results
Thermal camera images (5)	Canvas	Submit document with illustrative pictures of building and 1–2 pages interpretation of results
kilowatt measurements and analysis (5)	Canvas	Compile kilowatt measurements into a table and provide analysis of results
Analysis of blower door and indoor air quality measurements (10)	Canvas	Compile graphical data and bullet point summary analysis
Draft posters (25)	Canvas	Draft poster for SRS
Final poster and presentations to class (50)	Canvas	Powerpoint slide of poster
Final project deliverable poster at SRS in April (50)	Canvas	Final poster presented at SRS and submitted to canvas
Team member evaluations (25)	Canvas	Submit forms evaluating team member contributions

The final poster must contain the sections described below.

The project grade (200 points) will be based on the following:

(1) Team member evaluations (25 points). If all your team members fail you for non-participation, you earn a zero for the project
(2) Draft posters (25 points)
(3) Final posters and presentation of poster in class (50 points)
(4) Final poster with all components and measurements and presented at SRS (50 points)
(5) Deliverables throughout the semester (50 points)

/25 points total poster evaluation

/2 Introduction: Some cleanup noted in the annotated document.
A transition toward the main body of the document. It should take an uninformed reader from a level of zero knowledge to a level in which the reader is able to understand the main body of the document. A good introduction must have:

- Motivation (i.e., why is it important?)
- Background (i.e., what is the history of this issue?)
- Objectives (i.e., what are you trying to accomplish?)
- Scope (i.e., what is the focus of your analysis?)
- Limitations (i.e., what constraints did you face?)
- Content (i.e., what is in the report?)
- Organization (i.e., how the report is organized?)
- References: Find, examine, and pull information from at least five sources

/2 Literature review:
Examine the existing literature regarding building energy efficiency and provide a 3-page summary of relevant information. Consult literature sources (books, journal articles, and online information from reputable science sources such as Leadership in Energy and Environmental Design). Discuss split incentives and how they are important for rental housing. Refs, a min of 10 with half of them from books, published articles, science lit rather than pop lit.

/3 Methods:
A description of the methodological framework you have used in the project. Here you can talk about the IR camera you used, including info on how they work. You should discuss the heat loss calculations as a means to estimate your home's energy efficiency as well as to properly size an HVAC system. This will require a little research (online, journals, books, professional societies like ASME, ACEEE, LEED).

/8 Analysis of results: Overall reasonable analysis. Many of the notes and input on the draft report and other deliverables were overlooked in the final report.
The section should start with a description of the property under analysis: age, construction type, basement/slab/crawlspace, pitched or flat roof, materials, type of windows/doors/other leak points, etc. A description of the results obtained and an analysis of the implications associated with main results. It must be supported by figures and tables to facilitate, not to confuse, the reader. The inclusion of a floor or room diagram as well as key IR images is useful. Include a table with major systems including furnace, A/C, swamp cooler, refrigerator, water heater, etc. Discuss three recommendations and estimated payback times for those upgrades. Scale the household energy and water use appropriately and compare on a common basis (per capita, per square foot, per use, or continuous use as appropriate).

Metrics to consider are: gallons/person/day, therms/month, kWh/month, therms/square foot/year, normalizing to number of residents as appropriate.

/2 **Conclusions:**
A summary of the major findings you have arrived at in the previous sections. "Conclusions" is not an analysis section. Most of the analysis should occur in the "Analysis" section, and the conclusions should be summarized in a couple of paragraphs.

/2 **Recommendations:**
Insights into the next steps you recommend to be taken for improved energy efficiency. This must be supported by the analysis and conclusions section of the report. Include a prioritized list of prudent recommendations for efficiency improvements, including costs after retrofits and payback period.

/2 **References:**
A listing of books and articles you have used, or consulted, for methodological and nonmethodological issues.

/2 **Figures and tables**:
They are intended to facilitate understanding of the document by presenting relevant information and data in an easy-to-understand way. *They can be integrated into the main body or included in a separate section.* **Include pictures of home and heat loss areas.**

/2 **Writing style**, **grammar, and presentation:**
Is the report written with clarity, proper mechanics, and a succinct writing style as well as an attractive layout?

Tips on a successful project report and/or poster:
- Please specify in the report who did what (measurements, compiling tables, each section that is written, overall editing and report construction, etc.).
- You will also be asked to anonymously and separately rate the performance of your other team members, separate from the report, which will be a part of the grade each person receives.
- Remember, it's only EXTERNAL surfaces that participate in heat loss for the Manual J calculations. Interior walls or a floor on one level that's the ceiling on another level of the house do not result in heat loss from the house to the exterior.
- What did the IR images show? Where were the most important areas to focus on for improving the efficiency of the home? Include the most useful IR images, the ones that you learned something.
- How do your heat loss calculations compare to the size of the furnace that is in your house?
- Besides the textbook, you should consult at least four other sources of info on energy audits and home efficiency and cite these in your references. These can be used in the intro material or in your analysis.
- Spelling, grammar, and report organization style all count. Make it look like something you would give to a client or boss trying to sell them a project. It doesn't have to be glossy and slick, just polished and effective. Hand-scrawled worksheets are not acceptable at this point; put them in your spreadsheet.
- An example of the materials to include in response to the worksheet questions (and beyond the 10 pages of text write-up and a printout of the calculation spreadsheet) is shown below for your guidance.

Appendix A Home energy use analysis project

Home heating analysis

The following is an example from the Manual J heat loss calculator available (the heating and cooling values are BTU/hour):

http://www.loadcalc.net/

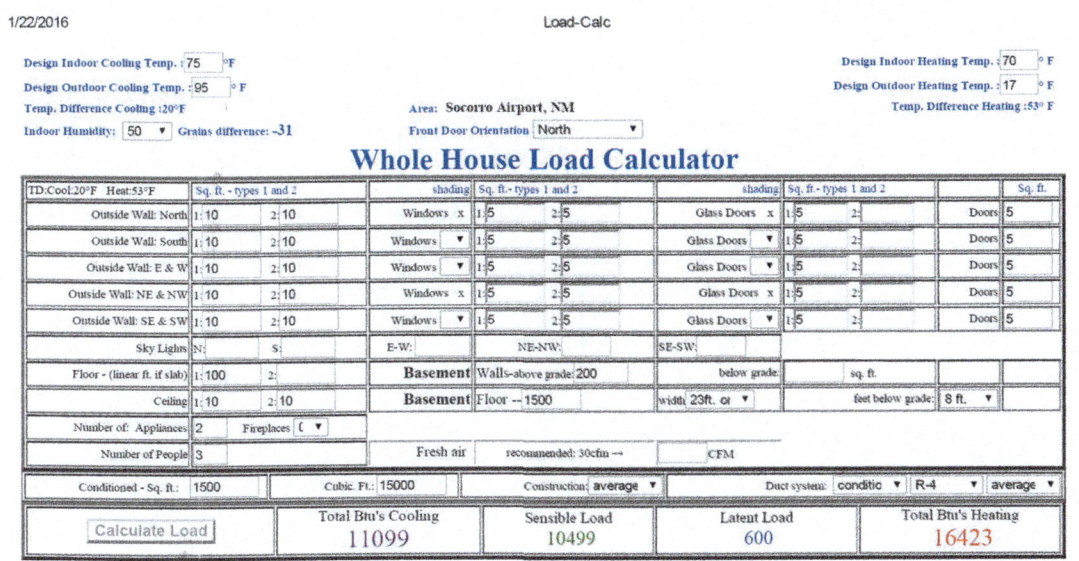

Example floor area layout (indicate windows and doors)

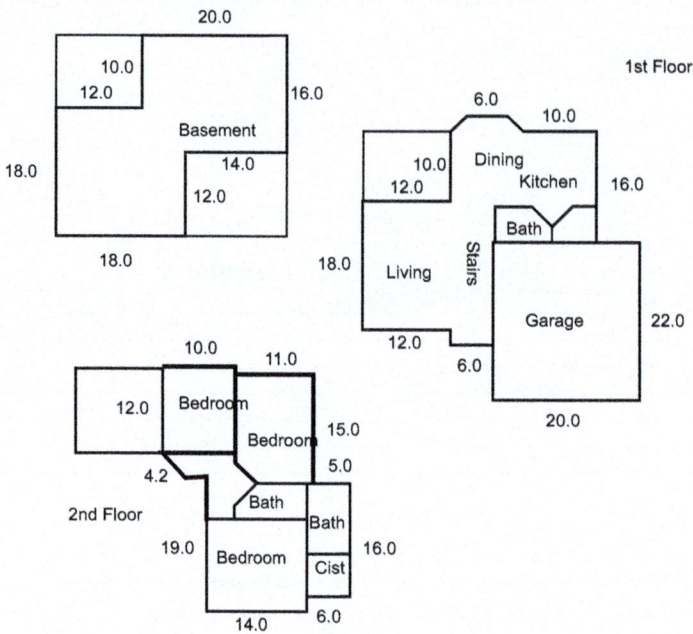

Appendix B. Green building project—Green doghouse

The project involves designing and constructing a model of a green doghouse. We will work as two teams as follows. In lieu of a report, you will have the doghouse and a poster for the SRS as deliverables.

Suggested project roles include but are not limited to (all students should be involved in all levels):

(1) **Project manager:** edits documents, makes sure sub-tasks are accomplished on schedule, makes sure deliverables are delivered on schedule, and interfaces with client to provide updates
(2) **Designer/constructor:** builds the model, works with hand and power tools, solves problems, examines documentation and parts, assesses completeness, and researches the needed parts of the system by examining the documentation
(3) **Data analyst report writer:** provides interpretation of data and writes the text of the report, takes draft of SOP from the lab and field tech to edit for accuracy and clarity, documents the project, plots data, looks for anomalies, and assembles into appropriate tables and figures
(4) **Presenter:** presents project to judges interview, prepares, and delivers a 15-minute poster presentation on the project to the client and to the SRS (Table 1)

Table 1 Project milestones

Task (total of 200 points)	Due date	Deliverable
Kickoff meeting: determine group member roles and provide synopsis of meeting and game plan (10)	2/7	List of roles for group members and brief synopsis of first meeting
Examine literature and provide 3-page background literature review for poster inclusion (10)	2/15	Poster section that is literature review
Initial design (on paper, 2-page description of design with at least 1 diagram and floor plan) (15)	2/26	Initial design document
Client review of construction progress (15)	3/13	Construction progress to date
Performance testing (25)	3/31	Monitor outside deployment and coordinate with performance testing
In-class poster presentation (25)	4/13	Presentation to the class and provide PowerPoint poster slide
Provide all documents compiled into a final poster presented at SRS and present doghouse at SRS (50)	4/17	Final Poster presented at SRS
Judge evaluations (25)	4/22	None
Provide team member evaluations (25)	4/26	Provide assessments of team member contributions—must receive your evals or you will not get credit

Design for the canine criteria:
- Functioning, portable outdoor doghouse for mid-size canine
- Maximize the use of recycled, reused, donated, and found materials
- Performance testing to include mock blower door test for air leakage and temperature stability (avoiding overheating during day and maintaining comfortable temperature at night)
- Excellence in floor plan design: The design submitted displays practical, livable, and comfortable space for a small to medium canine
- Excellence in exterior design: The design submitted displays well-thought-out landscaping, rooflines, window/door arrangements, proportions, and material selections
- Functional and aesthetic doghouse
- Performance testing metrics
- Displays knowledge of materials and construction methods and assemblies
- Innovative design points
- Judging will be done by an expert panel of judges, including the department chair

Judging panel criteria (scored 0 to 5; to be addressed in the model, presentation, and report)
 (1) Detailed design drawings description detail, quality, and representation
 (2) Detailed design description detail, quality, and representation
 (3) Example wall and roof sections for examination (drawing)
 (4) Model built following plan: The model submitted matches the plan submitted
 (5) Innovation (what distinguishing features are present in the design)
 (6) Performance: Maintain comfortable temperature uniformity day and night
 (7) Performance: Minimize infiltration as measured by a Mock Blower door test

(8) Performance: Stand up to the elements (wind and water)
(9) Aesthetics—Would you want to live in this house or want your dog to live in it? Is it marketable?
(10) Materials use reuse of material, recycled materials, found materials (discuss)
(11) Quality of model construction: The level of workmanship of the model submitted, as well as the choice of materials, textures, and colors
(12) Oral presentations on details of design approach and construction
(13) Embodied energy of materials—calculate an embodied energy of the construction materials as designed, including as a table.
(14) Estimate r-values of dog house and do heat loss calculations.
(15) Make an estimate: could a dog's internally generated waste heat keep the structure comfortable on a typical winter day?
(16) Interior light levels due to daylighting-performance test (consider)
(17) Incorporate knowledge gained from course
(18) Discuss how the building would be located and installed on a site

Rules:
(1) Budget: $100 or less and any additional out-of-pocket expenses are limited to another group combined $100 if you would like (receipts required)
(2) You may use any scrap material available in my lab and home (wood, tile, metal, etc.)
(3) Donations of materials and funds are fair game (document)
(4) Size limits: 8 feet combined dimensions, 25 kg mass
(5) You can build your model at home, or we can provide some lab space for its construction
(6) At least 1 accessible entry door. Window included at your discretion (daylighting is desirable)
(7) Passive house, no power available for mechanicals for heating or cooling except self-generated

The project grade (150 points) will be based on the following:

(1) Team member evaluations. If all your team members fail you for non-participation, you earn a zero for the project.
(2) Final poster presentation with all components and measurements and supporting documents
(3) Final presentation

Performance metrics (these will be tested on or about April 16, 2023). Metrics should include but are not limited to:

(1) Mock blower door test (what metric to use?)
(2) Maintain comfortable temperature regardless of conditions (will measure at least once in the daytime and once in the evening)
(3) Moderate indoor RH (midrange 30%–50%)
(4) IR pictures for heat loss
(5) Temperature uniformity in time and space
(6) Indoor light and sound levels?
(7) Weather resistance (mock rainstorm, windstorm?)
(8) Performance test: leaf blower for mechanical stability and weather resistance
(9) Real-life dog desirability testing

Appendix C. Semester project eco-concrete (Profs. Carrico and Morris)

A client is interested in building with concrete that has a lower environmental impact. You and your team (to be chosen in class) are going to produce concrete for strength testing with minimum materials and carbon footprint. As part of this, you will be given recovered material from an end-of-life solar photovoltaic panel and recycled concrete for inclusion in your new concrete mix. We will have the basic concrete materials (cement, aggregates, sand) available, and the rest of the mix design is up to you (supplementary cementitious materials, other recycled content).

Objectives
1. Produce a high compressive strength concrete
2. Calculate and minimize the virgin and total materials used in producing your concrete samples
3. Calculate and minimize the materials used in producing your concrete samples, providing a table quantifying the materials and their final disposal
4. Calculate and minimize the carbon footprint per kg of finished concrete

Details
- Students will split into teams of approximately five members (at least one environmental engineering student per team).
- Team member roles will be self-assigned based on team consensus and should include concrete mixologist, report writer, calculation master, material interaction consultant, and environmental impact consultant. Note that the assigned roles are for the *lead* in that area; all members are expected to contribute. For example, the report writer could assign, collate, and submit everyone's contribution to the report.
- Teams will produce four (4) 3″ × 6″ standard concrete cylinders for compression strength testing.
- NMT will provide the Portland cement, the required amount of recycled PV panel and concrete, molds, and options for sand and aggregate.
- Admixtures and other alternatively sourced materials (sand, aggregate, etc.) may be used, but must not incur any costs. These materials must be included in the analysis and mix design.
- Students will be given mix design guidelines.

Requirements
- Make at least four (4) uniform 3″ diameter cylinder samples that can be tested
- Compression test and report strength of (2) samples at 14 days
- Quantity (fraction weight percent) of recycled PV panel material-glass and PV cells. You must use the entire mass of PV panel sample given to you in your mix. You may grind the PV glass to your desired consistency but you may not add or remove anything from it without accounting for it in the mix design/carbon footprint
- Following the initial concrete design, you may make 2 design changes to your mix up to and including the mixing session. You must detail the changes, describe what you learned, provide justifications of those changes in your report, and redo your carbon footprint calculations.
- Groups can replace their allotted materials with substitutes provided that:
  - They are not virgin materials
  - Their mass contribution is quantified in the mix
  - They are zero-dollar cost
  - Their carbon footprint can be quantified and documented

- Final poster and presentation of project, including the Eco-crete project and your additional experiment and results will serve as the final for the course.

Competition

Students' concrete will be evaluated based on its strength and minimization of environmental impacts. **The best overall concrete project all things considered will be showcased at the C&EE department's Student Research Symposium!!** Other fantastic prizes may be awarded for metrics that will include:

- Lowest carbon footprint of mix
- Highest strength concrete (compression)
- Highest strength concrete (splitting tension)
- Highest strength-to-weight ratio
- Lowest carbon footprint concrete
- Most aesthetic concrete (visualize using for building exterior structure)
- Best in show—overall best concrete and most desirable package of deliverables
- One overall/final metric chosen by your team (can be a combination of those above or others)

Things to think about in your concrete

- Mixing and uniformity of concrete
- All cylinders will be cured in equivalent conditions in the lab for 14 days
- Interactions and expected effects of mix design choices and material choices
- Comment on feasibility of costs and scale-up of producing your concrete in the report
- Access and cite outside materials used in helping make your decisions

Mix design

- For the mix design, please submit **using the template provided** detailing your mix design requirements. Clearly present the total mass in kilograms of each component of your mix and a table showing the materials, sources, sizes, and quantities of the mix. Also specify the processing for your recycled PV panel materials inclusion (for example, what sizes are you going to obtain?). Parts of your mix design may change, but you will not be guaranteed more than your submitted quantities of Portland cement and virgin sand and aggregates. The template is shown below.

Poster rubric:

Your poster must address every item below. Follow the rubric below or you will have points deducted. You will turn in a single final pdf of your poster detailing team's concrete specs, mix design, competition metrics and results, and other attributes with figures, citations, and tables as appropriate.

- (10%) Poster Intro, background, and literature review
- (10%) Mix designs
- (10%) Calculations—carbon footprint (kg CO_2/kg of concrete)
- (10%) Minimization of waste: weighing and documentation of the mass of materials used in your test cylinders and the mass of material landfilled
- (10%) Report results of your mix design and testing
- (5%) Potential impacts of mix for scaling up production, costs, etc.
- (5%) Additional combined scoring metric and justification
- (10%) Lessons learned and what you might do differently next time (in mix design, mixing, material preparation, etc.)

Appendix C Semester project eco-concrete (Profs. Carrico and Morris)

- (10%) Team member contributions evaluations
- (10%) Report format and aesthetics—no problems with citations, formatting, and organization
- Bonus 10%! for innovative techniques, presentation, or environmental attributes

Carbon footprint

The total contribution of concrete to global greenhouse gas emissions is one of the largest industrial contributions outside of energy-related combustion of fossil fuels and contributes 4%–8% of global CO_2 emissions. Emissions come from both the energy use for concrete production, particularly the Portland cement production which involves fuel combustion for high-temperature calcining, as well as the curing of the cement where calcium carbonate forms lime, producing CO_2. Though the CO_2 per unit of material for cement and concrete is not particularly large (Fig. 1), the overall amount of concrete used globally is staggering, driving its importance to greenhouse gas emissions and hence climate. The approximate carbon footprint of concrete is ~100 kg CO_2/ton concrete overall, with some variability based on sources of materials, material processing, etc.

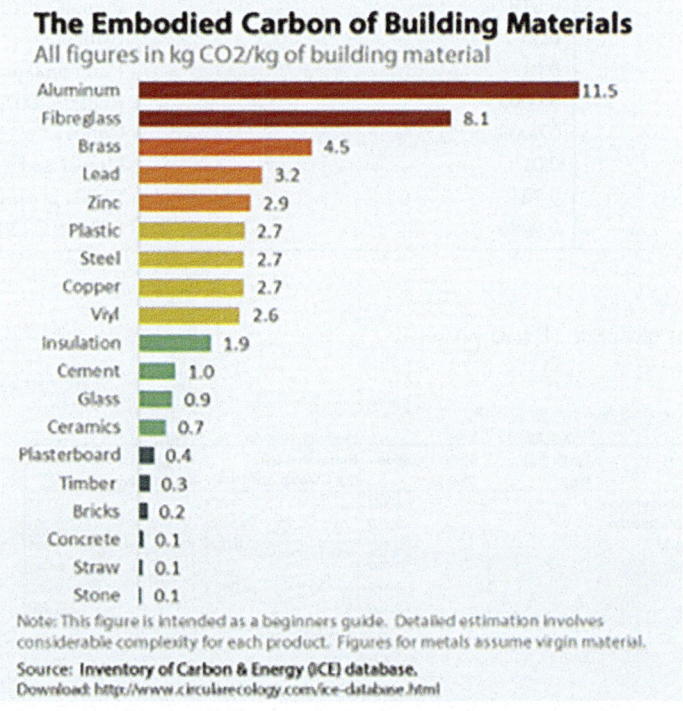

FIG. 1

Embodied carbon (kg CO_2/kg of material) for various components of concrete (ICE database).

For our purposes, we will use the following data for the carbon footprint of the components that comprise concrete. Substituted materials will need to be quantified similarly and references given. Different analyses can give different results; for our purposes, use the value provided in the template if multiple studies give varying values. The template that students use will be similar to that provided in Table 1. **Although virgin glass has a carbon footprint, since the PV glass and recycled concrete are reuse materials, they will be assigned a zero carbon footprint.**

Table 1 Embodied carbon (kg CO_{2e}/kg of material) for various components of concrete (Adesina, 2020).

Material	Embodied carbon (kg CO_{2e}/kg)	Source
Portland cement	0.83	Hammond and Jones (2008)
	0.83	Chiaia et al. (2014)
	0.82	Reiner (2007)
River sand	0.0050	Hammond and Jones (2008)
	0.0025	Chiaia et al. (2014)
Crushed aggregate	0.0062	Nisbet et al. (2002)
Fly ash	0.010	Hammond and Jones (2008)
	−0.0018	Reiner (2007)
	0.009	Purnell (2013)
Slag	0.070	Hammond and Jones (2008)
	0.019	Purnell (2013)
Metakaolin	0.33	Estimated
Silica fume	0.014	Hammond and Jones (2008)
Water	0.0003	Reiner (2007)
	0.0003	Chiaia et al. (2014)
Superplasticizer	0.01	Flower and Sanjayan (2007)
	0.72	Chiaia et al. (2014)
Air entrainers	0.0086	EFNARC (2005)

Carbon footprint template (Excel posted)

		V Input these V	fixed	fixed emission factors	fixed		
	Material	Mass (kg) (kg)	Mass Fraction (fraction)	Embodied CO2 (kg CO2/kg material)	Weighted Contribution (kg CO2/kg concrete)	reference	
	Portland cement	1	0.167	0.83	0.138		
	Crushed aggregate	1	0.167	0.0062	0.001		
	River sand	1	0.167	0.005	0.001		
	Fly Ash	1	0.167	0.01	0.002		
	Water	1	0.167	0.0003	0.000		
	PV glass	1	0.167	0	0.000		
Fill in others -->							
Fill in others -->							
Fill in others -->							
Fill in others -->						TOTAL EMBODIED CO2	
Fill in others -->						Sum of Contributions	
	TOTAL MASS	6 (kg)	1 (sums to 1)		0.142 (kg CO2/kg concrete)		

Appendix C Semester project eco-concrete (Profs. Carrico and Morris)

Mix design guide
Basic proportions of components based on the nominal maximum size of coarse aggregate for non-air-entrained concrete (adapted from Tables 7.11 and 7.12, Mamlouk and Zaniewski, 2017):

Nominal max. size of coarse aggregate mm (in.)	By weight (e.g., per 1 kg or 100 lbs of concrete)				By volume (e.g., per 1 m^3 or 1 ft^3 of concrete)			
	Cement	Fine aggregate	Coarse aggregate	Water	Cement	Fine aggregate	Coarse aggregate	Water
9.5 (3/8)	0.200	0.407	0.317	0.076	0.182	0.455	0.272	0.091
12.5 (1/2)	0.185	0.363	0.377	0.075	0.167	0.417	0.333	0.083
19 (3/4)	0.170	0.320	0.442	0.068	0.153	0.385	0.385	0.077
25 (1)	0.161	0.302	0.470	0.067	0.148	0.370	0.408	0.074

Notes:
1. Mix design guidelines above show the proportions of each component. For example, if you need 100 lbs of fresh concrete with 9.5 mm aggregate, you would use 20 lb cement, 40.7 lb fine aggregate, 31.7 lb coarse aggregate, and 7.6 lb water in your mix.
2. The volume of mixed concrete is approximately 2/3 the sum of the original bulk volumes. Voids in the volume of aggregates get filled in by the wet cement mixture.
3. Due to the shape of the aggregates and workability of the mix, if crushed stone or angular type aggregates are used, decrease coarse aggregate by 50 kg (3 lb) and increase fine aggregate by 50 kg (3 lb) for each cubic meter (cubic foot) of concrete.
4. Provided coarse aggregate is ⅜" nominal max. size.
5. Aggregates that are angular, smooth, wet, or in other conditions may affect workability of the mix; that is, your mixed concrete may be "too wet" or "too dry."
6. To use supplementary cementitious materials (SCM), the "cement" column given above refers to the total cementitious materials or the sum of the portions of cement and other SCM. For example, if you're using 15 wt% fly ash and the mix is supposed to have 20 lbs of total cements, then you'd use 3 lb fly ash and 17 lb cement.
7. The water-to-cement ratio (W/C) controls most properties of the cured concrete. Higher W/C mixes are easier to work with, but have higher porosity and lower strength. Lower W/C mixes are more difficult to mix and place, but have lower porosity and higher strength.[a] To find the W/C of a mix, divide the weight of the water by the total weight of all the cementitious materials. W/C less than 0.4 and greater than ~0.55 are not recommended.
8. W/C ratios for the mixes above assume that the aggregates are in the "saturated surface dry" condition; this means that they have "dry" surfaces but the pores may contain water. That is, the aggregates in the mixes above contribute a small amount of water that brings the W/C for the mix over 0.4.

---

[a]Admixtures are chemical mixtures that change the viscosity, reaction speed, and other characteristics of the mixed concrete to improve workability or other construction requirements without increasing the W/C or decreasing the strength of the cured concrete.

References

Abatzoglou, J.T., Williams, A.P., 2016. Impact of anthropogenic climate change on wildfire across western US forests. Proc. Natl. Acad. Sci. USA 113, 11770–11775.

Abbafati, C., et al., 2020. Global burden of 369 diseases and injuries in 204 countries and territories, 1990-2019: a systematic analysis for the Global Burden of Disease Study 2019. Lancet 396, 1204–1222.

Abramson, E., Dane, M., Brown, J., 2020. Transport Infrastructure for Carbon Capture and Storage: Whitepaper on Regional Infrastructure for Midcentury Decarbonization. Great Plains Institute, 2020, pp. 71.

Aguilera, R., Corringham, T., Gershunov, A., Benmarhnia, T., 2021. Wildfire smoke impacts respiratory health more than fine particles from other sources: observational evidence from Southern California. Nat. Commun. 12, 8.

Ahern, J., 2011. From fail-safe to safe-to-fail: sustainability and resilience in the new urban world. Landsc. Urban Plan. 100, 341–343.

Akhtar, A., Sarmah, A.K., 2018. Construction and demolition waste generation and properties of recycled aggregate concrete: a global perspective. J. Clean. Prod. 186, 262–281.

Al Horr, Y., Arif, M., Kaushik, A., Mazroei, A., Katafygiotou, M., Elsarrag, E., 2016. Occupant productivity and office indoor environment quality: a review of the literature. Build. Environ. 105, 369–389.

Altomonte, S., Schiavon, S., 2013. Occupant satisfaction in LEED and non-LEED certified buildings. Build. Environ. 68, 66–76.

Andreae, M., Andreae, T., Annegarn, H., Beer, J., Cachier, H., le Canut, P., et al., 1998. Airborne studies of aerosol emissions from savanna fires in southern Africa: 2. Aerosol chemical composition. J. Geophys. Res. Atmos. 103, 32119–32128.

Andrew, R.M., 2018. Global CO_2 emissions from cement production. Earth Syst. Sci. Data 10, 195–217.

Andrew, R.M., 2019. Global CO_2 emissions from cement production, 1928-2018. Earth Syst. Sci. Data 11, 1675–1710.

Andrews, J., Jelley, N., 2022. Energy Science: Principles, Technologies, and Impacts. Oxford University Press, Oxford, UK.

Anthony, K.R.N., Kline, D.I., Diaz-Pulido, G., Dove, S., Hoegh-Guldberg, O., 2008. Ocean acidification causes bleaching and productivity loss in coral reef builders. Proc. Natl. Acad. Sci. USA 105, 17442–17446.

Armstrong, S., Higbee, E., Anderson, D., Bailey, D., Kabat, T., 2021. A Pocket Guide to All-Electric Retrofits of Single-Family Homes. Redwood Energy, p. 90.

ASHRAE, 2020. ASHRAE Position Document on Infectious Aerosols American Society of Heating, Refrigerating, and Air Conditioning Engineers. ASHRAE, Atlanta, GA, p. 20.

ASHRAE, 2023. ANSI/ASHRAE Standard 55: Thermal Environmental Conditions for Human Occupancy. American Society of Heating, Refrigerating and Air Conditioning Engineers (ASHRAE).

Azapagic, A., 2018. Environmental systems analysis. In: Kutz, M. (Ed.), Handbook of Environmental Engineering. Wiley, Hoboken, NJ, USA, pp. 1–12.

Azimi, P., Stephens, B., 2013. HVAC filtration for controlling infectious airborne disease transmission in indoor environments: predicting risk reductions and operational costs. Build. Environ. 70, 150–160.

Azimi, P., Zhao, D., Stephens, B., 2014. Estimates of HVAC filtration efficiency for fine and ultrafine particles of outdoor origin. Atmos. Environ. 98, 337–346.

Bailes, A.A.I., 2022. A House Needs to Breathe…Or Does It? An Introduction to Building Science. BrightCommunications.net, Unspecified location.

Bains, P., Psarras, P., Wilcox, J., 2017. CO_2 capture from the industry sector. Prog. Energy Combust. Sci. 63, 146–172.

Barnett, T.P., Pierce, D.W., Hidalgo, H.G., Bonfils, C., Santer, B.D., Das, T., et al., 2008. Human-induced changes in the hydrology of the western United States. Science 319, 1080–1083.

Barnosky, A.D., Matzke, N., Tomiya, S., Wogan, G.O.U., Swartz, B., Quental, T.B., et al., 2011. Has the earth's sixth mass extinction already arrived? Nature 471, 51–57.

Barnosky, A.D., Hadly, E.A., Bascompte, J., Berlow, E.L., Brown, J.H., Fortelius, M., et al., 2012. Approaching a state shift in earth's biosphere. Nature 486, 52–58.

Barreira, E., Almeida, R., Moreira, M., 2017. An infrared thermography passive approach to assess the effect of leakage points in buildings. Energy Build. 140, 224–235.

Batjes, N.H., 2014. Total carbon and nitrogen in the soils of the world. Eur. J. Soil Sci. 65, 10–21.

Beckwith, A.L., Borenstein, J.T., Velasquez-Garcia, L.F., 2022. Physical, mechanical, and microstructural characterization of novel, 3D-printed, tunable, lab-grown plant materials generated from Zinnia elegans cell cultures. Mater. Today 54, 27–41.

Bell, M.L., Davis, D.L., 2001. Reassessment of the lethal London fog of 1952: novel indicators of acute and chronic consequences of acute exposure to air pollution. Environ. Health Perspect. 109, 389–394.

Bell, M.L., Davis, D.L., Fletcher, T., 2004. A retrospective assessment of mortality from the London smog episode of 1952: the role of influenza and pollution. Environ. Health Perspect. 112, 6–8.

Bianchini, A., Bangga, G., Baring-Gould, I., Croce, A., Cruz, J.I., Damiani, R., et al., 2022. Current status and grand challenges for small wind turbine technology. Wind Energ. Sci. 7, 2003–2037.

Bistline, J., Abhyankar, N., Blanford, G., Clarke, L., Fakhry, R., McJeon, H., et al., 2022. Actions for reducing US emissions at least 50% by 2030. Science 76 (6596), 922–924.

Bolan, N.S., Kirkham, M.B., Halsband, C., Nugegoda, D., Ok, Y.S. (Eds.), 2020. Particulate Plastics in Terrestrial and Aquatic Environments. CRC Press.

Borrelle, S.B., Ringma, J., Law, K.L., Monnahan, C.C., Lebreton, L., McGivern, A., et al., 2020. Predicted growth in plastic waste exceeds efforts to mitigate plastic pollution. Science 369, 1515.

Bourouiba, L., 2020. Turbulent gas clouds and respiratory pathogen emissions potential implications for reducing transmission of COVID-19. JAMA 323, 1837–1838.

Bowen, D., Li, G., 2017. Having a lot isn't enough: trends in upsizing houses and shrinking lots. Fed Notes, Board of Governors of the Federal Reserve System, Washington, DC.

Brager, G.S., de Dear, R.J., 1998. Thermal adaptation in the built environment: a literature review. Energy Build. 27, 83–96.

Braham, A., Casillas, S., 2021. Fundamentals in Sustainability in Civil Engineering. CRC Press, Boca Raton, FL, USA.

Brahney, J., Hallerud, M., Heim, E., Hahnenberger, M., Sukumaran, S., 2020. Plastic rain in protected areas of the United States. Science 368, 1257.

Brand, S., 1994. How Buildings Learn: What Happens After They Are Built. Viking Penguin, New York, NY, USA.

Bui, M., Adjiman, C.S., Bardow, A., Anthony, E.J., Boston, A., Brown, S., et al., 2018. Carbon capture and storage (CCS): the way forward. Energy Environ. Sci. 11, 1062–1176.

Burke, M., Childs, M., 2023. Wildfires are worsening air quality in the United States. Nature 2. https://doi.org/10.1038/d41586-023-02794-0 (Epub ahead of print). PMID: 37730773.

Burke, M., Childs, M.L., de la Cuesta, B., Qiu, M., Li, J., Gould, C.F., et al., 2023. The contribution of wildfire to PM(2.5) trends in the USA. Nature, 761–766.

Butchart, S.H.M., Walpole, M., Collen, B., van Strien, A., Scharlemann, J.P.W., Almond, R.E.A., et al., 2010. Global biodiversity: indicators of recent declines. Science 328, 1164–1168.

Cabeza, L.F., Rincon, L., Vilarino, V., Perez, G., Castell, A., 2014. Life cycle assessment (LCA) and life cycle energy analysis (LCEA) of buildings and the building sector: a review. Renew. Sustain. Energy Rev. 29, 394–416.

Cabeza, L.F., Bai, Q., Bertoldi, P., Kihila, J.M., Lucena, A.F.P., Mata, É., et al., 2022. Buildings. In: JS, P.R.S., Slade, R., Al Khourdajie, A., van Diemen, R., McCollum, D., Pathak, M., Malley, J. (Eds.), IPCC, 2022: Climate Change 2022: Mitigation of Climate Change. Contribution of Working Group III to the Sixth Assessment Report of the Intergovernmental Panel on Climate Change. Cambridge, UK and New York, NY, USA.

Caesar, L., Rahmstorf, S., Robinson, A., Feulner, G., Saba, V., 2018. Observed fingerprint of a weakening Atlantic Ocean overturning circulation. Nature 556, 191–196.

Carlton, E.J., Barton, K., Shrestha, P.M., Humphrey, J., Newman, L.S., Adgate, J.L., et al., 2019. Relationships between home ventilation rates and respiratory health in the Colorado Home Energy Efficiency and Respiratory Health (CHEER) study. Environ. Res. 169, 297–307.

Carpenter, S.R., Caraco, N.F., Correll, D.L., Howarth, R.W., Sharpley, A.N., Smith, V.H., 1998. Nonpoint pollution of surface waters with phosphorus and nitrogen. Ecol. Appl. 8, 559–568.

Carrico, C., Kus, P., Rood, M., Quinn, P., Bates, T., 2003. Mixtures of pollution, dust, sea salt, and volcanic aerosol during ACE-Asia: radiative properties as a function of relative humidity. J. Geophys. Res.-Atmos. 108 (D23), 8650. https://doi.org/10.1029/2003JD003405.

Carrico, C., Prenni, A., Kreidenweis, S., Levin, E., McCluskey, C., DeMott, P., et al., 2016. Rapidly evolving ultrafine and fine mode biomass smoke physical properties: comparing laboratory and field results. J. Geophys. Res.-Atmos. 121, 5750–5768.

Ce, L., Be, R., Baker, T., Brash, J., Birge-Liberman, P., Corner, J., et al., 2017. Deconstructing the High Line: Postindustrial Urbanism and the Rise of the Elevated Park. Rutgers University Press, New Jersey, USA.

Ceballos, G., Ehrlich, P.R., Barnosky, A.D., García, A., Pringle, R.M., Palmer, T.M., 2015. Accelerated modern human-induced species losses: entering the sixth mass extinction. Sci. Adv. 1 (5), e1400253. https://doi.org/10.1126/sciadv.1400253. PMID: 26601195. PMCID: PMC4640606.

CFSS, 2020. Sustainability Factsheets: Consumption Patterns, Impacts & Solutions. Center for Sustainable Systems, University of Michigan. Collection http://css.umich.edu/factsheets.

CGD, 2016. More than a lightbulb: five recommendations to make modern energy access meaningful for people and prosperity. A Report of the Energy Access Targets Working Group, Center for Global Development, Washington, DC, USA, p. 35.

Chafe, Z.A., Brauer, M., Klimont, Z., Van Dingenen, R., Mehta, S., Rao, S., et al., 2014. Household cooking with solid fuels contributes to ambient PM2.5 air pollution and the burden of disease. Environ. Health Perspect. 122, 1314–1320.

Chan, W.Y.R., Joh, J., Sherman, M.H., 2013. Analysis of air leakage measurements of US houses. Energy Build. 66, 616–625.

Chapin, F.S., Zavaleta, E.S., Eviner, V.T., Naylor, R.L., Vitousek, P.M., Reynolds, H.L., et al., 2000. Consequences of changing biodiversity. Nature 405, 234–242.

Chapra, S.C., 2018. Water quality. In: Kutz, M. (Ed.), Handbook of Environmental Engineering. Wiley, Hoboken, NJ, USA, pp. 333–350.

Chen, I.C., Hill, J.K., Ohlemueller, R., Roy, D.B., Thomas, C.D., 2011. Rapid range shifts of species associated with high levels of climate warming. Science 333, 1024–1026.

Chen, Y., Sherwin, E.D., Berman, E.S.F., Jones, B.B., Gordon, M.P., Wetherley, E.B., et al., 2022. Quantifying regional methane emissions in the New Mexico Permian Basin with a comprehensive aerial survey. Environ. Sci. Technol. 56, 4317–4323.

Chenari, B., Carrilho, J.D., da Silva, M.G., 2016. Towards sustainable, energy-efficient and healthy ventilation strategies in buildings: a review. Renew. Sustain. Energy Rev. 59, 1426–1447.

Childs, M.L., Li, J., Wen, J., Heft-Neal, S., Driscoll, A., Wang, S., Gould, C.F., Qiu, M., Burney, J., Burke, M., 2022. Daily local-level estimates of ambient wildfire smoke PM2.5 for the contiguous US. Environ. Sci. Technol. 56 (19), 13607–13621. https://doi.org/10.1021/acs.est.2c02934 (Epub 2022 Sep 22). PMID: 36134580.

Chu, S., Cui, Y., Liu, N., 2017. The path towards sustainable energy. Nat. Mater. 16, 16–22.

Clack, C.T.M., Qvist, S.A., Apt, J., Bazilian, M., Brandt, A.R., Caldeira, K., et al., 2017. Evaluation of a proposal for reliable low-cost grid power with 100% wind, water, and solar. Proc. Natl. Acad. Sci. USA 114, 6722–6727.

Cohen, J., Zhang, X., Francis, J., Jung, T., Kwok, R., Overland, J., et al., 2020. Divergent consensuses on Arctic amplification influence on midlatitude severe winter weather. Nat. Clim. Chang. 10, 20.

Cole, W., Mai, T., Bistline, J., Young, D., 2021. The Current State of Renewable Energy for Electricity. EM: The Magazine for Environmental Managers. May 2021. Air & Waste Management Association. Online.

Collignan, B., Lorkowski, C., Ameon, R., 2012. Development of a methodology to characterize radon entry in dwellings. Build. Environ. 57, 176–183.

Cook, J., Akar, S., Chang, D., Fensch, A., Nissen, K., Eric, O.'.S., 2024. SolarAPP+ Performance Review (2023 Data). US DOE National Renewable Energy Laboratory, Golden, CO, USA.

Cooper, C.D., 2015. Introduction to Environmental Engineering. Waveland Press, Long Grove, IL.

Cooper, C.D., Alley, F.C., 2011. Air Pollution Control: A Design Approach. Waveland Press, Long Grove, IL, USA.

Copeland, C., Carter, N.T., 2017. Energy-Water Nexus: The Water Sector's Energy Use. 7-5700, Congressional Research Service, Washington, DC, USA.

Correa, S.M., Arbilla, G., Marques, M.R.C., Oliveira, K.M.P.G., 2012. The impact of BTEX emissions from gas stations into the atmosphere. Atmos. Pollut. Res. 3, 163–169.

Coventry, P.A., Brown, J.E., Pervin, J., Brabyn, S., Pateman, R., Breedvelt, J., Gilbody, S., Stancliffe, R., McEachan, R., White, P.L., 2021. Nature-based outdoor activities for mental and physical health: systematic review and meta-analysis. SSM Popul. Health 16, 100934. https://doi.org/10.1016/j.ssmph.2021.100934. PMID: 34646931. PMCID: PMC8498096.

Creutzig, F., Breyer, C., Hilaire, J., Minx, J., Peters, G.P., Socolow, R., 2019. The mutual dependence of negative emission technologies and energy systems. Energy Environ. Sci. 12, 1805–1817.

CRS, 2024. In: Service UCR (Ed.), Introduction to the National Flood Insurance Program (NFIP), p. 33. R44593, Washington, DC, USA.

Cuce, E., Cuce, P.M., Wood, C.J., Riffat, S.B., 2014. Toward aerogel based thermal superinsulation in buildings: a comprehensive review. Renew. Sustain. Energy Rev. 34, 273–299.

Cunningham, C.X., Williamson, G.J., Bowman, D.M.J.S., 2024. Increasing frequency and intensity of the most extreme wildfires on earth. Nat. Ecol. Evol. 8 (8), 1420–1425. https://doi.org/10.1038/s41559-024-02452-2. (Epub 2024 Jun 24). PMID: 38914710.

Damtoft, J.S., Lukasik, J., Herfort, D., Sorrentino, D., Gartner, E.M., 2008. Sustainable development and climate change initiatives. Cem. Concr. Res. 38, 115–127.

Dane, A.E., Brown, J., 2020. Transport Infrastructure for Carbon Capture and Storage: Whitepaper on Regional Infrastructure for Midcentury Decarbonization. Great Plains Institute, p. 71.

Daszak, P., Cunningham, A.A., Hyatt, A.D., 2000. Wildlife ecology—emerging infectious diseases of wildlife—threats to biodiversity and human health. Science 287, 443–449.

Davis, M., 2019. Water and Wastewater Engineering: Design Principles and Practice. McGraw-Hill.

Davis, M., Cornwell, D., 2022. Introduction to Environmental Engineering. McGraw Hill, New York, NY, USA.

Davis, S.J., Lewis, N.S., Shaner, M., Aggarwal, S., Arent, D., Azevedo, I.L., et al., 2018. Net-zero emissions energy systems. Science 360, 1419.

De Alwis, D., Limaye, V.S., 2021. The Costs of Inaction: The Economic Burden of Fossil Fuels and Climate Change on Health in the United States Medical Society Consortium on Climate and Health Natural Resources Defense Council. Wisconsin Health Professionals for Climate Action, Ge, p. 13.

de Dear, R.J., Akimoto, T., Arens, E.A., Brager, G., Candido, C., Cheong, K.W.D., et al., 2013. Progress in thermal comfort research over the last twenty years. Indoor Air 23, 442–461.

DeLucia, E.H., Gomez-Casanovas, N., Greenberg, J.A., Hudiburg, T.W., Kantola, I.B., Long, S.P., et al., 2014. The theoretical limit to plant productivity. Environ. Sci. Technol. 48, 9471–9477.

Denison, E., Beech, N., 2019. How to Read Skyscrapers: A Crash Course in High-Rise Architecture. The Ivy Press.

Dessler, A., 2015. Introduction to Modern Climate Change. Cambridge University Press, Cambridge, UK.

Dimitroulopoulou, C., 2012. Ventilation in European dwellings: a review. Build. Environ. 47, 109–125.

Dixit, M.K., Fernandez-Solis, J.L., Lavy, S., Culp, C.H., 2012. Need for an embodied energy measurement protocol for buildings: a review paper. Renew. Sustain. Energy Rev. 16, 3730–3743.

Doan, D.T., Ghaffarianhoseini, A., Naismith, N., Zhang, T.R., Tookey, J., 2017. A critical comparison of green building rating systems. Build. Environ. 123, 243–260.

Dockery, D., Pope, C., Xu, X., Spengler, J., Ware, J., Fay, M., et al., 1993. An association between air-pollution and mortality in 6 United States cities. N. Engl. J. Med. 329, 1753–1759.

Dounis, A.I., Caraiscos, C., 2009. Advanced control systems engineering for energy and comfort management in a building environment—a review. Renew. Sustain. Energy Rev. 13, 1246–1261.

Dudka, S., Adriano, D.C., 1997. Environmental impacts of metal ore mining and processing: a review. J. Environ. Qual. 26, 590–602.

Duncombe, J., 2023. Potentially good news for solar energy during wildfires. Eos 104. https://doi.org/10.1029/2023EO230015.

Dunlap, R.A., 2019. Sustainable Energy, SI ed. Cengage Learning, Boston, MA.

Dykstra, A., 2016. Green Construction: An Introduction to a Changing Industry. Kirshner Publishing Company, Santa Rosa, CA, USA.

Edwards, R., Thomson, G., Wilson, N., Waa, A., Bullen, C., O'Dea, D., Gifford, H., Glover, M., Laugesen, M., Woodward, A., 2008. After the smoke has cleared: evaluation of the impact of a new national smoke-free law in New Zealand. Tob Control. 17 (1), e2. https://doi.org/10.1136/tc.2007.020347. PMID: 18218788.

Eigenbrod, C., Gruda, N., 2015. Urban vegetable for food security in cities. A review. Agron. Sustain. Dev. 35, 483–498.

EPRI, 2013. Electricity Use and Management in the Municipal Water Supply and Wastewater Industries. Electric Power Research Institute, Water Research Foundation, Palo Alto, CA, USA, p. 194.

Eriksen, M., Lebreton, L.C., Carson, H.S., Thiel, M., Moore, C.J., Borerro, J.C., Galgani, F., Ryan, P.G., Reisser, J., 2014. Plastic pollution in the world's oceans: more than 5 trillion plastic pieces weighing over 250,000 tons afloat at sea. PLoS One 9 (12), e111913. https://doi.org/10.1371/journal.pone.0111913.

Erten-Unal, M., Andrews, M., 2018. Environmental law for engineers. In: Kutz, M. (Ed.), Handbook of Environmental Engineering. Wiley, Hoboken, NJ, USA, pp. 119–136.

Esperon-Rodriguez, M., Tjoelker, M.G., Lenoir, J., Baumgartner, J.B., Beaumont, L.J., Nipperess, D.A., et al., 2022. Climate change increases global risk to urban forests. Nat. Clim. Chang. 12, 950.

Evangeliou, N., Grythe, H., Klimont, Z., Heyes, C., Eckhardt, S., Stohl, A., et al., 2020. Atmospheric transport is a major pathway of microplastics to remote regions. Nat. Commun. 11 (1), 1–11. https://doi.org/10.1038/s41467-020-17201-9.

Fahrig, L., 2003. Effects of habitat fragmentation on biodiversity. Annu. Rev. Ecol. Evol. Syst. 34, 487–515.

Farmer, D.K., Vance, M.E., Abbatt, J.P.D., Abeleira, A., Alves, M.R., Arata, C., et al., 2019. Overview of HOMEChem: house observations of microbial and environmental chemistry. Environ. Sci. Process Impacts 21, 1280–1300.

Feldman, D., Ramasamy, V., Fu, R., Ramdas, A., Desai, J., Margolis, R., 2021. U.S. solar photovoltaic system and energy storage cost benchmark: Q1 2020. Technical Report NREL/TP-6A20-77324, National Renewable Energy Laboratory, Golden, CO, USA.

FEMA (Ed.), 2020. National Risk Index Primer. FEMA Editor, Washington, DC, USA, p. 50.

FEMA, 2022. Paradise, California: Rebuilding Resilient Homes After the Camp Fire. FEMA Interagency Recovery Coordination Case Study. Federal Emergency Management Agency (FEMA), Washington, DC, USA, p. 11.

FERC, 2021. The February 2021 Cold Weather Outages in Texas and the South Central United States. Federal Energy Regulatory Commission, National Energy Reliability Corporation, p. 316.

Fischer, E.M., Knutti, R., 2015. Anthropogenic contribution to global occurrence of heavy-precipitation and high-temperature extremes. Nat. Clim. Chang. 5, 560–564.

Fischer, E.M., Sippel, S., Knutti, R., 2021. Increasing probability of record-shattering climate extremes. Nat. Clim. Chang. 11, 689–695.

Flores, H.C., 2006. Food Not Lawns: How to Turn Your Yard Into a Garden and Your Neighborhood Into a Community. Chelsea Green Publishing.

Foley, J.A., DeFries, R., Asner, G.P., Barford, C., Bonan, G., Carpenter, S.R., et al., 2005. Global consequences of land use. Science 309, 570–574.

France, R.M., Geisz, J.F., Song, T., Olavarria, W., Young, M., Kibbler, A., et al., 2022. Triple-junction solar cells with 39.5% terrestrial and 34.2% space efficiency enabled by thick quantum well superlattices. Joule 6 (5), 1121–1135.

Francis, J.A., Vavrus, S.J., 2015. Evidence for a wavier jet stream in response to rapid Arctic warming. Environ. Res. Lett. 10 (1), 014005. https://doi.org/10.1088/1748-9326/10/1/014005.

Francis, J.A., Vavrus, S.J., Cohen, J., 2017. Amplified Arctic warming and mid-latitude weather: new perspectives on emerging connections. Wiley Interdiscip. Rev. Clim. Chang. 8 (5). https://doi.org/10.1002/wcc.474.

Francis, J.A., Skific, N., Vavrus, S.J., 2018. North American weather regimes are becoming more persistent: is Arctic amplification a factor? Geophys. Res. Lett. 45, 11414–11422.

Frontczak, M., Wargocki, P., 2011. Literature survey on how different factors influence human comfort in indoor environments. Build. Environ. 46, 922–937.

Fuller, T., 2022. A super building for fragile times. N. Y. Times. (nytimes.com), New York, NY, USA.

Georgescu, M., Morefield, P.E., Bierwagen, B.G., Weaver, C.P., 2014. Urban adaptation can roll back warming of emerging megapolitan regions. Proc. Natl. Acad. Sci. USA 111, 2909–2914.

Gerring, D., 2023. Renewable Energy Systems for Building Designers: Fundamentals of Net Zero and High Performance Design. Routledge, New York, NY, USA.

Grieshop, A.P., Marshall, J.D., Kandlikar, M., 2011. Health and climate benefits of cookstove replacement options. Energy Policy 39, 7530–7542.

Griffith, S., 2021. Electrify: An Optimist's Playbook for Our Clean Energy Future. The MIT Press, Massachusetts Institute of Technology, Cambridge, MA, USA.

Griscom, B.W., Adams, J., Ellis, P.W., Houghton, R.A., Lomax, G., Miteva, D.A., et al., 2017. Natural climate solutions. Proc. Natl. Acad. Sci. USA 114, 11645–11650.

Guo, H., Shi, C.J., Guan, X.M., Zhu, J.P., Ding, Y.H., Ling, T.C., et al., 2018. Durability of recycled aggregate concrete—a review. Cem. Concr. Compos. 89, 251–259.

Guo, H.S., Aviv, D., Loyola, M., Teitelbaum, E., Houchois, N., Meggers, F., 2020. On the understanding of the mean radiant temperature within both the indoor and outdoor environment, a critical review. Renew. Sustain. Energy Rev. 117, 15.

Gupta, A.K., Hall, C.A.S., 2012. Energy cost of materials: materials for thin-film photovoltaics as an example. In: Ginley, D.S., David, C. (Eds.), Fundamentals of Materials for Energy and Environmental Sustainability. Cambridge University Press, Cambridge, UK, pp. 48–60.

Haddad, S., Zhang, W., Rea, P., 2024. Quantifying the energy impact of heat mitigation technologies at the urban scale. Nat. Cities 1, 62–72.

Hall, D.K., O'Leary, D.S., DiGirolamo, N.E., Miller, W., Kang, D.H., 2021. The role of declining snow cover in the desiccation of the Great Salt Lake, Utah, using MODIS data. Remote Sens. Environ. 252, 112106, ISSN 0034-4257. https://doi.org/10.1016/j.rse.2020.112106.

Halliday, S., 2008. Sustainable Construction. Elsevier, Oxford, UK.

Halverson, J.B., 2024. An Introduction to Severe Storms and Hazardous Weather. Routledge, Abingdon, UK and New York, NY, USA.

Hammer, M.J.S., Hammer, M.J.J., 2011. Water and Wastewater Technology. Pearson Prentice Hall, Upper Saddle River, NJ, USA.

Hammond, M.J., Chen, A.S., Djordjevic, S., Butler, D., Mark, O., 2015. Urban flood impact assessment: a state-of-the-art review. Urban Water J. 12, 14–29.

Hanninen, O.O., Lebret, E., Ilacqua, V., Katsouyanni, K., Kunzli, F., Sram, R.J., et al., 2004. Infiltration of ambient PM2.5 and levels of indoor generated non-ETS PM2.5 in residences of four European cities. Atmos. Environ. 38, 6411–6423.

Hansen, J.E., Sato, M., Simons, L., Nazarenko, L.S., Sangha, I., Kharecha, P., Zachos, J.C., von Schuckmann, K., Loeb, N.G., Osman, M.B., Jin, Q., Tselioudis, G., Jeong, E., Lacis, A., Ruedy, R., Russell, G., Cao, J., Li, J., 2023. Global warming in the pipeline. Oxford Open Clim. Change 3 (1), kgad008. https://doi.org/10.1093/oxfclm/kgad008.

Harrod, J., Shapiro, I.M., 2021. Preventing refrigerant leaks in heat pump systems. ASHRAE J., 36–42. ashrae.org.

Hawes, J.K., Goldstein, B.P., Newell, J.P., Dorr, E., Caputo, S., Fox-Kämper, R., et al., 2024. Comparing the carbon footprints of urban and conventional agriculture. Nat. Cities 1, 164–173.

Hemenway, T., 2009. Gaia's Garden: A Guide to Home-Scale Permaculture. Chelsea Green Publishers, White River Junction, VT, USA.

Hepburn, C., Adlen, E., Beddington, J., Carter, E.A., Fuss, S., Mac Dowell, N., et al., 2019. The technological and economic prospects for CO2 utilization and removal. Nature 575, 87–97.

Heredia-Velasquez, A.M., Giraldo-Silva, A., Nelson, C., Bethany, J., Kut, P., Gonzalez-de-Salceda, L., Garcia-Pichel, F., 2023. Dual use of solar power plants as biocrust nurseries for large-scale arid soil restoration. Nat. Sustainability 6 (8), 955–964. https://doi.org/10.1038/s41893-023-01106-8.

Hertwich, E.G., Peters, G.P., 2009. Carbon footprint of nations: a global, trade-linked analysis. Environ. Sci. Technol. 43, 6414–6420.

Hilaire, R.S., Arnold, M.A., Wilkerson, D.C., Devitt, D.A., Hurd, B.H., Lesikar, B.J., et al., 2008. Efficient water use in residential urban landscapes. HortScience 43, 2081–2092.

Hinrichs, R.A., Kleinbach, M.H., Wade, R., 2023. Energy: Its Use and the Environment. Cengage Learning, Boston, MA, USA.

Hoegh-Guldberg, O., Mumby, P.J., Hooten, A.J., Steneck, R.S., Greenfield, P., Gomez, E., et al., 2007. Coral reefs under rapid climate change and ocean acidification. Science 318, 1737–1742.

Hoekstra, A.Y., Wiedmann, T.O., 2014. Humanity's unsustainable environmental footprint. Science 344, 1114–1117.

Hooper, D.U., Chapin, F.S., Ewel, J.J., Hector, A., Inchausti, P., Lavorel, S., et al., 2005. Effects of biodiversity on ecosystem functioning: a consensus of current knowledge. Ecol. Monogr. 75, 3–35.

Hu, M., 2024. Green Building Costs: The Affordability of Sustainable Design. Routledge, London, UK and New York, NY, USA.

Hughes, T.P., Baird, A.H., Bellwood, D.R., Card, M., Connolly, S.R., Folke, C., et al., 2003. Climate change, human impacts, and the resilience of coral reefs. Science 301, 929–933.

Hughes, T.P., Kerry, J.T., Alvarez-Noriega, M., Alvarez-Romero, J.G., Anderson, K.D., Baird, A.H., et al., 2017. Global warming and recurrent mass bleaching of corals. Nature 543, 373.

Hund, K., La Porta, D., Fabregas, T.P., Laing, T., Drexhage, J., 2020. Minerals for Climate Action: The Mineral Intensity of the Clean Energy Transition. International Bank for Reconstruction and Development/The World Bank, Washington, DC, USA, p. 110.

Hurd, B.H., St Hilaire, R., White, J.M., 2006. Residential landscapes, homeowner attitudes, and water-wise choices in New Mexico. HortTechnology 16, 241–246.

IEA, 2021. The Role of Critical Minerals in Clean Energy Transitions. International Energy Agency, Paris, France, p. 283.

IPCC, 2005. IPCC special report: carbon dioxide capture and storage. In: Metz, B., Davidson, O., de Coninck, H.C., Loos, M., Meyer, L.A. (Eds.), Prepared by Working Group III of the Intergovernmental Panel on Climate Change. Intergovernmental Panel on Climate Change, Cambridge, UK and New York, NY, USA, p. 432.

IPCC, 2013. Climate change 2013: the physical science basis. In: Stocker, T.F., Qin, D., Plattner, G.-K., Tignor, M., Allen, S.K., Boschung, J., et al. (Eds.), Contribution of Working Group I to the Fifth Assessment Report of the Intergovernmental Panel on Climate Change. Intergovernmental Panel on Climate Change, Cambridge, UK and New York, NY, USA, p. 1535.

IPCC, The Core Writing Team, Pachauri, R.K., Meyer, L.A. (Eds.), 2014. Climate change 2014: synthesis report. Contribution of Working Groups I, II and III to the Fifth Assessment Report of the Intergovernmental Panel on Climate Change. IPCC, Geneva, Switzerland, p. 151.

IPCC, 2019. Climate change and land. In: JS, P.R.S., Buendia, E.C., Masson-Delmotte, V., DCR, H.-O.P., Zhai, P., Slade, R., Malley, J. (Eds.), An IPCC Special Report on Climate Change, Desertification, Land Degradation,

Sustainable Land Management, Food Security, and Greenhouse Gas Fluxes in Terrestrial Ecosystems. Intergovernmental Panel on Climate Change, p. 864.

IPCC, 2021a. Climate change 2021: impacts, adaptation, and vulnerability. In: The Working Group II Contribution to the Sixth Assessment Report. Intergovernmental Panel on Climate Change, Cambridge, UK and New York, NY, USA.

IPCC, 2021b. Climate change 2021: the physical science basis. In: Masson-Delmotte, V., Zhai, P., Pirani, A., Connors, S.L., Péan, C., Berger, S., Zhou, B. (Eds.), Contribution of Working Group I to the Sixth Assessment Report of the Intergovernmental Panel on Climate Change. Intergovernmental Panel on Climate Change, Cambridge, United Kingdom and New York, NY, USA.

IPCC, 2022. Climate resilient development pathways. In: Pörtner, H.-O., Roberts, D.C., Tignor, M., Poloczanska, E.S., Mintenbeck, K., Alegría, A., Singh, P.K. (Eds.), Climate Change 2022: Impacts, Adaptation and Vulnerability. Contribution of Working Group II to the Sixth Assessment Report of the Intergovernmental Panel on Climate Change. Intergovernmental Panel on Climate Change, Cambridge, UK and New York, NY, USA, pp. 2655–2807.

IPCC, Core Writing Team HLaJRe (Eds.), 2023. Summary for policymakers. Climate Change 2023: Synthesis Report. Contribution of Working Groups I, II and III to the Sixth Assessment Report of the Intergovernmental Panel on Climate Change. IPCC, Geneva, Switzerland, p. 34.

Irvine, L., Andre, C., 2023. Pet loss in an urban firestorm: grief and hope after Colorado's marshall fire. Animals 13, 16.

Irvine, P., Emanuel, K., He, J., Horowitz, L.W., Vecchi, G., Keith, D., 2019. Halving warming with idealized solar geoengineering moderates key climate hazards. Nat. Clim. Chang. 9, 295–299.

Jackson, R.E., 2019. Earth Science for Civil and Environmental Engineers. Cambridge University Press, Cambridge, UK and New York, NY, USA.

Jackson, J.B.C., Kirby, M.X., Berger, W.H., Bjorndal, K.A., Botsford, L.W., Bourque, B.J., et al., 2001. Historical overfishing and the recent collapse of coastal ecosystems. Science 293, 629–638.

Jackson, R.B., Le Quéré, C., Andrew, R.M., Canadell, J.G., Korsbakken, J.I., Liu, Z., et al., 2019. Global energy growth is outpacing decarbonization. A special report for the United Nations Climate Action Summit. Global Carbon Project, International Project Office, Canberra, Australia.

Jacob, D.J., Winner, D.A., 2009. Effect of climate change on air quality. Atmos. Environ. 43, 51–63.

Jacobson, M.Z., 2012. Air Pollution and Global Warming: History, Science, and Solutions. Cambridge University Press, Cambridge, UK.

Jacobson, M.Z., 2019. The health and climate impacts of carbon capture and direct air capture. Energy Environ. Sci. 12, 3567–3574.

Jacobson, M.Z., 2020. 100% Clean Renewable Energy and Storage for Everything. Cambridge University Press, Cambridge, UK.

Jambeck, J.R., Geyer, R., Wilcox, C., Siegler, T.R., Perryman, M., Andrady, A., et al., 2015. Plastic waste inputs from land into the ocean. Science 347, 768–771.

Johnson, S., 2006. The Ghost Map: The Story of London's Most Terrifying Epidemic and How It Changed Science, Cities, and the Modern World. Riverhead Books, New York, NY.

Johnson, N., Gross, R., Staffell, I., 2021. Stabilisation wedges: measuring progress towards transforming the global energy and land use systems. Environ. Res. Lett. 16, 064011. https://doi.org/10.1088/1748-9326/abec06.

Jones, A.P., 1999. Indoor air quality and health. Atmos. Environ. 33, 4535–4564.

Joshi, S., Mittal, S., Holloway, P., Shukla, P.R., Gallachoir, B.O., Glynn, J., 2021. High resolution global spatiotemporal assessment of rooftop solar photovoltaics potential for renewable electricity generation. Nat. Commun. 12.

Juliano, T.W., Lareau, N., Frediani, M.E., Shamsaei, K., Eghdami, M., Kosiba, K., et al., 2023. Toward a better understanding of wildfire behavior in the wildland-urban interface: a case study of the 2021 Marshall Fire. Geophys. Res. Lett. 50, 11.

Karlen, D.L., Mausbach, M.J., Doran, J.W., Cline, R.G., Harris, R.F., Schuman, G.E., 1997. Soil quality: a concept, definition, and framework for evaluation. Soil Sci. Soc. Am. J. 61, 4–10.

Kaya, Y., Yokobori, K., 1998. Environment, Energy, and Economy: Strategies for Sustainability. United Nations University Press, Tokyo, Japan.

Keys, B.J., Mulder, P., 2020. Neglected no more: housing markets, mortgage lending, and sea level rise. In: Research NBoE (Ed.), NBER Working Paper No. 27930, Cambridge, MA.

Kibert, C.J., 2016. Sustainable Construction: Green Building Design and Delivery. John Wiley & Sons, Hoboken, NJ.

King, B., 2017. The New Carbon Architecture. New Society Publishers, Gabriola Island, BC Canada.

Kleidon, A., 2021. What limits photosynthesis? Identifying the thermodynamic constraints of the terrestrial biosphere within the Earth system. Biochim. Biophys. Acta Bioenerg. 1862 (1), 148303. https://doi.org/10.1016/j.bbabio.2020.148303. (Epub 2020 Sep 11). PMID: 32926862.

Kolbert, D., Maines, M., Mottram, E., Briley, C., 2022. Pretty Good House: A Guide to Creating Better Homes. Taunton Press, Newtown, CT, p. 256.

Krarti, M., 2011. Energy Audit of Building Systems: An Engineering Approach. CRC Press, Taylor and Francis Group, Boca Raton, FL.

Krausmann, F., 2012. Global materials flow. In: Ginley, D.S., Cahen, D. (Eds.), Fundamental of Materials for Energy and Environmental Sustainability. Cambridge University Press, Cambridge, UK, pp. 81–89.

Krayenhoff, E.S., Moustaoui, M., Broadbent, A.M., Gupta, V., Georgescu, M., 2018. Diurnal interaction between urban expansion, climate change and adaptation in US cities. Nat. Clim. Chang. 8, 1097–1103.

Kreider, M.R., Higuera, P.E., Parks, S.A., Rice, W.L., White, N., Larson, A.J., 2024. Fire suppression makes wildfires more severe and accentuates impacts of climate change and fuel accumulation. Nat. Commun. 15, 2412.

Kroos, K.A., Potter, M.C., 2014. Thermodynamics for Engineers, SI ed. Cengage.

Kruger, A., Seville, C., 2012. Green Building: Principles and Practice in Residential Construction. Cengage Learning, Delmar.

Kutscher, C.F., Milford, J.B., Kreith, F., 2019. Principles of Sustainable Energy Systems. CRC Press, Taylor and Francis Group, Boca Raton, FL, USA.

Lal, R., 2004. Soil carbon sequestration impacts on global climate change and food security. Science 304, 1623–1627.

Lancaster, B., 2019. Rainwater harvesting for drylands and beyond. In: Guiding Principles to Welcome Rain Into Your Life and Landscape, third ed. vol. 1. Rainsource Press.

Le Quéré, C., Jackson, R.B., Jones, M.W., et al., 2020. Temporary reduction in daily global CO_2 emissions during the COVID-19 forced confinement. Nat. Clim. Chang. 10, 647–653.

Lebel, E.D., Lu, H.S., Speizer, S.A., Finnegan, C.J., Jackson, R.B., 2020. Quantifying methane emissions from natural gas water heaters. Environ. Sci. Technol. 54, 5737–5745.

Lebreton, L., Slat, B., Ferrari, F., Sainte-Rose, B., Aitken, J., Marthouse, R., et al., 2018. Evidence that the Great Pacific Garbage Patch is rapidly accumulating plastic. Sci. Rep. 8.

Lehne, J., Preston, F., 2018. Making concrete change: innovation in low-carbon cement and concrete. Chatham House Reports, London, UK, The Royal Institute of International Affairs, p. 122.

Lenton, T.M., Rockström, J., Gaffney, O., Richardson, S., Steffen, K., Schellnhuber, W., Joachim, H., 2019. Climate tipping points—too risky to bet against. Nature 575, 592–595.

Leung, D.Y.C., Caramanna, G., Maroto-Valer, M.M., 2014. An overview of current status of carbon dioxide capture and storage technologies. Renew. Sustain. Energy Rev. 39, 426–443.

Lewis, T., 2013. Divided Highways: Building The Interstate Highways, Transforming American Life. Cornell University Press, Ithaca, NY, USA and London, UK.

Lewis, J.J., Pattanayak, S.K., 2012. Who adopts improved fuels and cookstoves? A systematic review. Environ. Health Perspect. 120, 637–645.

Li, Y., Leung, G.M., Tang, J.W., Yang, X., Chao, C.Y.H., Lin, J.Z., et al., 2007. Role of ventilation in airborne transmission of infectious agents in the built environment—a multidisciplinary systematic review. Indoor Air 17, 2–18.

Liu, C., Chen, R., Sera, F., Vicedo-Cabrera, A.M., Guo, Y.M., Tong, S.L., et al., 2019. Ambient particulate air pollution and daily mortality in 652 cities. N. Engl. J. Med. 381, 705–715.

Loehrlein, M., 2021. Sustainable Landscaping: Principles and Practices. CRC Press, Taylor & Francis Group, LLC, Boca Raton, FL, USA.

Ludwig, A., 2015. The New Create an Oasis With Greywater 6th Ed: Integrated Design for Water Conservation, Reuse, Rainwater Harvesting, and Sustainable Landscaping. Oasis Design, Santa Barbara, CA, USA.

Mack, M.C., Walker, X.J., Johnstone, J.F., Alexander, H.D., Melvin, A.M., Jean, M., et al., 2021. Carbon loss from boreal forest wildfires offset by increased dominance of deciduous trees. Science 372, 280–283.

Mackenzie, W.R., Hoxie, N.J., Proctor, M.E., Gradus, M.S., Blair, K.A., Peterson, D.E., et al., 1994. A massive outbreak in milwaukee of cryptosporidium infection transmitted through the public water supply. N. Engl. J. Med. 331, 161–167.

Magwood, C., 2017. Essential Sustainable Home Design: A Complete Guide to Goals, Options, and the Design Process. New Society Publishers.

Magwood, C., Mack, P., Therrien, T., 2005. More Straw Bale Building: A Complete Guide to Designing and Building With Straw. New Society Publishers.

Malm, W.C., Sisler, J.F., Huffman, D., Eldred, R.A., Cahill, T.A., 1994. Spatial and seasonal trends in particle concentration and optical extinction in the United States. J. Geophys. Res.-Atmos. 99, 1347–1370.

Manoli, G., Fatichi, S., Schlapfer, M., Yu, K.L., Crowther, T.W., Meili, N., et al., 2019. Magnitude of urban heat islands largely explained by climate and population. Nature 573, 55.

Manoli, G., Fatichi, S., Bou-Zeid, E., Katul, G.G., 2020. Seasonal hysteresis of surface urban heat islands. Proc. Natl. Acad. Sci. USA 117, 7082–7089.

Martin, W., Sumanasooriya, M., Kaye, N.B., Putman, B., 2018. Design of porous pavements for improved water quality and reduced runoff. In: Kutz, M.B. (Ed.), Environmental Engineering Handbook. Wiley, Hoboken, NJ, USA, pp. 425–451.

Mazria, E., 2003. It's the architecture, stupid! Solar Today, 48–51.

McClure, C.D., Jaffe, D.A., 2018. US particulate matter air quality improves except in wildfire-prone areas. Proc. Natl. Acad. Sci. USA 115, 7901–7906.

McPherson, E.G., Berry, A.M., van Doorn, N.S., 2018. Performance testing to identify climate-ready trees. Urban For. Urban Green. 29, 28–39.

Me, K., 2018. Handbook of Environmental Engineering. Wiley, Hoboken, NJ, USA.

Mehta, A.J., Dean, R.G., Montague, C.L., Hayter, E.J., Khare, Y.P., 2013. Understanding sea level rise and coastal hazards. In: Watts, R.G. (Ed.), Engineering Response to Climate Change. CRC Press, Taylor and Francis Group, Boca Raton, FL, USA, pp. 139–178.

Mehta, M., Scarborough, W., Armpriest, D., 2018. Building Contruction: Principles, Materials and Systems. Pearson Education, New York, NY.

Merlino, K.R., 2018. Building Reuse. University of Washington Seattle, Seattle, WA, USA.

Meyer, C., 2009. The greening of the concrete industry. Cem. Concr. Compos. 31, 601–605.

Middleton, R.S., Yaw, S.P., Hoover, B.A., Ellett, K.M., 2020. SimCCS: an open-source tool for optimizing CO_2 capture, transport, and storage infrastructure. Environ. Model Softw. 124, 104560. https://doi.org/10.1016/j.envsoft.2019.104560.

Mihelcic, J.R., Zimmerman, J.B., 2014. Environmental Engineering: Fundamentals, Sustainability, Design. John Wiley and Sons, Hoboken, NJ.

Milford, J., 2018. Environmental law for engineers. In: Kutz, M. (Ed.), Handbook of Environmental Engineering. Wiley, Hoboken, NJ, USA, pp. 45–66.

Milton, D.K., Fabian, M.P., Cowling, B.J., Grantham, M.L., McDevitt, J.J., 2013. Influenza virus aerosols in human exhaled breath: particle size, culturability, and effect of surgical masks. PLoS Pathog. 9 (3): e1003205. https://doi.org/10.1371/journal.ppat.1003205 (Epub 2013 Mar 7). PMID: 23505369. PMCID: PMC3591312.

Mirletz, H., Hieslmair, H., Ovaitt, S., Curtis, T., Barnes, T., 2023. Unfounded concerns about photovoltaic module toxicity and waste are slowing decarbonization. Nat. Phys. 19, 1376–1378. https://doi.org/10.1038/s41567-023-02230-0.

Morawska, L., Thomas, S., Gilbert, D., Greenaway, C., Rijnders, E., 1999. A study of the horizontal and vertical profile of submicrometer particles in relation to a busy road. Atmos. Environ. 33, 1261–1274.

Morawska, L., He, C.R., Hitchins, J., Gilbert, D., Parappukkaran, S., 2001. The relationship between indoor and outdoor airborne particles in the residential environment. Atmos. Environ. 35, 3463–3473.

Morawska, L., He, C.R., Hitchins, J., Mengersen, K., Gilbert, D., 2003. Characteristics of particle number and mass concentrations in residential houses in Brisbane, Australia. Atmos. Environ. 37, 4195–4203.

Morawska, L., Tang, J.W., Bahnfleth, W., Bluyssen, P.M., Boerstra, A., Buonanno, G., Cao, J., Dancer, S., Floto, A., Franchimon, F., Haworth, C., Hogeling, J., Isaxon, C., Jimenez, J.L., Kurnitski, J., Li, Y., Loomans, M., Marks, G., Marr, L.C., Mazzarella, L., Melikov, A.K., Miller, S., Milton, D.K., Nazaroff, W., Nielsen, P.V., Noakes, C., Peccia, J., Querol, X., Sekhar, C., Seppänen, O., Tanabe, S.I., Tellier, R., Tham, K.W., Wargocki, P., Wierzbicka, A., Yao, M., 2020. How can airborne transmission of COVID-19 indoors be minimised? Environ. Int. 142, 105832. https://doi.org/10.1016/j.envint.2020.105832. (Epub 2020 May 27) PMID: 32521345. PMCID: PMC7250761.

Morseletto, P., 2020. Targets for a circular economy. Resour. Conserv. Recycl. 153, 12.

Moss, K.J., 2007. Heat and Mass Transfer in Buildings. Taylor and Francis, New York, NY.

Nardell, E.A., Keegan, J., Cheney, S.A., Etkind, S.C., 1991. Airborne infection—theoretical limits of protection achievable by building ventilation. Am. Rev. Respir. Dis. 144, 302–306.

NASEM, 2021. The Future of Electric Power in the United States. National Academies of Sciences, Engineering, and Medicine, Washington, DC.

NASEM, 2022. In: National Academies of Sciences, Engineering, and Medicine (Ed.), Resilience for Compounding and Cascading Events. The National Academies Press, Washington, DC, p. 59.

NASEM, 2023. Committee on Accelerating Decarbonization in the United States: Technology, Policy, and Societal Dimensions. National Academies of Sciences, Engineering, and Medicine, Washington, DC, USA, p. 652.

NASEM, 2024. Modernizing Probable Maximum Precipitation Estimation. The National Academies Press, Washington, DC.

Nazaroff, W.W., Weschler, C.J., 2004. Cleaning products and air fresheners: exposure to primary and secondary air pollutants. Atmos. Environ. 38, 2841–2865.

NCEP N NOAA National Centers for Environmental Information (NCEI) U.S., 2024. Billion-Dollar Weather and Climate Disasters., https://doi.org/10.25921/stkw-7w73. https://www.ncei.noaa.gov/access/billions/.

Norton, B.A., Coutts, A.M., Livesley, S.J., Harris, R.J., Hunter, A.M., Williams, N.S.G., 2015. Planning for cooler cities: a framework to prioritise green infrastructure to mitigate high temperatures in urban landscapes. Landsc. Urban Plan. 134, 127–138.

Novak, C.A., Van Giesen, G.E., DeBusk, K.M., 2014. Designing Rainwater Harvesting Systems: Integrating Rainwater Into Building Systems. John Wiley & Sons, Hoboken, NJ, USA.

NREL, 2012. Rebuilding It Better Greensburg, Kansas: High Performance Buildings Meeting Energy Savings Goals. NREL, Washinton, DC, p. 12.

Ohlsen, E., 2023. The Regenerative Landscaper: Design and Build Landscapes That Repair the Environment. Synergistic Press, Santa Fe, NM, USA.

Oliveira, C.M.d., Albergaria De Mello Bandeira, R., Vasconcelos Goes, G., Schmitz Gonçalves, D.N., D'Agosto, M.D.A., 2017. Sustainable vehicles-based alternatives in last mile distribution of urban freight transport: a systematic literature review. Sustainability 9, 1324. https://doi.org/10.3390/su9081324.

Olsson, J., Hellström, D., Pålsson, H., 2019. Framework of last mile logistics research: a systematic review of the literature. Sustainability 11 (24), 7131. https://doi.org/10.3390/su11247131.

Omer, A.M., 2008. Energy, environment and sustainable development. Renew. Sustain. Energy Rev. 12, 2265–2300.

Orsini, F., Marrone, P., 2019. Approaches for a low-carbon production of building materials: a review. J. Clean. Prod. 241, 14.

Pacala, S., Socolow, R., 2004. Stabilization wedges: solving the climate problem for the next 50 years with current technologies. Science 305, 968–972.

Pall, P., Aina, T., Stone, D.A., Stott, P.A., Nozawa, T., Hilberts, A.G.J., et al., 2011. Anthropogenic greenhouse gas contribution to flood risk in England and Wales in autumn 2000. Nature 470, 382–385.

Parmesan, C., Yohe, G., 2003. A globally coherent fingerprint of climate change impacts across natural systems. Nature 421, 37–42.

Paul, W.L., Taylor, P.A., 2008. A comparison of occupant comfort and satisfaction between a green building and a conventional building. Build. Environ. 43, 1858–1870.

Paulikas, D., Katona, S., Ilves, E., Ali, S., 2020. Life cycle climate change impacts of producing battery metals from land ores versus deep-sea polymetallic nodules. J. Clean. Prod. 275, 123822. https://doi.org/10.1016/j.jclepro.2020.123822.

Pauliuk, S., Heeren, N., Berrill, P., Fishman, T., Nistad, A., Tu, Q., et al., 2021. Global scenarios of resource and emission savings from material efficiency in residential buildings and cars. Nat. Commun. 12, 5097.

Pe, H., 2017. Drawdown: The Most Comprehensive Plan Ever Proposed to Reverse Global Warming. Penguin Books, New York, NY, USA.

Pearce, A.R.A., Han, Y., Hoesa, H.K.C., 2018. Sustainable Building and Infrastructure. Routledge, Oxford, UK and New York, NY, USA.

Pearson, J.K., Derwent, R., 2022. Air Pollution and Climate Change the Basics. Routledge, Taylor and Francis Group, London, UK.

Pecl, G.T., Araújo, M.B., Bell, J.D., Blanchard, J., Bonebrake, T.C., Chen, I.C., Clark, T.D., Colwell, R.K., Danielsen, F., Evengård, B., Falconi, L., Ferrier, S., Frusher, S., Garcia, R.A., Griffis, R.B., Hobday, A.J., Janion-Scheepers, C., Jarzyna, M.A., Jennings, S., Lenoir, J., Linnetved, H.I., Martin, V.Y., McCormack, P.C., McDonald, J., Mitchell, N.J., Mustonen, T., Pandolfi, J.M., Pettorelli, N., Popova, E., Robinson, S.A., Scheffers, B.R., Shaw, J.D., Sorte, C.J., Strugnell, J.M., Sunday, J.M., Tuanmu, M.N., Vergés, A., Villanueva, C., Wernberg, T., Wapstra, E., Williams, S.E., 2017. Biodiversity redistribution under climate change: impacts on ecosystems and human well-being. Science 355 (6332), eaai9214. https://doi.org/10.1126/science.aai9214. PMID: 28360268.

Perlin, J., 2013. Let It Shine: The 6,000-Year Story of Solar Energy. New World Library, Novato, CA, USA.

Petrokofsky, G., Harvey, W.J., Petrokofsky, L., Ochieng, C.A., 2021. The importance of time-saving as a factor in transitioning from 2 woodfuel to modern cooking energy services: a systematic map. Forests 12 (9), 1149. https://doi.org/10.3390/f12091149.

Pope, C., Ezzati, M., Dockery, D., 2009. Fine-particulate air pollution and life expectancy in the United States. N. Engl. J. Med. 360, 376–386.

Potting, J., Hekkert, M., Worrell, E., Hanemaaijer, A., 2017. Circular Economy: Measuring Innovation in the Product Chain: Policy Report. PBL Netherlands Environmental Assessment Agency, The Hague, Netherlands.

Potts, S.G., Biesmeijer, J.C., Kremen, C., Neumann, P., Schweiger, O., Kunin, W.E., 2010. Global pollinator declines: trends, impacts and drivers. Trends Ecol. Evol. 25, 345–353.

Psarras, P.C., Comello, S., Bains, P., Charoensawadpong, P., Reichelstein, S., Wilcox, J., 2017. Carbon capture and utilization in the industrial sector. Environ. Sci. Technol. 51, 11440–11449.

Qadir, M., Smakhtin, V., Koo-Oshima, S., Guenther, E.E., 2022. Unconventional Water Resources. Springer.

Radeloff, V.C., Mockrin, M.H., Helmers, D., Carlson, A., Hawbaker, T.J., Martinuzzi, S., et al., 2023. Rising wildfire risk to houses in the United States, especially in grasslands and shrublands. Science 382 (6671), 702–707. https://orcid.org/0000-0002-0090-3352.

Reddy, T.A., Kreider, J.F., Curtiss, P.S., Rabl, A., 2017. Heating and Cooling of Buildings: Principles and Practice of Energy Efficient Design. CRC Press, Taylor and Francis Group, Boca Raton, FL, USA.

Reddy, K.R., Cameselle, C., Adams, J.A., 2019. Sustainable Engineering: Drivers, Metrics, Tools, and Applications. John Wiley and Sons.

Reed, K.A., Wehner, M.F., Zarzycki, C.M., 2022. Attribution of 2020 hurricane season extreme rainfall to human-induced climate change. Nat. Commun. 13, 1905.

Reid, C., Brauer, M., Johnston, F., Jerrett, M., Balmes, J., Elliott, C., 2016. Critical review of health impacts of wildfire smoke exposure. Environ. Health Perspect. 124 (124), 1334–1343.

Richards, L., Brew, N., Lizzie, S., 2020. 2019–20 Australian bushfires—frequently asked questions: a quick guide. In: Services DoP (Ed.), Research Paper Series 2019–20. Commonwealth of Australia, Parliament of Australia, Canberra, Australia, p. 10.

Rillig, M.C., Lehmann, A., 2020. Microplastic in terrestrial ecosystems. Science 368, 1430–1431.

Ritchie, H., 2020. What Are the Safest and Cleanest Sources of Energy? OurWorldInData.org.

Ritchie, H., Rosado, P., Roser, M., 2023. CO_2 and Greenhouse Gas Emissions. Published online at OurWorldInData.org. Retrieved from: https://ourworldindata.org/co2-and-greenhouse-gas-emissions.

Robock, A., Marquardt, A., Kravitz, B., Stenchikov, G., 2009. Benefits, risks, and costs of stratospheric geoengineering. Geophys. Res. Lett. 36 (19), L19703. https://doi.org/10.1029/2009GL039209.

Robock, A., MacMartin, D.G., Duren, R., Christensen, M.W., 2013. Studying geoengineering with natural and anthropogenic analogs. Clim. Chang. 121, 445–458.

Rocca, M.E., Brown, P.M., MacDonald, L.H., Carrico, C.M., 2014. Climate change impacts on fire regimes and key ecosystem services in Rocky Mountain forests. For. Ecol. Manag. 327, 290–305.

Rochman, C.M., Hoellein, T., 2020. The global odyssey of plastic pollution. Science 368, 1184–1185.

Rodrigue, J.-P., 2020. The Geography of Transport Systems. Routledge.

Rosa, L., Reimer, J.A., Went, M.S., D'Odorico, P., 2020. Hydrological limits to carbon capture and storage. Nat. Sustainability 3, 658–666.

Roser, M., Ritchie, H., 2021. Burden of Disease. Retrieved from: https://ourworldindata.org/burden-of-disease. OurWorldinData.org, Published online at.

Rupp, R.F., Vasquez, N.G., Lamberts, R., 2015. A review of human thermal comfort in the built environment. Energy Build. 105, 178–205.

Saidani, M., Yannou, B., Leroy, Y., Cluzel, F., Kendall, A., 2019. A taxonomy of circular economy indicators. J. Clean. Prod. 207, 542–559.

Samet, J.M., Dominici, F., Curriero, F.C., Coursac, I., Zeger, S.L., 2000. Fine particulate air pollution and mortality in 20 US cities, 1987-1994. N. Engl. J. Med. 343, 1742–1749.

Schendler, A., Udall, R., 2005. LEED is Broken: Let's Fix It, LEED is Broken - Aspen/Snowmass (yumpu.com)., p. 20.

Schewe, P.F., 2007. The Grid: A Journey Through the Heart of Our Electrified World. Joseph Henry Press, Washington, DC, USA.

Schlickman, E., Milligan, B., 2023. Design by Fire: Resistance, Co-Creation, and Retreat in the Pyrocene. Routledge, New York, NY.

Schoen, L.J., 2020. Guidance for building operations during the COVID-19 pandemic. ASHRAE J. (May), 72–74.

Schug, F., Bar-Massada, A., Carlson, A.R., Cox, H., Hawbaker, T.J., Helmers, D., Hostert, P., Kaim, D., Kasraee, N.K., Martinuzzi, S., Mockrin, M.H., Pfoch, K.A., Radeloff, V.C., 2023. The global wildland-urban interface. Nature 621 (7931), 94–99. https://doi.org/10.1038/s41586-023-06320-0 (Epub 2023 Jul 19). PMID: 37468636. PMCID: PMC10482693.

Schwarzenbach, R.P., Escher, B.I., Fenner, K., Hofstetter, T.B., Johnson, C.A., von Gunten, U., et al., 2006. The challenge of micropollutants in aquatic systems. Science 313, 1072–1077.

Seddon, N., Chausson, A., Berry, P., Girardin, C.A.J., Smith, A., Turner, B., 2020. Understanding the value and limits of nature-based solutions to climate change and other global challenges. Philos. Trans. R. Soc., B-Biol. Sci. 375, 12.

See, S.W., Balasubramanian, R., 2008. Chemical characteristics of fine particles emitted from different gas cooking methods. Atmos. Environ. 42, 8852–8862.

Seinfeld, J.H., Pandis, S.N., 1998. Atmospheric Chemistry and Physics. John Wiley and Sons, New York, NY, USA.

Seinfeld, J.H., Pandis, S.N., 2016. Atmospheric Chemistry and Physics: From Air Pollution to Climate Change. Wiley, New Jersey, USA.

Sepulveda, N.A., Jenkins, J.D., de, Sisternes FJ, Lester RK., 2018. The role of firm low-carbon electricity resources in deep decarbonization of power generation. Joule 2, 2403–2420.

Seymour, L.M., Maragh, J., Sabatini, P., Di Tommaso, M., Weaver, J.C., Masic, A., 2023. Hot mixing: mechanistic insights into the durability of ancient Roman concrete. Sci. Adv. 9, 13.

Sfakianaki, A., Pavlou, K., Santamouris, M., Livada, I., Assimakopoulos, M.N., Mantas, P., et al., 2008. Air tightness measurements of residential houses in Athens, Greece. Build. Environ. 43, 398–405.

Sharma, S.K., Zote, K.K., 2010. Mulberry—a multi purpose tree species for varied climate. Range Manage. Agrofor. 31, 97–101.

Shiklomanov, I., 1993. World fresh water resources. In: Gleick, P.H. (Ed.), Water in Crisis: A Guide to the World's Fresh Water Resources. Oxford University Press, New York, pp. 13–24.

Shindell, D.T., Faluvegi, G., Koch, D.M., Schmidt, G.A., Unger, N., Bauer, S.E., 2009. Improved attribution of climate forcing to emissions. Science 326, 716–718.

Shu, E.G., Porter, J.R., Hauer, M.E., Sandoval Olascoaga, S., Gourevitch, J., Wilson, B., et al., 2023. Integrating climate change induced flood risk into future population projections. Nat. Commun. 14, 7870.

Siddique, R., Cachim, P., 2018. Waste and Supplementary Cementitious Materials in Concrete: Characterisation, Properties and Applications. A volume in Woodhead Publishing Series in Civil and Structural Engineering, Elsevier, p. 621.

Silberstein, J.M., Mael, L.E., Frischmon, C.R., Rieves, E.S., Coffey, E.R., Das, T., et al., 2023. Residual impacts of a wildland urban interface fire on urban particulate matter and dust: a study from the Marshall Fire. Air Qual. Atmos. Health 16, 1839–1850.

Singh, A., Syal, M., Grady, S.C., Korkmaz, S., 2010. Effects of green buildings on employee health and productivity. Am. J. Public Health 100, 1665–1668.

Smith, A.B., Katz, R.W., 2013. US billion-dollar weather and climate disasters: data sources, trends, accuracy and biases. Nat. Hazards 67, 387–410.

Smith, A.B., Matthews, J.L., 2015. Quantifying uncertainty and variable sensitivity within the US billion-dollar weather and climate disaster cost estimates. Nat. Hazards 77, 1829–1851.

Snow, J., 1936. Snow on Cholera. Humphrey Milford, Oxford University Press, New York.

Solomon, S., Plattner, G.K., Knutti, R., Friedlingstein, P., 2009. Irreversible climate change due to carbon dioxide emissions. Proc. Natl. Acad. Sci. USA 106, 1704–1709.

Sonter, L.J., Dade, M.C., Watson, J.E.M., Valenta, R.K., 2020. Renewable energy production will exacerbate mining threats to biodiversity. Nat. Commun. 11 (1), 4174. https://doi.org/10.1038/s41467-020-17928-5. PMID: 32873789. PMCID: PMC7463236.

Sorvig, K., Thompson, W.J., 2018. Sustainable Landscape Construction: A Guide to Green Building Outdoors. Island Press.

Spracklen, D.V., Mickley, L.J., Logan, J.A., Hudman, R.C., Yevich, R., Flannigan, M.D., Westerling, A.L., 2009. Impacts of climate change from 2000 to 2050 on wildfire activity and carbonaceous aerosol concentrations in the western United States. J. Geophys. Res 114, D20301. https://doi.org/10.1029/2008JD010966.

Stangler, C., 2008. The Craft and Art of Bamboo: 30 Eco-Friendly Projects to Make for the Home & Garden. Lark Books, New York, NY, USA.

Steffen, W., Rockstrom, J., Richardson, K., Lenton, T.M., Folke, C., Liverman, D., et al., 2018. Trajectories of the earth system in the anthropocene. Proc. Natl. Acad. Sci. USA 115, 8252–8259.

Sternberg, J., Pilla, S., 2023. Chemical recycling of a lignin-based non-isocyanate polyurethane foam. Nat. Sustain. 6 (3). https://doi.org/10.1038/s41893-022-01022-3.

Sternfeld, J., 2023. Walking the Highline, Revised ed. Steidl.

Stott, P.A., Christidis, N., Otto, F.E.L., Sun, Y., Vanderlinden, J.P., van Oldenborgh, G.J., et al., 2016. Attribution of extreme weather and climate-related events. Wiley Interdiscip. Rev. Clim. Chang. 7, 23–41.

Straube, J., 2017. Building Enclosure Fundamentals: A Concise Introduction. University of Waterloo and RDH Building Science, p. 29.

Strom, S., Nathan, K., Woland, J., 2013. Site Engineering for Landscape Architects. John Wiley and Sons, Hoboken, NJ, USA.

Suaria, G., Achtypi, A., Perold, V., Lee, J.R., Pierucci, A., Bornman, T.G., Aliani, S., Ryan, P.G., 2020. Microfibers in oceanic surface waters: a global characterization. Sci. Adv. 6 (23), eaay8493. https://doi.org/10.1126/sciadv.aay8493. PMID: 32548254. PMCID: PMC7274779.

Sundell, J., 2004. On the history of indoor air quality and health. Indoor Air 14, 51–58.

Sundell, J., Levin, H., Nazaroff, W.W., Cain, W.S., Fisk, W.J., Grimsrud, D.T., et al., 2011. Ventilation rates and health: multidisciplinary review of the scientific literature. Indoor Air 21, 191–204.

Supertall, A.S., 2022. How the World's Tallest Buildings Are Reshaping Our Cities and Our Lives. W. W. Norton & Company.

Sweeney, J.L., 2016. Energy Efficiency: Building a Clean, Secure Economy. Hoover Institution Press, Stanford, CA, USA.

Tayebi-Khorami, M., Edraki, M., Corder, G., Golev, A., 2019. Re-thinking mining waste through an integrative approach led by circular economy aspirations. Minerals 9, 13.

Thomas, C.D., Cameron, A., Green, R.E., Bakkenes, M., Beaumont, L.J., Collingham, Y.C., et al., 2004. Extinction risk from climate change. Nature 427, 145–148.

Tosun, I., 2015. Thermodynamics: Concepts and Applications. World Scientific Publishing, Singapore.

Trenberth, K.E., Fasullo, J.T., Shepherd, T.G., 2015. Attribution of climate extreme events. Nat. Clim. Chang. 5, 725–730.

Turns, S.R., Pauley, L.L., 2020. Thermodynamics: Concepts and Applications. Cambridge University Press, Cambridge, UK.

UNEP, 2014. Sand, Rarer Than One Thinks. United Nations Environment Programme, p. 15.

UNEP, 2016. In: Schandl, H., Fischer-Kowalski, M., West, J., Giljum, S., Dittrich, M., Eisenmenger, N., et al. (Eds.), Global Material Flows and Resource Productivity: Assessment Report for the UNEP International Resource Panel. United Nations Environment Programme, Paris, France, p. 197.

UNEP, 2019. The Search for Sustainable Sand Extraction Is Beginning. News and Stories-Ecosystems and Diversity. United Nations Environment Programme, Web.

UNEP, 2021. Global Status Report for Buildings and Construction: Towards a Zero-Emission, Efficient and Resilient Buildings and Construction Sector. 2021 United Nations Environment Programme, Nairobi, Kenya, p. 104.

UNEP, 2022. Spreading Like Wildfire: The Rising Threat of Extraordinary Landscape Fires. A UNEP Rapid Response Assessment. United Nations Environment Programme (UNEP), Nairobi, Kenya, p. 124.

USEPA, 2019. Advancing Sustainable Materials Management: 2017 Factsheet. United States Environmental Protection Agency, Office of Land and Emergency Management (5306P), Washington, DC, p. 21.

USFS, 2015. The Rising Cost of Wildfire Operations: Effects on the Forest Service's Non-Fire Work. United States Department of Agriculture, United States Forest Service, p. 16.

Vecellio, D.J., Wolf, S.T., Cottle, R.M., Kenney, W.L., 2022. Evaluating the 35°C wet-bulb temperature adaptability threshold for young, healthy subjects (PSU HEAT Project). J. Appl. Physiol. 132 (2), 340–345 (1985). https://doi.org/10.1152/japplphysiol.00738.2021. (Epub 2021 Dec 16). PMID: 34913738. PMCID: PMC8799385.

Vethaak, A., Legler, J., 2021. Microplastics and human health knowledge gaps should be addressed to ascertain the health risks of microplastics. Science 371, 672–674.

Vitousek, P.M., Mooney, H.A., Lubchenco, J., Melillo, J.M., 1997. Human domination of earth's ecosystems. Science 277, 494–499.

Walker, X.J., Baltzer, J.L., Cumming, S.G., Day, N.J., Ebert, C., Goetz, S., et al., 2019. Increasing wildfires threaten historic carbon sink of boreal forest soils. Nature 572, 520.

Wasley, E., Dahl, T.A., Simpson, C.F., Fischer, L.W., Helgeson, J.F., Kenney, M.A., et al., 2023. Adaptation. In: Crimmins, A.R., Avery, C.W., Easterling, D.R., Kunkel, K.E., Stewart, B.C., Maycock, T.K. (Eds.), Fifth National Climate Assessment. U.S. Global Change Research Program, Washington, DC USA, https://doi.org/10.7930/NCA5.2023.CH31 (Chapter 31).

Waters, C.N., Zalasiewicz, J., Summerhayes, C., Barnosky, A.D., Poirier, C., Galuszka, A., et al., 2016. The Anthropocene is functionally and stratigraphically distinct from the Holocene. Science 351, 137.

Watts, R.J., Teel, A.L., Gardner, C.M., 2022. Hazardous Wastes: Assessment and Remediation. Wiley, New York, NY, USA.

Weinstein, G., 1999. Xeriscape Handbook. Fulcrum Publishing, Golden, CO, USA.

Westerling, A., Hidalgo, H., Cayan, D., Swetnam, T., 2006. Warming and earlier spring increase western US forest wildfire activity. Science 313, 940–943.

Westerling, A., Bryant, B., Preisler, H., Holmes, T., Hidalgo, H., Das, T., et al., 2011. Climate change and growth scenarios for California wildfire. Clim. Chang. 109, 445–463.

Wiedmann, T.O., Schandl, H., Lenzen, M., Moran, D., Suh, S., West, J., et al., 2015. The material footprint of nations. Proc. Natl. Acad. Sci. USA 112, 6271–6276.

Williams, A.P., Cook, E.R., Smerdon, J.E., Cook, B.I., Abatzoglou, J.T., Bolles, K., et al., 2020. Large contribution from anthropogenic warming to an emerging North American megadrought. Science 368, 314.

Williams, A.P., Cook, B.I., Smerdon, J.E., 2022. Rapid intensification of the emerging southwestern North American megadrought in 2020–2021. Nat. Clim. Chang. 12, 232–234.

Wiser, R., Millstein, D., Rand, J., Donohoo-Vallett, P., Gilman, P., Mai, T., 2021. Halfway to Zero: Progress Towards a Carbon-Free Power Sector. U.S. DOE Office of Energy Efficiency and Renewable Energy, Berkeley, CA, USA, p. 27.

Wolkoff, P., Kjaergaard, S.K., 2007. The dichotomy of relative humidity on indoor air quality. Environ. Int. 33, 850–857.

Worm, B., Barbier, E.B., Beaumont, N., Duffy, J.E., Folke, C., Halpern, B.S., et al., 2006. Impacts of biodiversity loss on ocean ecosystem services. Science 314, 787–790.

Wright, S.L., Thompson, R.C., Galloway, T.S., 2013. The physical impacts of microplastics on marine organisms: a review. Environ. Pollut. 178, 483–492.

Wuebbles, D.J., 2013. Adapting to climate change. In: Watts, R.G. (Ed.), Engineering Response to Climate Change. CRC Press, Taylor and Francis Group, Boca Raton, FL, USA, pp. 391–412.

Wurtsbaugh, W.A., Miller, C., Null, S.E., DeRose, R.J., Wilcock, P., Hahnenberger, M., et al., 2017. Decline of the world's saline lakes. Nat. Geosci. 10, 816.

Wynes, S., Nicholas, K.A., 2017. The climate mitigation gap: education and government recommendations miss the most effective individual actions. Environ. Res. Lett. 12 (7), 074024. https://doi.org/10.1088/1748-9326/aa7541.

Xi, F.M., Davis, S.J., Ciais, P., Crawford-Brown, D., Guan, D.B., Pade, C., et al., 2016. Substantial global carbon uptake by cement carbonation. Nat. Geosci. 9, 880.

Yang, L., Yan, H.Y., Lam, J.C., 2014. Thermal comfort and building energy consumption implications—a review. Appl. Energy 115, 164–173.

Yeh, N.C., Chung, J.P., 2009. High-brightness LEDs-energy efficient lighting sources and their potential in indoor plant cultivation. Renew. Sustain. Energy Rev. 13, 2175–2180.

Zanaga, D., Van De Kerchove, R., De Keersmaecker, W., Souverijns, N., Brockmann, C., Quast, R., et al., 2021. ESA WorldCover 10 m 2020 v100., https://doi.org/10.5281/zenodo.5571936.

Zhang, W., Zhou, T., Wu, P., 2024. Anthropogenic amplification of precipitation variability over the past century. Science 385, 427–432.

Zhao, L., 2018. Urban growth and climate adaptation. Nat. Clim. Chang. 8, 1034.

Zhao, W., Zhu, B., Davis, S.J., Ciais, P., Hong, C., Liu, Z., et al., 2023. Reliance on fossil fuels increases during extreme temperature events in the continental United States. Commun. Earth Environ. 4, 473.

Zib, L., Byrne, D.M., Marston, L.T., Chini, C.M., 2021. Operational carbon footprint of the U.S. water and wastewater sector's energy consumption. J. Clean. Prod. 321, 128815.

Index

Note: Page numbers followed by *f* indicate figures, *t* indicate tables and *b* indicate boxes.

A

Accreditation Board for Engineering and Technology, Inc. (ABET) accreditation program, 10–11
Adaptation, 149–150, 381–384
   to altered climate change, 124–125
   coastal, 387
Aerosols, 80
Agricultural and forestry system (AFOLU), 117–118
Air
   conditioning, 216
   leakage, thermal envelope, 206–208, 206*f*
   pollutants, 302
   pollution, ambient and indoor, 72–81
   quality, healthy indoor environments, 301–304, 303*f*
   sealing, 186, 200–201
   source heat pumps, 223–224
Air quality index (AQI), 73–74, 74*f*, 77*f*, 80*t*
Algae bloom, 257, 257*f*
Ambient and indoor air pollution, 72–81
American Society for Heating, Refrigeration, and Air Conditioning Engineers (ASHRAE), 368
American Society of Civil Engineers (ASCE), 5, 62*t*, 366–367
Antarctic ice mass balance, 26*f*
Anthropogenic climate change and climate crisis, 81–101
Antoine equation, 32, 32*f*, 33*b*
Architecture 2030, 367
Arctic amplification, 379
Asphalt, 169, 170*f*
Assisted migration, 399
Attics, 199–200
Attributes, materials' mass, 154–155
Auditing, 149

B

Babcock Ranch Community in Florida, 390–391, 391*f*
Bamboo, 172–173, 174*f*
Basements, 200–201
Battery storage, 17*b*, 341
Beneficial electrification approach, 117
Biodiversity, 71–72
Bioenergy
   with CCS (BECCS), 122
   current costs and scaling, 123*b*
Biofuels, 338, 339*b*, 340
Biohazard, 306–309
Biomass energy, 338–340
Black and graywater systems, 262–263
Blower door testing, thermal envelope, 207*b*
Body temperature, 23*f*
British thermal units (BTU), 14
Brundtland Commission, 5
Buildings
   codes, 149, 358, 372
   heat and mass transfer in, 186–187
Built environment
   extreme weather events, 376–381
   solutions to climate change, 117

C

Calgary District Heating facility, 231, 232*f*
Carbon capture use and sequestration (CCUS), 118–122, 119–120*f*
   US federal tax credits, 121*t*
   viability, 121–122*b*
Carbon cycle and human fingerprint, 88–91, 89*f*
Carbon footprint, 114
Carbon-sequestering materials, 155
Carnot COP of refrigerator, 30*b*, 33–34
Ceiling fan, 228–229, 229*t*
Cement, concrete and, 163–169
Chemical energy in food, 20–21*b*
Cholera outbreaks, 256
Circular economy, 110
Clausius-Clapeyron relationship, 32
Cliff dwellings, 3*f*
Climate change
   adaptation to altered, 124–125
   anthropogenic, 92–96
   built environment solutions to, 117
   decarbonization of energy infrastructure, 112–115
   EPA Office of Pollution Prevention, 110, 111*f*
   extremes with incremental solutions, 384*t*
   hierarchy of choices, 111*f*
   life cycle analysis, 112
   migrations, 399
   mitigations, 109–112
   risks, 379, 383*t*
   urgency of anthropogenic, 109
Climate crisis, anthropogenic climate change and, 81–101
Climate disruption, 96–98

439

Climate mitigation, 387
   and adaptation, 380
Climate, radiative forcing, 85–87
Climate system, physics, 81–85
Coal-fired power plant, 53f, 54–55b
Coastal construction, 387–389
Coefficient of performance (COP), 29
CO_2 emissions, 163
   factors, 114, 114t
   fleet replacement and, 116b
   from largest emissions sources, 91f
   timeline by nation, 90f
Color rendering index (CRI), 300
Combined heat and power (CHP) systems, 231f, 233–234
Combustion, 27–29
   high-efficiency, 222f
   and reacting systems, 30
Commercial space, 278–279
Commissioning, 135–136, 149
Community space, 278–279
Concrete
   and cement, 163–169, 164t
   staircases, 166–167b
Condensation, 299b
Conduction, 22–23
Construction, 131, 134, 136, 138
   and demolition waste and debris (C&D), 69, 158–159, 159f
   methods, 174–176
   scheduling, 143
   waste, 67–69
Contaminant concentrations, 250–251
Convection, 22–23
Cool roof technologies, 199–200
Crawlspaces, 200–201
Cross laminated timber (CLT), 170–172
Cryosphere, 377

D

Darcy's law flow rate, 66b
DC motor ceiling fan, 228–229, 229t
Decarbonization
   building-centric, 115–116
   of energy infrastructure, 112–115
Demand management, 240–241
Demolition waste, 67–69
Design, 134–138
Doors, 196–199
Drought risks, 392
Dual-use solar, 327

E

Earthship architecture, 161f
Electrical appliances and phantom loads, 239, 239f

Electrical generating unit (EGU), 51
Electrical relationships, 17
Electric arc furnaces (EAFs), 169
Embodied carbon, 155–157, 157f
Embodied energy, 155–157, 157f
Empire State Building (ESB), 179, 180f, 181
Energy, 56–57
   conservation, 115–116
   curtailment, 115–116
   efficiency, 115–116
   forms of, 14, 15t
   as fundamental physical quantity, 13–14
   and mass balance, 26–27
   and power, 17b
   scales and transformations, 21f
   units, 14t
Energy Information Association (EIA), 2, 313, 314f
Energy recovery ventilators (ERV)., 230
Energy Star program, 368
Energy systems in buildings, 215
   climate control systems, conditioned spaces, 216–232, 217f
   commercial buildings, 216f
Energy-water nexus, 64
Enthalpy (h), 18
Environmental engineering, 1
Environmental Impact Statement (EIS), 103
Environmental justice, 102–103
Environmental policy, 102–103
Environmental quality, healthy indoor environments, 300
Environmental stresses, 61–62
Envision program, 366
Equity, Sustainability, and Governance (ESG) focus, 5–6
Evaporative coolers, 225–226, 226b
Extreme weather events, 376–381
   risks to buildings and infrastructure, 378t
   structural damage, 381f

F

Federal Emergency Management Agency, 358
Federal policy and financial incentives, 357–358
Fertilizer runoff, 251
Fine-mode aerosols, 76
Firewise program, 393–394, 394f
First Law of Thermodynamics, 27
Floating residences in Netherlands, 389f
Flooding and mitigation, 389–390
Fluid flow, 30
Food onsite, 288
Forestry Stewardship Council (FSC), 118
Fossil fuels, 50
Freshwater resources, 64, 249–250, 260–261
Fuel and iron redevelopment in Pueblo, 146–147

G

Geoengineering approaches, 125–126, 125f
Geothermal system, 320, 333–334, 334f
Ghirardelli Square, San Francisco, California, 139, 140f
Global population and environmental impacts, 42–44
Global warming potential, 219–221
Government Services Agency (GSA), 358
Graywater, 256–257, 262–263, 289–290
Green attributes, 154–155
Green building, 1, 176–178, 370–372
    certification programs, 365, 367–370, 368–370t, 372–373
Green Building Certification Incorporated (GBCI), 132
Greenhouse gas emissions, 43–44b, 163, 164t
    with building-related emissions, 2–3, 2f
    sources, 92f
Greenhouse gases, 85f, 100–101
    radiative forcing efficiency of, 87t, 88b
Green infrastructure, 382, 404f
Green roofs, 292, 293f
Grid
    resiliency, 395–397
    stability, 396
Groundwater resources, 250, 262f

H

Habitat loss, 71–72
Hazardous waste, 67–69, 162f
Healthy indoor environments
    air quality, 301–304, 303f
    biohazard concerns, 306–309, 310f
    environmental quality, 300
    human comfort, 297–299
    hygro-thermal properties, 297–299
    interior furnishings, 310–311
    lighting characteristics, 300
    psychometric chart, 299f
    sound characteristics, 300–301
Heat
    capacity, 24–26, 24t
    engines, 27–29, 28f, 51
    and mass transfer in buildings, 186–187
    and work, 19–21, 22b
Heating systems, 216
Heat loss
    conductive, 194b
    through thermal envelope, 193–194
Heat pump, 27–29, 28f, 219f
Heat recovery ventilators (HRV), 230
Heat transfer mechanisms, 22–23
Hellstrip landscaping, 286–287, 286f, 287t
Hemp straw bale, 172–173, 172f
High-efficiency combustion systems, 221–223, 222f
Home Energy Rating System (HERS) rates, 368

Horizontal Axis Wind Turbine (HAWT), 337
Housing costs, 141b
Human comfort, healthy indoor environments, 297–299
Human Development Index (HDI), 41–42, 42f, 363, 363f
Human fingerprint and carbon cycle, 88–91, 89f
Human population and exponential growth, 39–42, 40f
HVAC systems, 217–218t
    sizing and ducting, 221
Hydrogen economy, 123–124, 124b
Hydrological resources, planet earth, 62–64, 63f
Hydrology, 249
Hygro-thermal properties, healthy indoor environments, 297–299

I

Ideal gas law, 30–31
Indoor air quality (IAQ), 302, 302f
    mitigation approaches, 304–305
Indoor environment
    air quality, 301–304, 303f
    biohazard concerns, 306–309, 310f
    environmental quality, 300
    human comfort, 297–299
    hygro-thermal properties, 297–299
    interior furnishings, 310–311
    lighting characteristics, 300
    psychometric chart, 299f
    sound characteristics, 300–301
Indoor environmental quality (IEQ), 300
    certification programs, 367
Indoors, 205f
Infrared interior, 199f
Inland flooding and mitigation, 389–390
Institute for Sustainable Infrastructure (ISI), 366–367
Insulating materials, 195–196
    and R-value in series, thermal envelope, 188–192, 188f, 189t, 191–192b
Integrated Climate and Land-Use Scenarios (ICLUS) project, 386
Integrated design process (IDP), 136
Intergovernmental Panel on Climate Change (IPCC), 86, 86–87f, 94–95
Interior furnishings, healthy indoor environments, 310–311
Internal energy (u), 18
IPAT equation, 43

J

Joule's law, 17

L

Laboratory building district chiller systems control, 232–233
Landscape design
    history and importance of, 273–275
    maintenance and equipment use, 292

Land use, 65–66, 65f
Latent heat, 25–26, 25f, 31–32
Leadership in Energy and Environmental Design (LEED) program, 131, 364, 365t, 366f
Levelized cost of energy (LCOE), 316f
Life-cycle analysis (LCA), 56, 372
Lighting, 235–238, 236f, 237b, 238f
Lighting characteristics, healthy indoor environments, 300
Linear park in New York city, 279–281, 280f
Load shifting, 240–241
Locating and Selecting Scenarios Online (LASSO) tool, 386
Low-carbon energy
Low impact development (LID), 135

M

Manufacturing, 174–176
Marine debris and microplastics, 69–70
Mass balance, 153
  on Antarctic ice mass, 26–27b
  and energy balance, 26–27
Mass timber, 170–172
Materials extraction, 65–66
  and processing, 67
Materials' mass
  attributes and sustainability, 154–155
  on Earth, 153
Material use and energy use, 53, 56f
Maximum contaminant level (MCL), 250
Mean sea level, 95f
Microgrids, 314, 340, 395–397, 397f
Microplastics, 251
  and marine debris, 69–70
Micropollutants, in water, 250
Mini-split systems, 223–224, 224f
Municipal solid waste (MSW), 158–162

N

National Ambient Air Quality Standards (NAAQS), 73t, 75f
National Association of Home Builders (NAHB), 154
National Flood Insurance Program (NFIP), 390
National Risk Index (NRI), 385f
Net present value (NPV), 357b
New Green Tall Tower in Austin, TX, 181–182, 182f
Noise pollution, 301, 301b

O

Ocean acidification, 95–96b
Ocean plastics, 70b
Office landscapes, 278–279

Ohm's law, 17
Open vs. closed systems, 19b
Oregon state treasury building, 398, 398f
Organic architecture, 274f
Orientation and passive solar considerations, 147–148
Outdoor environments, 287–288

P

Particulate matter (PM), 73–74, 100–101
Permaculture, 281, 288
Permeable concretes, 165
Petroleum
  refinery, 113f
  resources, 50, 50b
Planck's Law, 318–319
Plumbing, 215
Potable water treatment, 252–255
Power (W), 15–17
  consumption in Watts, 16t
Power density/flux (W/m2), 15–17
Problem-solving, 6
  and dimensional homogeneity, 8–9
Pyrocene, 393

R

Radiation, 22–23, 24b
  emission of, 319f
Radon evacuation system, 306
Rain exclusion, 265
Rainwater, 289–290
  collection, 290b
Redevelopment, 131–132, 144, 145f, 146–147
Refrigerants, 219–221
Refrigerated air systems, 216
  advent of, 218
Regenerative landscaping, 288
Relative humidity, 31–32
Renewable energy systems
  behind the meter renewable systems, 320
  constraints and opportunities, 343–345, 343t
  definition, 313–316
  mid-century schoolhouse/church repurposed into a home, 345–347, 347f
  Planck's Law, 318–319
  solar energy and electromagnetic energy, spectrum of, 317–319, 317f
  solar photovoltaic cells and PV systems, 320–327
  solar PV land use, 319b
  time history of global, 316f
  US National Renewable Energy Laboratory map, 318f
Renovation, 131, 138–139, 139f, 143
  of deteriorating concrete staircases, 166–167b

Index

Resiliency, 376, 381–384
    in mid-century schoolhouse/church into home, 400–401
    organizations and tools, 385–387
Resource and Recovery Act (RCRA), 158
Retreat, managed, 399
Reynold Number (Re), 31
Roofs, 199–200
R-value
    of parallel configuration, 193*b*
    parallel pathway, 192
    in series, insulating materials and, 192*b*
    of vermiculite-filled CMU, 192*b*

S

Saltwater resources, 251
Sand constraints, 168*f*, 169*b*
Sea level rise, 387–389, 387*f*
Sensible heat, 13, 24–26
SI system, 14
Site water management systems, 289–290
Smart grid, 395–397
Societal energy, 314
Socioeconomic progress, 360
    efficiency, 362–363
    environmental justice issues and building industry, 360–363, 362*f*
    federal policy and financial incentives, 357–358
    gross domestic product (GDP) per capita, 363
    holistic business model, 363
    human development indicator, 363, 363*f*
    life cycle analysis, 364
    utilities and rate structures, 359–360
    zoning and building codes, 358
Soil, 65–66
    properties, 282–283
    quality, 66
    sequestration, 118–119, 127
Solar Automated Permitting Process, 327
Solar energy and electromagnetic energy, spectrum of, 317–319, 317*f*
Solar photovoltaic cells and PV systems
    active solar thermal systems, 331–332, 332*b*
    biofuels, 339*b*
    biomass energy, 338–340
    electricity generation with, 323–324, 324–325*f*
    energy storage systems and transmission, 340–343, 341*f*, 342*b*
    geothermal energy systems, 333–334, 334*f*
    grid-tied solar photovoltaic system, 323*f*
    micro hydro applications, 340
    residential solar photovoltaic power systems, 327*f*
    solar cell physics and efficiency, 320–322, 322*f*
    solar daylighting and passive solar space heating, 328–330, 328–329*f*
    solar PV costs, 324–327
    wind power systems, 335–338, 335–337*f*, 338*b*
Solar Rating and Certification Corporation (SRCC), 368
Solid waste, 67–69
Sound characteristics, healthy indoor environments, 300–301
Species extinction, 71–72
Steel, 169
Stefan-Boltzmann constant, 23
Stormwater runoff, 277–278
    management of, 132, 133*f*
Stormwater systems, 250, 257*f*
    to greywater reuse, 263
    low impact development (LID), 261
    management, 258–260, 268
    runoff of, 260
Straight payback, 356, 356*b*
Stratospheric aerosol injection (SAI), 125–126
Stratospheric ozone depletion, 219–221
Sustainability, 5, 56–57, 109–112
    materials' mass, 154–155
Sustainable design, 140–142, 141*f*
    site development, 131–134
Sustainable global energy use, 50–51
Sustainable SITES Initiative, 132, 133–134*t*
Sustainable sites program, 276, 277*f*
Sustainable transportation, 144–145, 145–146*f*
Swamp coolers. *See* Evaporative coolers

T

Tall buildings, 176–178, 177*b*
Tankless on-demand systems, 233–234
Thermal bridging, 201–203, 202*f*, 202–203*b*
Thermal envelope
    air leakage in, 206–208, 206*f*
    blower door testing, 207*b*
    definition, 185–186
    heat loss through, 193–194
    insulating materials and R-value in series, 188–192, 188*f*, 189*t*, 191–192*b*
    tightening, 188*b*
Thermal mass, 201–203
Thermal resistance, 190
Thermal transmittance, 190
Thermodynamic properties, 18–19, 18*f*, 19*t*
3D-printed residential building, 175, 175–176*f*
Tightening, thermal envelope, 188*b*
Trombe wall, 330, 330*f*

U

Ultimate insulator, 188
United States and global energy use, 44–49, 45f, 47f
United States Forest Service (USFS), 100
UN sustainable development goals, 135t
Urban agriculture (UA), 288
Urban heat island effect (UHI), 379
US Department of Energy (DOE), 154
US Environmental Protection Agency (EPA), 72
US Green Building Council (USGBC), 39, 131
US National Oceanic and Atmospheric Administration, 378f
US Resiliency Council (USRC), 386

V

Ventilation, 309
    for space cooling and heating, 226–229, 229f

W

Wastewater
    components of, 256f
    generation, 267
    industrial source, 250
    treatment plant, 250, 252, 256–258, 256–257f, 262–263, 267f, 267b
Water droplet lifetime, 307f
Water-energy nexus, 265–266, 265–266b
Water heating
    costs, 234–235b
    systems-storage tank, 233–234
Water quality
    management on building scale, 263–264, 264f, 267f
    parameters and pollution, 249–252
Water resources, 64, 64–65b
    freshwater, 249–250
    global, 265
    groundwater, 250
Water treatment
    components of, 256f
    discharge point of, 250
    energy and vice versa, 265–266
    industrial, 261
    municipal, 253
    onsite, 253
    potable, 252–255, 254f
    terminal settling velocity, 255b
Water usage
    landscape, 274–275
    for residence, 274–275, 275f
Water vapor, 31–32
Wetlands, constructed, 289–290
Wider World of green certification programs, 367–370
Wildland fire, 98–101
    and built environment, 392–395, 392f
Wildland-urban interface (WUI), 393
Windows, 196–199
    improvements, 197
    performance of, 197t
    profiler, 198f
    types, 198t
Wind power systems, 335–338, 335–337f, 338b
Wood products, 170–172
Work, heat and, 19–21, 22b
Wright, Frank Lloyd, 273–275, 274f

X

Xeriscape, 283–285
    design and plantings, 284
    residential landscapes, 291, 291f